WIRELESS INTERNET AND MOBILE COMPUTING

BICENTENNIAL
1807
WILEY
2007
BICENTENNIAL

THE WILEY BICENTENNIAL—KNOWLEDGE FOR GENERATIONS

*E*ach generation has its unique needs and aspirations. When Charles Wiley first opened his small printing shop in lower Manhattan in 1807, it was a generation of boundless potential searching for an identity. And we were there, helping to define a new American literary tradition. Over half a century later, in the midst of the Second Industrial Revolution, it was a generation focused on building the future. Once again, we were there, supplying the critical scientific, technical, and engineering knowledge that helped frame the world. Throughout the 20th Century, and into the new millennium, nations began to reach out beyond their own borders and a new international community was born. Wiley was there, expanding its operations around the world to enable a global exchange of ideas, opinions, and know-how.

For 200 years, Wiley has been an integral part of each generation's journey, enabling the flow of information and understanding necessary to meet their needs and fulfill their aspirations. Today, bold new technologies are changing the way we live and learn. Wiley will be there, providing you the must-have knowledge you need to imagine new worlds, new possibilities, and new opportunities.

Generations come and go, but you can always count on Wiley to provide you the knowledge you need, when and where you need it!

WILLIAM J. PESCE
PRESIDENT AND CHIEF EXECUTIVE OFFICER

PETER BOOTH WILEY
CHAIRMAN OF THE BOARD

WIRELESS INTERNET AND MOBILE COMPUTING
Interoperability and Performance

Yu-Kwong Ricky Kwok
Colorado State University

Vincent K. N. Lau
The Hong Kong University of Science and Technology

IEEE

IEEE PRESS

BICENTENNIAL
1807
WILEY
2007
BICENTENNIAL

**WILEY-
INTERSCIENCE**

A JOHN WILEY & SONS, INC., PUBLICATION

Library of Congress Cataloging-in-Publication Data:

Kwok, Yu-Kwong Ricky.
   Wireless Internet and mobile computing : interoperability and performance /
Yu Kwong Ricky Kwok and Vincent Lau.
       p. cm.
    ISBN 978-0-471-67968-4 (cloth)
1. Wireless communication systems.   2. Mobile computing.   3. Internetworking
(Telecommunication)   I. Lau, Vincent K. N.   II. Title.
TK5103.2.K95 2007
621.382—dc22

                                                                    2007001709

Printed in the United States of America.

10 9 8 7 6 5 4 3 2 1

To our wives and kids:
Fion, Harold, Amber,
Elvina, and Sze-Chun

CONTENTS

PREFACE

The Theme:

The all-mighty Internet has extended its reach to the wireless realm. With this exciting development, we have an ambitious goal in this book—to help the reader to build up sound technical understanding of what is going on in a large-scale networking system as depicted graphically in Figure 0.1. Using a bottom-up approach, we would like the reader to understand how it is feasible, for instance, for a cellular device user (top left corner in the figure) to communicate, via the all-purpose TCP/IP protocols, with a wireless notebook computer user (bottom left corner in the figure), traversing all the way through a base station in a cellular wireless network (e.g., GSM, CDMA), a public switched telephone network (PSTN), the Internet, an intranet, a local area network (LAN), and a wireless LAN access point. In traveling through this long path, the information bits are processed by numerous disparate communication technologies (in slightly more technical terms, processed by many different protocol stacks and wireless air interfaces). We also describe the technologies involved in infrastructureless (i.e., ad hoc) wireless networks (the bottom right corner in the figure), which are widely envisioned to be the most popular form of mobile computing infrastructure, pertaining to many interesting applications: most notably, wireless network games. Our focus is on how these distinctive technologies can work together

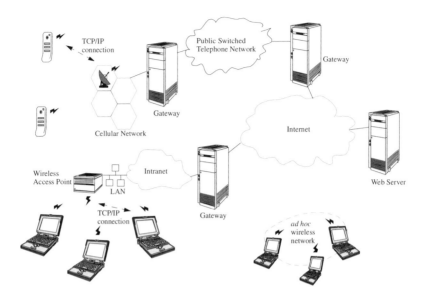

Figure 0.1   The Internet extends into the wireless realm.

through various adaptation methods. Specifically, throughout this book we use two major technical challenges to delineate and motivate our discussion—*interoperability* and *performance*.

We believe that interoperability is the very first concern in integrating so many different components to realize a seamless communication path for users with disparate wireless devices. Indeed, many interesting features of intelligent communication technologies emerge out of the need to address the interoperability issues. Performance is another major challenge in a resource-limited wireless environment. Combined, these two challenges have baffled researchers and engineers for years, and there are many technical imperfections yet to be tackled.

Cutting vertically through the protocol stacks at various junctions in the distributed wireless networking environment shown in Figure 0.1, we divide the book into six parts:

I. Essentials of Wireless Communications: In this part we provide the reader with a solid foundation about wireless communications in a concise, yet complete manner. We do not aim to repeat all the minute details that can be found in classical digital communications theory books, but we explicate the overall theory in a unified manner.

II. Cellular Wireless Technologies: In this part we describe in an evolutionary manner the developments of the various practical cellular wireless technologies: the last wireless mile of the Internet, highlighting the ingenious adaptation schemes designed for satisfying the interoperability requirement while enhancing performance continuously from one generation to another.

III. Short-Range Wireless Technologies: In this part we describe and explain the various short-range wireless technologies: the last 100 m of the Internet, including Bluetooth, wireless LAN, and infrared technologies. Unlike the cellular counterpart, which is driven by voice communication-oriented application requirements, short-range wireless technologies are driven by computing oriented application requirements, and as such, entail a number of different system-level issues. In particular, the coexistence issue between the two most popular technologies, Bluetooth and IEEE 802.11b WLAN, is very acute, as it critically affects the performance perceived by users.

IV. Protocol Adaptations for Wireless Networking: In this part we describe the various ingenious adaptation techniques for our all-purpose HTTP/TCP/IP protocol suite, which is a vital component in many parts of the distributed wireless networking environment, as shown in Figure 0.1. We start with the IP-Mobile IP and IPv6. We then describe the infamous wireless application protocol (WAP) and its design rationales, which suggest a complete reworking of the TCP/IP stack. We move on to the TCP adaptations, which are important in providing stable application-level performance.

V. Wireless Resources Management: In this part we discuss management of the scarce resources in a wireless environment for different layers: physical layer, multiple access control layer, network layer, and application layer. Specifically, we describe power management, packet scheduling policies, routing, data caching, and security issues. From the interoperability and performance angles, the interplay between the wireless resources management schemes and the various protocol adaptation technologies is also our focus throughout this part.

VI. Mobile Computing Applications: In this part we move on to the application level in that we describe the various adaptation techniques, interoperability issues, and performance aspects in realizing mobile computing services, including voice, video, and file system abstraction. To achieve efficient anytime, anywhere wireless information services, location-dependent applications are envisioned as the core underlying delivery mode. Finally, we also look into the future by discussing the ultimate form of truly peer-to-peer wireless ad hoc wireless networking and wireless sensor networking.

Blending Practice and Theory:

Throughout the book we use practical real-life case studies, derived from our own research and development experience in both the industry and academia, to serve as motivating examples for helping the reader to gain in-depth understanding of the technologies taught. These examples are designed to bring out the dynamics of the underlying systems. For example, in addition to describing various subsystems (e.g., soft handover, power control, etc.) in cellular systems, we illustrate the interactions of various subsystems involved in several user application scenarios in a dynamic manner.

The technologies we describe in the book have evolved from several areas: digital communications, information theory, signal processing, wireless networking, Internet protocol advancements, and mobile computing applications. We firmly believe that these technological areas will synergize with each other, and the ultimate goal is to achieve optimized performance at the user level. For instance, while video processing techniques have traditionally been the outcome of signal processing research, recent advancements are the fruits of research projects that take into account wireless networking and physical layer design results.

We strongly feel that a comprehensive book, developed via the interoperability and performance perspective, and cutting through the various protocol layers—from the physical layer to the application layer—is in pressing need by both researchers and practitioners in the field. Indeed, we have conducted various lectures on the subject for senior undergraduate and master-level students in the past few years. It is difficult to find a comprehensive textbook and reference book on the market that covers both the scope and depth of the subject. This motivated us to develop a book that will allow people to better understand the engineering details, followed by a firm grasp of development trends at all the protocol layers which are interdependent, and then to form an opinion as to when and how the technologies can be applied in various settings, be they commercial or educational.

Our target audience of this book includes senior undergraduate students, graduate students, and engineers in the field. The prerequisite knowledge required includes basic communication engineering concepts and basic understanding of internetworking protocols. We expect our readers to walk through the technological descriptions and most important, the practical illustrative examples, and then map them to their own academic and/or practical experience. We would describe the book as being on an advanced level, but not a research-oriented specialist treatment.

Road-Map:

- **Part I: Essentials of Wireless Communications**

  - **Chapter 1: The Mobile Radio Propagation Channel** This chapter aims at providing the reader with a solid understanding of the properties of the wireless transmission medium. Such a background will be useful for the reader to understand the various adaptation and optimization approaches devised to enhance system performance. To strengthen the reader's understanding, we provide a detailed description of the wireless channel propagation characteristics in the popular GSM (narrowband), CDMA (wideband), and IEEE 802.11a/b/g (wideband) environments.

  - **Chapter 2: Modulation Techniques** The core of this chapter is the evolution of different digital modulation techniques, which are critical in achieving high bandwidth efficiency. The basic trade-off of bit rate, power/bandwidth, and quality of communication is illustrated both qualitatively and quantitatively. Design considerations and issues with respect to various channel models described in Chapter 1 (narrowband and wideband) are illustrated. To allow the reader to appreciate the intelligent design of popular modulation methods, we use the GSM, CDMA, and IEEE 802.11a/b/g as examples to illustrate the concepts.

– **Chapter 3: Multiuser Communications** While Chapter 2 deals with point-to-point communications, we extend the framework in this chapter to consider multiuser communications. The chapter begins with an information-theoretical introduction on multiuser communications. Next, we introduce various commonly adopted resource partitioning techniques, together with a quantitative comparison of their strengths and weaknesses. We then illustrate the practical design choices in the GSM, CDMA, and IEEE 802.11b systems.

– **Chapter 4: Diversity Techniques** Diversity is becoming an important problem dimension that can be exploited to further enhance system capacity and performance. In this chapter we discuss various diversity parameters that are considered as practicable performance enhancement approaches. We describe the practical implementation methodology in realizing the various diversity techniques (e.g., RAKE receiver, open-loop transmit diversity, closed-loop transmit diversity). To give the reader a solid understanding of the efficacy of the diversity concept, we use the UMTS and IEEE 802.11a systems as real-life examples.

• **Part II: Cellular Wireless Technologies**

– **Chapter 5: Overview and Evolution of Cellular Technologies** In this chapter we aim to provide a comprehensive review of the rationale, theory, and practical issues of modern cellular wireless communications technologies. Motivated by the efficient frequency reuse features, cellular networks are designed and engineered to be highly cost-effective but also entail a number of system issues such as mobility management and co-channel interference. We give a brief overview of the evolution of cellular systems from 1G to 3G+ and compare the paths in Europe, the United States, and Japan. We discuss the technical challenges in realizing 3G services. Finally, we describe briefly how GPRS/EDGE/3G1x/UMTS overcomes these technical challenges.

– **Chapter 6: CDMA (IS-95)** Based on the theoretical foundation given in Chapter 4, we describe in detail the design goals and system architecture of a practical CDMA (IS-95) cellular infrastructure. To prepare for the 3G topics in Chapter 10, we present the details of various subsystems [such as the air interface, the radio resource management (soft handoff, power control), and mobility management]. To highlight the dynamic interactions of various subsystems, we describe several real-life call-processing scenarios.

- **Chapter 7: GSM** GSM is by far the most successful cellular communication technologies, driven perhaps not only by technical superiority but by economic forces as well. In a retrospective angle, we describe its original design goals, followed by a detailed presentation of its practical architecture. Besides describing the design of various subsystems (such as the air interface, the protocol architecture, radio resource management, mobility management), we shall explain the design based on the theories described in Part I. To help the reader appreciate the efficacy of the GSM system, we describe the interactions of various subsystems in a dynamic way through the use of detailed real-life call-processing examples.

- **Chapter 8: Wideband CDMA and Beyond** To cope with bandwidth-hungry applications (e.g., multimedia messaging services), there is no choice but to let the cellular network systems march into the broadband era. Here, interoperability is a prime concern, as operators have to provide a seamless migration of the huge user population and install base to next-generation systems. Thus, there are a plethora of engineering chores that must be tackled with high efficiency in the actual implementations of the so-called 3G systems. In this chapter we describe the most common modes of IMT2000 umbrella: the CDMA2000 and UMTS systems. The focus is on packet-switched services. High-data-rate services are then described in detail. Some advanced topics such as macroscopic scheduling and microscopic scheduling, are also discussed. Besides describing various subsystems, we also explain the design philosophy behind the standards. Finally, we use a number of practical wireless Internet access scenarios to illustrate the salient features of 3G systems.

- **Part III: Short-Range Wireless Technologies**

  - **Chapter 9: IEEE 802.11x WLAN Standards** Due to its local cost (commoditized technologies and free wireless spectrum), IEEE 802.11x-based wireless LAN systems have proliferated. Based on the mature MAC design principles in the original IEEE 802 architecture, the IEEE 802.11x standards employ several different physical layer designs (FHSS, DSSS, OFDM, IrDA, etc.), providing a hierarchy of data rates. We describe these system issues in detail and provide a mobile computing scenario based on IEEE 802.11b LANs, using both the point coordination function (PCF) and distributed coordination function (DCF).

  - **Chapter 10: Bluetooth WPAN** Bluetooth attracted high-level media attention when it was introduced around the beginning of the new millennium. Also based on the free ISM frequency band, Bluetooth turned out to be much delayed in its deployment for various engineering and economic reasons. Nevertheless, Bluetooth's compact design—the TDD TDMA master-slave communication model—has eventually made it pervasive in a wide range of practical gadgets. We discuss the design rationale of the many interesting features of Bluetooth, including the piconet and scatternet modes, the latter yet to be realized. We provide several users' application scenarios to let the reader know the unique characteristics of Bluetooth.

- **Chapter 11: Coexistence Issues** The free ISM frequency band is now occupied by a crowd of short-range wireless technologies, most notably Bluetooth and IEEE 802.11b. The consequence is that when devices of different technologies come into each other's range, interference inevitably occurs, rendering significant performance degradation unavoidable. Thus, various commercial efforts and IEEE task forces have been dispatched to investigate possible solutions of the coexistence problem. We first illustrate the collision problem quantitatively and then describe various suggested solutions and their rationales. Among the many suggestions, adaptive frequency hopping (e.g., to be used in Bluetooth systems) seems to be most practicable. We provide a solid description of such technologies, followed by a real-life illustrative example.

- **Chapter 12: Competing Technologies** Short-range wireless communication has traditionally been the territory of IrDA. We describe the functionality and features of IrDA-based systems and its applications. Recent competitors, other than Bluetooth and IEEE 802.11x, include several commercial efforts, such as HomeRF and HIPERLAN. We also discuss in detail the strengths and weaknesses of these new technologies. We then illustrate the usefulness of IrDA using a few real-life application examples.

- **Part IV: Protocol Adaptations for Wireless Networking**

  - **Chapter 13: Mobile IP** In this chapter we begin our discussion of higher-layer protocols. Specifically, we motivate the reader regarding the need to devise intelligent adaptation strategies to make the classical wireline protocols (e.g., IP, TCP, HTTP) work in a wireless environment to satisfy the interoperability requirement. We describe various ingenious designs in making a mobile version of IP interoperable with the wireline version. We use several practical scenarios to illustrate the functionality of Mobile-IP.

  - **Chapter 14: Ipv6** IPv6, designed to tackle a number of problems in wireline Internet, is also very important in the wireless environment. Indeed,

IPv6 is designed deliberately to have a number of features that make IP independent of the underlying physical medium. We describe such features in detail and compare them against those discussed in Chapter 15. We use similar practical situations to illustrate similarities and differences in relation to the two different IP adaptation approaches.

– **Chapter 15: WAP** The Wireless Application Protocol also caught intense public attention when it was first published. Designed for seamless integration of wireline Internet and cellular wireless networks, WAP possesses a number of special features to optimize the transportation of heavy weight Internet contents while satisfying the interoperability requirements. We use a real-life scenario that is found in a commercial environment to illustrate the usefulness of WAP as well as to show the weakness of WAP, thereby letting the reader know why it has not become as popular as it was initially touted to be.

– **Chapter 16: TCP over Wireless** TCP has a number of salient features, making it the hard core of many Internet services for over three decades. Among these features, congestion control has been receiving great attention in the wireline Internet research community, and as such, many intelligent new techniques are devised to make the congestion control performance better. However, such techniques are not suitable in a wireless environment, in which packet reception errors (e.g., missing packets within a TCP window) may be due not only to congestion but to channel error as well. We describe such phenomena in detail to motivate the reader about the need to design new techniques for the adaptation of TCP in a wireless environment. We describe several more mature approaches using numerous examples. We then use several practical scenarios to illustrate how these methods are being used in a real-life situation.

- **Part V: Wireless Resources Management**

  – **Chapter 17: Wireless Packet Scheduling** This chapter kicks off our discussion of resources management, which is of utmost importance in a wireless computing environment, wherein resources (power and bandwidth) are desperately deficient compared with the connected wireline counterpart. We talk about packet scheduling in this chapter by walking through several popular scheduling approaches, designed based on the fundamental fairness notions. We discuss in detail the performance evaluation criteria for scheduling techniques. We then use the scheduling approach deployed in practical cdma2000 systems to show the reader that a good scheduler can indeed enhance the utilization and performance of a wireless infrastructure.

  – **Chapter 18: Power Management** Power is always a limiting factor in a wireless environment. We describe quantitatively how we quantify power consumption. We then discuss general principles in conserving power. We describe the performance issues related to these power-saving strategies. Using the popular IEEE 802.11b standard as an example, we illustrate how far we can go in saving power in a mobile computing system.

– **Chapter 19: Ad Hoc Routing** Finding a stable and good (in the sense that the bandwidth is higher) route in an infrastructureless network (i.e., ad hoc network) is critical for two devices that are out of each other's range to communicate. Leveraging the legacy techniques designed in the 1980s for packet radio networks, many new algorithms are devised for efficient routing in ad hoc networks. We describe in detail the system model, performance metrics, and protocol features. We then use an IEEE 802.11b-based ad hoc environment to illustrate the use of such ad hoc routing protocols.

– **Chapter 20: Wireless Data Caching** Contrary to an infrastructureless environment, in today's mobile computing systems, there usually are centralized servers to provide wireless access and information services to the tetherless client devices. Among the many system services provided by the server, the cooperative caching (cooperation between the clients and the server) model is one of the most important features. Specifically, the server, usually acting as a proxy, will periodically broadcast invalidation reports to alert clients that some data cached in the clients may be obsolete. We discuss in detail the application scenarios, system architecture, and performance issues of the caching problem. We then describe several interesting approaches to tackling this problem. We use an IEEE 802.11b-based infrastructure to illustrate the practicality of the caching techniques.

– **Chapter 21: Wireless Security** Wireless security is an important but largely immature technical area. In this chapter we aim to provide a general overview of the issues involved and the current approaches used. We divide our discussion into cellular long-range wireless security and short-range wireless security. We use an IEEE 802.11b network to illustrate the use of security enhancement techniques and the difficult problems faced by security protocol designers.

• **Part VI: Mobile Computing Applications**

– **Chapter 22: VoIP on Wireless** This chapter begins the discussion of various higher-level mobile computing issues. We talk about an alternative approach in providing voice services over wireless-based VoIP. We discuss about the advantages and weaknesses of this approach, through a detailed description of the use of VoIP over wireless. We then describe the various engineering efforts required to fulfil the performance and interoperability requirements. We use an IEEE 802.11b-based system to illustrate the actual practice of VoIP over wireless.

– **Chapter 23: Wireless Video** Video is beyond doubt a very important media in the use of wireless infrastructure. Due to the heavy resource requirements of video, sophisticated approaches are devised to carry video efficiently over a wireless link. Most notably, many joint source channel coding techniques have been suggested in the research community. We provide a unified overview of such techniques. We use an IEEE 802.11b-based system to highlight the difficulties involved and the solutions designed for handling video data.

– **Chapter 24: Wireless File Systems** A file system is a diehard abstraction that any meaningful information system must efficiently provide. A wireless mobile computing environment is no exception. However, the frequent disconnection problem and the time-varying channel quality make providing such an abstraction rather difficult. We discuss in detail several effective techniques in tackling the various file system problems, such as consistency, integrity, and security. We use the infamous Coda file system as an illustrative example.

– **Chapter 25: Location-Dependent Services** It is widely conceived that providing suitable services in a suitable environment at a suitable time is the most desirable to a wireless device user, due to the needs for conserving power, simplifying user input, and maximizing system utilization. However, providing truly seamless location-dependent services entails a number of difficult engineering issues, including location management, intelligent information pushing, and efficient remote query processing. We discuss these issues in detail and then we use a practical real-life example to illustrate the functionality of location-dependent advertising.

– **Chapter 26: Trust Bootstrapping in Wireless Sensor Networks** We discuss a recently hot research topic about wireless sensor networks in this chapter—the establishment of mutual trust among sensors. Without such mutual trust, it is infeasible for the wireless sensors to carry out secure data exchange. However, a major constraint is that usually we have no control about the network topology of a wireless sensor network. Consequently, we usually do not know the neighboring relationship among sensors so as to set up encryption keys in an a priori manner. We summarize representative major research findings in this interesting topic.

– **Chapter 27: Peer-to-Peer (P2P) Computing over Wireless Networks** In this chapter we aim to help the reader to look ahead into the future about what mobile computing will look like. We provide our personal opinions based on our own research as well as on various radical ideas from the research community. In particular, we describe in detail to the reader what we believe is the ultimate form of wireless mobile information processing—selfish wireless computing.

- **Chapter 28: Incentives in Peer-to-Peer Computing** In this chapter we follow up Chapter 27 by discussing a fundamental requirement of P2P computing—to provide adequate incentives for autonomous and possibly selfish peers to cooperate. Several types of incentive schemes have been proposed recently by world-renowned researchers. The reader should find this chapter's concepts and techniques interesting, yet intriguing.

RICKY AND VINCENT

Hong Kong

ACKNOWLEDGMENTS

The authors would like to thank our families for their love and support in this project. We sincerely thank Mr. Tyrone Kwok and Mr. Carson Hung (both are Ph.D. students of Y.-K. Kwok) for their invaluable help in compiling the materials and illustrations in the book. In particular, Tyrone has provided various criticisms and comments on various parts of the book. We also thank Mr. Ray Lau (a close friend of Y.-K. Kwok), who made important contributions in the design of the book cover. The authors also thank all the people who have helped in the preparation of this book in one way or the other. Thanks are due to Ms. Rachel Witmer and Ms. Angioline Loredo for their professional advice and assistance. Indeed, we owe a deep sense of gratitude to our friends and families, especially now that this project has been completed.

Y.-K.

ACRONYMS

ACK	Acknowledgment
ACL	Asynchronous Connectionless Link
AFH	Adaptive Frequency Hopping
AOA	Angle of Arrival
AODV	Ad Hoc On-Demand Distance Vector
ARQ	Automatic Repeat Request
AVSG	Accessible Volume Storage Group
AWGN	Additive White Gaussian Noise
BPSK	Binary Phase-Shift Keying
CCK	Complementary Code Keying
CDMA	Code-Division Multiple Access
CRC	Cyclic Redundancy Check
CSMA/CA	Carrier-Sense Multiple Access with Collision Avoidance
CSMA/CD	Carrier-Sense Multiple Access with Collision Detection

CTS	Clear to Send
D-CDMA	Deterministic CDMA
DCF	Distribution Coordination Function
DHT	Distributed Hash Table
DMT	Discrete Multitone
DSDV	Destination-Sequenced Distance-Vector
DSL	Digital Subscriber Line
DSR	Dynamic Source Routing
DSSS	Direct Sequence Spread Spectrum
FDMA	Frequency-Division Multiple Access
FHSS	Frequency-Hopping Spread Spectrum
GGSN	Gateway GPRS Support Node
GMSK	Gaussian Minimum Shift Keying
GPRS	General Packet Radio Service
GPS	Global Positioning System
GSM	Global System for Mobile Communications
HLR	Home Location Register
IFS	Interframe Space
IMSI	International Mobile Subscriber Identity
ISI	Intersymbol Interference
ISM	Industrial, Scientific, Medical
IV	Initialization Vector
L2CAP	Logical Link Control and Adaptation Protocol
LMP	Link Manager Protocol
LMU	Location Measurement Unit
MAC	Multiple Access Control
MDMS	Master Delay MAC Scheduling
MSC	Mobile Switching Center
NAP	Network Access Point
NAT	Network Address Translation
OBEX	Object Exchange
OFDM	Orthogonal Frequency-Division Multiplexing

OMA	Open Mobile Alliance
OVSF	Orthogonal Variable Spreading Factor
P2P	Peer-to-Peer
PCF	Point Coordination Function
PCS	Personal Communication System
PPM	Pulse Position Modulation
PPP	Point-to-Point Protocol
PSMM	Pilot Strength Measurement Message
PSTN	Public Switched Telephone Network
PSNR	Peak Signal-to-Noise Ratio
QAM	Quadrature Amplitude Modulation
QoS	Quality of Service
QPSK	Quadrature Phase-Shift Keying
R-CDMA	Random CDMA
RLP	Radio Link Protocol
RREP	Route Reply
RREQ	Route Request
RSSI	Received Signal Strength Indication
RTS	Request to Send
SACK	Selective ACK
SCO	Synchronous Connection-Oriented
SDMA	Space-Division Multiple Access
SDP	Service Discovery Protocol
SGSN	Serving GPRS Support Node
SIM	Subscriber Identification Module
SMLC	Serving Mobile Location Center
SRES	Signed Response
TCP	Transport Control Protocol
TKIP	Temporal Key Integrity Protocol
TLS	Transport Layer Security
TMSI	Temporary Mobile Subscriber Identity
TODA	Time Difference of Arrival

UDP	User Datagram Protocol
VLR	Visitor Location Register
WAE	Wireless Application Environment
WAP	Wireless Application Protocol
WDP	Wireless Datagram Protocol
WEP	Wired Equivalent Privacy
WLAN	Wireless Local Area Network
WML	Wireless Markup Language
WPA	Wi-Fi Protected Access
WSP	Wireless Session Protocol
WTA	Wireless Telephony Application
WTAI	Wireless Telephony Application Interface
WTLS	Wireless Transport Layer Security
WTP	Wireless Transaction Protocol
XML	Extensible Markup Language

PART I

ESSENTIALS OF WIRELESS COMMUNICATIONS

CHAPTER 1

THE MOBILE RADIO PROPAGATION CHANNEL

The beaten path is the safest. (*Via trita est tutissima.*)

—Latin proverb

1.1 INTRODUCTION

A communication system consists of a transmitter and a receiver connected by a *channel*. The channel is a *black box* representation of the actual physical medium where a signal at the channel input will produce a corresponding channel output after suffering from various *distortions*. Examples of a physical medium include telephone wire, coaxial cable, radio frequency (RF), and optical fiber. Different meda have different signal propagation characteristics and require different designs. In this chapter we focus on understanding the signal propagation characteristics of wireless channels.

Wireless Internet and Mobile Computing. By Yu-Kwong Ricky Kwok and Vincent K. N. Lau
Copyright © 2007 John Wiley & Sons, Inc.

The simplest wireless channel is the *additive white Gaussian Noise* (AWGN) channel where the output signal from the channel is given by:

$$y(t) = x(t) + z(t)$$

where $x(t)$ is the channel input and $z(t)$ is the white Gaussian channel noise. In this channel, the received signal composes of an undistorted transmit signal contaminated with channel noise. In Chapter 2, we discuss the design and performance of basic digital communication systems using AWGN channel as a simple example. Note that AWGN channel is quite an accurate model for deep space communications between earth station and geostationary satellites or spacecrafts.

On the other hand, for terrestrial wireless communications, the channel model is much more complicated due to the time-varying nature as well as the multipath nature of the wireless channel [271]. For instance, multipath refers to the situation that there are more than one propagation path between a transmitter and a receiver. The received signals from the multipaths are superimposed with each other. Depending on the phase difference (or path difference) of the multipaths, the superposition may be *constructive* or *destructive*. When the superposition is constructive (path difference around $n\lambda$ for some integer n where λ is the carrier wavelength), the received signal will be strong. On the other hand, when the superposition is destructive (path difference around $n\lambda/2$ for some integer n), the received signal will be quite weak. This is illustrated in Figure 1.1

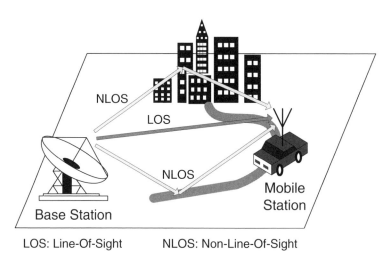

Figure 1.1   Illustration of multipath in terrestrial wireless communications.

Typically, the signal envelop as a result of multipath superposition can fluctuate over 30dB or more over a span of a few λ in distance. Suppose the carrier frequency is 3GHz, the wavelength λ is around 10cm and the large fluctuation of signal strength (35 dB) can be observed with a span of ~30cm which is quite short.

On the other hand, time variation of the wireless channel results from the mobility of the transmitter, the receiver or the environment (e.g., scatterer). Hence, the number of multipaths, the strength of multpaths as well as the delays of the multipaths can be time varying.

To capture different time scale of the distortions introduced by the wireless channels, we introduce a *three-level model*. The first level model is called the *large scale path loss model*. The path loss model focuses on the study of the long-term or large scale variations on the average received signal strength due to the variation of distance from the transmitter and the receiver. The second level model is called the *shadowing model* which focuses on the study of the medium term variation on the medium-term average (over hundreds of λ) received signal strength due to local changes of terrain features or man-made obstacles (such as blockage or coverage shadows). The third level model is called the *microscopic fading model* which focuses on the study of the short-term received signal strength (over a few λ) due to constructive and destructive interference of the multipath. Figure 1.2 illustrates the three level models of wireless channels.

The path loss model is usually used for system planning, coverage analysis and link budget [271]. On the other hand, the shadowing model is usually used for power control analysis, second order interference analysis as well as more detailed coverage and link budget analysis [152]. Finally, the microscopic fading model is usually used for the detail design of the physical layer transmitter and receiver such as coding, modulation, interleaving, etc. [320]. In this chapter, we elaborate the three level models using a qualitative approach in the following sections.

1.2 LARGE SCALE PATH-LOSS

Path loss model focuses on the study of the large scale variation on the average received signal strength due to variation in distance between the transmitter and the receiver [203]. The simplest path loss model is the *free space path loss model* where the averaged received signal strength is inversely proportional to the square of the distance between the transmitter and the receiver. For instance, the received signal power P_r is given by:

$$P_r = \frac{P_t G_t G_r c^2}{16\pi^2 d^2 f^2}$$

where c is the speed of light, P_t is the transmitted power, G_t and G_r are the transmit and receive antenna gains, d is the distance separation between the transmitter and the receiver and f is the carrier frequency of the transmitted signal. In dB form, the received signal power is given by:

$$P_r(dB) = P_t(dB) + K - 20\log_{10} d - 20\log_{10} f.$$

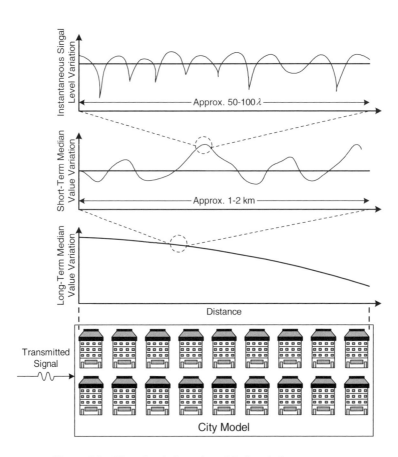

Figure 1.2   Three level channel models for wireless systems.

Define the path loss as:

$$PL(dB) = P_t(dB) - P_r(dB).$$

Hence, the free space path loss equation[23] is given by:

$$PL(dB) = -K + 20\log_{10} d + 20\log_{10} f$$
$$= PL(d_0) + 20\log_{10} \frac{d}{d_0} \tag{1.1}$$

for some reference distance d_0. The path loss equation indicates how fast the received signal strength drops with respect to the change in distance. For example, the received signal strength is reduced by 4 times if we double the distance between transmitter and the receiver or double the carrier frequency. The inverse square law is a direct result of the *Maxwell wave equation* [23] in free space and can be used to model the signal loss due to *line-of-sight* (LOS) propagation mode where the transmitter and receiver can *see* each other.

The free space path loss equation is not accurate to model the path loss in terrestrial wireless communications with multipaths from the transmitter to the receiver. In general, the path loss model can be represented by the equation:

$$PL(dB) = PL(d_0) + 10n\log_{10}(d/d_0)$$

where n is the *path loss exponent* which depends on the environment. Table 1.1 illustrates the values of path loss exponents under various environment.

Table 1.1   Path loss exponents.

Environment	Path Loss Exponent
In building line-of-sight	1.6–1.8
Free space	2
Obstructed in factories	2–3
Urban area cellular	2.7–3.5
Shadowed urban cellular	3–5
Obstructed in building	4–6

The higher the path loss exponent is, the faster the signal strength drops with respect to the increase in distance. For example, a commonly assumed path loss exponent in non line-of-sight environment is $n = 4$. In this case, doubling the distance separation between the transmitter and the receiver will result in 16 times reduction in received signal strength.

In some more complicated propagation environments such as irregular terrain and cities, there is no simple analytical path loss model. Empirical models based on extensive channel measurements are used to model the path loss versus distance in such complex environments. Examples are the Okumura model [261], Hata model and COST 231 extension to the Hata model [316] for cellular systems simulations and link budget analysis.

1.3   SHADOWING EFFECTS

In the shadowing model, we are interested to study the medium term variation in received signal strength when the distance between the transmitter and receiver is fixed. For example, image a mobile station is circling around a base station with radius r. As the mobile moves, one expects some medium term fluctuations in the received signal strength but the variation is not due to the path loss component because the distance between the transmitter and the receiver is not changed. This signal variation is due to the variations in terrain profile such as variation in the blockage due to trees, buildings, hills, etc. This effect is called shadowing. Consider a signal undergoing multiple reflections (each with an attenuation factor a_i) and multiple diffractions (each with an attenuation factor b_i) as illustrated in Figure 1.3

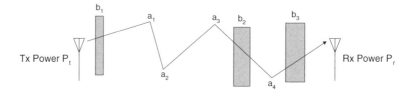

Figure 1.3   Shadowing model—variations in path loss predication due to buildings, trees, hills, etc.

The received signal strength is given by:

$$P_r = P_t \Pi_{i=1}^{N_r} a_i \Pi_{i=1}^{N_d} b_i \tag{1.2}$$

where N_r and N_t denote the number of obstacles with reflecting and diffracting the signal respectively. Expressing the received power in dB, we have:

$$P_r(dB) = P_t(dB) + \sum_{i=1}^{N_r} a_i(dB) + \sum_{j=1}^{N_d} b_i(dB)$$

$$= P_t(dB) + \sum_i \alpha_i(dB)$$

where α_i is the attenuation coefficient due to reflections or diffractions. Each of the term $\alpha_i(dB)$ represents a random and statistically independent attenuation. As the number of reflectors and diffractors increase, by central limit theorem, the sum

$S(dB) = \sum_i \alpha_i(dB)$ approaches a Gaussian random variable. Hence, the received signal power can be expressed as:

$$P_r(dB) = S(dB) + P_t(dB)$$
$$= \mu_S(dB) + X(dB) + P_t(dB) \tag{1.3}$$

where $S(dB)\,\mathcal{N}(\mu_S, \sigma^2)$ and $X(dB)\,\mathcal{N}(0, \sigma^2)$. The mean shadowing μ_S is usually included into the path loss model and that is why the path loss exponent can be larger than 2. Hence, without loss of generality, we move the term $\mu_S(dB)$ to the path loss model and consider only the shadowing effect $X(dB)$. Expressing the received power in linear scale, we have:

$$P_r = P_t 10^{X/10} = A_s P_t$$

where $A_s = 10^{X/10}$ is the power attenuation due to shadowing and is modeled by the *log-normal distribution* with standard derivation σ (in dB).

Combining the path loss model and the shadowing model, the overall path loss is given by:

$$PL(dB) = PL_{av}(dB) + X(dB)$$

where PL_{av} is the path loss component obtained from the large scale path loss model and $X(dB)$ is the shadowing component which is modeled as a zero-mean Guassian random variable with standard derivation σ in dB.

1.4 SMALL SCALE MULTIPATH FADING EFFECTS

The third component of variation of received signal strength is called the small scale fading or *microscopic fading*. In the small scale fading component, the signal strength fluctuates over 30dB within a very short time scale of the order of milli-second as illustrated in Figure 1.4. There are two independent dimensions in microscopic fading, namely the *multipath dimension* and the *time-varying dimension*. They are elaborated below.

Figure 1.5 illustrates the concept of multipath dimension and time varying dimension of microscopic fading. For example, a narrow pulse transmitted at $t = t_0$ results in one resolvable received pulse. The received pulse in general has a different shape compared with the transmitted pulse because of multipaths propagation. At $t = t_1$, the same transmit pulse results in two resolvable pulses with the first pulse stronger than the second pulse. At $t = t_2$, the same transmit pulse results in two resolvable pulses with the first pulse weaker than the second pulse. Note that the multipath dimension and the time-varying dimension are two independent dimensions. We quantify the two dimensions below.

Multipath dimension: To quantify the multipath dimension in microscopic fading, we can either look at the *delay spread* or *coherence bandwidth*. Delay spread

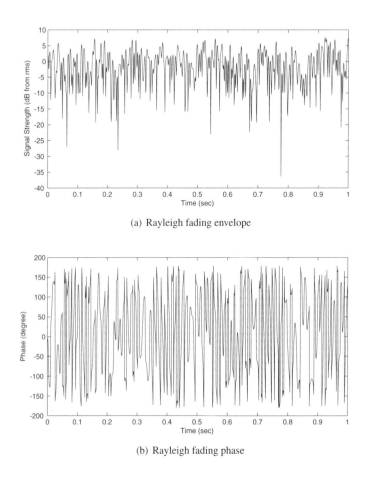

(a) Rayleigh fading envelope

(b) Rayleigh fading phase

Figure 1.4   Illustration of the envelope and phase variations in microscopic fading.

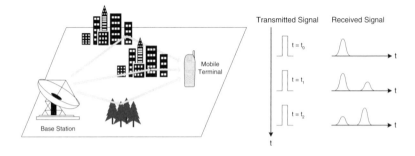

Figure 1.5   Illustration of the multipath and time-varying dimensions in microscopic fading.

is defined as the range of multipath components with *significant* power when an impulse is transmitted as illustrated in Figure 1.6 In Figure 1.6(a), a narrow pulse (in time) at the transmitter results in a spread of energy across the delay dimension. The graph of received power versus delay is called the *power-delay profile* and is used to measure the delay spread experimentally. Typically, the delay spread in enclosed indoor environment is smaller than the outdoor environment because the range of delays in the multipaths are more contained in the indoor environment.

On the other hand, to quantify the multipath dimension [167], an equivalent parameter, namely the *coherence bandwidth*, can be used. Figure 1.6(b) illustrates the concept of *coherence bandwidth*. Suppose at the transmitter, we transmit two single tone signals (impulse in frequency domain) with unit power and frequencies f_1 and f_2 respectively. At the receiver, the two single tones will be received with received powers α_1 and α_2 respectively. α_1 and α_2 represent the random channel attenuation (or fading) experienced by the two single-tone signals. It is found that if $|f_1 - f_2| > B_c$, the attenuation α_1 and α_2 are uncorrelated. Otherwise, the attenuation α_1 and α_2 are correlated. Hence, the minimum separation of frequency such that uncorrelated fading is resulted is called the *coherence bandwidth*, B_c. It is found that for 0.5 correlation coefficient between α_1 and α_2, the coherence bandwidth is related to the delay spread σ_τ by:

$$B_c \approx \frac{1}{\sigma_\tau}$$

Typical values of the coherence bandwidth for indoor and outdoor environment are 1MHz and 100kHz respectively. Note that delay spread and coherence bandwidth are two sides of the same coin to characterize the multipath dimension of microscopic fading.

Time Varying dimension: To characterize the time variation dimension, we can have either the *Doppler spread* or *coherence time*. Doppler spread [167] refers to the spread of frequency introduced by the fading channel when we transmit a single tone signal (narrow pulse in frequency domain) as illustrated in Figure 1.7(a). The spread in frequency is due to the mobility between the transmitter and the receiver or the mobility of the surrounding obstacles. The maximum Doppler spread is given by:

$$f_d = \frac{v}{\lambda}$$

where λ is the wavelength of the signal and v is the maximum speed between the mobile and the base station. Typical values of Doppler spread for pedestrian users (5km/hr) and vehicular users (100km/hr) at carrier frequency of 2.4 GHz are given by 14 Hz and 300 Hz respectively.

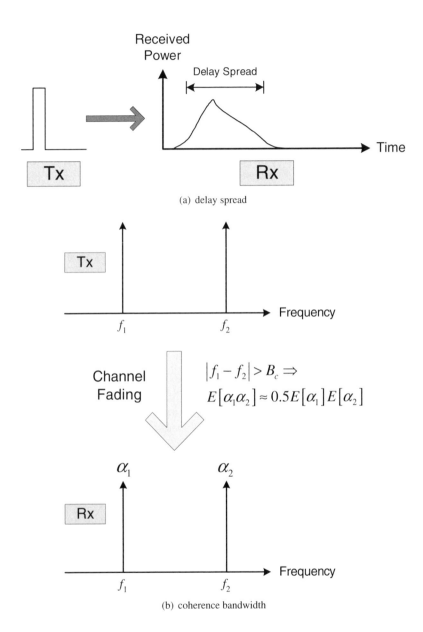

Figure 1.6   Illustration of the delay spread and coherence bandwidth.

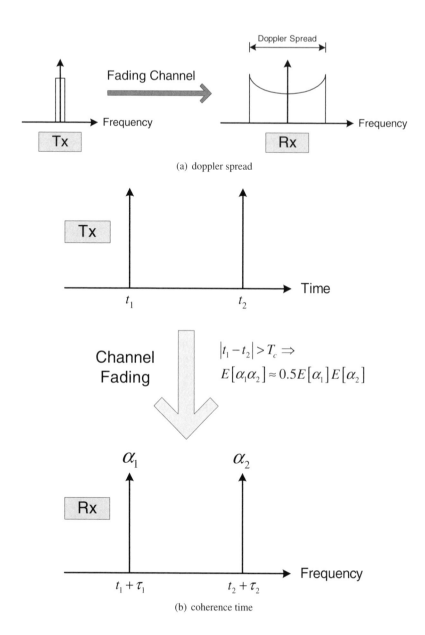

(a) doppler spread

Channel Fading

$$|t_1 - t_2| > T_c \Rightarrow$$
$$E[\alpha_1\alpha_2] \approx 0.5E[\alpha_1]E[\alpha_2]$$

(b) coherence time

Figure 1.7   Illustration of the doppler spread and coherence time.

Similarly, an equivalent parameter to quantify the time variation dimension of microscopic fading is given by the *coherence time*. Figure 1.7(b) illustrates the concept of coherence time. Suppose we transmit two narrow pulses in time domain at times $t = t_1$ and $t = t_2$ with unit energy. After channel fading, the corresponding signals will be received at $t = t_1 + \tau_1$ and $t = t_2 + \tau_2$ with energies α_1 and α_2 respectively. Similar to coherence bandwidth, if we separate the two transmissions sufficiently far in time, ($|t_1 - t_2| > T_c$), the channel fading α_1 and α_2 are uncorrelated. Hence, the coherence time is defined as the minimum time separation at the transmitter in order to have uncorrelated fading at the receiver. It is found that for 0.5 correlation coefficient between α_1 and α_2, the coherence time T_c is related to the Doppler spread f_d by:

$$T_c \approx \frac{1}{f_d}$$

Note that Doppler spread and coherence time are two sides of the same coin to characterize the time variation dimension of microscopic fading.

1.4.1   Flat Fading vs Frequency Selective Fading

Based on the delay spread and the transmitted symbol duration, (or equivalently, the coherence bandwidth and the transmitted bandwidth), the multipath fading can be classified as *frequency flat fading* or *frequency selective fading* [287]. A fading channel is said to be *frequency flat* if $\sigma_\tau < T_s$ or equivalently, $B_c > W_{tx}$ where T_s and W_{tx} are the symbol duration and the transmitted signal bandwidth, σ_τ and B_c are the delay spread and the coherence bandwidth, respectively. Otherwise, the fading channel is said to be *frequency selective*. Note that whether the transmitted signal will experience frequency flat fading or frequency selective fading depends on both the *channel parameters* (such as σ_τ, B_c) as well as the *transmitted signal parameters* (such as T_s and W_{tx}).

When a transmitted signal experience frequency flat fading, the received signal consists of superposition of multipath signals but the multipaths have delays much smaller than the symbol duration T_s and hence, the multipaths are *unresolvable*. As a result of the unresolvable multipath, the received pulse shape will be different from the transmitted pulse shape but the received pulse appears as one single pulse at the receiver. The received signal is modeled as:

$$y(t) = h(t)x(t - \tau) + z(t)$$

where $h(t)$ is the time-varying channel fading which is modeled as complex Gaussian random variable with zero mean and unit variance, $z(t)$ is the complex AWGN noise and $x(t)$ is the low-pass equivalent complex transmitted signal.

On the other hand, when a transmitted signal experience frequency selective fading, the received signal consists of multipaths with delay » symbol duration (T_s). Hence, the received signal consists of pulses at *resolvable delays* (integral multiples of T_s). In general, the number of resolvable multipaths is given by:

$$L_p = \lceil \frac{\sigma_\tau}{T_s} \rceil = \lceil \frac{W_{tx}}{B_c} \rceil$$

The received signal for frequency selective fading channel is modeled as:

$$y(t) = \sum_{i=1}^{L_p} h_i(t)x(t - iT_s) + z(t)$$

where $h_i(t)$ is the time varying channel fading for the i-th resolvable multipaths which is modeled as complex Gaussian random variable with zero mean and unit variance.

1.4.2   Fast Fading vs Slow Fading

Similarly, based on the Doppler spread and the transmitted bandwidth (or equivalently the coherence time and the symbol / frame duration), the fading channel is classified as *fast fading* or *slow fading*. A fading channel is classified as *fast fading* if the symbol duration T_s is larger than the coherence time T_c. Otherwise, the fading channel is called *slow fading*. Note that whether a transmitted signal experiences fast fading or slow fading depends on both the channel parameters (such as the Doppler spread or coherence time) and the transmitted signal parameters (such as the transmitted bandwidth or the symbol duration). Hence, for slow fading channels, the channel fading coefficients $h_i(t)$ remains to be quasi-static within a symbol duration T_s.

Note that sometimes, we define slow fading to be the case when T_c (coherence time) is larger than T_f (frame duration) instead of symbol duration.

1.5   PRACTICAL CONSIDERATIONS

In this section, we briefly discuss the implications of path loss, shadowing and microscopic fading components on system design. In fact, the design and performance of communication systems depends heavily on the underlying channels. For instance, path loss exponent determines how fast the received signal strength attenuates with respect to distance separation between the transmitter and receiver. For a point to point digital link, higher path loss exponents results in faster signal attenuation and therefore is undesirable. This is because for the same transmitter and receiver design, a higher path loss exponent results in shorter communication range (for the same transmit power) or higher required transmit power (for the same distance) to overcome the higher path loss. This holds in general for *noise-limited systems*. However, if we consider multicell systems such as cellular networks like GSM, or CDMA systems, higher path loss exponent results in more confined interference (as the interference signal cannot propagate very far) and this translates into higher system capacity because more aggressive resource reuse can be realized. In general, higher path loss exponent is desirable for *interference-limited systems*. We will elaborate the cellular system designs in Chapter 3.

Shadowing introduces randomness in the coverage of cellular systems. For instance, if the standard derivation of the shadowing component is large, the average received signal power at the cell edge will have large fluctuations. To achieve a certain Quality of Service (QoS) such as 90% probability that the received signal strength at the cell edge is above a target threshold, higher power or shorter cell radius is needed to allow for some shadowing margins to satisfy the QoS target. Hence, shadowing with larger standard derivation is undesirable.

Finally, the effects of microscopic fading have high impact on the physical layer design of communication systems. For instance, by common sense, it is more challenging to setup a wireless link to transmit high quality video then audio signals. This is because for the same environment, transmitting a video signal generally involves a higher transmit bandwidth, W_{tx}, relative to an audio signal. From Section 1.4.1, it is very likely that the video signal will experience frequency selective fading while the audio signal will experience frequency flat fading. As we elaborate in Chapter 4, frequency selective fading will introduce *intersymbol interference* (ISI) and this induces irreducible error floor. Hence, complex equalization at the receiver is needed. On the other hand, for audio signal, since $W_{tx} << B_c$, the signal will experience flat fading only and no equalization at the receiver is needed. Hence, this results in more simple design. Another common-sense example is that it is easier to setup a wireless link for indoor environment than an outdoor environment to transmit the same signal. This again can be analyzed by the frequency selective or frequency flat fading channels. For the same transmit signal, the indoor environment in general has a larger coherence bandwidth and hence, the number of resolvable multipaths in indoor environment will be small. On the other hand, the coherence bandwidth for outdoor environment in general will be small and this results in larger number of resolvable multipaths (frequency selective fading). Hence, more complicated designs are needed in the latter case.

In general, it is imperative to understand the target operating environment in order to optimize the transmitter and receiver design of the communication systems. There is no such thing as a *universal communication systems* that is effective for all communication environments. For examples, both wireless LAN systems and 3G networks are designed to offer high bit rate. For IEEE 802.11g system, the highest bit rate is 54Mbps. For UMTS systems (Rel 99), the highest bit rate is just 2Mbps. Obviously, the cost of the UMTS system is much higher than the wireless LAN systems. Hence, one might query about why we still require UMTS system when wireless LAN systems cost only less than US$100. Figure 1.8 illustrates a more comprehensive comparison between Wi-Fi and UMTS systems. We do not just look at the maximum bit rate offered by both systems but we also have to compare the target environment of operations. For instance, while Wi-Fi systems offer higher bit rate, they are designed to operate at pedestrian mobility and indoor coverage. For 3G/4G cellular systems, although their maximum bit rate is lower than Wi-Fi systems, they are designed to operate at very high mobility and macroscopic coverage.

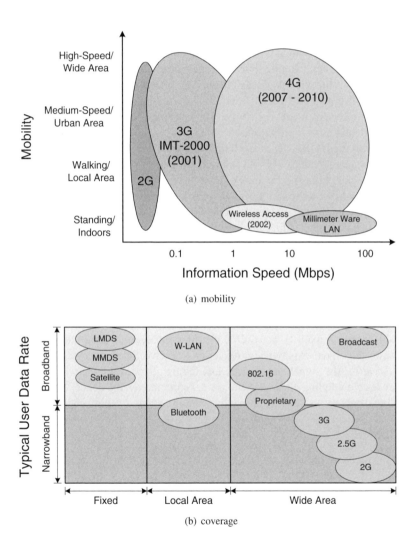

Figure 1.8 A comparison between WiFi systems and 3G cellular systems.

1.6 SUMMARY

In this chapter, we have elaborated on the modeling of wireless channels. We have introduced the 3 levels of channel models, namely the *large scale path loss model*, the *medium scale shadowing model* as well as the *microscopic scale fading model*. The path loss model deals with the variation of received signal strength with respect to variation of distance between the transmitter and the receiver and is focused on the time scale of seconds. The path loss between a transmitter and a receiver is characterized by a path loss exponent. In free space or line of sight propagation condition, the path loss exponent is 2 meaning that the signal power will be reduced by 4 times for every double in the distance separation. In non line-of-sight propagation environment, the path loss exponent can reach 4 or above. The larger the path loss exponent, the faster the signal attenuates as it propagates.

Shadowing model deals with medium scale variation of received signal strength when the distance is fixed. This is contributed by the variation in the terrain profile such as obstacles, hills and buildings. Due to law of large number, the shadowing effects (received signal strength variations) can be modeled by *log-normal shadowing* component and parameterized by the standard derivation $\sigma(dB)$.

Finally, microscopic fading deals with small scale variation of received signal strength due to constructive and destructive multipath superpositions. The time scale of interests can be of millisecond order. Microscopic fading can be parameterized by the *delay spread* (*coherence bandwidth*) for the multipath dimension as well as the *Doppler spread* (*coherence time*) for the time variation dimension. Note that the multipath dimension and the time variation dimension are two *independent* dimensions. The number of resolvable multipaths is given by $L_p = \lceil W_{tx}/B_c \rceil$. When there is only one resolvable multipath ($W_{tx} < B_c$), the signal experiences flat fading channels. Otherwise, the signal experiences frequency selective fading channels. Similarly, when $T_s > T_c$, the signal experiences fast fading channels. Otherwise, the signal experiences slow fading channels. Figure 1.9 summarizes the concept of frequency flat fading, frequency selective fading, fast fading and slow fading channels.

PROBLEMS

1.1 A wideband signal is more difficult to transmit than a narrowband signal because of the inter-symbol interference problem faced by the wideband signal. [True/False]

1.2 In flat fading channels, there is no multipath effect at all as there is only a single resolvable echo at the receiver. [True/False]

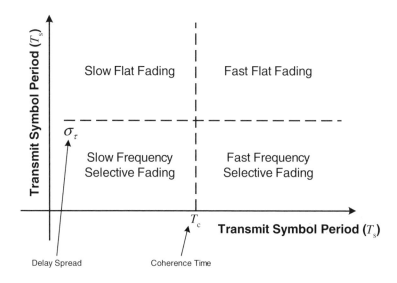

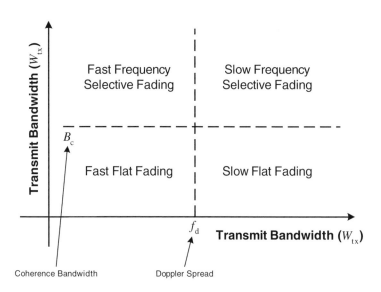

Figure 1.9 Summary of classifications of fading channels.

CHAPTER 2

MODULATION TECHNIQUES

The whole earth was of one language, and of one speech.

—Genesis XI, 1, c. 700 B.C.

2.1 INTRODUCTION

In this chapter, we introduce a generic model for digital modulation and demodulation design based on the concept of signal space. We focus on the point-to-point communications where there is a single transmitter, a single receiver, and a channel connecting the transmitter and receiver, as illustrated in Figure 2.1

Performance:

In Chapter 1, we have discussed various channel models for wireless media. In general, the design of the transmitter and the receiver has to match the characteristics of the channels. There does not exist a *universal communication system* that is optimized in all different operating environments. In this chapter, we elaborate on the design of digital modulator and digital demodulator for various wireless channels.

Wireless Internet and Mobile Computing. By Yu-Kwong Ricky Kwok and Vincent K. N. Lau
Copyright © 2007 John Wiley & Sons, Inc.

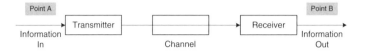

Figure 2.1   Point-to-point communication link.

To study the digital communication design in Figure 2.1, we discuss and quantify the tradeoff between the performance dimensions and the cost dimensions. Specifically, the performance dimension of a digital communication link refers to *bit rate* and *bit error rate* (or packet error rate). Bit rate, which is defined as the number of bits delivered to the receiver per second (bits / sec), is a measure of how fast information bits are transferred in the communication link. Obviously, from a user perspective, we would like the bit rate to be as high as possible. Bit error rate (BER), on the other hand, is a measure of the *reliability* of the communication link. BER is defined as the number of bits in error over the total number of bits transmitted. From a user perspective, we would like the BER to be as low as possible.

On the other hand, the cost dimension of a digital communication link refers to the *transmit power* and *channel bandwidth*. Obviously, a communication link that requires a higher power or larger channel bandwidth will be most costly to operate. Hence, a very important focus of this chapter is to understand and study the detail dynamics behind the tradeoff of the *performance* and *cost* parameters. The tradeoff relationship is summarized in Figure 2.2

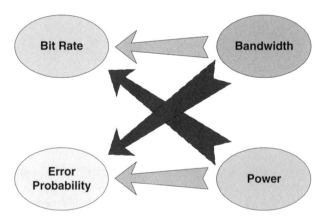

Figure 2.2   Tradeoffs of the performance and cost in digital communication systems.

We first introduce the concept of signal space[287] and how to represent time domain signals using vectors in the signal space[147] in Section 2.2. We then discuss modulation techniques in Section 2.3. Next, we consider the design and performance of detection techniques over a simple *additive white Gaussian noise* channel (AWGN) in Section 2.4. Finally, we give some practical illustrations on the application of digital modulations in real-life systems in Section 2.6. We conclude with a brief summary of main points in Section 2.7.

2.2 SIGNAL SPACE AND GEOMETRIC REPRESENTATION OF SIGNALS

While the physical transmitted signals and received signals are all in continuous time domain, it is quite difficult to gain proper design insights by working in the original forms. For example, in digital communication receiver, the primary task is to deduce what has been transmitted at the transmitter given a noise observation (or received signal) $y(t)$. It is not very obvious on what should be the optimal structure to extract the information bits carried by the signal. However, as we illustrate, if we represent the time domain signals as points or *vectors* on a *signal space*, the design of the digital receiver will be much easier. Before we discuss the detail operation of the digital modulator and digital receiver, we try to introduce the concept of *signal space* and how to represent continuous time signals as *points* or *vectors* in the signal space.

If we want to discuss the coordinate of a point, say $(1, 2)$, in Euclidean space, we have to first establish what is the *frame of reference* or equivalently, what is the x-axis and y-axis. In other words, before knowing the coordination reference, it is meaningless to discuss the coordinate of a point. Similarly, when we say we want to represent a time domain signal $x(t)$ by a point $\vec{x}$ in the signal space, we mean expressing the signal $x(t)$ into a coordinate $\vec{x} = (x_1, \ldots, x_D)$ where D is the dimension of the signal space. However, for similar reasons, before we can discuss the coordinate of a point, we have to first establish the *frame of reference* or *axis* of the signal space.

In fact, signal space and Euclidean space are two specific examples of vector space. Recall that a vector space is a collection of objects that satisfies a number of *axioms*[258]. Hence, signal space is simple a vector space where the *objects* or the vectors are time domain signals. We define the *dot-product* or *inner product* between two signals $x(t)$ and $y(t)$ (with duration T_s) in the signal space as:

$$< x(t), y(t) >= \int_0^{T_s} x(t)y^*(t)dt. \tag{2.1}$$

To specify a $D-$dimensional signal space, we need D orthonormal basis functions $\{\phi_1(t), \ldots, \phi_D(t)\}$ where

$$\langle \phi_i(t), \phi_j(t) \rangle = \begin{cases} 1 & \text{if } i = j \\ 0 & \text{otherwise} \end{cases}$$

Note that the D *basis functions* in signal space play the same role as the D *axis* in Euclidean space.

Suppose we would like to represent any M time domain signals $\{x_1(t), \ldots, x_M(t)\}$ with finite symbol duration T_s into M vectors $\{\vec{x}_1, \ldots, \vec{x}_M\}$ in a signal space. We have to go through the following two steps.

Step 1. Find out the basis functions: The first step is to find out the basis functions or frame of reference of the underlying signal space that contains the M signals. It can be shown that given any M time domain signals $\{x_1(t), \ldots, x_M(t)\}$ with finite duration T_s, there exists a D dimensional signal space with $D \leq M$ that can contain these M signals. The corresponding orthonormal basis $\{\phi_1(t), \ldots, \phi_D(t)\}$ for the D-dimensional signal space can be obtained by the *Gram-Schmidt Orthogonal Procedure*[283] as illustrated below.

(i): The first basis function $\phi_1(t)$ is given by:

$$\phi_1(t) = x_1(t)/\sqrt{E_1}$$

where $E_1 = \int_0^{T_s} |x_1(t)|^2 dt$.

(ii): The second basis function $\phi_2(t)$ is obtained by first computing the projection of $x_2(t)$ onto $\phi_1(t)$, which is:

$$c_{1,2} = \int_0^{T_s} x_2(t)\phi_1^*(t)dt.$$

Then $c_{1,2}\phi_1(t)$ is subtracted from $x_2(t)$ to yield

$$f_2(t) = x_2(t) - c_{1,2}\phi_1(t).$$

Note that $f_2(t)$ is orthogonal to $\phi_1(t)$ because $< f_2(t), \phi_1(t) > = 0$. Normalizing $f_2(t)$ to unit energy signal, we have:

$$\phi_2(t) = f_2(t)/\sqrt{E_2}$$

where $E_2 = \int_0^{T_s} |f_2(t)|^2 dt$.

(iii): In general, the k-th orthogonal basis $\phi_k(t)$ is given by:

$$\phi_k(t) = f_k(t)/\sqrt{E_k}$$

where

$$f_k(t) = x_k(t) - \sum_{i=1}^{k-1} c_{i,k}\phi_i(t)$$

and

$$c_{i,k} = < x_k(t), \phi_i(t) > = \int_0^{T_s} x_k(t)\phi_i^*(t)dt.$$

The orthogonalization process continues until all the M signals $\{x_1(t), \ldots, x_M(t)\}$ have been exhausted and $D \leq M$ orthonormal basis functions have been constructed.

Step 2. Find out the coordinates of the M points: After the D orthonormal basis $\{\phi_1(t), \ldots, \phi_D(t)\}$ have been found from Step 1, we can determine the coordinates of the M vectors. For instance, let $\vec{x}_m = (x_{m1}, x_{m2}, \ldots, x_{mD})$ be the coordinates of the $m-$th point. The j-th coordinate of the m-th point is given by:

$$x_{mj} = <x_m(t), \phi_j(t)> = \int_0^{T_s} x_m(t)\phi_j^*(t) \tag{2.2}$$

Example 1 (Geometric Representation of 4 time domain signals) *Let $x_1(t), x_2(t), x_3(t), x_4(t)$ be 4 time domain signals with duration $T_s = T$ as illustrated in Figure 2.3*

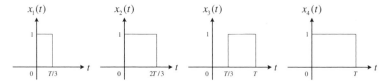

Figure 2.3   Illustration of the time domain waveforms of the 4 signals in example 1.

Following the approach in step 1, we requires a 3-dimensional signal space to contain all the 4 signals. The basis functions of the signal space $\{\phi_1(t), \phi_2(t), \phi_3(t)\}$ are illustrated in Figure 2.4

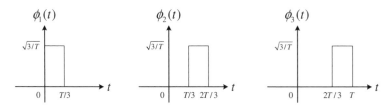

Figure 2.4   Basis functions of the signal space in example 1.

Following the step 2, the coordinates of the 4 time domain signals are given by $\vec{x}_1 = (\sqrt{T/3}, 0, 0)$, $\vec{x}_2 = (\sqrt{T/3}, \sqrt{T/3}, 0)$, $\vec{x}_3 = (0, \sqrt{T/3}, \sqrt{T/3})$ *and* $\vec{x}_4 = (\sqrt{T/3}, \sqrt{T/3}, \sqrt{T/3})$ *as illustrated in Figure 2.5*

Hence, given any M time domain signals, we can find a signal space of at most M dimension to represent these M signals as M points. Notation-wise, we denote such process as

$$x_m(t) \longrightarrow \vec{x}_m = (x_{m1}, \ldots, x_{mD}).$$

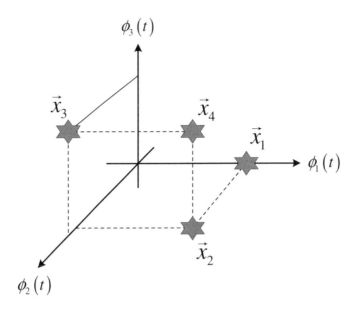

Figure 2.5   Geometric representation of the 4 signals in example 1.

One point to emphasis is that nomatter how we describe the M signals (either in time domain or in geometric domain), the physical nature of the signal is not changed. For example, the energy of a signal is defined in time domain as:

$$E = \int_0^{T_s} |x(t)|^2 dt \tag{2.3}$$

However, by the definition of the *dot-product* in the signal space in (2.1), we have

$$< x(t), x(t) >= \int_0^{T_s} |x(t)|^2 dt = E = \|\vec{x}\|^2. \tag{2.4}$$

In other words, the energy of a signal $x(t)$ is also given by the square of the length of the vector $\vec{x}$. A signal point further away from the origin implies higher energy. Similarly, the power of a signal in time domain is defined as:

$$P = \frac{1}{T_s} \int_0^{T_s} |x(t)|^2 dt \tag{2.5}$$

Using the definition of dot-product, we have the power of the signal also given by:

$$P = \frac{1}{T_s} \|\vec{x}\|^2. \tag{2.6}$$

2.3   MODULATION DESIGN AND SIGNAL CONSTELLATIONS

In Section 2.2, we have introduced the concept of representing time domain signals as *points* or *vectors* on a signal space. We apply this *geometric representation* of signals to model the digital communication link design. In this section, we elaborate on the digital transmitter (or modulator). A digital modulator[296] can be modeled by a *black box* with information bits $\{b_1, b_2, \ldots\}$ as inputs and modulation symbols $s(t)$ as output. The digital modulator is parameterized by:

Symbol Duration T_s: The symbol duration T_s specifies how often the modulator produces a *modulation symbol*. The *baud rate* or *symbol rate* of a digital modulator is defined as the number of modulation symbols per second, which is given by:

$$R_{baud} = \frac{1}{T_s} \tag{2.7}$$

The baud rate or symbol rate is a very important parameter because it determines the required channel bandwidth of a digital modulator.

Signal Set S: The signal set of a digital modulator is a collection of M time domain signals given by:

$$S_M = \{s_1(t), \ldots, s_M(t)\} \tag{2.8}$$

for some integer M. Each of the signal element in the signal set S is labeled with a unique bit pattern which will be explained below.

The operation of a $M - ary$ digital modulator is very simple. For simplicity, we consider the *memoryless* modulator where the encoding of information bits is done on a *symbol-by-symbol* basis. At the n-th symbol duration, the digital modulator accepts $\log_2(M)$ bits as input. Depending on the bit pattern of the input bits, one of the M modulation signals (with exactly the same label as the input bit pattern) in the signal set S_M is chosen as the modulation symbol output. For example, when $M = 2$ (binary modulator), the modulator accepts one single bit on every symbol duration. The signal set $S_2 = \{s_1(t), s_2(t)\}$ has the bit label as $s_1(t)$ associated with bit value = 0 and $s_2(t)$ associated with bit value = 1. When the input bit is 0, $s_1(t)$ will be produced at the modulator output during the current symbol duration and vice versa. After *modulation process* is completed for the current symbol duration, we proceed to the next symbol duration and fetch a new information bit for modulation. Figures 2.6(a) and 2.6(b) illustrate the output waveforms of two binary modulators with different signal sets. In both binary modulators, the symbol durations are $T_s = 1$. In the first binary modulator, the signal set is given by $S_2(A) = \{s_1^{(A)}(t), s_2^{(A)}(t)\}$ which is illustrated in Figure 2.7(a). In the second binary modulator, the signal set is given by $S_2(B) = \{s_1^{(B)}(t), s_2^{(B)}(t)\}$ which is illustrated in Figure 2.7(b).

Note that for binary modulators, the bit rate R_b is given by:

$$R_b = \frac{1}{T_b} = R_{baud} \tag{2.9}$$

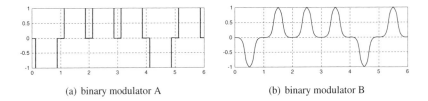

(a) binary modulator A   (b) binary modulator B

Figure 2.6 Illustration of Binary Modulations.

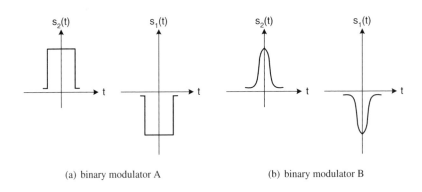

(a) binary modulator A   (b) binary modulator B

Figure 2.7 Illustration of the Signal Sets of the Binary Modulators A and B.

where T_b denotes the bit duration. In other words, the bit rate is the same as baud rate for binary modulators. Hence, it is obvious that increasing the channel bandwidth W_{tx} can result in an increase of information bit rate as indicated in the tradeoff diagram in Figure 2.2

On the other hand, for 4-ary modulator ($M = 4$), the signal set $\mathcal{S}_4 = \{s_1(t),\ldots,s_4(t)\}$ consists of 4 distinct signal elements which are labeled as $\{00, 01, 10, 11\}$. During each symbol duration, 2 bits of information is fetched and modulated. Depending on the specific bit pattern observed, the signal with the corresponding label will be produced at the output. Similarly, we can extend the operation of a general M-ary modulator by the *black box* model. For $M-$ary modulator, since one symbol can be considered as carrying $\log_2 M$ information bits, the bit rate is given by:

$$R_b = \frac{1}{T_b} = \log_2 M \frac{1}{T_s}. \tag{2.10}$$

On the other hand, the transmission bandwidth of a digital modulator with signal set $\mathcal{S}$ is in general given by [287]:

$$W_{tx} \propto \text{bandwidth} \left[\frac{1}{M} \sum_{m=1}^{M} |S_m(f)|^2 \right] \tag{2.11}$$

where $|S_m(f)|$ is the amplitude spectrum of the m-th signal $s_m(t)$ in the signal set and W_{tx} denotes the required channel bandwidth. For example, if the signal set $\mathcal{S}$ consists of *baseband signals* (all with pulse duration T_s), the required transmission bandwidth (one-sided lowpass bandwidth) is given by:

$$W_{tx} = (1 + \alpha)\frac{1}{2T_s} \tag{2.12}$$

where $\alpha \geq 0$ is a constant determined by the *pulse shape* of the modulation symbol. If the signal set $\mathcal{S}$ consists of *bandpass signals* (all with pulse duration T_s), the required transmission bandwidth (one-sided bandpass bandwidth) is given by:

$$W_{tx} = (1 + \alpha)\frac{1}{T_s} \tag{2.13}$$

Finally, the average transmit power of the modulator (assuming equiprobable input information bit) is given by:

$$P_{tx} = \frac{1}{MT_s}\|\vec{s}_m\|^2 = \frac{1}{MT_s}\int_0^{T_s} |s_m(t)|^2 dt.$$

Hence, intuitively, the mean square distance of the constellation points from the origin determines the average transmit power of the digital modulator.

From Equation (2.10), it seems that given any fixed channel bandwidth W_{tx} (and hence a fixed baud rate or symbol duration T_s), one can increase the bit rate to an arbitrarily high value by increasing M. The analogy is that modulation symbols are like mini-bus. The bandwidth or baudrate determines the *frequency* of mini-buses (number of mini bus per second). However, the M determines the number of *seats* in a mini bus. Hence, given a fixed mini bus frequency, one can still increase the bit rate (number of passengers per second) by increasing the number of seats in a mini bus ($\log_2 M$). This is obviously too good to be true as otherwise, bandwidth will be worthless. We will revisit this point when we discuss the operation of the demodulator to get a complete tradeoff picture.

Finally, as we have mentioned, to specify the behavior and operation of a modulator, we just need to specify the symbol duration and the signal set. For the signal set, we can choose to represent it using time domain approach (draw out the waveforms of the M signal elements). Equivalently, we can represent the signal set using *geometric domain* approach. In that case, the M signal elements are represented as M points on the signal space $\mathcal{S}_M = \{\vec{s}_1, \ldots, \vec{s}_M\}$. If we draw out the M points on the signal space, this is called the *constellation diagram* of the digital modulator. Figure 2.8 illustrates the time domain (signal waveforms) and geometric domain (signal constellation) of a quaternary modulator.

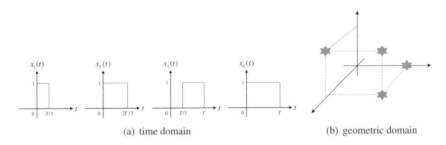

(a) time domain   (b) geometric domain

Figure 2.8   Illustration of the time domain and geometric domain representations of the signal set.

Example 2 (Signal Constellation of BPSK and BASK)   *Consider two binary modulators, namely the Binary Phase Shift Keying (BPSK) and the Binary Amplitude Shift Keying (BASK) modulators. The signal outputs and the signal constellations are illustrated in Figures 2.9 and 2.10*

2.4 DEMODULATION DESIGN AND OPTIMAL DETECTION IN AWGN CHANNELS

In Section 2.3, we focused on the modelling of the digital modulator. In order to have a complete picture of the tradeoff in Figure 2.2, we discuss the design of the digital demodulator in this section. The primary task of the digital demodulator is

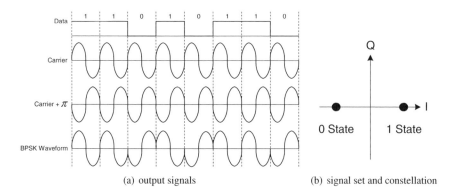

(a) output signals   (b) signal set and constellation

Figure 2.9 Illustration of BPSK modulation and the constellation.

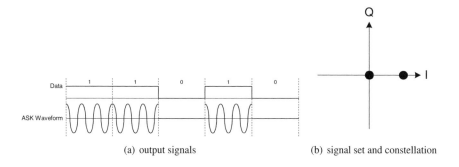

(a) output signals   (b) signal set and constellation

Figure 2.10 Illustration of BASK modulation and the constellation.

to detect what was transmitted at the digital modulator based on noisy or distorted observations $y(t)$. Depending on the specific media, the received signal $y(t)$ will be distorted in different ways as discussed in Chapter 1. In the following, we consider a simple example of Additive White Gaussian Noise (AWGN) channel. The received signal $y(t)$ is given by:

$$y(t) = x(t) + z(t)$$

where $x(t)$ is the transmitted signal and $z(t)$ is the white Gaussian noise. Figure 2.11 illustrates a segment of the received signal for a binary modulator after transmission over the AWGN channel.

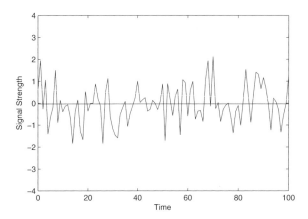

Figure 2.11   Illustration of received signal $y(t)$ of a binary modulator after transmission over the AWGN channel.

From the time domain perspective, there is little insight on what should the demodulator do in order to extract the information bits from the noisy observation $y(t)$. Moreover, it is difficult to deduce what should be the *optimal* detection rule for the demodulator. In the followings, we illustrate the demodulation process using geometric domain perspective.

Assume that symbol synchronization[1] has been achieved before passing the received signal to the demodulator. Using the concept of signal space, the received signal during the n-th symbol duration can be expressed as:

$$\mathbf{y}_n = \mathbf{x}_n + \mathbf{z}_n$$

where $\mathbf{x}_n$ is the transmitted signal vector with respect to the signal space $\mathcal{X}$ that defines the constellation of the modulator, $\mathbf{z}_n$ is the noise vector of the channel

---

[1]Symbol synchronization refers to the timing synchronization with respect to the symbol boundary. In other words, after achieving symbol synchronization, the demodulator is able to partition the received signal into multiple symbols, each with duration T_s.

noise $z(t)$ projected[2] onto the signal space $\mathcal{X}$ and $\mathbf{y}_n$ is the received signal vector of $y(t)$ projected onto the signal space $\mathcal{X}$. For Gaussian noise, the noise vector $\mathbf{z}_n$ is a random vector with uniformly distributed orientation and Rayleigh distributed length. Since the square of the length of the signal vector represents signal energy, hence, the stronger the channel noise is, the longer the average length of the noise vector will be. Without loss of generality, consider a quaternary modulator with signal set $\mathcal{S} = \{\mathbf{s}_1, \ldots, \mathbf{s}_4\}$. Assume that $\mathbf{x}_n = \mathbf{s}_1$ is the actual transmitted signal. If the channel is noiseless, the received signal $\mathbf{y}_n$ must always be given by $\mathbf{s}_1$. However, due to the random nature of the noise vector $\mathbf{z}_n$, the received signal $\mathbf{y}_n$ will fall within a *noise cloud* about the actual transmitted signal $\mathbf{s}_1$. This is illustrated in Figure 2.12.

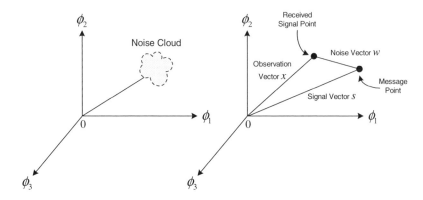

Figure 2.12   Illustration of noise cloud on the received signal.

Suppose now, the received signal $\mathbf{y}_n$ is at the position as illustrated by Figure 2.13 Note that due to quaternary modulator, the demodulator realizes that only one of the four possible positions, $\{\mathbf{s}_1, \ldots, \mathbf{s}_4\}$, could have been transmitted. Given that the current received vector is $\mathbf{y}_n$, a simple and intuitive scheme is to pick the decoded signal $\hat{\mathbf{x}}$ as the point that is *closet* to the received vector $\mathbf{y}_n$. In the example illustrated in Figure 2.12, the demodulator should pick the point $\mathbf{s}_1$ as the detected symbol $\hat{\mathbf{x}}_n$. This detection scheme is called the *minimum distance detection*.

In general, the minimum distance detector for a M-ary modulator is summarized below.

Minimum distance detector:

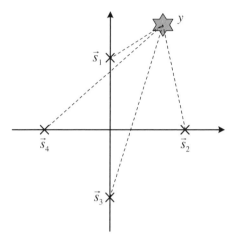

Figure 2.13   An example of the received signal $\mathbf{y}_n$ on a two-dimensional signal space.

Step 1: Convert the time domain received signal $y(t)$ to the received vector $\mathbf{y}_n$.
As mentioned in the previous section, suppose the constellation signal space $\mathcal{X}$ (with D dimension) has basis $\{\phi_1(t), \ldots, \phi_D(t)\}$. The j-th coordinate of the received vector $\mathbf{y}_n$ is given by:

$$\mathbf{y}_n(j) = \, < y(t), \phi_j(t) > = \int_{(n-1)T_s}^{T_s} y(t)\phi_j^*(t)dt \qquad (2.14)$$

for $j = \{1, 2, \ldots, D\}$.

Step 2: Determine the M distances. Given the received vector $\mathbf{y}_n$ as determined in step 1, we measure M distances between the received vector and the M possible *hypothesis points* $\{\mathbf{s}_1, \ldots, \mathbf{s}_M\}$. That is,

$$
\begin{aligned}
d(\mathbf{y}_n, \mathbf{s}_m)^2 &= \|\mathbf{y}_n - \mathbf{s}_m\|^2 = \, < y(t) - s_m(t), y(t) - s_m(t) > \\
&= \int_{(n-1)T_s}^{nT_s} |y(t) - s_m(t)|^2 dt \qquad (2.15)
\end{aligned}
$$

Step 3: Determine the minimum distance. The detected symbol $\hat{\mathbf{x}}_n$ is given by:

$$\hat{\mathbf{x}}_n = \mathbf{s}_{m^*} \qquad (2.16)$$

where

$$m^* = \arg\min_m d(\mathbf{y}_n, \mathbf{s}_m)^2 \qquad (2.17)$$

The three steps are illustrated in Figure 2.14 Note that the first step can be implemented by a correlator or a *matched filter*. The minimum distance detector is simple and easy to understand intuitively. Moreover, it can be shown (see Appendix A) that it is the optimal detector (best possible detector) for Gaussian noise.

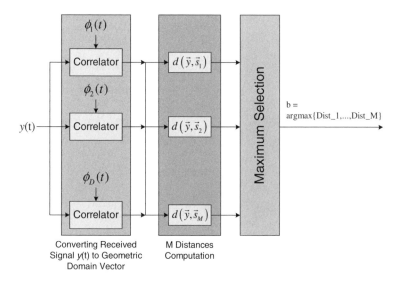

Figure 2.14 A block diagram of the minimum distance detector.

In fact, the 3-stage minimum distance detector can be further simplified. For instance, Equation (2.15) can be expressed as:

$$d(y(t), s_m(t))^2 = \|\mathbf{y}\|^2 + \|\mathbf{s}_m\|^2 - 2 <\mathbf{y}, \mathbf{s}_m>$$
$$= \|\mathbf{y}\|^2 + E_m - 2 \int_{(n-1)T_s}^{nT_s} y(t)s_m^*(t)dt \qquad (2.18)$$

where $E_m = \|\mathbf{s}_m\|^2$ is the energy of the m-th signal element $s_m(t)$. Since the key of the minimum distance detector is to select the signal point that gives the minimum distance and $\|\mathbf{y}\|^2$ is independent of m, hence, the minimum distance detection rule is equivalent to:

$$m^* = \arg\max_m \int_{(n-1)T_s}^{nT_s} y(t)s_m^*(t)dt - E_m/2 \qquad (2.19)$$

Hence, the minimum distance detector structure can be simplified as illustrated in Figure 2.15

Note that the first stage consists of M correlators or matched filters between the received signal $y(t)$ and $\{s_1(t), \ldots, s_M(t)\}$.

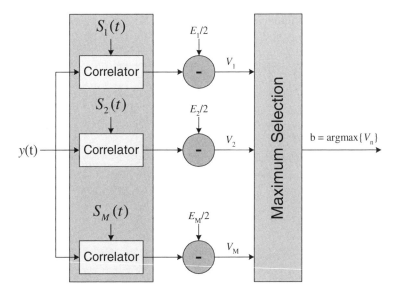

Figure 2.15   A simplified structure of the minimum distance detector.

2.5   PERFORMANCE AND TRADEOFFS

In the previous sections, we have introduced the concept of geometric representation of signals (signal space), the geometric models of digital modulators and digital demodulators. In this section, we complete the tradeoff big picture as stipulated in Figure 2.2

To understand the tradeoffs, we have to first understand the mechanism that a demodulator makes error on the detection process. Suppose the optimal demodulator, namely the *minimum distance detector*, is used and assume that $\vec{s}_1$ is actually transmitted. Figure 2.16(a) illustrates the situation when the channel noise power is small. As illustrated, the received signals $\vec{y}_n$ will fall within the noise cloud most of the time and since the noise cloud is small, the demodulator (minimum distance detector) will make the right decisions most of the time.

On the other hand, Figure 2.16(b) illustrates the situation when the channel noise power is large. When the received vector $\vec{y}_n$ falls in the area A, the minimum distance detector will pick $\vec{s}_1$ as the detected output and no error is made. However, when the received vector $\vec{y}_n$ falls in the area B, the minimum distance detector will pick $\vec{s}_2$ as the detected symbol and this results in decoding error. In general, there is a noise cloud around each constellation point at the receiver side and the minimum distance detector will make errors whenever the noise clouds between adjacent constellation points overlaps as illustrated in Figure 2.16(b). The probability of making detection error (symbol error probability) is given by the area of overlapping between constellation

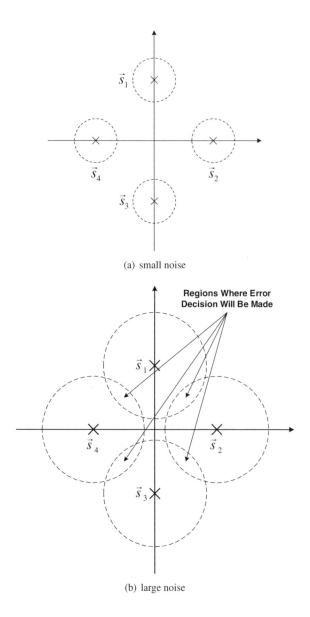

(a) small noise

(b) large noise

Figure 2.16 Illustration of detection errors in minimum distance detector.

points. Obviously, the symbol error probability is limited by the minimum distance between any two constellation points in the signal constellation.

$$P_e \approx Q\left(\sqrt{\frac{d_{min}^2}{\eta_0}}\right) \qquad (2.20)$$

where $d_{min} = \min_{i,j} \|\vec{s}_i - \vec{s}_j\|$, η_0 is the noise power spectral density and $Q(x)$ is the Gaussian Q-function. Hence, to reduce the error probability P_e, we have to maximum the minimum separations between constellation points.

Next, we apply the concept of noise cloud to complete the tradeoff picture. This is elaborated below.

Increasing transmit power can reduce error probability. Recall that P_e is related to the minimum distance between any two constellation points, d_{min}. Increasing the transmit power is equivalent to pulling all the constellation points further away from the origin. In this way, the minimum separation between any two constellation points will be increased and therefore, the error probability will be decreased without decreasing the bit rate.

Increasing transmit power can indirectly increase bit rate. On the other hand, one method to increase the bit rate without increasing the bandwidth requirement is to increase the modulation level M. For example, the bit rate can be doubled by using 16QAM constellation instead of QPSK constellation. However, doing so will increase the density of the constellation points unless the transmit power is increased. In that case, more constellation points can be squeezed into the signal space (higher bit rate) and at the same time, maintaining a similar *minimum distance separation between constellation points* (similar error probability) at the expense of higher transmit power.

Increasing bandwidth can increase bit rate. Recall that bandwidth is directly proportional to the baud rate or symbol rate of a digital communication link. Hence, increasing the bandwidth will directly increase the baud rate. Hence, the bit rate, which is given by $\log_2 M \times$ baud rate, will be increased.

Increasing bandwidth can indirectly reduce error probability. Finally, increasing bandwidth (while keeping bit rate unchanged) can increase the *dimension* of the signal space. For instance, say the dimension of the signal space is increased from 2 to 4 due to an increase in bandwidth. For the same number of signal constellation points and the same average transmit power (mean square distance of the points from the origin), the inter-constellation point spacing will be increased if the dimension of the signal space is increased. This results in a decrease of error probability without increasing power (at the expense of higher bandwidth).

The qualitative model of digital modulator and demodulator enable the design to compare the performance of different digital modulator designs without going through complicated math calculations. This is illustrated by a few examples in the following.

Example 3 (Design of Quarternary Modulator) *Consider a quarternary modulator with 4 points on the 2 dimensional signal space. What is the optimal transmit constellation given a fixed power constraint?*

Since the energy of a signal point $\vec{s}_i$ is given $\|\vec{s}_i\|^2$, the average power of the quarternary modulator with points $\{\vec{s}_1, \ldots, \vec{s}_4\}$ is given by:

$$P_{av} = \sum_{i=1}^{4} \frac{\|\vec{s}_i\|^2}{4T_s}$$

Suppose the four points have different lengths to the origin. To keep the average power P_{av} constant, if we pull any two points $\{\vec{s}_1, \vec{s}_2\}$ away from the origin (so as to increase the separation of the two constellation points), the remaining two points $\{\vec{s}_3, \vec{s}_4\}$ have to move closer to the origin. Since the error probability is limited by the minimum distance between any two constellation points, the bottleneck of the constellation will be on $\vec{s}_3$ and $\vec{s}_4$. Hence, by symmetry, the four points should be equi-distance from the origin (or lie on a circle). Again by symmetry, the four points should be equi-distance from each other (i.e., lie on the four corners of a square) as illustrated in Figure 2.17

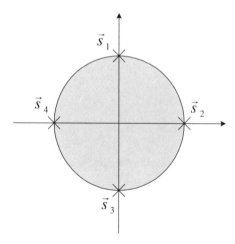

Figure 2.17   Optimal constellation of quarternary modulator on a 2-D signal space.

Example 4 (Comparison of ASK and BPSK modulators) *Consider a ASK and BPSK modulators as illustrated in Figure 2.9 Which of the two binary modulator designs is more energy efficient?*

*From the time domain waveforms, it is hard to say which of the two binary modu-
lators has a better performance. Consider the signal constellation of the two binary
modulators. Figure 2.18 illustrates the signal constellations of the BASK and BPSK
modulators. Both designs have two constellation points in the signal set but the BASK
design has one signal point at the origin. On the other hand, the BPSK design has
the two signal points equi-distance from the origin with one on the positive x-axis
and the other on the negative x-axis. Let the BPSK and BASK have the same error
probability. This means that the distance separation between the two constellation
points in both cases are the same (normalize to 2). For BASK, the average power is
given by*

$$P_{av}(BASK) = \frac{1}{2T_s} \left(\|\vec{s}_0\|^2 + \|\vec{s}_1\|^2 \right) = \frac{2}{T_s}$$

On the other hand, the average power of the BPSK design is given by:

$$P_{av}(BPSK) = \frac{1}{2T_s} \left(\|\vec{s}_0\|^2 + \|\vec{s}_1\|^2 \right) = \frac{1}{T_s}$$

*Hence, we have $P_{av}(BASK) = 2P_{av}(BPSK)$. That is, the BASK design requires
3dB more power than the BPSK design for the same error probability.*

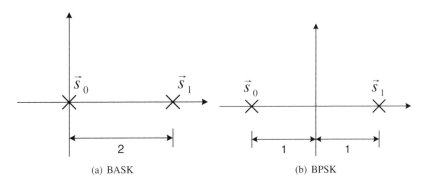

Figure 2.18   Constellations of the BPSK and BASK modulators.

2.6   PRACTICAL ILLUSTRATIONS: DIGITAL MODULATION

In this section, we illustrate how digital modulations are applied in practical systems.
From the previous sections, the power and bandwidth are two important parameters
to characterize the *resource space*. Given the transmit power P_{tx} and bandwidth
resource W, there is a maximum limit on the supported bit rate at a given error
probability. For instance, there are two extreme operating regions in a digital communication system. They are namely systems with *high bandwidth efficiency* and

systems with *high bandwidth expansion*. When a system operates in the *high band-width efficiency* region, the system is bandwidth limited (rather than power limited). In this case, the digital modulation design relies on $M-$ary modulation scheme to increase the bit rate (at the expense of higher power). Examples are GSM, GPRS and EDGE cellular systems[141]. On the other hand, when a system operates in the *high bandwidth expansion* region, the system is power limited (rather than band-width limited). In this case, the digital modulation design relies on the expansion of signal space dimensions in order to reduce the error probability. Examples are satellite communications, deep space communications and spread spectrum systems. We elaborate how various practical systems operate in these two different regions.

High bandwidth efficiency. Consider the IS54 (Digital AMPS) system, which is the US digital cellular system. In IS54, the digitalized speech is protected using convolutional encoding and modulated using Offset-Quadrature Phase Shift Keying (O-QPSK) modulation. In regular QPSK system, the signal constellations is given by $\{\pi/4, -\pi/4, 3\pi/4, -3\pi/4\}$. Hence, the four time domain signals can be expressed as:

$$s(t) = A\cos(\omega_c t + \theta)$$
$$= \frac{A}{\sqrt{2}} d_I(t) \cos(\omega_c t) + \frac{A}{\sqrt{2}} d_Q(t) \sin(\omega_c t) \qquad (2.21)$$

where $d_I(t)$ is given by:

$$d_I(t) = \cos(\theta_n) = \{\pm\frac{1}{\sqrt{2}}\}$$

and $d_Q(t)$ is given by:

$$d_Q(t) = \sin(\theta_n) = \{\pm\frac{1}{\sqrt{2}}\}.$$

Hence, the QPSK modulator can be regarded as an independent I-Q modulator with the information bit streams $\{d_0, d_1, d_2, \ldots\}$ split into two independent bit streams $\{d_0, d_2, \ldots\}$ driving the I-th modulator $d_I(t)$ and $\{d_1, d_3, \ldots\}$ driving the Q-th modulator $d_Q(t)$. Figure 2.19(a) illustrates the QPSK modulation from the I-Q perspective.

Figure 2.19(b) illustrates the QPSK time domain waveform. The carrier phase change once every $2T_b$ duration where T_b is the bit duration. If from one bit interval to the next, neither the I or the Q bit stream changes sign, then the carrier phase remains unchanged. If either the I or the Q component changes sign, a phase change of $\pi/2$ occurs. However, if both the I and Q components change signs, a phase change of π occurs. For practical consideration, the QPSK signal with a phase change of π will go through zero momentarily after

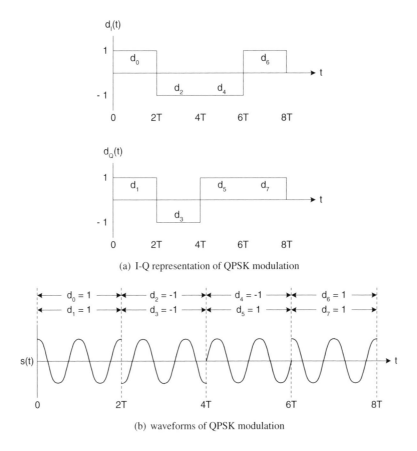

(a) I-Q representation of QPSK modulation

(b) waveforms of QPSK modulation

Figure 2.19 Illustration of regular QPSK modulation.

pulse filtering. This results in non-constant envelop in the QPSK waveform or in general, a higher Peak-to-Average (PAR) ratio[299]. Signals with high PAR is notorious of *spectral regrowth* in the side lobes after non-linear process (such as the class B power amplifier). Hence, we need to operate the power amplifier with a higher *backoff factor* to avoid the saturation of the power amplifier during some peaks of the envelop. Therefore, the regular QPSK is not desirable from a spectral regrowth perspective.

Offset QPSK (O-QPSK) is designed to overcome the high PAR problem of the QPSK signal. For instance, the I and Q streams of information $d_I(t)$ and $d_Q(t)$ are offset by $T_b/2$ in time as illustrated in Figure 2.20(a).

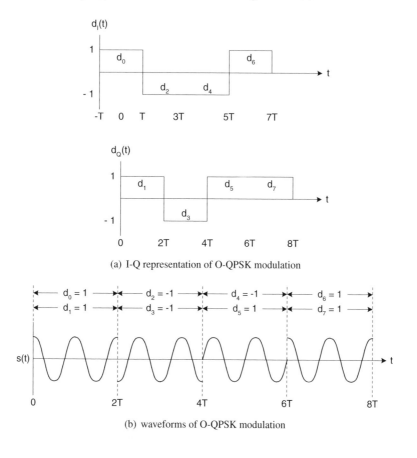

(a) I-Q representation of O-QPSK modulation

(b) waveforms of O-QPSK modulation

Figure 2.20   Illustration of Offset-QPSK modulation in IS54 systems.

Thus, the possible range of phase transition becomes 0 to $\pi/2$. The phase change of π is eliminated by the $T_b/2$ time offset. While the amplitude fluctuations still occur in O-QPSK waveform, it fluctuates over a smaller amplitude range because the amplitude will not pass through a zero point anymore. Hence, the O-QPSK signal has a smaller PAR[306] and therefore, smaller power back-off is required at the power amplifier to avoid spectral regrowth problem. This enhances the efficiency of the power amplifier and therefore, the O-QPSK was adopted in the IS-54 standard. Note that the O-QPSK modulation has the same error performance and bit rate performance as the regular QPSK modulation. The only difference is the O-QPSK systems will have a lower PAR and this favors the robustness towards the non-linear spectral regrowth effects.

Similarly, the modulation system of GSM system, which is the second generation cellular system, follows a similar consideration. Speech signal is first digitalized using speech compression technique (vocoder) to produce a digital stream of source data at a rate of 13 kbps. The source data is protected by convolutional encoder and modulated using Gaussian Minimum Shift Keying (GMSK) and transmitted over the air to the receiver. The GMSK is derived from the O-QPSK design by replacing the rectangular pulse in the I and Q waveforms with a *Gaussian pulse*. The time domain equation of a Gaussian pulse is given by:

$$g(t) = \frac{1}{2T_b} \left[Q\left(2\pi B \frac{t - T_b/2}{\sqrt{\log_e 2}}\right) - Q\left(2\pi B \frac{t + T_b/2}{\sqrt{\log_e 2}}\right) \right] \qquad (2.22)$$

where $BT_b \in [0, \infty)$ and $Q(t)$ is the Q-function defined as:

$$Q(t) = \int_t^\infty \frac{1}{\sqrt{2}} \exp - \left(-x^2/2\right) dx$$

and B is the low-pass bandwidth of the Gaussian pulse. Figure 2.21 illustrates the Gaussian pulse with $BT_b = 0.5$.

We not go into the details of GMSK modulation here and the interested reader could refer to [306] for a more quantitative treatment. Similar to O-QPSK, the design of GMSK is to reduce the spectral regrowth as a result of non-linear processing in the transmitter chain.

Consider one more example in the cellular system. EDGE (Enhanced Data rate for GSM System) system is regarded as a 2.5G technology designed to enhance the data rate of the GSM / GPRS air interface. It increases the maximum data rate from 100kbps of GPRS system to around 384kb/s in EDGE systems. The first order enhancements in the EDGE system is the application of higher order modulation (BPSK and 8PSK) as well as the use of *adaptive modulation*. For instance, as we have discussed above, the 8PSK modulation offers a higher bit rate at the expense of higher required power for same error performance. The BPSK modulation offers a more robust performance (error probability) at the expense of lower bit rate. It is found that the received Signal-to-Noise ratio

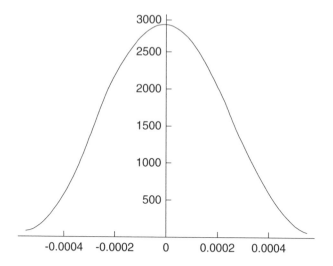

Figure 2.21   Illustration of Gaussian pulse shape in GMSK modulation with $BT_b = 0.5$.

(SNR) of mobiles is a random variable with a certain mean value. Sometime, the instantaneous received SNR of a mobile is high and sometimes, it is small. If the transmitter knows about the instantaneous received SNR, it can adaptively change the modulation constellations according to the SNR level. Say if the current SNR is high, the transmitter should transmit information using 8PSK modulation to boost the throughput. On the other hand, when the current SNR is low, the transmitter should transmit using BPSK modulation for robust error performance at the expense of bit rate. In EDGE, the transmitter can adaptively select the modulation level on a *block level* (a block consists of several frames). This adaptive modulation approach substantially enhances the average bit rate of the mobile users.

High bandwidth expansion. On the other hand, some practical systems are designed to operate in high bandwidth expansion mode. For example, in CDMA systems such as IS-95, CDMA2000 or W-CDMA systems, the modulation design is based on high-bandwidth expansion approach. For instance, in IS95 systems, *time-orthogonal modulation* is used in the reversed link (mobile to base station). Six information bits $\{d_0, d_1, \ldots, d_5\}$ are grouped together to select one out of 64 *modulation symbols* $\{\vec{s}_1, \ldots, \vec{s}_{64}\}$. Hence, it is a 64-ary modulation but the constellation (with 64 points) are positioned in a 64-dimensional signal space. Moreover, the 64 modulation symbols are *orthogonal* to each other. That is:

$$< \vec{s}_i, \vec{s}_j >= 0$$

where $< x, y >$ denotes the vector dot-product. Hence, the modulation scheme is also called 64-ary orthogonal modulation. The 64-dimensional signal space are derived from the *time-spreading operation* which will be discussed in detail in Chapter 3. The design philosophy in IS95 orthogonal modulation is to expand the required bandwidth to trade for robustness in error performance in the presence of strong interference (multi-user interference, intersymbol interference as well as multi-cell interference). This robustness is also called the *interference suppression capability* in CDMA systems. In other words, it is operating in the tradeoff direction of using extra bandwidth to reduce the error probability.

2.7 SUMMARY

In this chapter, we have elaborated the fundamental of point-to-point digital communication system design. We have elaborated a *black box model* for the digital modulator and the digital demodulator. For instance, a digital modulator can be characterized by (i) the symbol duration, and (ii) the signal set $\mathcal{S} = \{s_1(t), \ldots, s_M(t)\}$. The required bandwidth is given by $BW = \alpha \frac{1}{T_s}$. The transmitted bit rate is given by $R_b = \log_2(M)\frac{1}{T_s}$. The signal set can be described by either the time domain approach or *geometric domain approach* (vectors in a signal space). Using the signal space concept, the operations of digital demodulator can be easily understood and the optimal digital demodulator with respect to the AWGN channel is the *minimum distance decoding rule*. The decoded symbol is chosen to be the one of the M possible points $\{\vec{s}_1, \ldots, \vec{s}_M\}$ that is closet to the received vector $\vec{y}$. The error probability is a function of the minimum distance between any two constellation points.

After the introduction of digital modulation and digital demodulation, we have elaborated on the big picture of system tradeoffs between the bandwidth, power, bit rate and error probability. For instance, using higher bandwidth can result in higher bit rate for obvious reasons. On the other hand, using a higher bandwidth can also result in lower error probability indirectly due to the expansion of the dimension of the signal space. Using a higher transmit power results in smaller error probability directly because signal constellations are separated further. On the other hand, using a higher transmit power can also result in a higher bit rate indirectly because it allows the use of higher level modulation (M) over the same dimension signal space. Higher power allows a similar separation between the constellation points after more constellation points are squeezed into the signal space.

Finally, we have illustrated the use of digital modulation in some commercial systems such as the digital AMPS, GSM, EDGE as well as CDMA systems. The digital AMPS, GSM, EDGE systems are designed to operate in high bandwidth efficiency manner where a higher power is required to tradeoff for higher bit rate. On the other hand, the CDMA systems are designed based on high bandwidth expansion modes where a higher bandwidth expansion is required to tradeoff for robustness against interference.

PROBLEMS

2.1   Suppose there is a scheme that makes use of a special traffic light to send digital information. There are 3 lights (with different colors: red, yellow and green). The traffic light changes pattern once every 10ms. Suppose we want to build a "modulator" that transmits a "symbol" based on the different light combinations (each light can be on or off but the three lights cannot be "on" simultaneously) and the modulation process is symbol-by-symbol (encoding across one 10ms symbol only). What are the baud rate and the bit rate of the system if we consider a basic symbol with duration 10ms?

2.2   One way to increase the information bit rate through such a setting is to encode information bits based on two "symbols" together. (i.e., a new "super-symbol" of duration 20ms is formed by two consecutive symbols and information bits are carried by the new "super-symbol"). What is the resulting bit rate?

2.3   What is the bit rate when grouping N symbols at a time during the modulation? How about the asymptotic bit rate as $N \rightarrow \infty$?

CHAPTER 3

MULTIUSER COMMUNICATIONS

Men, like all other animals, naturally multiply in proportion to the means of their subsistence.
—Adam Smith: *The Wealth of Nations*, I, 1776

3.1 INTRODUCTION

In Chapter 2, we have introduced a generic model for digital modulation and demodulation design based on the concept of signal space. We have also discussed various channel coding and decoding techniques. All these focused on the point-to-point communications where there is a single transmitter and a single receiver. However, in many situations, communications may involve multiple users. In this chapter, we focus on both the theoretical and application aspects of techniques for multi-user communications.

We consider a system with single base station and multiple mobile users. In general, there are two directions of multi-user communications, namely the uplink direction and the downlink direction. Uplink refers to the direction from mobile users

Wireless Internet and Mobile Computing. By Yu-Kwong Ricky Kwok and Vincent K. N. Lau
Copyright © 2007 John Wiley & Sons, Inc.

to the base station as illustrated in Figure 3.1(a). This is a *many-to-one* communication scenario and is called the *multi-access problem*.

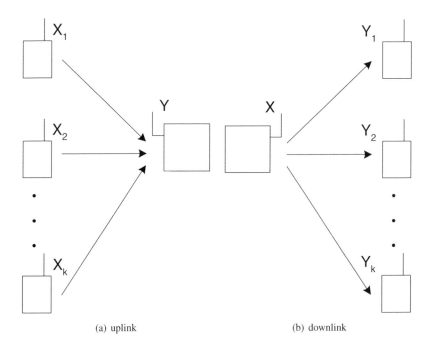

(a) uplink   (b) downlink

Figure 3.1   Illustration of two directions of multi-user communications.

On the other hand, downlink refers to the communication direction from the base station to the mobile users as illustrated in Figure 3.1(b). This is a *one-to-many* communication scenario and is called the *broadcast problem*. The two problems were treated separately for the past few decades but lately, it is shown that the two problems are dual of each other. In Section 3.2, we give an overview on the information theoretical framework for the two problems. In Section 3.3, we discuss various commonly used techniques to partition resources in both the uplink and downlink directions in an orthogonal manner. For example, commonly used techniques such as FDMA (Frequency Division Multiple Access), TDMA (Time Division Multiple Access), Deterministic CDMA (Code Division Multiple Access) belong to this category. In Section 3.4, we discuss various commonly used techniques for resource partitioning in a non-orthogonal manner. For example, random CDMA as well as space division multiple access belong to this category. In Section 3.5, we compare the pros and cons of various technologies in single cell and multi-cell scenarios. In Section 3.6, we describe how these technologies are applied in practical systems such as GSM, CDMA and IEEE 802.11b systems. Finally, we give a summary of the main points and concepts in Section 3.7.

3.2 INFORMATION THEORETIC OVERVIEW

In the context of point to point communications, there are two approaches to attack the design problem. One approach, namely on the *signal processing approach*, focused on the specific design of modulation and error control coding schemes. Obviously, different specific design will have different physical layer performance (as elaborated in Chapter 2) in terms of supported bit rate, required transmit power, required bandwidth and resulting error probability. A natural question to ask is *"could we do a better physical layer design"?*. The signal processing approach could not offer a complete answer to this question.

Another approach, namely on the *information theoretical approach*, focused on characterizing the *channel capacity* [78]. Specifically, the theoretical framework isolates the best performance from the specific implementation details of modulation and error control coding. It gives the performance bound achieved by the most optimal design of error control coding and modulation schemes. For example, given a fixed transmit power and a fixed bandwidth, the maximum data rate [78] that could be transmitted *reliably*[1] over the *Additive White Gaussian Noise* (AWGN) channel is given by:

$$C = W \log_2 \left(1 + \frac{P}{\eta_0 W} \right) \tag{3.1}$$

where W is the channel bandwidth, P is the received power and η_0 is the channel noise density. In other words, there is no channel coding and modulation scheme that could *reliably* deliver a data rate larger than C and C is called the *channel capacity* or *Shannon's capacity*. This approach has the beauty that it gives a complete answer to the previous question because it presents the optimal solution. Unfortunately, this approach gives no indication on the structure of the optimal channel coding design. Therefore, there are still plenty of different researchers adopting these two approaches to address the point-to-point communication link design.

Similarly, there are two approaches to address the multi-user communication system design, namely the *signal processing approach* and the *information-theoretical approach*. We introduce major information-theoretical results with respect to multi-user communications in this section. However, unlike the point-to-point situation where the channel is characterized by a scalar called *channel capacity*, the multi-user

---

[1]Reliable transmission of information bit to the receiver refers to the asymptotic situation that the error probability could be made arbitrarily small at arbitrarily large block length.

$$\lim_{N \longrightarrow \infty} P_e = 0$$

channel is characterized by a K−dimensional region $\mathcal{R}$ where K is the number of users involved in communications.

3.2.1 Multi-Access Problem

3.2.1.1 General Problem Formulation   As illustrated in Figure 3.1(a), let $X_k \in \mathcal{C}$ be the transmitted symbol out of user k in the uplink and $Y \in \mathcal{C}$ be the received symbol at the base station where $\mathcal{C}$ represent a set of complex number. The multiaccess channel is in general characterized by a probabilistic function (or *channel transition probability*) $p(\mathbf{Y} = \mathbf{y}|\mathbf{X}_1 = \mathbf{x}_1, \ldots, \mathbf{X}_K = \mathbf{x}_K)$ where $\mathbf{Y} = [Y(1), \ldots, Y(n), \ldots, Y(N)]$ is the block of received symbols, $\mathbf{X}_k = [X_k(1), \ldots, X_k(N)]$ is the block of transmitted symbols from user k and n is the time index. If the channel transition probability could be expressed as

$$
\begin{aligned}
p(\mathbf{Y} &= \mathbf{y}|\mathbf{X}_1 = \mathbf{x}_1, \ldots, \mathbf{X}_K = \mathbf{x}_K) \\
&= \Pi_{n=1}^{N} p(Y(n) = y(n)|X_1(n) = x_1(n), \ldots, X_K(n) = x_K(n)) \quad (3.2)
\end{aligned}
$$

the multiaccess channel is called *memoryless*. Otherwise, the mutliaccess channel is called *memory*. For simplicity, we assume a simple multiaccess channel, namely the multiaccess channel with Additive White Gaussian Noise. The AWGN channel is a memoryless multiaccess channel and the received signal is given by:

$$
Y = \sum_{k=1}^{K} X_k + Z \qquad\qquad (3.3)
$$

where Z is the white Gaussian noise with variance σ_z^2. Putting the channel model into the form of *channel transition probability*, we have:

$$
\begin{aligned}
p(Y(n) &= y(n)|X_1(n) = x_1(n), \ldots, X_K(n) = x_K(n)) \\
&= \frac{1}{\sqrt{2\pi}\sigma_z} \exp\left(-\frac{|y(n) - \sum_k x_k(n)|^2}{2\sigma_z^2}\right) \qquad (3.4)
\end{aligned}
$$

We first have a few definitions.

Definition 1 *A multiaccess coding scheme with block length N for K users is a collection of codebooks $\mathcal{B} = \{\mathcal{B}_1, \ldots, \mathcal{B}_K\}$, where the codebook for the k−th user is a mapping from the message set $\mathcal{M}_k$ to the block of transmitted symbols $\mathbf{X}$. That is: $\mathcal{B}_k : \mathcal{M}_k \longrightarrow \mathcal{C}^N$ where $\mathcal{M}_k$ is a set of positive integers with cardinality $|\mathcal{M}_k|$.*

Figure 3.2(a) illustrates the concept of 2-user multi-access coding scheme as an example.

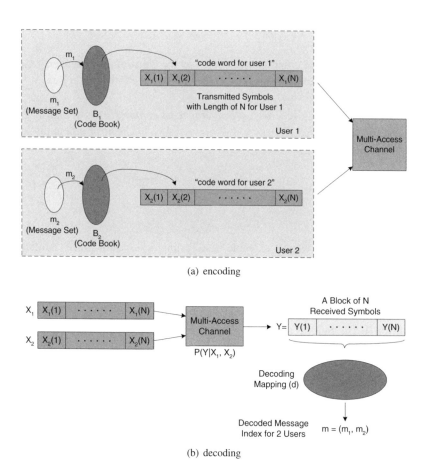

(a) encoding

(b) decoding

Figure 3.2 Illustration of two-user multiaccess encoding and decoding scheme.

Definition 2 *The multiaccess coding scheme is said to satisfy* power constraints, $(P_1, \ldots, P_K)$, *if for each* $k \in [1, K]$ *and each* $m_k \in \mathcal{M}_k$, *we have*

$$\frac{1}{N} \sum_{n=1}^{N} |X_k(n)|^2 \leq P_k$$

The power constraint could be interpreted as the constrain on the average transmitted power out of user k where the average is over the entire coding frame.

Definition 3 *A* decoding scheme *for the multiaccess coding scheme is characterized by a mapping* $d : \mathcal{C}^N \longrightarrow [\mathcal{M}_1] \times [\mathcal{M}_2] \times \ldots \times [\mathcal{M}_K]$. *The* error probability *of the decoding scheme is given by:*

$$P_e = \Pr\left[d(\mathbf{y}) \neq (m_1, \ldots, m_K)\right]$$

Figure 3.2(b) illustrates an example of decoding scheme for 2-user multiaccess code. In other words, the task of the decoder at the base station is to separate the *embedded information* from all the K users based on the observation $\mathbf{Y}$ over the frame of N symbols.

Definition 4 *The* rate vector *of the multiaccess coding scheme is defined by* $\mathbf{r} = (r_1, \ldots, r_K)$ *where:*

$$r_k = \frac{1}{N} \log_2(|\mathcal{M}_k|)$$

The multiaccess code is called achievable *at a rate vector* $\mathbf{R}$ *if there exists a codebook* $\mathcal{B}$ *such that* $\lim_{N \to \infty} P_e = 0$ *and* $\lim_{N \to \infty} \mathbf{r}(\mathcal{B}) \geq \mathbf{R}$. *Note that* $(r_1, r_2) \geq (R_1, R_2)$ *if and only if* $r_1 \geq R_1$ *and* $r_2 \geq R_2$.

Definition 5 *The* capacity region *of a multiaccess channel with* K *users satisfying the power constraints* $\{P_1, \ldots, P_K\}$ *is defined as the* convex closure *of all achievable rate vectors.*

With the above definitions, the problem of multiaccess communication is to find out the *capacity region*[78] given a multiaccess channel model (or the channel transition probability) and power constraints. Therefore, the *channel capacity* in the context of point-to-point communication link is simply a special case of the *capacity region* for multiaccess problem.

3.2.1.2   *General Solution to the Multiaccess Problem*   The general solution of the multiaccess problem [78] is summarized by the following theorem.

Theorem 1 *The capacity region of a multi-access channel is given by the region defined by the following inequalities.*

$$\mathbf{r}(S) \leq I\left(Y; (X_k)_{k \in S} | (X_j)_{j \notin S}\right) \quad \forall S \subset [1, \ldots, K]$$

for some input distributions $p(X_1)p(X_2)\ldots(X_K)$ where $\mathbf{r}(S) = \sum_{k \in S} r_k$ and $I(X;Y|B)$ denotes the conditional mutual information between X and Y given B.

In other words, there exists *achievable* multiaccess codes if the rate vector lies within the above region. On the other hand, if a rate vector lies outside the above region, all multiaccess codes will be *non-achievable*. The proof of this theorem is omitted as this is outside the scope of the book.

For example, consider a 2-user multiaccess system with AWGN channel. Similar to the point-to-point situation, the *capacity achieving input distribution* in this case is also Gaussian. Hence, the capacity region is given by:

$$r_1 \leq I(X_1; Y|X_2) = \log_2\left(1 + \frac{P_1}{\sigma_z^2}\right) \tag{3.5}$$

$$r_2 \leq I(X_2; Y|X_1) = \log_2\left(1 + \frac{P_2}{\sigma_z^2}\right) \tag{3.6}$$

$$r_1 + r_2 \leq I(X_1, X_2; Y) = \log_2\left(1 + \frac{P_1 + P_2}{\sigma_z^2}\right) \tag{3.7}$$

The 2-user AWGN capacity region is illustrated in Figure 3.3

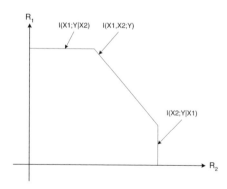

Figure 3.3   2-user multiaccess capacity region.

All the points could be realized by *successive interference cancellation* and/or *time sharing* citeVerdu1998. For instance, point A could be realized by decoding information for user 2 first, subtract X_2 from Y and then decode information for user 1. Point B could be realized by decoding user 1 followed by user 2. Point C could be realized by time sharing between the 2 decoding order.

3.2.2 Broadcast Problem

Broadcast problem[77] refers to the communication directions from the base station to the K mobile users.

3.3 ORTHOGONAL RESOURCE PARTITIONING

While Section 3.2 highlights the main results in multiuser communication theory, the theory itself simply gives the optimal performance bound. It does not provide information on how to realize the expected performance. In practice, the optimal schemes as suggested by theories have high computational complexity[2] and are usually infeasible. Yet, the theoretical results are important in the sense that they could give us a reference on the best achievable results. Based on the optimal performance reference, we could compare between various practical designs and evaluate how far they are from the optimal schemes.

In the next few sections, we address the multiuser communication problem based on the second approach, *signal processing approach.* We introduce various schemes for multiuser communications commonly used in practice due to their simplicity and compare their relative performance in single cell and multi cell scenarios. In this section, we focus on the *orthogonal schemes* such as FDMA, TDMA and deterministic CDMA. In the next section, we discuss on the *non-orthogonal schemes* such as random CDMA and SDMA. In Section 3.5, we compare their performance.

Assuming AWGN channel, the received signal $Y(t)$ (uplink direction) or the transmitted signal (downlink direction) at the base station is given by:

$$Y(t) = \sum_{k=1}^{K} X_k(t) + Z(t) \tag{3.8}$$

where $Z(t)$ is the AWGN channel noise, $X_k(t)$ is the transmitted signal from (or to) the $k-$th user. In the uplink direction, the task of the *multiuser detector* is to separate information related to the K users. In the downlink direction, the task of the *multiuser transmitter* is to make sure user k receives the signal specific to it ($X_k(t)$) without severe degradation. In other words, K *user channels* have to be formed between the base station and the K users. In general, some means of *user coordination* is needed in order to reduce the complexity of base station and mobile station processing. One

---

[2]For example, the optimal multiaccess detection algorithm is shown to have exponential order of complexity with respect to number of users.

obvious and simple approach is to partition the resource space based on *orthogonality principle*. That is, K orthogonal user channels are formed as a result of the resource partitioning schemes. By orthogonal, we mean that the *processed signal* for user k_1 do not interfere with the *processed signal* for user k_2. We have 3 popular orthogonal multiuser communication schemes, namely the frequency division multiple access (FDMA), time division multiple access (TDMA) and *deterministic* time division multiple access (CDMA). The FDMA scheme partitions the K user channels based on the frequency dimension of the signal space. The TDMA scheme partitions the K user channels based on the time dimension of the signal space. Finally, deterministic CDMA partitions the user channels based on the *code domain* of the signal space. These three schemes are applicable to both the multiaccess problem and the broadcast problem and they will be explained in details below.

3.3.1   Frequency Division Multiple Access (FDMA)

FDMA partitions the communication resource in the *frequency dimension* as illustrated in Figure 3.4(a).

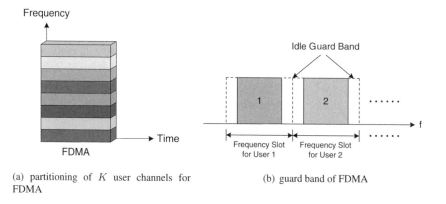

(a) partitioning of K user channels for FDMA

(b) guard band of FDMA

Figure 3.4   Principle of FDMA.

Specifically, the allocated spectrum with bandwidth W is divided into K frequency slots to form K user channels. Each channel has a bandwidth of W/N. Different mobiles are assigned to transmit (or receive) at different frequency slots and hence, there is no interference of signals between different users. The K channels are therefore *orthogonal channels*. Users could transmit (or receive) at the assigned frequency slot(s) over the entire time frame.

Example: Suppose a spectrum of 15 MHz is allocated to a mobile operator. Let the modulation throughput be 1 bit per symbol and the required data rate of individual user be 25 kbps. Find out how many users could be supported based on the FDMA scheme.

Solution With binary modulation, the required bandwidth per user to support an average bit rate of 25kbps is 25kHz. Hence, the number of frequency slots available over a spectrum of 15MHz is given by:

$$N_{FDMA} = 15M/25k = 600.$$

For the uplink multiaccess communication, signal transmitted from user k ($X_k(t)$) could be extracted from the received signal using a bandpass filter tuned to the frequency range of the assigned channel. For the downlink communications, signal to user k ($X_k(t)$) is modulated to the assigned frequency slot. However, in practice, *perfect bandpass filtering* is not realizable. Practical filters have finite slope in the frequency response. To facilitate simple processing at the base station, *frequency guard band* is maintained between channels as illustrated in Figure 3.4(b). Another design consideration is that the number of bandpass filters or RF units required at the base station is equal to the number of mobile users in the system. Since these components are analog circuits, they are usually quite bulky compared with digital components.

3.3.2   Time Division Multiple Access (TDMA)

On the other hand, TDMA partition the K user channels based on the time dimension of the signal space as illustrated in Figure 3.5(a).

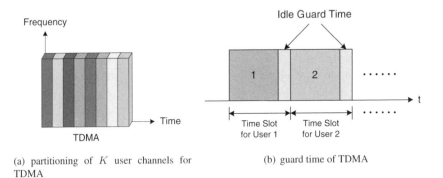

(a) partitioning of K user channels for TDMA

(b) guard time of TDMA

Figure 3.5   Principle of TDMA.

Specifically, all the K user channels occupy the same entire bandwidth W. However, each user takes turn to use the bandwidth in a coordinated manner. Different mobiles are assigned to transmit (or receive) at different *time slots* and therefore, there is no interference of signals between different users. The resulting K user channels are therefore orthogonal channels.

Example: Suppose a spectrum of 15 MHz is allocated to a mobile operator. Let the modulation throughput be 1 bit per symbol and the required average data rate of individual user be 25 kbps. Find out how many users could be supported based on the TDMA scheme.

Solution: Let T_p be the TDMA frame period and T_k be the time slot duration assigned to user k. When user k transmits in the assigned time slot, the peak bit rate is given by 15Mbps (because the modulation throughput is 1 bit per symbol). However, user k does not transmit during the rest of the time on T_p. Hence, the average bit rate of user k is given by:

$$R_b = 15M \times (T_k/T_p) = 25k$$

On the other hand, the number of time slots available for assignment is given by T_p/T_k. Hence, the capacity of TDMA is given by:

$$N_{TDMA} = T_p/T_k = 15M/25k = 600$$

For the TDMA scheme to work, all the K users have to be *synchronized* so that they have a common timing reference. However, in practice, it is not possible to have all the K users having exactly the same timing reference as the base station due to the variation in propagation delays of the K users. To allow certain tolerance as a result of user distribution, *guard time* is used between time slots as illustrated in Figure 3.5(b). During the guard time interval, no signal is transmitted and therefore, it does not interfere with other channels even if the user timing is slightly offset from the base station timing. In fact, the guard time imposes a limit on the maximum cell radius. Another characteristics of TDMA is the burstiness of physical data rate per user. A user will transmit at the *peak rate* (using the entire bandwidth) during the assigned time slot and remain silence at other time slots. This presents two potential problems. Firstly, it is more difficult to transmit the wideband signal (at the allocated time slot) because multipath fading will be seen by the transmitted signal. Complex channel equalization is therefore required at the receiving end to reduce the *inter-symbol-interference* (ISI). Secondly, the rapid switching on and off of the power amplifier at the TDMA transmitter will generate high level of *electromagnetic interference*. On the other hand, one advantage of TDMA scheme is the requirement on a single RF transceiver at the base station regardless of the number of TDMA channels. Therefore, in general, operators seldom partition the system resource based on TDMA entirely.

3.3.3 Deterministic Code Division Multiple Access (D-CDMA)

While FDMA and TDMA partition the system resource based on the frequency dimension and time dimension of the signal space, CDMA [339] partition the system resource based on the *code-dimension* of the signal space. In other words, all the K

user signals share the entire bandwidth and the entire time duration but differentiate among themselves based on different *code channels*. This extra code dimension is introduced through the *spreading process* as illustrated in Figure 3.6(a).

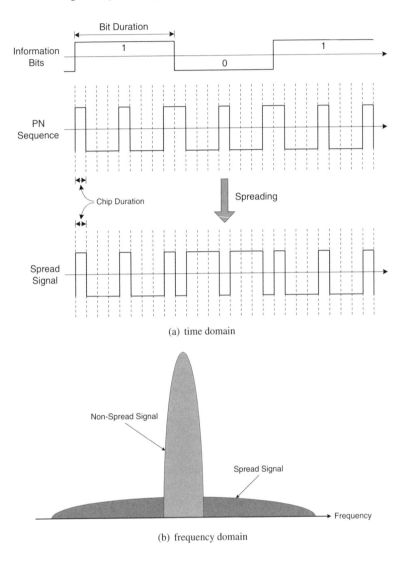

(a) time domain

(b) frequency domain

Figure 3.6   Principle of CDMA spreading.

Specifically, consider the transmit processing of the signal for user k. Individual information bit (or modulation symbol) is *spread* with a code sequence $c_k(t) \in$

$\{+1, -1\}$ to form the transmit signal. The code sequence is a time sequence with the smallest time unit of transition called a *chip*. The time duration of a chip is called *chip duration* T_c. Hence, effective, a modulation symbol is *chopped* into many tiny pieces by the code sequence. The ratio between the modulation symbol duration T_s and the chip duration T_c is called the *spreading factor* (SF). That is:

$$SF = \frac{T_s}{T_c} \tag{3.9}$$

As a result of the spreading process, the transmit bandwidth is increased by SF times. That is:

$$W_{tx} = W_s \times SF \tag{3.10}$$

where W_s is the signal bandwidth without spreading. Figure 3.6(b) illustrates the spreading process in frequency domain.

Each of the K transmit signals from (or to) the K users are spread by the code sequence $\{c_1(t), \ldots, c_K(t)\}$ respectively. Note that these K signals all share the same time and same bandwidth and they are separated based on the distinct code sequence only. For the uplink direction, the received signal at the base station is given by:

$$Y(t) = \sum_{k=1}^{K} S_k(t)c_k(t) + Z(t)$$

where $S_k(t)$ is the information signal for the $k-$th user, $c_k(t)$ is the spreading code sequence for the $k-$th user and $Z(t)$ is the channel noise. Since the transmitted signal all share the same time domain and frequency domain, we have to exploit on the characteristics of code sequence $c_k(t)$ in order to separate the K signals. Specifically, we apply *despreading* processing at the receiving end. To extract signal for user k, we multiply the received signal $Y(t)$ with the corresponding code sequence $c_k(t)$ as illustrated in Figure 3.7(a).

The despread signal is given by:

$$W_k(t) = Y(t) \times c_k(t) = \underbrace{S_k(t)}_{\substack{\text{Signal} \\ \text{Component}}} + \underbrace{\sum_{j \neq k} S_j(t)c_j(t)c_k(t)}_{\text{MultiuserInterference}} + \underbrace{Z(t)c_k(t)}_{\substack{\text{Channel} \\ \text{Noise}}} \tag{3.11}$$

due to the fact that $c_k(t) \times c_k(t) = 1$. Hence, we could see that the desired signal component for user k, $S_k(t)$, is *constructed* after the despreading operation. However, in general, in addition to the channel noise component, there is another *multiuser interference* component. If the code sequence $\{c_1(t), \ldots, c_K(t)\}$ are chosen in such a way that the multiuser interference is zero, the CDMA scheme is called *deterministic CDMA* and similar to FDMA or TDMA, the K user channels become orthogonal.

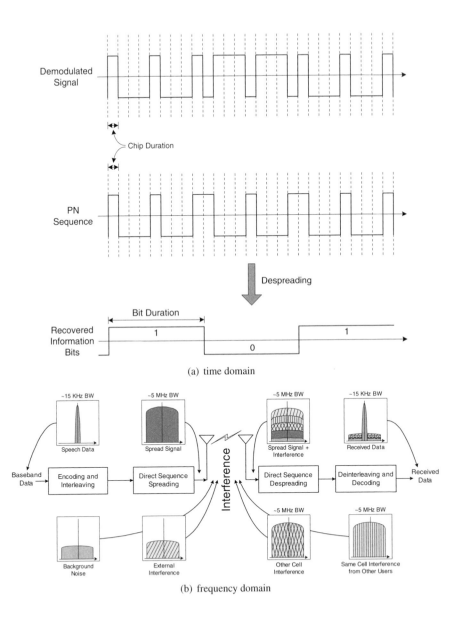

(a) time domain

(b) frequency domain

Figure 3.7   Principle of CDMA despreading.

We derive the condition on code sequence design in order to achieve deterministic CDMA. If the code sequence (periodic with period T_s/T_c) is expressed in discrete time domain $c_k(t) = \sum_{n=0}^{T_s/T_c-1} c_{k,n} u(t - nT_c)$ where $c_{k,n} \in \{+1, -1\}$ and $u(t)$ is a rectangular unit pulse with duration T_c, the received signal is expressed as:

$$W_k(t) = S_k(t) + \sum_{j \neq k} S_j(t) \sum_{n=0}^{T_s/T_c-1} c_{k,n} c_{j,n} u(t - nT_c) + Z(t) \sum_{n=0}^{T_s/T_c-1} c_{k,n} u(t - nT_c)$$

$$(3.12)$$

Theorem 2 *If $T_c \ll T_s$ and the code sequence is chosen such that*

$$R_{k,j} = \frac{1}{SF} \sum_{n=0}^{SF-1} c_{k,n} c_{j,n} = 0 \qquad (3.13)$$

where $SF = \frac{T_s}{T_b}$ is the spreading factor, *the interference component in (3.12) becomes 0 and the resulting CDMA scheme achieves* deterministic CDMA.

Proof: Converting $W_k(t)$ to a signal vector $\mathbf{W}_k = (W_{k,1}, \ldots, W_{k,D})$ in a D−dimensional signal space with basis $\{\phi_1(t), \ldots, \phi_D(t)\}$, we have:

$$\mathbf{W}_k = (< W_k(t), \phi_1(t) >, \ldots, < W_k(t), \phi_D(t) >)$$

where $< W_k(t), \phi_d(t) >$ denotes the *dot-product* of the signal space and is given by:

$$< W_k(t), \phi_d(t) >= \int_0^{T_s} W_k(t) \phi_d^*(t) dt$$

$$= \underbrace{\int_0^{T_s} S_k(t) \phi_d^*(t) dt}_{Desired Signal} + \underbrace{\sum_{j \neq k} \sum_{n=0}^{T_s/T_c-1} c_{k,n} c_{j,n} \int_0^{T_s} S_j(t) \phi_d^*(t) u(t - nT_c) dt}_{Multiuser interference}$$

$$+ \underbrace{\sum_{n=0}^{T_s/T_c-1} c_{k,n} \int_0^{T_s} Z(t) \phi_d^*(t) u(t - nT_c) dt}_{Noise}$$

In general, since $T_c \ll T_s$, we have $f(n) = \int_0^{T_s} S_j(t) \phi_d^*(t) u(t - nT_c) dt = \int_{nT_c}^{(n+1)T_c} S_j(t) \phi_d^*(t) u(t - nT_c) dt$ is *approximately* independent of n. Hence, the multi-user interference term could be expressed as $f \sum_{j \neq k} c_{k,n} c_{j,n}$. If $R_{k,j}(0) = 0$, the multi-user interference term vanishes.

Hence, the despread signal vector of user k is given by $\mathbf{W}_k = \mathbf{S}_k + \mathbf{Z}_k$ where $\mathbf{S}_k$ is the desired signal vector and $\mathbf{Z}_k$ is the channel noise vector.

Similar to TDMA and FDMA, the capacity of the resource partitioning is given by the number of available orthogonal channels. In deterministic CDMA, the number of available channels is equal to the number of code sequences with spreading factor $SF = T_s/T_c$ that satisfy the *orthogonality requirement* in (3.13). In other words, the capacity of deterministic CDMA system is *code limited*. Since the code dimension is created by the spreading operation, the degree of freedom of the code dimension is equal to the spreading factor $SF = T_s/T_c$ and therefore, the number orthogonal code sequences available in the system is given by the spreading factor SF. Furthermore, complete orthogonality of code sequence requires perfect synchronization between $c_k(t)$ and $c_j(t)$. In general, it is easier to achieve this condition in the downlink direction. This is because signals for the K users originates from the same point (the base station). As long as the transmitted codes are synchronized at the transmitting point, the same timing relationship could be maintained at the mobile and therefore, deterministic CDMA is feasible in the downlink direction. On the other hand, it is more difficult to achieve this synchronous state between code sequence in the uplink direction. This is because the code sequence $c_k(t)$ and $c_j(t)$ originates from user k and user j respectively and they are in general at different locations and uncoordinated. Even if the two codes start at the same system time, this timing relationship could not be maintained at the receiving point (the base station) because they have different propagation delays. Hence, deterministic CDMA is not applied in the uplink direction.

Example: Suppose a spectrum of 15 MHz is allocated to a mobile operator. Let the modulation throughput be 1 bit per symbol and the required average data rate of individual user be 25 kbps. Find out how many users could be supported based on the deterministic CDMA scheme.

Solution: With binary modulation, the maximum chip rate based on a bandwidth of 15 MHz is 15 Mcps. Hence, the spreading factor of the D-CDMA system is $SF = T_s/T_c = 15M/25k = 600$. Hence, there are 600 code channels in D-CDMA system.

Note that the maximum number of spreading code is limited by the *spreading factor* $SF = T_s/T_c$. Given a fixed bit rate (and hence a fixed bit duration T_b), the symbol duration $T_s = \log_2(M)T_b$ is a function of the modulation level M. For example, $T_s(QPSK) = 2T_b$ and $T_s(16QAM) = 4T_b$. Hence, the number of orthogonal channels could be expanded by using higher order modulation. The tradeoff is possibly higher required bit-energy to noise ratio as a result of higher order modulation.

D-CDMA shares the same advantages as the TDMA system in the sense that single RF transceiver is needed regardless the number of users in the system because all the K user channels share the same RF channel. On the other hand, since despreading operation require re-construction of the desired signal component by multiplying the corresponding code sequence to the received signal, accurate timing synchronization is required so that the despreading sequence $c_k(t)$ is time aligned with the spreading sequence $c_k(t)$ embedded in the received signal.

Similar to TDMA, the transmitted signal has high bandwidth due to the spreading operation. Hence, multipath fading will be present and in general, expensive channel equalization is needed at the receive to alleviate the inter-symbol interference (ISI). However, due to property of code sequence, a simple equalization structure exists for CDMA and this is called the *RAKE receiver*.

Finally, in terms of practical implementation, the implementation bottleneck of D-CDMA is usually on the maximum supportable chip rate $1/T_c$. In particular, the D-CDMA receiver front end operates at a speed of the chip rate order and there is a natural limitation on how fast the integrated circuit could operate. Therefore, in general, operators seldom partition the channel based entirely on D-CDMA.

3.3.4   Hybrid Design of FDMA, TDMA and D-CDMA

As introduced in the previous sections, FDMA, TDMA and Deterministic CDMA are all orthogonal resource partitioning schemes. As illustrated in the examples, the number of channels as a result of the 3 partitioning methods are the same for single cell system. (We elaborate the comparison more in Section 3.5) and this is in general always the case. Therefore, all the 3 schemes are equally effective theoretically. However, each of the schemes have their own implementation limitations. For example, FDMA system suffers a disadvantage of requiring one RF element per user. TDMA system suffers from a disadvantage of bursty transmission with very high peak rate. CDMA system suffers from a disadvantage of high speed signal processing at the chip rate level. In general, system resource is partitioned by either hybrid FDMA/TDMA or hybrid FDMA/CDMA schemes and they are introduced below through two examples.

Example: Suppose a spectrum of 15 MHz is allocated to a mobile operator. Let the modulation throughput be 1 bit per symbol and the required average data rate of individual user be 25 kbps. Assume that the channel coherence bandwidth is 200kHz. Suggest how the operator should partition the system resource based on FDMA and TDMA so that no equalizer is needed at the receiver.

Solution: Figure 3.8 illustrates the hybrid partitioning scheme.

Since the channel coherence bandwidth is 200kHz, we constrain the maximum bandwidth of signal transmission to be 200kHz so that expensive equalization is not required at the receiving side. Hence, the 15 MHz spectrum is first partitioned into $15M/200k = 75$ carriers based on FDMA as illustrated in the figure. Each carrier is then further partitioned into N_t channels using TDMA. Similar to the example described earlier, the number of time slots (channels) per carrier is given by:

$$N_t = 200k/25k = 8.$$

The total number of channels in the system is given by $N_t N_f = 600$ which is the same as partitioning based on FDMA or TDMA alone. Using the hybrid scheme, the number of RF transceivers required at the base station is reduced to 75 instead of 600 (for pure FDMA). On the other hand, the transmitted bandwidth of user signal

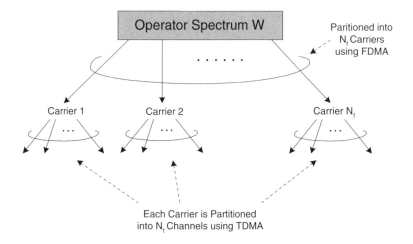

Figure 3.8   Hybrid FDMA and TDMA.

is reduced to $200k$Hz (instead of 15MHz as in pure TDMA) and therefore, avoiding expensive channel equalization.

Example: Suppose a spectrum of 15 MHz is allocated to a mobile operator. Let the modulation throughput be 1 bit per symbol and the required average data rate of individual user be 25 kbps. Assume that maximum supportable chip rate is 1.5Mcps. Suggest how the operator should partition the system resource based on FDMA and D-CDMA so that no equalizer is needed at the receiver.

Solution: Figure 3.9 illustrates the hybrid partitioning scheme.

Since the maximum supportable chip rate is 1.5 Mcps and the modulation throughput is 1 bits per symbol, the D-CDMA signal requires 1.5MHz transmission bandwidth. Hence, the 15 MHz spectrum is first partitioned into $15M/1.5M = 10$ carriers based on FDMA as illustrated in the figure. Each carrier is then further partitioned into N_c channels using D-CDMA. The maximum spreading factor per carrier becomes $T_s/T_c = 1.5M/25k = 60$ and hence, the number of code channels per carrier is given by:

$$N_c = 60.$$

The total number of channels in the system is given by $N_c N_f = 600$ which is the same as partitioning based on FDMA or D-CDMA alone. Using the hybrid scheme, the number of RF transceivers required at the base station is reduced to 75 instead of 600 (for pure FDMA). On the other hand, the maximum chip rate is reduced to 1.5Mcps (instead of 15Mcps as in pure D-CDMA).

In both of the hybrid schemes, FDMA is used to partition the spectrum into a few *carriers* of smaller bandwidth. Each of the carrier is further partitioned based on either TDMA or D-CDMA schemes to a number of *user channels*. In this way,

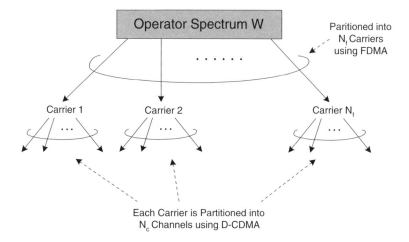

Figure 3.9   Hybrid FDMA and D-CDMA.

we could limit the number of RF units required due to FDMA and at the same time limit the peak rate transmission (for TDMA) or maximum supportable chip rate (for D-CDMA) of individual carriers. The total number of channels available after the hybrid partitioning are the same.

3.4   NON-ORTHOGONAL RESOURCE PARTITIONING

In the previous section, we introduce a few common resource partitioning schemes based on the orthogonality principle. The resulting channels are orthogonal to each other, meaning that users on different channels do not interfere with each other. However, as we could see in Section 3.5, the orthogonal resource partitioning schemes are in fact not the optimal schemes. They are chosen because of low implementation complexity. In this section, we introduce a few other resource partitioning scheme that are *non-orthogonal*. In other words, users on different channels do interfere with each other but in a controlled manner.

3.4.1   Random Code Division Multiple Access (R-CDMA)

Similar to deterministic CDMA, each user is assigned a spreading code sequence $c_k(t) = \sum_n c_{k,n} u(t - nT_c)$ where $c_{k,n} \in \{+1, -1\}$ and $u(t - nT_c)$ is a unit pulse with interval $t \in [nT_c, (n + 1)T_c]$. If we choose the code sequence such that the orthogonality condition in (3.13) is satisfied, the system is limited by the number of codes. On the other hand, if we relax the orthogonality condition slightly to tolerate small cross-correlation (instead of zero), the system is no longer limited by the number of codes but limited by the interference. Such a system is called *random CDMA* [339].

Specifically, the criteria of the code sequence $\{c_k(t)\}$ satisfies:

Auto-correlation: The auto-correlation of the spreading code:

$$R_{k,k}(\tau) = \frac{1}{T_s} \int_0^{T_s} c_k(t)c_k(t+\tau)dt$$

has sharp fall off to a small *residual value* at $\tau \geq T_c$. This is illustrated in Figure 3.10

Time-averaged Cross-correlation: The cross-correlation of the spreading code:

$$R_{k,j}(\tau) = \frac{1}{T_s} \int_0^{T_s} c_k(t)c_j(t+\tau)dt$$

has small value for all τ.

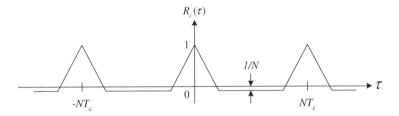

Figure 3.10   Requirement of autocorrelation property of spreading code sequence.

The transmitted signal of user k is given by $S_k(t)c_k(t)$ and the received signal as a result of K users transmission through the AWGN channel is given by:

$$Y(t) = \sum_{k=1}^{K} S_k(t)c_k(t) + Z(t)$$

There are several ways to extract information signal for user k. One simple receiving technique, which has been widely applied in today's CDMA systems, is called *isolated decoding*. Specifically, signals due to other users are treated as interference and the desired information signal of user k is extracted through despreading with respect to $c_k(t)$. That is:

$$W_k(t) = Y(t)c_k(t) = \underbrace{S_k(t)}_{\substack{\text{Signal} \\ \text{Component}}} + \underbrace{\sum_{j\neq k} S_j(t)c_j(t)c_k(t)}_{\text{MultiuserInterference}} + \underbrace{Z(t)c_k(t)}_{\substack{\text{Channel} \\ \text{Noise}}}$$

Expressing $W_k(t)$ as a vector in a signal space in a similar way as in the previous section, we have:

$$\mathbf{W}_k = \mathbf{S}_k + \sum_{j \neq k} \sum_n c_{j,n} c_{k,n} \mathbf{V}_{j,n} + \mathbf{Z}_k \tag{3.14}$$

where $\mathbf{S}_k$ is the information signal vector of user k, $\mathbf{Z}_k$ is the noise vector projected onto the signal space and $\mathbf{V}_{j,n}(d) = \int_{nT_c}^{(n+1)T_c} S_j(t)\phi_d^*(t)dt$ represents the d–th component of the multi-user interference vector contributed by user j. Unlike deterministic CDMA, the *multi-user interference* term $\sum_{j \neq k} \sum_n c_{j,n} c_{k,n} \mathbf{V}_{j,n}$ is non-zero and is a random process. In general, the power of the interference component depends on the specific choice of code sequences. Detail description of the code design[288] and performance is outside the scope of the book. To capture the performance of random CDMA based on isolated detection, we employ a statistical approach.

Assume that the code sequences for the K users $\{c_1(t), \ldots, c_K(t)\}$ are *random sequences* given by:

$$c_k(t) = \sum_n c_{k,n} u(t - nT_c) \tag{3.15}$$

where $c_{k,n}$ is an identical and independent distributed (i.i.d.) random variable drawn from $\{+1, -1\}$ with equal probability. Hence, we have $\mathcal{E}[c_{k,n}] = 0$, $\mathcal{E}[c_{k,n}^2] = 1$, $\mathcal{E}[c_{k,n} c_{k,m}] = 0$ and $\mathcal{E}[c_{k,n} c_{j,m}] = 0$.

Since the received signal vector $\mathbf{W}_k$ in (3.14) is a *sufficient statistic* for the information signal $\mathbf{S}_k$, the maximal likelihood (ML) detection rule on $\mathbf{S}_k$ is given by:

$$\text{argmax}_{\mathbf{S}_k} [p(\mathbf{W}_k | \mathbf{S}_k)] \tag{3.16}$$

For sufficiently large number of users, the multi-user interference becomes Gaussian. Hence, together with the Gaussian channel noise vector, the decision vector $\mathbf{W}_k$ is a Gaussian vector. To evaluate the performance of the ML detection, we have to evaluate the mean and co-variance of the received signal vector $\mathbf{W}_k$.

Conditional Mean of $\mathbf{W}_k$:

Given $\mathbf{S}_k$, the mean of $\mathbf{W}_k$ is given by:

$$\mathcal{E}[\mathbf{W}_k | \mathbf{S}_k] = \mathbf{S}_k$$

The co-variance of the Noise Term:

For AWGN channel, the noise vector $\mathbf{Z}_k$ is a zero-mean Gaussian random vector with covariance matrix given by $\mathcal{E}[\mathbf{Z}_k \mathbf{Z}_k^*] = \eta_0 \mathbf{I}_D$ where η_0 is the noise spectral density of $Z(t)$.

The co-variance of the Multiuser Interference Term:

When the number of users K is sufficiently large, the interference term $\mathbf{V}_I = \sum_j \sum_n c_{j,b} c_{k,n} \mathbf{V}_{j,n}$ is approximately Gaussian. We have the mean given by:

$$\mathcal{E}[\mathbf{V}_I | \mathbf{S}_k] = \mathbf{0} \tag{3.17}$$

because $\mathcal{E}[R_{j,k}(mT_c)] = 0$.

and the covariance matrix given by:

$$\mathcal{E}[\mathbf{V}_I \mathbf{V}_I^* | \mathbf{S}_k]$$

$$= \sum_{j \neq k} \sum_n \mathcal{E}[|c_{k,n}|^2] \mathcal{E}[|c_{j,n}|^2] \left| \int_{nT_c}^{(n+1)T_c} S_j(t)\phi_d^*(t)dt \right|^2 \mathbf{I}$$

$$\leq \sum_{j \neq k} \sum_n \int_{nT_c}^{(n+1)T_c} |S_j(t)|^2 dt \int_{nT_c}^{(n+1)T_c} |\phi_d^*(t)|^2 dt \mathbf{I}$$

$$= \frac{T_c}{T_s} \sum_{j \neq k} \int_0^{T_s} |S_j(t)|^2 dt \mathbf{I}$$

$$= \sum_{j \neq k} E_s(j) \mathbf{I}$$

where $E_s(j) = \int_0^{T_s} |S_j(t)|2dt$ is the *symbol energy* due to user j and the inequality is due to Cauchy-Swartz inequality.

Hence, the log-likelihood detection rule in (3.16) is equivalent to the *minimum distance decoding*:

$$\mathrm{argmax}_{\mathbf{S}_k}[\|\mathbf{W}_k - \mathbf{S}_k\|^2] \qquad (3.18)$$

Let $E_b(k) = P_k T_b$ be the bit energy of user k and $E_s(j) = P_j T_s$ be the symbol energy of user j. The error performance of such a decoder is usually expressed as a function of *bit-energy to variance ratio* $\gamma = \frac{E_b}{\sigma^2}$ where σ^2 is the total variance of the channel noise and multiuser interference. Hence, γ of random CDMA is given by:

$$\gamma_k = \frac{E_b}{\eta_0 + \frac{T_c}{T_s}\sum_{j \neq k} E_s(j)} = \frac{P_k}{\eta_0 W + \sum_{j \neq k} P_j} \frac{T_b}{T_c} \qquad (3.19)$$

where $W = \frac{1}{T_c}$ and P_j is the transmitted power of user j. From this equation, we could see that the γ is given by the SNIR (*signal-to-noise plus interference* power ratio of the $k-$th user) multiplied by a factor $\frac{T_b}{T_c}$. This factor is called the *processing gain* of random CDMA.

$$PG = \frac{T_b}{T_c} \qquad (3.20)$$

Hence, the key characteristic of random CDMA is its ability to suppress interference power by a factor of PG. Hence, signals from multiple users could be transmitting using the same time and frequency dimensions. The *associated interference* power as a result of transmitting in the same frequency and the same time is reduced by the action of spreading and despreading. Note that the interference is *suppressed* but not eliminated. Let us consider an example to illustrate the operation of random CDMA.

Example: In a 3-user uplink system, signals from user 1, 2, 3 are received at power levels of 20dBm, 18 dBm and 23dBm respectively at the base station. All users use BPSK modulation and transmit at a bit rate of 30 kbps. Random CDMA is applied on the uplink with a chip rate of 3M cps. That is: user k is spread with a code sequence $c_k(t)$ according to equation (3.15). Find out the BER performance of the 2nd user where the BER performance of BPSK modulation in AWGN channel is given by $Q\left(\sqrt{\left(\frac{2E_b}{\eta_0}\right)}\right)$.

Solution: The processing gain is given by $PG = \frac{T_b}{T_c} = W/R_b = 100$. The transmit bandwidth = 3MHz and hence, the noise power (in linear scale) is given by $\eta_0 W = 10^{-13.9}$. The received power levels for the 3 users (in linear scale) are 0.1, 0.063 and 0.2 respectively. For the 2nd user, the bit energy to noise plus interference density (γ_2) is given by:

$$\gamma_2 = 100 \frac{P_2}{\eta_0 W + P_1 + P_3} = 21$$

Hence, the BER performance of the user 2 is given by $Q(\sqrt{(2\gamma_2)}) \approx 10^{-8}$.

Unlike deterministic CDMA where the maximum number of channels is limited by the number of available orthogonal code sequence, the random CDMA system is *interference limited*. In other words, since we relaxed the requirement on the cross-correlation of different code sequence to be non-zero, the number of code sequence satisfying this mild criteria is abundant. Therefore, the system is no longer *code limited*. However, channels become *non-orthogonal* and every time a new user is added to the system, the overall interference level to all other users increases. Eventually, the system capacity is limited by the maximal tolerable interference level. This point is illustrated by an example below.

Example: Suppose a spectrum of 15 MHz is allocated to a mobile operator. Assume that binary modulation is used and the required average data rate of individual user be 25 kbps. Find out how many channels could be partitioned based on random CDMA scheme alone if the maximal *tolerable* E_b/σ is given by 7dB (assuming mobiles are transmitting at infinity power).

Solution: Given a bandwidth of 15MHz and a bit rate of 25kbps, the processing gain is given by: $PG = T_b/T_c = W/R_b = 600$. From (3.19), the asymptotic number of channels as $P_j \to \infty$ is given by:

$$N_{R-CDMA,\infty} = 1 + \frac{600}{\gamma_{\text{target}}} = 1 + \frac{600}{5} = 121.$$

Note that PG is the operational parameter of R-CDMA. Furthermore, *spreading factor* and *processing gain* are two *different* parameters for deterministic CDMA and random CDMA respectively. The spreading factor governs the maximum number of orthogonal spreading code sequence. On the other hand, the *processing gain* determines the ability to *suppress* multiuser interference. When binary modulation is used, $T_s = T_b$ and therefore, the two parameters are equivalent. On the other hand, when M−ary modulation is used, the two parameters is different. For instance, given a fixed bit rate (and therefore a fixed T_b), number of orthogonal channels supported by deterministic CDMA could be increased by using higher order modulations (e.g., 16QAM) at the expense of higher required γ. Hence, there is a natural tradeoff between the number of orthogonal channels and the required γ for deterministic CDMA using higher order modulation.

However, the processing gain of random CDMA (and therefore, the interference suppression capability) is independent of the modulation level. In other words, it is always sub-optimal to apply higher level of modulation in random CDMA. That is why most commercial systems operating at random CDMA have either BPSK or QPSK modulation but not 8PSK, 16QAM. This point is better illustrated using the following example.

Example: Suppose a spectrum of 15 MHz is allocated to a mobile operator. Possible modulation schemes include BPSK, QPSK, 16QAM modulations. The required average data rate of individual user be 25 kbps. At a given target BER of 10^{-3}, the required $\gamma_{dB} = 10 \log_{10}(E_b/\sigma^2)$ for the 3 modulation schemes are 7 dB, 7 dB, 17 dB respectively. Which modulation is the best choice for R-CDMA system if we assume that all users transmit at infinite power?

Solution: The *processing gain* of R-CDMA system is given by $PG = T_b/T_c = W/R_b = 15M/25k = 600$. The *asymptotic number of channels* available in R-CDMA with such processing gain is given by:

$$N_{R-CDMA,\infty} = 1 + \frac{600}{\gamma_{\text{target}}}$$

Obviously, using 16 QAM is a sub-optimal choice because the required γ_{dB} is the largest. On the other hand, it is equally effective to use either QPSK or BPSK in R-CDMA system because they require the same bit-energy to noise plus interference density ratio.

3.4.2   Space Division Multiple Access (SDMA)

Another new techniques of non-orthogonal resource partitioning is to utilize the spatial dimension through multiple antenna technologies. Specifically, SDMA serves different users by utilizing the fact that each users are located in different "positions" and are therefore *separable*. Hence, different users could transmit signals based on the same frequency and the same time duration but via different *spatial channels*. For line-of-sight propagation environment, one could visualize the process as forming different narrow beams to focus the transmission to individual users situated in different spatial positions as illustrated in Figure 3.11(a). However, the concept of beams may not apply in a more general environment, namely the *scattering environment* where there is no line of sight propagation. In this case, a more general concept, namely *spatial filtering*, is more appropriate.

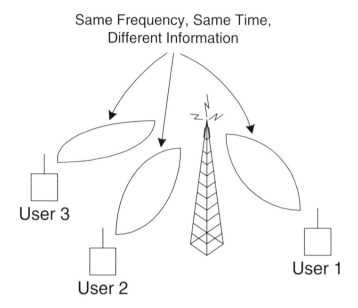

(a) illustration of SDMA using concepts of beam-forming

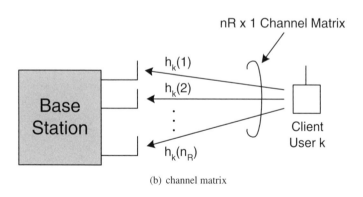

(b) channel matrix

Figure 3.11   Illustration of SDMA using concepts of beam-forming.

We illustrate the concept using uplink as an example. Assume that the base station has n_R antennas and the client station has single antenna. Let $S_k(t)$ be the transmitted signal from user k and $\mathbf{Y}(t)$ be the $n_R \times 1$ dimensional received signal vector at the base station. Each client user is linked up to the base station through an $n_R \times 1$ dimension complex channel matrix, $\mathbf{h}_k$ as illustrated in Figure 3.11(b). The channel matrix is in general a complex random process and the distribution depends on the propagation environment. In one extreme case when we have complete line of sight, $\mathbf{h}_k$ becomes a deterministic quantity whose amplitude is 1 and phase value depends on the relative position between the client user and the base station. In another extreme when we have complete scattering environment, the channel matrix is Rayleigh distributed in amplitude and uniformly distributed in phase. In some intermediate cases, the channel matrix is a mixture between the two ends and it could be characterized by Rician distribution. In any case, the channel matrix is normalized such that

$$\mathcal{E}[\mathbf{h}_k \mathbf{h}_k^*] = \mathbf{I}_{n_R}$$

The received signal vector is given by:

$$\mathbf{Y}(t) = \mathbf{h}_k S_k(t) + \sum_{j \neq k} \mathbf{h}_j S_j(t) + Z(t) \tag{3.21}$$

where $\mathbf{h}_k \neq \mathbf{h}_j$ for all $k \neq j$. The first term represents the desired signal and the middle term represents the multi-user interference. The last term represents the channel noise. Similar to R-CDMA, all signals are transmitted at the same frequency range and the same time but have different *spatial signature* $\mathbf{h}_k$ and hence, they are separable. Hence, there is interference between the user signals and therefore, the SDMA is a non-orthogonal resource partitioning scheme.

There are various ways to separate signals from the K users. The optimal scheme is given by the joint-detection and successive interference scheme as suggested by information theory. However, this implementation complexity is exponential in K and is not feasible in practice. Similar to R-CDMA, we consider a simple and effective method to separate signals from the K users, namely *linear spatial filtering*. Specifically, to extract information from user k, we apply a $n_R \times 1$ dimension *linear complex weight* $\mathbf{w}_k$ to the received signal vector. That is:

$$X_k(t) = \mathbf{w}_k^* \mathbf{Y}(t) = \mathbf{w}_k^* \mathbf{h}_k S_k(t) + \sum_{j \neq k} \mathbf{w}_k^* \mathbf{h}_j S_j(t) + \mathbf{w}_k^* Z(t)$$

The problem becomes how to choose an appropriate weight $\mathbf{w}_k$. In general, there could be several design objectives, such as *interference nulling* and *minimum mean-square error*. For interference nulling, the problem is: Choose weight $\mathbf{w}_k$ such that

$$\mathbf{w}_k^* \mathbf{h}_k = 1.$$

$$\mathbf{w}_k^* \mathbf{h}_j = 0 \quad \forall j \neq k$$

For minimum mean-square error (MMSE), the optimization problem becomes: Choose weight $\mathbf{w}_k$ such that the Mean-Square-Error (MSE) is minimized.

$$\text{argmin}_{\mathbf{w}_k} \left(\mathcal{E} \left[|\mathbf{w}_k^* \mathbf{Y}(t) - S_k(t)|^2 \right] \right)$$

For interference nulling, we are solving a system of linear equation and in order to have solution, we have:

$$K \leq n_R \tag{3.22}$$

Hence, the maximum number of *spatial channel* as a result of interference nulling spatial filtering is the number of antennas at the base station (n_R). Note that interference nulling gives n_R orthogonal spatial channels at the expense of enhanced noise power.

On the other hand, for MMSE spatial filtering, the optimal weight is given by:

$$\mathbf{w}_k = \mathbf{\Phi}^{-1} \mathbf{p}_k$$

where $\mathbf{\Phi} = \mathcal{E}[\mathbf{Y}(t)\mathbf{Y}^*(t)]$ and $\mathbf{p}_k = \mathcal{E}[\mathbf{Y}(t)S_k^*(t)]$. If we look at the MMSE filtered output for user k, it consists of the desired signal for user k, the multi-user interference (*suppressed* by MMSE filtering) and the residual channel noise. Therefore, the spatial channels are not *orthogonal* with each other and they suffer from multi-user interference but with a suppressed power due to spatial filtering. Similar to R-CDMA, the number of spatial channels available in the system is *interference limited* and depends on the maximal tolerable interference level.

3.5   SPECTRAL EFFICIENCY AND PERFORMANCE ISSUES

In the previous sections, we have introduced an information theoretical framework for multi-user communications. We have also introduced various common resource partitioning schemes, resulting in both orthogonal and non-orthogonal channels. In this section, we give a performance comparison of the various schemes in both single cell and multi-cell scenarios.

3.5.1   Single Cell Comparison

We first assume that the system consists of a single cell and K mobile users. In this scenario, a very important multiuser performance metric is the system capacity given by $\sum_k R_k$ (R_k) is the data rate of user k. On the other hand, when all users have the same data rate (such as voice applications), the system capacity could be expressed as the number of users (or number of channels) that could be supported. As we illustrate below, all the orthogonal resource partitioning schemes (namely FDMA, TDMA and D-CDMA) are *equally effective*. On the other hand, the performance of non-orthogonal resource partitioning schemes depends heavily on the decoding methods employed. For linear complexity signal separation as we have illustrated in the previous section, the performance is *interference limited* and is in general worst than that of the orthogonal schemes. However, when more sophisticated interference cancellation schemes are employed (such as multi-user joint detection), the performance of non-orthogonal schemes are actually better than that of the orthogonal schemes.

Performance of FDMA System:

Let W be the spectrum available for resource partitioning, and R_b be the required average bit rate per user. The transmission bandwidth per user is given by:

$$W_k = R_b/\log_2(M)$$

where M is the modulation level in terms of bits per modulation symbol. Since the bit rate per user is the same for all users, the number of channels partitioned by FDMA is given by:

$$N_{FDMA} = W/W_k = (W/R_b)\log_2(M)$$

Performance of TDMA System:

Let T_p be the frame duration of a TDMA frame and T_k be the width of the time slot assigned to user k. The peak bit rate of user k during the assigned time slot is given by $R_p = W\log_2(M)$. Hence, the average bit rate per user k is given by:

$$R_b = P_p(T_k/T_p) = W\log_2(M)T_k/T_p$$

The total number of time slots available is thus given by:

$$N_{TDMA} = T_p/T_k = (W/R_b)\log_2(M)$$

Performance of D-CDMA Systems:

The modulation symbol duration is given by $T_s = T_b\log_2(M) = \log_2(M)/R_b$. With a bandwidth of W, the chip duration $T_c = 1/W$. Hence, the *spreading factor* of D-CDMA is given by:

$$SF = T_s/T_c = \log_2(M)(W/R_b)$$

Hence, the total number of channels available for D-CDMA is given by:

$$N_{D-CDMA} = \log_2(M)(W/R_b)$$

As could be seen from the above results, we observe that $N_{FDMA} = N_{TDMA} = N_{D-CDMA}$. In other words, they are equally effective in terms of system capacity. Furthermore, observe that the system capacity could be enhanced by increasing the modulation level M. Yet, the price to pay is increased signal to noise ratio to maintain the BER target.

Comparing the orthogonal resource partitioning schemes with the information theoretical (optimal) results, we could see that the orthogonal schemes are in fact, sub-optimal. This is illustrated in the *capacity region* of the multiple access (uplink) and broadcast (downlink) problems in Figures 3.12(a) and 3.12(b) respectively.

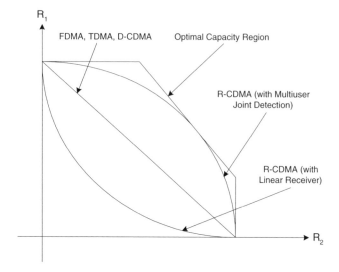

(a) multiples access region

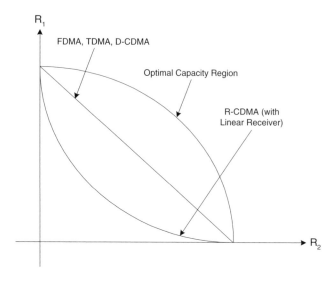

(b) broadcast region

Figure 3.12   Comparison of FDMA, TDMA, D-CDMA and R-CDMA with respect to the information theoretical capacity region.

Performance of R-CDMA:

For R-CDMA with *isolated decoding*, the system capacity is *interference limited* instead of code limited. In other words, as more and more users are added to the system, the background interference level grows and individual existing users have to further increase their transmit powers to offset the rise in background interference (so that the target SNR could be maintained.). There reaches a point when no more users could be added and every existing users are transmitting at infinite power levels. When this happens, we said the capacity of the R-CDMA system reaches the *pole capacity*. This concept is illustrated below.

Consider the reverse link as a simple illustration of R-CDMA. As illustrated in the previous section, the received bit-energy-to noise plus interference level of user k, γ_k, is given by:

$$\gamma_k = \frac{P_k}{\eta_0 W + \sum_{j \neq k} P_j} \frac{T_b}{T_c} \tag{3.23}$$

where P_j is the received signal power from user j. Assume that to maintain a target BER for user k, the required γ_k has to be at least γ_{target}. This translates into the requirement:

$$\frac{P_k}{\eta_0 W + \sum_{j \neq k} P_j} \frac{T_b}{T_c} \geq \gamma_{\text{target}}$$

For voice applications, the source rate is quite steady and the real time speech frames could not be queued. Therefore, by symmetry, the optimal operating point (with respect to the number of users that could be accommodated) is when all users signal powers are equal and given by P. In this case, we have:

$$\frac{P}{\eta_0 W + (K - 1)P} \frac{T_b}{T_c} \geq \gamma_{\text{target}}$$

Therefore, the capacity of R-CDMA is given by:

$$N_{R-CDMA} \leq \left(\frac{T_b/T_c}{\gamma_{\text{target}}} - \frac{\eta_0 W}{P} + 1 \right)$$

Figure 3.13 illustrates the required received power P versus the number of users in the system N_{R-CDMA}. Observe that in the *lightly loaded* range, as more users are added to the R-CDMA system, there is a proportional increase in the received signal power P to compensate for the rise in background interference. However, as the system gets more and more *heavily loaded*, the corresponding increase in the received signal power required rise in a dis-proportional manner and *diminishing return* is observed.

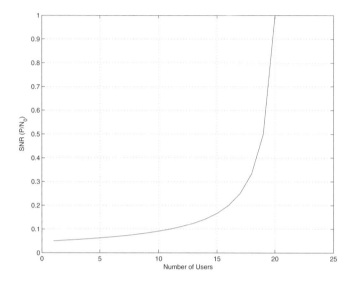

Figure 3.13   Received power versus the system capacity in R-CDMA (isolated decoding). $T_b/T_c = 100$, target $\gamma = 5$.

In the limit when $P \to \infty$, we have the *pole capacity*[3] given by:

$$N_{R-CDMA}(\infty) = 1 + \frac{T_b/T_c}{\gamma_{\text{target}}} = 1 + \frac{W/R_b}{\gamma_{\text{target}}} \qquad (3.24)$$

In general, the required γ_{target} is a function of the modulation level and coding schemes employed. In most cases, the required γ_{target} would be of the order 7 dB for a BER of 10^{-3}. Hence, we could easily see that the pole capacity N_{R-CDMA} is less than the system capacity for FDMA, TDMA or D-CDMA. The performance of R-CDMA (with isolated decoding) with respect to the information theoretical performance is illustrated in Figure 3.12(a). On the other hand, when *multiuser joint-detection* is used, the performance of R-CDMA is substantially improved. This is because with joint detection, multi-user interference could be cancelled between successive decoding stages. Yet, the implementation complexity of such receiver is in general very high and it is outside the scope of the book to discuss the topics on multiuser detection.

Another point to make is on the relationship between the R-CDMA capacity N_{R-CDMA} and the modulation level. Unlike the orthogonal schemes (FDMA, TDMA, D-CDMA) where the number of channels is increased by increasing the

---

[3]It is called the pole capacity because to achieve $P \to \infty$, one way is to have all the users transmitting at finite power but situated arbitrarily close to the base station antenna. Therefore, it is called the pole capacity.

modulation level M, the system capacity of R-CDMA depends solely on the required γ. In general, using higher modulation level (for bandwidth efficient modulation schemes such as 8PSK, 16QAM, 64QAM) requires higher γ to maintain the same BER, which is detrimental to the system capacity. Therefore, we seldom use anything beyond QPSK modulation for R-CDMA systems.

3.5.2 Multi-Cell Comparison

The previous section compares the system performance of various resource partitioning schemes in single cell environment. Here, we extend the comparison to consider multi-cell environment[203] where the service area is covered by multiple cells. Although FDMA/TDMA/D-CDMA achieves better performance compared to R-CDMA (isolated detection) in single cell environment, we illustrate that R-CDMA (isolated detection) performs better than the orthogonal resource partitioning schemes in multi-cell environment.

Resource Reuse in Orthogonal Resource Partitioning

Without loss of generality, assume that we have N orthogonal channels in the system. These N orthogonal channels could be partitioned using FDMA, TDMA, D-CDMA or hybrids of these schemes but we do not try to differentiate the detail partitioning here. In multi-cell scenarios, we have to assign the N orthogonal channels to the multi-cells. On one hand, we want to increase the *channel reuse efficiency* so that the overall system capacity could be multiplied based on the N channels. On the other hand, we have to make sure that *co-channel interference*[4] as a result of *channel reuse* is controlled at an acceptable level. There are many ways to do that. A simple and commonly used scheme is called *fixed channel assignment* (FCA).

With FCA, the N orthogonal channels are partitioned into N_c *clusters*. These N_c clusters are assigned to the multi-cells according to a reuse rule that the distance separation between the *co-cluster* cells are maximized. The corresponding system capacity is therefore given by:

$$N_{orthogonal} = (N/N_c) \times B \qquad (3.25)$$

where B is the total number of cells in the system. For hexagonal cell shape, the condition on N_c for tessellating the cells is that $N_c = i^2 + ij + j^2$ for all non-negative integers i, j. Figure 3.14 illustrates various clusterization patterns.

In one extreme case, we could have $N_c = 1$ where this is equivalent to *complete reuse*. In this case, the system capacity from (3.25) will be maximized. However, the side effect is the co-channel interference. In this case, since co-channel cells are adjacent to each other, the interference level is very high and therefore, the quality of the communication link is very bad.

---

[4]Co-channel interference is the interference between multi-users as a result of re-using a channel in different cells.

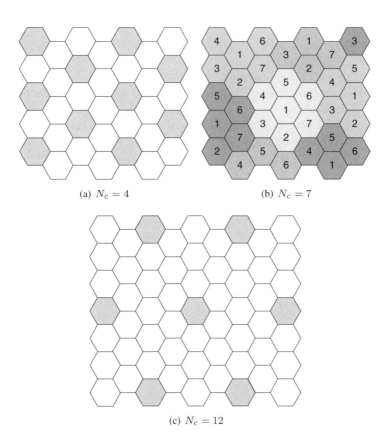

(a) $N_c = 4$

(b) $N_c = 7$

(c) $N_c = 12$

Figure 3.14 Illustration of several cluster reuse patterns in FCA.

In another extreme case, we could have a very large N_c and this is equivalent to *no reuse*. In this case, the system capacity is simply equal to N but the communication quality is very good because there is no co-channel users.

Obviously, these two extremes are not desirable operating point and the desirable operating point should be a minimum cluster size N_c where the co-channel interference is just below the maximal tolerable level. In fact, it can be easily verified that the signal to interference level (SIR) as a result of channel reuse is given by:

$$SIR \approx \sqrt{3N_c}^{\alpha}$$

where α is the path loss exponent.

Example: Find the multi-cell capacity of AMPS system where the minimum required SIR is 18dB and the path loss exponent is 4.

Solution: Since the minimum SIR is 18 dB, the lowest cluster size $N_c = 7$. Therefore, we could at best operate the channel assignment based on 7-cluster pattern. With this setting, the system capacity is given by:

$$N = (N_{AMPS}/7)B \tag{3.26}$$

where B is the total number of cells in the system and N_{AMPS} is the total number of channels in AMPS system.

Example: Find the multi-cell capacity of Digital-AMPS system where the minimum required SIR (as a result of digital transmission, error correction and sectorization) is 12dB.

Solution: Since the minimum SIR is 12 dB, the lowest cluster size $N_c = 4$. Therefore, we could at best operate the channel assignment based on 4-cluster pattern. With this setting, the system capacity is given by:

$$N = (N_{DAMPS}/4)B = (W/R_b)(B/4) \tag{3.27}$$

where B is the total number of cells in the system, N_{DAMPS} is the total number of channels in AMPS system, W is the bandwidth of the operator spectrum and R_b is the required bit rate per user. Note that due to digital speech compression, $N_{DAMPS} = 3N_{AMPS}$ for the same available spectrum. Hence, the capacity gain of 2G digital systems with respect to 1G analog systems are contributed by two factors, namely *higher number of channels as a result of speech compression* and *better reuse efficiency as a result of smaller cluster size*.

Since received signal strength is inversely proportional to d^{α} where $\alpha \approx 4$ in terrestrial propagation environment, the co-channel interference as a result of reuse is reduced by the path loss effect. In a way, the increased path loss exponent in terrestrial propagation helps the channel reuse in multi-cell because signals are more *contained* within their own cells.

Multi-cell Performance of R-CDMA

For R-CDMA, the extension to multicell scenario is relatively easy. This is because the intra-cell channels are non-orthogonal already and there is not much difference between intra-cell interference and inter-cell interference. To account for the additional intercell interference, we have

$$I_{oc} = \beta I_{sc} = \beta(K - 1)P\mu$$

where I_{oc} denotes the other cell interference power, I_{sc} denotes the same cell interference power and μ denotes the voice activity factor of users. Typical values of β and μ are $[0.4, 0.7]$ (depending on the path loss exponent) and 0.5 respectively. Hence, the received bit-energy to noise plus interference ratio of user k γ_k is given by:

$$\gamma_k = \frac{P}{\eta_0 W + (1 + \beta)\mu(K - 1)P} \frac{T_b}{T_c} \geq \gamma_{\text{target}}$$

The pole capacity of a cell is given by $N_{R-CDMA,cell} \approx 1 + \frac{T_b/T_c}{(1+\beta)\gamma_{\text{target}}}$. Since all the adjacent cells use the same frequency and the system is not code limited, the total system capacity in multi-cell scenario is multiplied by B and is given by:

$$N_{R-CDMA,sys} = B \left(1 + \frac{W/R_b}{(1 + \beta)\mu\gamma_{\text{target}}} \right) \tag{3.28}$$

where B is the total number of cells in the system.

Example: Find the multi-cell capacity of IS 95 system where the minimum required SIR (as a result of digital transmission, error correction and sectorization) is 7dB. The voice activity factor μ is 0.5 and the co-channel interference factor β is 0.4.

Solution: Since the minimum SIR is 7 dB (which is approximately 5 in linear scale), the system *pole capacity* is given by:

$$N_{IS95,sys} \approx B \left(1 + \frac{W/R_b}{3.5} \right)$$

Observe that R-CDMA multi-cell capacity is larger than the multi-cell capacity of the orthogonal resource partitioning schemes (FDMA/TDMA/D-CDMA) because of the complete channel reuse. Figure 3.15 illustrates the system capacity comparison of the orthogonal and non-orthogonal partitioning schemes. It is shown that the larger the number of cells in the system, the greater the capacity difference is.

3.6 PRACTICAL ILLUSTRATIONS: GSM, CDMA AND WIRELESS LAN

In this section, we explain how different practical systems apply the resource partitioning techniques to enable multi-user communications. We look at GSM (2G cellular system), CDMA (2G cellular system) and Wireless LAN (short-range wireless system) [129, 274] as examples.

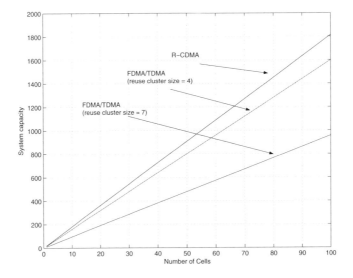

Figure 3.15 Comparison of multicell system capacity between the orthogonal schemes (FDMA,TDMA,D-TDMA) and the non-orthogonal schemes (R-CDMA). $W = 1.5$ MHz, $R_b = 25$kbps.

3.6.1 Multiuser resource partitioning in GSM

GSM is one of the major second generation cellular standards with over 70% world-wide penetration based on TDMA. We leave it to Part II for a detail elaboration of the GSM architecture and design but in this section, we focus on the multi-access aspects in GSM system[141]. In GSM, the spectrum bandwidth is partitioned into radio channels based on hybrid FDMA / TDMA approach. The radio spectrum is first partitioned into a number of *carriers* based on FDMA where each carrier has a bandwidth of 200kHz. Each 200kHz carrier is further partitioned into 8 *time slots* based on TDMA. Voice traffic and control traffic are delivered to and from the mobiles via the timeslots. For example, a GSM operator may be allocated a paired spectrum of 15MHz on the uplink and downlink respectively. The 15MHz spectrum is first partitioned into 75 *carriers* (each carrier has a bandwidth of 200kHz) based on FDMA. Each of the 200kHz carrier is partitioned into 8 timeslots based on TDMA [143]. Hence, the total number of *full duplex channels* available to the operator is given by $75 \times 8 = 600$. The choice of 200kHz carrier bandwidth is a compromise between two factors. For instance, since the end-user channel is partitioned based on TDMA, the user on the assigned timeslot will be transmitting over the entire carrier bandwidth. If the carrier bandwidth is too large (say 2MHz instead of 200kHz), the number of resolvable multipaths as seen by the GSM mobile will be quite high because the typical outdoor coherence bandwidth is of the order of 100kHz. Hence, complex equalization is required at the GSM mobile which is highly undesirable. We elaborate more on the effect of multi-path fading in Chapter 4. On the other hand, if the bandwidth of the carrier is too small (say 50kHz instead of 200kHz), the number of carriers derived from the 15MHz spectrum will be 300 instead of 75. Hence, a base station (fully equipped with all the 600 channels) will require 300 pairs of RF units which is very bulky and expensive. Therefore, the choice of 200kHz carrier bandwidth is a compromise between the equalization complexity at the GSM mobile and the base station complexity.

In fact, the 600 channels available to the GSM operator in the example is a small number. Imagine the operator has to serve the entire city with only 600 channels from the 15MHz paired spectrum. This is highly insufficient and obviously, multi-cell spatial reuse has to be exploited in order to scale up the total number of channels in the GSM system as the number of base stations increases. Strictly speaking, since a user channel in GSM is labeled by both the timeslot number and the carrier number, two user channels in GSM systems are regarded as *co-channel* users if the two user channels have the same time slot and carrier number coordinate. This is because if either the time slot number or the carrier number of two user channels differs, the two user channels will not be interfering with each other. For example, two user channels with (time slot number, carrier number) coordinate $= (1, 3)$ and $(1, 4)$ are not co-channel users. Two user channels with coordinate $= (1, 3)$ and $(2, 3)$ are also not co-channel user. Hence, strictly speaking, the *user channels* on adjacent cells can be separated based on either the time slot dimension as well as the carrier dimension. However, in GSM systems (FDD), the base stations are only *coarsely synchronized* in time with each other. Furthermore, due to the uncertainty of propagation delay between mobiles in different cells, it is technically too complicated if we try to separate adjacent cell users based on time slots. In GSM systems, the adjacent cells are assigned with different carriers in a similar way as traditional frequency planning (hence, channels from adjacent cells are separated based on the carrier dimension only). The carrier is reused as long as co-channel cells are sufficiently separated. For example, if base station 1 and base station 2 are next to each other, they will be assigned two different carriers. Within each of the assigned carriers, the base station can utilize all the 8 time slots without worrying about mutual interference. Due to the digital transmission and the error correction in the speech frame, the required SIR in GSM system is around 13dB. Together with sectorization, the frequency reuse factor can be reduced from $K = 7$ to $K = 4$. This represents substantial system capacity gain compared to the 1st generation systems.

3.6.2   Multiuser resource partitioning in IS95

Similarly, we leave to Part II for a more detail elaboration of the system architecture and operations of IS95 (another example of 2G digital cellular systems based on CDMA technology). In this section, we just focus on the multi-user aspects of IS95 systems. In IS95 systems, user channels are partitioned from the allocated spectrum based on hybrid FDMA/CDMA. The allocated spectrum is first partitioned into 1.5MHz *carrier*. Each of the 1.5MHz carrier is further partitioned into user channels based on CDMA. For example, suppose the operator has been allocated 15MHz of paired spectrum. We have $15M/1.5M = 10$ full duplex CDMA carriers after FDMA partitioning. Each of the 1.5MHz CDMA carrier is further partitioned into a number of code channels based on CDMA. Similar to the GSM system, the choice of 1.5MHz is a compromise between CDMA receiver complexity and the efficiency of CDMA. For instance, if the CDMA carrier bandwidth is too high, the chip rate will be high and this will induce higher complexity in the CDMA receiver processing (chip synchronization, searching and despreading). On the other hand, if the CDMA carrier bandwidth is too low, the spreading factor and the processing gain will be too low and this reduces the efficiency of CDMA to suppress various interference.

Unlike the GSM example where we can precisely specify the total number of full duplex channels from the 15MHz pair spectrum, we cannot precisely calculate the total number of channels here. This is because in IS95 downlink, D-CDMA (with spreading factor of 64) is used to partition the 1.5MHz carrier into orthogonal code channels. Hence, there are a total of $64 \times 10 = 640$ downlink channels. However, on the uplink, R-CDMA (with processing gain of 64) is used to partition the 1.5MHz carrier into *non-orthogonal* code channels. As we have introduced, the number of channels R-CDMA can support is not code limited but interference limited. Hence, we have *soft capacity* instead of the *hard capacity*. Nonetheless, in IS95, the R-CDMA capacity is around 14-16 concurrent users per 1.5MHz CDMA carrier.

Similarly, the total number of channels have to be able to scale in the multicell network as we increase the number of base stations. As we have illustrated, CDMA systems allowed complete frequency reuse due to the inherit interference suppression capability. Adjacent cells in IS95 networks are assigned with the same CDMA carriers and the co-channel interference will be suppressed by the inherit processing gain of 64. When the capacity of the 1.5MHz CDMA carrier has been used up in some hot spots, additional CDMA carrier can be added. In other words, each CDMA carrier can be viewed as a *layer* and multiple carriers are *overlaid* with each other in the IS95 multicell system.

3.6.3   Multiuser resource partitioning in Wireless LAN

In cellular systems such as GSM or IS95 systems, the channels are partitioned and coordinated at the base station. This refers to the centralized approach. On the other hand, wireless LAN (Wi-Fi) systems [129] are designed based on a *distributed* manner. There is no centralized controller (similar to the base station) in a wireless LAN network. The key design challenge in Wi-Fi systems is robust operations under distributed and uncoordinated situations. In Part III, we elaborate more on the architecture and the operation of Wi-Fi systems. In this section, we just focus on the multi-user resource aspects in Wi-Fi.

Wi-Fi system is designed to operate in ISM band (2.4GHz or 5GHz) which is an unlicensed spectrum. For instance, the ISM band in 2.4GHz is partitioned into 11 *channels*. However, each Wi-Fi transmission has a bandwidth of 20MHz which occupies 5 consecutive ISM channels. Hence, in the 11 ISM channels, there can only be three non-overlapping Wi-Fi channels, namely channel 1, channel 6 and channel 11. The key to Wi-Fi systems concerning multi-user communications is based on the *carrier sense multiple access with collision avoidance* (CSMA/CA). For instance, before a Wi-Fi user can transmit a packet, it has to listen to the channel and make sure there is no existing transmissions. In other words, in normal operations, there will be no concurrent transmissions among multiple Wi-Fi users operating in the same channel. Occasionally, when there is concurrent transmissions among Wi-Fi users, this results in *collision* and none of the packets will get through. Hence, in the steady state, the Wi-Fi users *coordinate* among themselves in a *distributed* manner to time share the channel dynamically. At any time instance, only one Wi-Fi user is allowed to use the channel.

Notice that Wi-Fi system such as 802.11b utilize spread spectrum to transmit the packet (using Barker code). The spreading factor of the Barker code is 11. Unlike IS95 systems, all the Wi-Fi users in the systems are assigned the same and unique Barker code. Hence, the Barker code is not used to separate other Wi-Fi users for simultaneous transmissions. The Barker code (spread spectrum) is used for interference suppression with respect to other ISM users.

On the multi-cell aspects, there is no need for coordination in Wi-Fi systems as well. For example, multiple access points in Wi-Fi systems can be configured to operate in the same frequency. The *co-channel interference* between adjacent access points is automatically taken care by the CSMA/CA protocol. If the two access points are within the *sensing range*, the Wi-Fi users associated with the two access points will time-share the channel dynamically using the CSMA/CA. On the other hand, if the two access points are separated sufficiently far apart that they are outside the *sensing range*, the Wi-Fi users associated with the two access points will simultaneously use the same channel without affecting each other. Hence, in Wi-Fi systems, the design of sensing range is critical to the scalability of the network capacity. For instance, if the sensing range is too large, the capacity scalability will be very poor and the total throughput supported by the entire network will not scale effectively as we increase the number of access points. If the sensing range is too small, there will be increased *hidden node* populations and this will dramatically reduce the network throughput due to collisions.

3.7 SUMMARY

In this chapter, we address the problem of multiuser communications using both the *information theoretical* approach and *practical design* approach. In the information theoretical context, we state the multiuser communication problem as the *multiaccess problem* (uplink) and *broadcast problem* (downlink) and offer a general solution for AWGN channel. In practical design approach, we have reviewed and compared the performance of various *orthogonal resource partitioning* schemes (such as FDMA, TDMA, D-CDMA) as well as *non-orthogonal resource partitioning* schemes (such as R-CDMA and SDMA). While orthogonal schemes offer better performance than non-orthogonal scheme (with isolated detection) in single cell scenario, the non-orthogonal schemes give better performance in multi-cell scenario due to the ability to have complete resource reuse. Practical illustrations on the use multiuser communication technologies in GSM, CDMA, and wireless LAN are given in the final section.

PROBLEMS

3.1   Suppose a digital voice user has a source bit rate of 8kbps (full duplex). Assuming QPSK is used as the common modulation scheme and the baud rate is equal to the required channel bandwidth. Consider a single cell system with 1.6 MHz bandwidth allocated for uplink and 1.6MHz for downlink respectively. Calculate the capacity (in turns of number of concurrent active voice users) achievable by the FDMA. [Assume that coherent phase reference is available at the receiver].

3.2   Using the same setting as in Question 1, what about the capacity of the single-cell TDMA system?

3.3   What about the forward link capacity of the single-cell CDMA system Hint: The number of distinct orthogonal spreading codes available is equal to the number of chips per modulation symbol? Which technology is more effective in terms of forward link capacity in a single cell system? Will there be any difference in multiple cell system?

CHAPTER 4

DIVERSITY TECHNIQUES

Diversity—The division of classes among a certain population.

—Wikipedia, the free encyclopedia

4.1 INTRODUCTION

Communication system consists of a transmitter and a receiver connected by a *channel*. In Chapter 1, we have elaborated the common channel models to capture the major propagation characteristics of mobile radio channels. In Chapter 2, we have discussed the operation and tradeoff picture of digital modulator and demodulator. However, we have used AWGN as a simple example when we discuss the performance of digital communication systems. In this chapter, we shall further elaborate and focus on the detrimental effects (and the possible solutions) of channel fading on digital communication systems.

We shall first start with flat fading channels and frequency selective fading channels and discuss their effects on the BER performance of the physical layer. Next, we shall

Wireless Internet and Mobile Computing. By Yu-Kwong Ricky Kwok and Vincent K. N. Lau
Copyright © 2007 John Wiley & Sons, Inc.

discuss the corresponding solutions to the detrimental effects, namely the *diversity techniques* and the *equalization effects*. We shall elaborate various means of obtaining diversity, namely the *time diversity*, *frequency diversity* and *spatial diversity* [289]. Following these, we shall illustrate the practical application of diversity techniques, namely the RAKE receiver [67], in CDMA systems (such as IS95, UMTS, 3G1X). Finally, we shall conclude with a brief summary of main points.

4.2 EFFECTS OF FLAT FADING ON BER PERFORMANCE

Recall in Chapter 1 that when the transmit signal bandwidth is less than the coherence bandwidth (narrowband transmission), the transmit signal experiences flat fading channels. The low-pass equivalent received signal through slow flat fading channels is given by:

$$y(t) = \alpha(t)x(t - \tau) + z(t)$$

where $x(t)$ is the transmitted signal, $\alpha(t)$ is the complex fading coefficient, τ is the propagation delay and $z(t)$ is the complex Gaussian channel noise with zero mean and power spectral density N_0. In the AWGN channel, the transmitted signal is affected by the channel noise only. However, in flat fading channels, the transmitted signal is *distorted* by the fading coefficient $\alpha(t)$ as well as the channel noise. Both the channel noise and the fading coefficient $\alpha(t)$ are random processes. Since we consider slow fading channels, $\alpha(t) \approx \alpha$ within a modulation symbol duration. The amplitude $|\alpha|$ represents the random attenuation introduced by flat fading. The phase $\angle\alpha$ represents the random phase shift introduced by flat fading. Since the complex fading coefficient α is Gaussian distributed with zero-mean and unit variance $\mathcal{E}[|\alpha|^2] = 1$, the amplitude $|\alpha|$ is Rayleigh distributed and the phase $\angle\alpha$ is uniformly distributed over $[0, 2\pi]$.

For simplicity, consider BPSK modulation where the signal set is given by $\mathcal{S} = \{s_1(t), s_2(t)\}$. Conditioned on α, the flat fading channel is the same as AWGN channel. As illustrated in Chapter 2, the conditional error probability of BPSK modulation in flat fading channels (conditioned on α) is determined by the Euclidean distance between the 2 constellation points $\|\alpha\vec{s}_1 - \alpha\vec{s}_2\|^2$ and is given by:

$$P_e(\alpha) = Q\sqrt{\left(\frac{\|\alpha\vec{s}_1 - \alpha\vec{s}_2\|^2}{2N_0}\right)} = Q\sqrt{\left(\frac{2|\alpha|^2 E_s}{N_0}\right)} \tag{4.1}$$

where $E_s = \int_0^{T_s} |s_i(t)|^2 dt$.

Hence, the average error probability of the BPSK modulation (averaged over the random fading coefficient α) is given by [287]:

$$P_e = \mathcal{E}[P_e(\alpha)] = Q\sqrt{\left(\frac{2\gamma E_s}{N_0}\right)} f_\gamma(\gamma) d\gamma \approx \frac{1}{4E_s/N_0} \tag{4.2}$$

where $\gamma = |\alpha|^2$, $f_\gamma(\gamma) = \exp(-\gamma)$ is the pdf of γ and the approximation holds for large SNR E_s/N_0. Consider plotting the BER (in log-scale) versus SNR (in dB), we have

$$10 \log_{10} P_e \approx -10 \log_{10}(E_s/N_0) + K.$$

Hence, the BER curve (in log scale) of BPSK modulation in flat fading channels has a slope of -1 at large SNR.

Figure 4.1 illustrates the effect of flat fading on the BER vs SNR curve of BPSK modulation. The error probability drops rapidly with respect to SNR for AWGN channel. However, for flat fading channels, the slope of the BER curve *flattened*. Hence, at a given target BER (say 10^{-3}), the required SNR is 40 dB which is 30 dB larger than that required for AWGN channel[287]. As illustrated in the figure, due to the flat fading coefficient α, the required SNR has to be increased substantially to maintain a certain target error probability. For point-to-point communications, the penalty of flat fading is a reduced communication range (for the same transmit power) or extra transmit power (for the same communication range). For multi-user communications (such as R-CDMA), the penalty of the flat fading is translated into a reduction of system capacity.

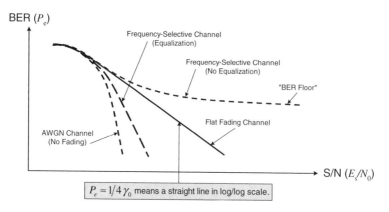

Figure 4.1   BER curves of BPSK in AWGN and flat fading channels.

4.3   EFFECTS OF FREQUENCY SELECTIVE FADING ON BER PERFORMANCE

When the transmit signal bandwidth is larger than the channel coherence bandwidth, the multipaths become *resolvable* and the transmit signal experiences frequency selective fading channels. The lowpass equivalent received signal $y(t)$, after passing through frequency selective fading channels, is given by:

$$y(t) = \sum_{l=0}^{L_p-1} \alpha_l(t)x(t - lT_s) + z(t)$$

where $\alpha_l(t)$ is the complex random channel fading coefficient of the $l-$th resolvable path, T_s is the symbol duration, $z(t)$ is the complex Gaussian channel noise with zero mean and power spectral density N_0, and $L_p = \lceil W_{tx}/B_c \rceil$. Due to the resolvable multipaths, the *influence* of a transmitted symbol at the n-th symbol duration spans over multiple symbol durations $(n - L_p + 1, n - L_p + 2, \ldots, n)$. In other words, the received symbols are mixed up with each other and this results in *intersymbol interference* (ISI). This is illustrated in Figure 4.2 The larger the number of resolvable multipaths (L_p), the more severe the ISI will be.

The effects of frequency selective fading channels is therefore more severe than the flattening of the slope of the BER curves. In addition, due to the ISI, the BER curves will exhibit *error floor* as illustrated in Figure 4.2 This can be explained by the following. The BER can be seen as a function of SINR (Signal to Noise + Interference Power).

$$\gamma = \frac{S}{N + I}$$

where S is the received signal power, N is the noise power, and I is the ISI power. At the $n-$th symbol duration, the ISI is contributed by all the past echos of the past symbols. Hence, the ISI power (conditioned on the fading coefficients) is given by:

$$I = S \sum_{l=1}^{L_p-1} |\alpha_l|^2$$

When the SNR (S/N) is small, $N > I$ and therefore, the BER decreases with increasing SNR. However, at large SNR, we have $I >> S$ and hence the SINR γ is given by:

$$\gamma \rightarrow S/I = \frac{1}{\sum_{l=1}^{L_p-1} |\alpha_l|^2} = \gamma_0$$

Hence, γ approaches a constant γ_0 and therefore, the BER does not decrease further with increasing SNR. Therefore, irreducible error floor[219] is resulted. Note that the error floor level depends on the residual SIR γ_0. Hence, when L_p is large, the error floor will be high.

4.4   DIVERSITY: A KEY TECHNIQUE TO COMBAT FLAT FADING CHANNELS

In this section, we shall elaborate a common and effective technique to combat the *flattening* of the slope of the BER curve as a result of flat fading. In Chapter 2, we

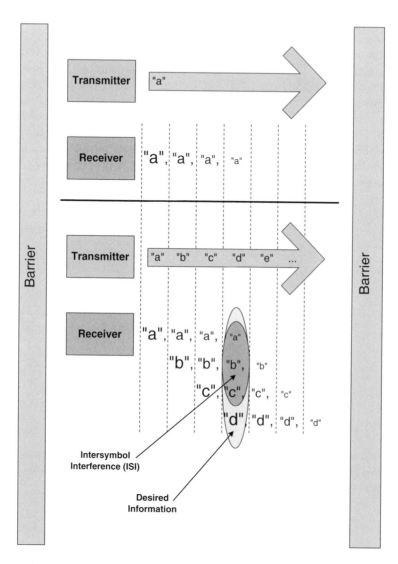

Figure 4.2   Illustration of intersymbol interference (ISI). The transmitter (transmitting letter "a") results in multiple echos at the receiver. Hence, when the transmitter transmits a sequence of letters "a,b,c,...", ISI will be resulted at the receiver where the desired information (letter 'd') is interfered by the past echos of letters "c", "b", "a".

have elaborated on the minimum distance detection design. For a given received symbol, the "best" guess on what has been transmitted is the one that is *closet* to the *observation*. Normally, the detection is made based on a single observation. However, if L *independent observations* about the same information are available at the receiver, the detection can be made based on all the L observations. In that case, the receiver is said to exploit L-th order diversity. With L-th order diversity, the probability that all the L independent copies of observations about the same information fade simultaneously is dramatically reduced.

There are in general three common ways of making use of L independent observations about the same information at the receiver. They are referred to *diversity combining techniques*, namely *selection diversity*, *equal gain combining* and *maximal ratio combining*. These methods, as illustrated in Figure 4.3, are elaborated in the following subsections.

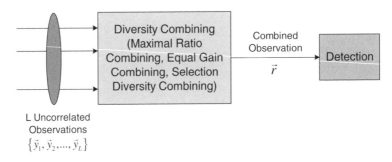

Figure 4.3   Diversity Combining.

4.4.1   Diversity Combining

For simplicity, consider a binary modulator with signal set $\mathcal{S} = \{\vec{s}_1, \vec{s}_2\}$ transmit over flat fading channels with L independent observations. The received vectors from the l-th observation is given by:

$$\vec{y}_l = \alpha_l \vec{x} + \vec{z}_l$$

for $l = [1, \ldots, L]$ where $\vec{z}_l$ is the complex Gaussian noise vector with zero mean and covariance matrix $\mathcal{E}[\vec{z}_l \vec{z}_l^*] = N_0 \mathbf{I}$. We shall consider three possible diversity combining rules below.

Maximal Ratio Combining: Since $\{\vec{y}_1, \ldots, \vec{y}_L\}$ contains the same information $\vec{x}$, the optimal detection should be based on the entire observations $\{\vec{y}_1, \ldots, \vec{y}_L\}$. The ML detection can be formulated as:

$$\vec{s} = \arg\max_{\vec{x} \in \mathcal{S}} p(\vec{y}_1, \ldots, \vec{y}_L | \vec{x}, \alpha_1, \ldots, \alpha_L) \tag{4.3}$$

Since the L observations are independent, $\alpha_1, \ldots, \alpha_L$ and $\vec{z}_1, \ldots, \vec{z}_L$ are independent with each other. Hence, the joint density $p(\vec{y}_1, \ldots, \vec{y}_L | \vec{x}, \alpha_1, \ldots, \alpha_L)$ can be expressed as:

$$
\begin{aligned}
p(\vec{y}_1, \ldots, \vec{y}_L | \vec{x}, \alpha_1, \ldots, \alpha_L) &= \Pi_{l=1}^{L} p(\vec{y}_l | \vec{x}, \alpha_l) \\
&= K \exp\{-\left(\frac{\sum_{l=1}^{L} \|\vec{y}_l - \alpha_l \vec{x}\|^2}{2\sigma_z^2}\right) \quad (4.4)
\end{aligned}
$$

for some constant K. Hence, the ML detection rule in (4.3) can be simplified as:

$$
\begin{aligned}
\vec{s} = \arg\min_{\vec{x} \in S} & \sum_{l=1}^{L} \|\vec{y}_l - \alpha_l \vec{x}\|^2 \\
= \arg\min_{\vec{x} \in S} & -\sum_l <\vec{y}_l, \alpha_l \vec{x}> - \sum_l <\alpha_l \vec{x}, \vec{y}_l> + \sum_l |\alpha_l|^2 \|\vec{x}\|^2 \\
= \arg\min_{\vec{x} \in S} & - <\sum_l \alpha_l^* \vec{y}_l, \vec{x}> - <\vec{x}, \sum_l \alpha_l^* \vec{y}_l> + \sum_l |\alpha_l|^2 \|\vec{x}\|^2 \\
= \arg\min_{\vec{x} \in S} & - <\vec{r}_{MRC}, \vec{x}> - <\vec{x}, \vec{r}_{MRC}> + \sum_l |\alpha_l|^2 \|\vec{x}\|^2 \quad (4.5)
\end{aligned}
$$

where

$$
\vec{r}_{MRC} = \sum_{l=1}^{L} \alpha_l^* \vec{y}_l \quad (4.6)
$$

Since the ML detection metric in 4.5 is a function of the *combined observation* $\vec{r}_{MRC}$, the observation $\vec{r}_{MRC}$ is called the *sufficient statistics* with respect to the information $\vec{x}$. In other words, there is no loss of information to build a detector based on the combined observation $\vec{r}_{MRC}$ relative to a detector based on the L independent observations $\{\vec{y}_1, \ldots, \vec{y}_L\}$. The combined observation $\vec{r}_{MRC}$ in (4.6) is called the *maximal ratio combined* observation as illustrated in Figure 4.4 and it is the *optimal* way of combining the L independent observations. However, to achieve maximal ratio combining, the receiver has to estimate the fading coefficients $\{\alpha_1, \ldots, \alpha_L\}$.

Next, we shall look at the BER performance of BPSK modulation with L-order diversity and maximal ratio combining. The observation obtained by maximal ratio combining $\vec{r}_{MRC}$ is given by:

$$
\vec{r}_{MRC} = \left(\sum_{l=1}^{L} |\alpha_l|^2\right) \vec{x} + \sum_{l=1}^{L} \alpha_l \vec{z}_l
$$

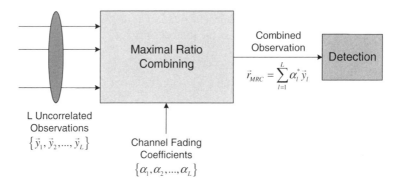

Figure 4.4 Illustration of Maximal Ratio Combining (MRC).

Conditioned on the fading coefficients $\{\alpha_1, \ldots, \alpha_L\}$, the BER is given by:

$$P_e(\alpha_1, \ldots, \alpha_L) = Q\left(\sqrt{\frac{2\gamma E_s}{N_0}}\right)$$

where $\gamma = \sum_{l=1}^{L} |\alpha_l|^2$. Comparing with the flat fading performance in (4.2), the maximal ratio combining alters the distribution of the term γ from *negative exponential distribution* to *central χ-square distribution* with $2L$ degrees of freedom. The average BER expression (average over the fading coefficients $\alpha_1, \ldots, \alpha_L$) is given by:

$$P_e = \mathcal{E}[P_e(\alpha_1, \ldots, \alpha_L)] \approx \binom{2L-1}{L} \frac{1}{(4E_s/N_0)^L}$$

for large SNR. Hence, the slope of the BER versus SNR curve (in log scale) is given by $-L$. The L-th order diversity increases the slope of the BER curve from -1 to $-L$ (steepens the BER curves).

Equal Gain Combining: For equal gain combining, the combined observation $\vec{r}_{EGC}$ is given by:

$$\vec{r}_{EGC} = \sum_{l=1}^{L} \exp(-\angle\alpha_l)\vec{y}_l = \vec{x}\left(\sum_{l=1}^{L} |\alpha_l|\right) + \sum_{l=1}^{L} \exp(-\angle\alpha_l)\vec{z}_l$$

This is illustrated in Figure 4.5

Hence, conditioned on the fading coefficients $\alpha_1, \ldots, \alpha_L$, the conditional BER is given by:

$$P_e(\alpha_1, \ldots, \alpha_L) = Q\left(\sqrt{\frac{2\gamma E_s}{2N_0}}\right)$$

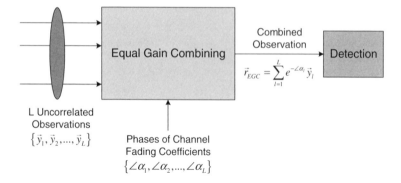

Figure 4.5   Illustration of Equal Gain Combining (EGC).

where $\gamma = (\sum_l |\alpha_l|)^2$. Hence, the average BER (average over fading coefficients $\alpha_1, \ldots, \alpha_L$) is given by:

$$P_e(EGC) = \mathcal{E}[P_e(\alpha_1, \ldots, \alpha_L)] \approx \frac{K_{EGC}}{(4E_s/N_0)^L}$$

for some constant K_{EGC} and large SNR. Hence, the slope of the average BER versus SNR curve is $-L$ with a shift of $10\log_{10}(K_{EGC}/4^L)$ dB. For equal gain combining, we need to know the phase of α_l but not the amplitude of α_l.

Selection Diversity Combining: Consider a third diversity combining technique, namely the selection diversity combining. Specifically, the combined observation $\vec{r}_{SDC}$ is given by:

$$\vec{r}_{SDC} = \vec{y}_{l^*} \tag{4.7}$$

where

$$l^* = \arg\max_l \{|\alpha_1|, \ldots, |\alpha_L|\}$$

This is illustrated in Figure 4.6

In other words, the combined observation is obtained by selecting the *best* observation from the L independent observations $\{\vec{y}_1, \ldots, \vec{y}_L\}$. Similarly, the conditioned BER is given by:

$$P_e(\alpha_1, \ldots, \alpha_L) = Q\left(\sqrt{\frac{2\gamma E_s}{2N_0}}\right)$$

where $\gamma = \max_l \beta_l$ and $\beta_l = |\alpha_l|^2$. The cdf of γ is given by $F_\gamma(\gamma) = [F_\beta(\gamma)]^L$. Hence, the pdf of γ is given by:

$$f_\gamma(\gamma) = \frac{dF_\gamma}{d\gamma} = L(1 - \exp(-\gamma))^{L-1} \exp(-\gamma)$$

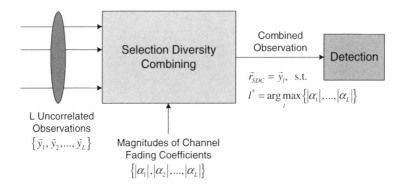

Figure 4.6   Illustration of Selection Diversity Combining (SDC).

The average BER (average over fading coefficients $\alpha_1, \ldots, \alpha_L$) is given by:

$$P_e(SDC) = \mathcal{E}[P_e(\alpha_1, \ldots, \alpha_L)] \approx \frac{K_{SDC}}{(4E_s/N_0)^L}$$

for some constant K_{SDC} and large SNR. Hence, with L-th order selection diversity, the slope of the average BER versus SNR curve is $-L$. For selection diversity combining, we do not need to know the phase of α_l and hence, this method is suitable for receivers without accurate channel estimation or equivalently, it is robust to channel estimation errors in $\alpha_1, \ldots, \alpha_L$.

In all the above combining methods with respect to the L-th order diversity, the slope of the average BER versus SNR curves are steepen from -1 to $-L$. The difference between the three combining schemes in BER performance is just the horizontal shifts as illustrated in Figure 4.7 As illustrated, the slope of the BER curves approach that of the AWGN curve as $L \to \infty$. The maximal ratio combining achieves the best BER performance, followed by the equal gain combining and the selection diversity.

4.4.2   Realization of Diversity

The key of diversity is to obtain L independent received observations *about the same information*. In general, there are three possible ways of realizing L independent received observations, namely the *time diversity* [351], *frequency diversity* [196] and the *spatial diversity* [87]. They are elaborated briefly below.

For time diversity, the L independent observations are obtained from multiple transmissions on the time dimension (separated sufficiently far in time). For example, if the same information symbol is transmitted twice in time domain, we obtain two observations (in time domain). As we have illustrated in Chapter 1, the channel fading experienced by the two observations separated in time by Δt is uncorrelated if $|\Delta t| > T_c$ where T_c is the coherence time of the fading channel. Hence, to realize

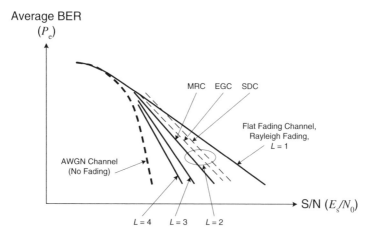

Figure 4.7   Diversity Combining.

L-th order diversity based on the time-domain, we must separate the L copies of observations by more than LT_c. In practice, this is realized by error correction coding over time domain and fast fading (channel fading coefficients are independent between modulation symbols). For slow fading channels where the channel fading coefficient remains quasi-static for an encoding frame, time diversity cannot be realized.

For frequency diversity, the L independent observations are obtained by transmission of the same information symbol at different frequencies. To achieve independent observations, the range of frequencies must be separated by more than LB_c where B_c is the coherence bandwidth of the fading channel. Hence, frequency diversity can be realized for wideband transmission over frequency selective fading channels. For narrowband transmission, the channel fading is quasi-static within the transmission bandwidth and therefore, no frequency diversity can be realized.

Finally, for spatial diversity, the L independent observations are obtained by either transmission of the same information symbol using different antennas at the transmitter or by receiving L observations using different antennas at the receiver. The former case is referred to the *transmit diversity* and the latter case is referred to the *receive diversity*. For independent observations, the antennas must be separated sufficiently far apart. For instance, the separation between any two antennas must be greater than D_c (coherence distance). Since at the base station side, the antennas are elevated at high altitude and hence, the angle spread as seen by the transmitter antennas is smaller compared with that as seen by the mobile antennas. Therefore, the coherence distance D_c at the base station side is larger than that at the mobile side. In general, D_c at the base station is of the order of 10λ whereas D_c at the mobile station is of the order of $\lambda/2$.

4.5 EQUALIZATION*

In Section 4.3, we have illustrated the effects of frequency selective fading channels (multiple resolvable multipaths) on the BER of the physical layer. Due to the inter-symbol interference (ISI), there will be two penalties on the BER vs SNR curve, namely the *flattening of BER curve* and the *irreducible error floor*. While the solution with respect to the former degradation is diversity, the fundamental solution[1] to the latter degradation is *equalization* [213, 287]. A channel equalizer [109] is a signal processing unit at the receiver side which is designed to estimate and cancel the ISI on the received signal. We shall briefly outline the principle of equalization below.

4.5.1 Maximal Likelihood Sequence Estimator (MLSE)

For simplicity, consider a digital modulator where the signal set $\mathcal{S}$ can be characterized by a two dimensional signal constellation such as QPSK, 16QAM, etc. The received signal after passing through a frequency selective fading channel with L_p resolvable multipaths is given by:

$$y_n = \sum_{l=0}^{L_p-1} h_l x_{n-l} + z_n$$

where $x_n \in \mathcal{S}$ is the transmitted modulation symbol, h_l is the complex Gaussian random variable (characterizing the l-th resolvable multipath fading coefficient) and z_n is the complex Gaussian additive channel noise. The optimal equalization process can be formulated as a *maximal likelihood sequence estimator* (MLSE). Consider a span of $N + 1$ observations $\mathbf{y}_0^N = \{y_0, \ldots, y_N\}$. The MLSE equalization is to deduce the most likely transmitted information sequence $\mathbf{x}_0^N = \{x_0, \ldots, x_N\}$ based on the observations. This is formulated mathematically as:

$$\hat{\mathbf{x}}_0^N = \arg\max_{\mathbf{x}_0^N} \log p(\mathbf{y}_0^N | \mathbf{x}_0^N, \mathbf{h}_0^{L_p-1})$$

$$= \arg\max_{\mathbf{x}_0^N} \sum_{n=0}^{N} \log p(y_n | \mathbf{x}_{n-L_p+1}^n, \mathbf{h}_0^{L_p-1}) \tag{4.8}$$

where the second equality is due to the fact that the current observation y_n depends on the past L_p transmitted modulation symbols only. Define the system state $S_n =$

---

[1]Strictly speaking, diversity with maximal ratio combining may alleviate the error floor as well. This is because with MRC, the overall SINR is enhanced by the independent observations as well. Hence, the error floor will be lowered. On the other hand, if the number of resolvable multipath L_p is large, the effect of selection diversity on the error floor is insignificant because the overall average SINR is virtually not affected.

$(x_{n-L_p+1}, \ldots, x_{n-1})$ and the optimal log-likelihood function as

$$F^*(m, 0, S_m) = \max_{\mathbf{x}_0^m} \sum_{n=0}^{m} \log p(y_n | S_n, x_n, \mathbf{h}_0^{L_p-1}) \tag{4.9}$$

The MLSE optimization problem can be expressed into the following recursive form:

$$
\begin{aligned}
F^*(m, 0, S_m) &= \max_{x_m} \left(\log p(y_m | S_m, x_m, \mathbf{h}_0^{L_p-1}) \right. \\
&\quad + \max \mathbf{x}_0^{m-1} \sum_{n=0}^{m-1} \log p(y_n | S_n, x_n, \mathbf{h}_0^{L_p-1}) \Big) \\
&= \max_{x_m} \left(\log p(y_m | S_m, x_m, \mathbf{h}_0^{L_p-1}) \right. \\
&\quad + F^*(m-1, 0, (S_m/x_m))) \tag{4.10}
\end{aligned}
$$

for $m = 0, 1, \ldots$ and $F^*(-1, 0, S) = 0$. Hence, the solution of the MLSE optimization problem can be divided and conquered based on *dynamic programming* approach. More precisely, the MLSE algorithm is identical to the conventional *Viterbi decoding algorithm* for convolutional code with a branch metric given by $\log p(y_m | S_m, x_m, \mathbf{h}_0^{L_p-1})$.

Note that the MLSE equalizer requires the knowledge of the frequency selective fading coefficients $\{h_0, \ldots, h_{L_p-1}\}$. This can be estimated from the pilot channels or pilot symbols. Furthermore, since there are M^{L_p} state sequence S_n in the trellis, the complexity of the MLSE equalizer is of exponential order in L_p. The performance analysis of the MLSE equalizer is quite involved and we shall just quote the results here. The conditional symbol error probability of M-ary modulator (conditioned on the channel fading coefficients $\{h_0, \ldots, h_{L_p-1}\}$) is given by:

$$P_e(h_0, \ldots, h_{L_p-1}) \leq KQ\left(\sqrt{\frac{6}{M^2-1}E_s/N_0 \delta_{min}^2(\mathbf{h}_0^{L_p-1})}\right)$$

for some constant K where $\delta_{min}^2(\mathbf{h}_0^{L_p-1})$ is given by:

$$\delta_{min}(\mathbf{h}_0^{L_p-1})^2 = \sum_i \sum_j \alpha_{ij} h_i h_j^*$$

for some constants α_{ij}. The term δ_{min}^2 represents the SNR loss due to the ISI. The average error probability is given by $\mathcal{E}[P_e(h_0, \ldots, h_{L_p-1})]$ where the expectation is taken over all possible realizations of $h_0, \ldots, h_{L_p-1}$. We observe that after the optimal equalization, there is no error floor anymore. In addition, the slope of the BER curve is steepened due to the inherit frequency diversity due to frequency selective fading coefficients $h_0, \ldots, h_{L_p-1}$. This is illustrated in Figure 4.8 and the steepening of the BER curve can be understood by noting that δ_{min}^2 consists of the terms $\sum_i \alpha_{ii} |h_i|^2$ which is equivalent to the MRC error probability.

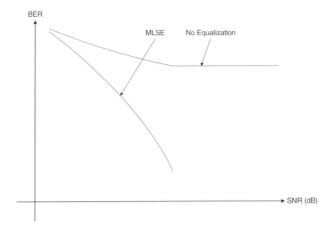

Figure 4.8   BER performance of MLSE equalizer in frequency selective fading channels.

4.5.2 Linear Equalization

While the MLSE equalization achieves the optimal performance, the complexity is of exponential order in L_p and hence, would be too complicated for moderate to large L_p. In this section, we consider a sub-optimal equalization algorithm[132], namely the minimum mean-square error (MMSE) equalization[210], that has linear order of complexity.

Figure 4.9 illustrates a generic block diagram of a linear equalizer for frequency selective fading channels. The received observations $\{y_n, y_{n-1}, \ldots, y_{n-2L}\}$ are processed by a *linear filter* with coefficients $\{w_0, \ldots, w_{2L}\}$ (where $2L$ is the length of impulse response of the linear filter) to produce a processed output r_n for detection. In other words, the processed output is given by:

$$r_n = \mathbf{w}^* \mathbf{y} \tag{4.11}$$

where $\mathbf{w} = [w_0, \ldots, w_{2L}]^*$, $\mathbf{y}_n = [y_n, \ldots, y_{n-2L}]^*$ and $(.)^*$ denotes conjugate transpose. The detection (demodulation) is performed on the processed output r_n to obtain $\hat{x}_n$.

Define the mean square error as:

$$\mathcal{E}[|e_n|^2] = \mathcal{E}[|r_n - x_n|^2]$$

The MMSE weight $\mathbf{w}$ is given by:

$$\mathbf{w}_{opt} = \arg\min_{\mathbf{w}} \mathcal{E}[|r_n - x_n|^2] \tag{4.12}$$

Substituting (4.11) into (4.12), the MMSE weight is given by:

$$\mathbf{w}_{opt} = \left(\mathcal{E}[\mathbf{y}_n \mathbf{y}_n^*]\right)^{-1} \mathcal{E}[\mathbf{y}_n x_n]$$

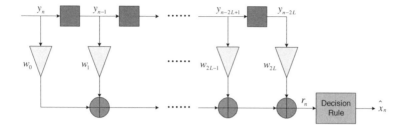

Figure 4.9   Block diagram of linear equalizer.

where

$$(\mathcal{E}[\mathbf{y}_n\mathbf{y}_n^*])_{ij} = \begin{array}{ll} x_{i-j} + N_0\delta_{ij} & |i-j| \le L_p \\ 0 & \text{otherwise} \end{array}$$

and

$$(\mathcal{E}[\mathbf{y}_n x_n])_i = \sigma_x^2 h_i^* \;\; \forall i \in [0, L_p]$$

where $\sigma_x^2 = \mathcal{E}[|x_n|^2]$.

While the MMSE equalizer has linear processing complexity with respect to L_p, there is residual ISI at the filtered outputs r_n. Hence, there will be residual error floor in the BER vs. SNR curve.

4.6   PRACTICAL ILLUSTRATION: RAKE RECEIVER

In this section, we shall illustrate how the RAKE receiver in spread spectrum systems (such as CDMA IS95 and 3G) combat frequency selective fading channels. The RAKE receiver can be considered as an equalizer with linear complexity with respect to L_p. Furthermore, the RAKE receiver allows L_p independent observations about the same information to be combined before detection on the information symbol is done. Hence, it allows L_p-order frequency diversity. The operation of RAKE receiver is explained below.

Figure 4.10 illustrates the processing of RAKE receiver in spread spectrum systems. For simplicity, consider a frequency selective fading channel with two multipaths. The complex channel fading coefficients of the two paths are $\{h_1, h_2\}$ and the corresponding delays are $\{\tau_1, \tau_2\}$. At the transmitter, we assume the information signal $s(t)$ (with symbol duration T_s) is spread by the sequence $c(t)$ (with chip duration T_c). The received signal is given by:

$$y(t) = \underbrace{h_1 s(t - \tau_1)c(t - \tau_1)}_{\text{path 1}} + \underbrace{h_2 s(t - \tau_2)c(t - \tau_2)}_{\text{path 2}} + z(t)$$

To extract the information for detection, the receiver has to perform despreading. Suppose the receiver knows the spreading sequence $c(t)$ and the path delays τ_1, τ_2,

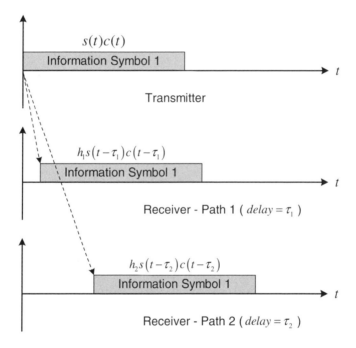

Figure 4.10 Illustration of the RAKE receiver processing.

there are two choices of despreading, namely despreading with respect to path 1 ($c(t - \tau_1)$) or despreading with respect to path 2 ($c(t - \tau_2)$). Consider the former case, the output after despreading is given by:

$$
\begin{aligned}
r_1(t) \quad &= \quad < c(t - \tau_1), y(t) >= h_1 s(t - \tau_1) < c(t - \tau_1), c(t - \tau_2) > \\
&\quad + h_2 < c(t - \tau_1), s(t - \tau_2) c(t - \tau_2) > + < c(t - \tau_1), z(t) > \\
&= \quad \underbrace{h_1 s(t - \tau_1)}_{\text{Desired signal}} + \underbrace{h_2 s(t - \tau_2) < c(t - \tau_1), c(t - \tau_2) >}_{\text{ISI}}
\end{aligned}
$$

$$
+ \quad \underbrace{\tilde{z}(t)}_{\text{channel noise}} \tag{4.13}
$$

where $< c(t - \tau_1), y(t) >= \frac{1}{T_s} \int_{\tau_1}^{\tau_1 + T_s} c(t - \tau_1) y^*(t) dt$, $< c(t - \tau_1), c(t - \tau_1) >= 1$ and the equivalent noise $\tilde{z}(t)$ has the same power as the original noise $z(t)$. Note that the second term represents the ISI term but the ISI power is weighted by the autocorrelation $< c(t - \tau_1), c(t - \tau_2) >= R_c(\tau_1 - \tau_2)$. It can be shown that the autocorrelation of a properly designed pseudo-random noise (PN) sequence is upper bounded by $R_c(\tau) \approx \sqrt{frac T_s T_c} = \sqrt{N_{SF}}$. Hence, the ISI power is suppressed by the spreading factor N_{SF} and is given by:

$$
ISI_1 = S/N_{SF}
$$

The SINR of the decision variable based on $r_1(t)$ is given by:

$$
SINR(1) = \frac{S}{N + S/N_{SF}}
$$

Compare with the system without spread spectrum, the error floor of the BER vs SNR curve will be lowered because the SIR is increased by N_{SF} times due to the processing gain of spread spectrum. Hence, the RAKE receiver acts as a simple equalizer with respect to frequency selective fading channels. Figure 4.11 illustrates the BER curve of the spread spectrum system based on despreading with respect to the first path.

Similarly, if the receiver despread with respect to the second path $c(t - \tau_2)$, the despread signal is given by:

$$
r_2(t) \quad = \quad < c(t - \tau_2), y(t) >== \underbrace{h_2 s(t - \tau_2)}_{\text{Desired signal}}
$$

$$
+ \underbrace{h_1 s(t - \tau_1) < c(t - \tau_2), c(t - \tau_1) >}_{\text{ISI}} + \quad \underbrace{\tilde{z}(t)}_{\text{channel noise}} \tag{4.14}
$$

Hence, the residual ISI power is given by S/N_{SF}. Observe that while the error floor is reduced due to the processing gain, the slope of the BER curve is still flattened.

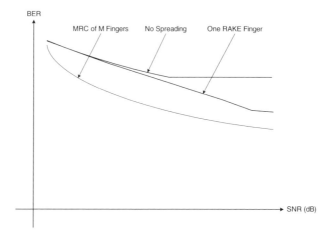

Figure 4.11   BER versus SNR of the RAKE receiver in frequency selective fading channels.

However, since $r_1(t)$ and $r_2(t)$ contain independent observations about the information signal, we can better exploit the inherit frequency diversity by combining the two observations before detecting the signal. In this way, we achieve two-order frequency diversity. The BER curve of the MRC combined output based on $r_1(t)$ and $r_2(t)$ is illustrated in Figure 4.11 Due to the two-order diversity, the slope of the BER curve is increased. Hence, the RAKE receiver kills two birds with one stone.

Figure 4.12 illustrates the block diagram of a typical RAKE receiver. The received signal $y(t)$ is feed into multiple fingers where the $m-$th finger is responsible for despreading with respect to the $m-$th resolvable multipath. The outputs from the M fingers $r_1(t), \ldots, r_M(t)$ are fed into diversity combining block where detection of the information is made on the diversity-combined signal. In general, when the frequency selective fading channel has L_p resolvable multipaths, the RAKE receiver can achieve L_p-th order frequency diversity. Furthermore, since the maximum number of RAKE fingers required is given by L_p, the complexity of the RAKE receiver is linear with respect to L_p

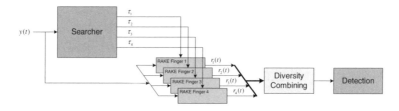

Figure 4.12   Block diagram of the RAKE receiver with $M = 4$ fingers.

In the RAKE processing, we have assumed that the RAKE receiver knows about the path delays $\tau_1, \ldots, \tau_M$. In general, the path delays are estimated by a block called the *searcher*. The searcher consists of a correlator which correlates the received signal with respect to a local spreading sequence $c(t)$. In spread spectrum systems, there is usually a pilot channel where the data symbol consists of all-ones or all-zeros sequence. Hence, the received signal from the pilot channel is given by:

$$y_p(t) = h_1 c(t - \tau_1) + h_2 c(t - \tau_2) + \ldots + h_{L_p} c(t - \tau_{L_p}) + z(t)$$

where $c(t)$ is the spreading sequence used in the pilot channel. Based on the pilot channel, the searcher obtains correlation with respect to the spreading sequence at various delays, i.e.,

$$r(\tau) = <y_p(t), c(t - \tau)> = h_1 R_c(\tau - \tau_1) + \ldots + h_{L_p} R_c(\tau - \tau_{L_p}) + \tilde{z}(\tau)$$

Hence, the correlation output value $r(\tau)$ will hit large values only when $\tau \approx \tau_m$ for $m \in [1, \ldots, L_p]$. Figure 4.13 illustrates a typical searcher output versus the delays. The paths with the correlation values exceeding some predefined thresholds are used to configure the RAKE fingers to perform despreading and RAKE processing.

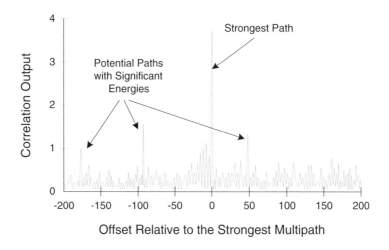

Figure 4.13   Illustration of searcher results—correlation values versus path delays.

4.7 SUMMARY

In this chapter, we have elaborated the effects of fading channels on the physical layer performance of a digital communication link. For instance, flat fading channels introduces the flattening of the BER vs SNR curve where the slope is reduced to -1.

Hence, for the same target BER, more power is required for digital communication over flat fading channels. In multi-user systems, the SNR loss is translated into capacity loss.

For frequency selective fading channels, there are two penalties on the BER curves. In addition to the flattening of the BER curves, we have an irreducible error floor. This error floor is caused by the intersymbol interference (ISI) in the frequency selective fading channels. The problem of ISI cannot be resolved by simply transmitting at a higher power because the ISI power will also increase with the transmit signal power.

A common solution for combating flat fading is diversity. Diversity refers to the technique of detecting information based on multiple independent observations with respect to the same information symbol. With L-th order diversity, the slope of the BER vs SNR curve becomes $-L$ and hence, in the limit of very large diversity order, the slope of the BER vs SNR curve approaches that of the AWGN performance. There are three common techniques for diversity combining, namely the maximal ratio combining, equal gain combining as well as selection diversity. The MRC requires accurate knowledge of the channel fading coefficients from the L independent observations but achieves the best performance. The EGC requires just the phases of the channel fading coefficients and achieves slightly inferior performance relative to MRC. Finally, the SDC requires knowledge of the amplitude of channel fading coefficients and achieve the worst performance of the three combining schemes. However, in all the three schemes, the slope of the BER curve is given by $-L$. The difference in the three schemes are just the *offsets in SNR* in the BER vs SNR curves. Diversity can be realized in time domain (through error correction coding over fast fading channels) or frequency domain (through wideband transmission over frequency selective fading channels) or in spatial domain (through multiple antenna transmission / receptions).

The fundamental solution for frequency selective fading channels is more than diversity. On the contrary, channel equalization at the receiver is needed to cancel the effects of the ISI and hence, alleviating the irreducible error floor. The optimal equalizer is called the MLSE which has exponential order of complexity with respect to the number of resolvable multipaths L_p. For moderate to large L_p, the MLSE becomes impractical in terms of implementation.

Finally, we illustrate the use of RAKE receiver to achieve frequency diversity as well as combating the ISI in CDMA systems. Due to the interference suppression capability of spread spectrum, the ISI power is reduced by the spreading factor N_{SF}. On the other hand, by despreading with respect to different path delays, each RAKE finger produces one independent observation about the information signal. Hence, by combining the L observations from the L RAKE fingers, we can achieve L-th order frequency diversity. In general, the maximum diversity order we can achieve in RAKE receiver is given by $L = L_p$ (number of resolvable multipaths). The complexity of the RAKE receiver is only linear with respect to the number of resolvable paths L_p.

PROBLEMS

4.1   Power control is used to overcome the inter-symbol interference problem in frequency selective fading channel [True/False].

4.2   Transmitting the same information at different time instances, separated smaller than the coherence time of the fading channel, is an effective means to combat frequency flat fading [True/False].

PART II

CELLULAR WIRELESS TECHNOLOGIES

CHAPTER 5

OVERVIEW AND EVOLUTION OF CELLULAR TECHNOLOGIES

Every cell comes from a cell. (*Omnis cellula e cellula.*)

—Rudolf Virchow: *Cellular-Pathologie*, I, 1858

5.1 INTRODUCTION

In this chapter, we shall apply the theories we have introduced in Part I and review the design of cellular systems. A cellular system [203] is a collection of entities. Figure 5.1 illustrates a generic architecture of modern cellular systems. A cellular system consists of three basic elements, namely the *mobile station* (mobile phone)[MS], the *base station* [BS], as well as the *mobile switching center* [MSC].

The mobile stations are connected to the base station over the *radio interface* while the base stations are connected to the mobile switching center (MSC) through fixed line transmission systems such as T1, E1 or ATM. The service area is covered by multiple base stations where the coverage of each base station is restricted to a limited range only. Mobile stations cannot communicate directly with each other and

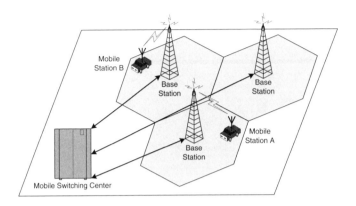

Figure 5.1   A generic architecture of cellular system.

all communications have to go through the base station and the switch. For example, suppose mobile station A establishes a connection with mobile station B. The end to end connection path between A and B consists of two radio segments [from mobile A to BS and from BS to mobile B] and two fixed-line segments [from BS to MSC and from MSC to BS].

In general, the bottleneck of the cellular system (say in terms of the maximum number of users that can be supported or system capacity) is on the radio interface. Let us say the capacity of the fixed line segment is used up. We can easily increase the capacity by installing an additional line. On the other hand, this is not possible in radio or wireless communications. This is because the cellular system has to operate in a license spectrum where the radio signal can only be transmitted in a given range of frequencies. Once the bandwidth in the allocated spectrum is used up, the radio capacity will be used up as well. Hence, various technologies such as FDMA, TDMA, CDMA or SDMA have been developed to utilize the limited spectrum more efficiently. Furthermore, the technologies of the radio interface form an important part in defining the standards of 1G, 2G, 3G as well as 4G systems. In Part I, we have reviewed basic communication theories from point-to-point design to multi-user and multi-cell designs. In Part II (starting from this chapter), we shall elaborate more on the specific designs of industrial standards on cellular systems. We shall first discuss the motivations of the evolution of the cellular technologies from 1G to 3G in the following sections.

5.2   EVOLUTION OF CELLULAR SYSTEMS

In this section, we shall outline the evolution of various cellular systems from 1st generation systems to the current 3rd generation systems. Figure 5.2 illustrates the timeline of the evolution of various cellular systems.

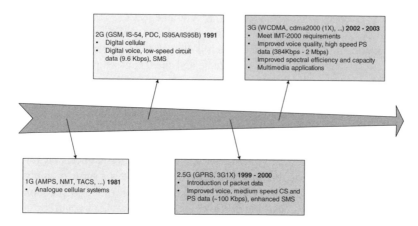

Figure 5.2   Timeline of the evolution of some common cellular systems.

The first generation (1G) cellular systems [202] are also called *analog cellular systems* because the voice is delivered using FM. Commercial deployment started in around 1981. Common examples are the Advanced Mobile Phone Systems (AMPS), the TACS as well as the NMT systems. The AMPS is the US standard and the TACS is the UK standard. Moreover, the 1G systems support voice communications only and the different standards are incompatible with each other. Hence, no roaming across different 1G systems is possible. Channels are partitioned based on FDMA approach. As we have illustrated in Chapter 3, the frequency reuse factor in 1G systems are quite high ($K = 7$) due to the high SIR requirement for FM transmission (SIR 18dB). Hence, the voice capacity (number of voice channels in the system) is in general a primary concern in the 1G systems because the multi-cell spectral efficiency of the radio interface is quite low.

On the other hand, the second generation (2G) cellular systems [324] are called the *digital cellular systems* because the voice transmission is based on digital communication technologies. The speech signal is first digitalized and compressed into digital bit streams by speech vocoder. Hence, the speech channel is *less bulky* compared with the 1G systems. That means for the same channel bandwidth, 2G systems can support more speech users compared to the 1G systems. Hence, the total number of voice channels that can be partitioned from the allocated spectrum is increased and this contributes to the capacity gain in 2G systems. Moreover, the digitalized speech is error protected and transmitted over the air using digital modulation techniques as introduced in Chapter 1. Due to the digital transmission techniques, the required SIR is generally lower (SIR $\sim$13dB or lower) and this allows a more aggressive frequency reuse. Hence, the voice capacity in 2G systems scales better as we increase the number of base stations compared with 1G systems. Common examples of 2G systems are GSM [324], IS54 (Digital AMPS) and IS95 (CDMA) [127]. 2G systems can support not only the voice service but also the digital data service.

In 2.5G systems [141], such as GPRS, EDGE and 3G1X, a new wireless data service, namely the *packet switched service*, is introduced. In 1G and 2G systems, the primary application is voice service which is supported by *circuit-switched connection*. By circuit switched connection, a dedicated channel is setup for the user during the entire communication session. The user has to pay for the *air time* for the radio resource assigned to it. This connection mode is well suited for voice communication because the source bit rate for voice applications is quite steady and time invariant. On the other hand, in packet switched connection, the dedicated radio resource assigned to the user will be dynamically adjusted based on the traffic requirement. The radio resource may be released to the system if the application source has been idle for some time. The resource will be re-established again when there is a surge of application packets to deliver. Hence, this mode of connection is well suited to accommodate bursty data applications such as video streaming, web browsing or email.

Finally, 3G systems, such as UMTS[152] and CDMA2000 [127] are based on wideband CDMA technology to offer high speed wireless internet access[232] in addition to the conventional voice service. They support a peak bit rate of 2Mbps with Quality of Service (QoS), improved spectral efficiency and multimedia services.

In general, a cellular system has to go through the following typical life cycle, namely the *initial deployment*, the *growth stage* and the *mature/saturated stage* [203]. Suppose an operator just bid for a licensed spectrum from a regulatory body in a city. The operator is about the setup the cellular system. At this stage, the bottleneck of the system is network coverage rather than network capacity. This is because the number of subscribers will not be too high during the initial stage. The primary concerns of the potential subscribers will be the coverage or the size of service area. Hence, the primary strategy of the operator is to maximize the network coverage with the smallest possible investments. This refers to the initial deployment stage where the operator will install a minimum number of base stations to provide a large network coverage. The operator would like to minimize the number of base station because it is quite expensive to setup a new base station. The cost include the hardware cost of the base station as well as the real estate cost to install the base station. In other words, the coverage of the base stations will be large and we called that *macro-cells* as illustrated in Figure 5.3

As the business of the operator evolves, more and more subscribers may join the network and this shall increase the *traffic loading* of the cellular network. For instance, in some densely populated areas, capacity hotspots may be developed where the total number of subscribers covered by a base station may far exceed the number of channels available in that base station. In that case, the network operator can

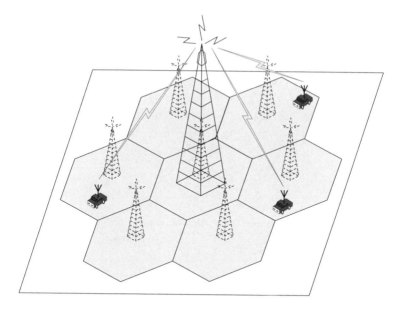

Figure 5.3   Illustration of macrocells during initial deployment stage of a cellular network.

alleviate the hotspot problem by *cell splitting*. Note that cell splitting is a system engineering approach which does not involve technology change or upgrade. By cell splitting, the original macrocell is split into multiple smaller cells. Each cell has a smaller coverage area compared to the original macrocell. With a smaller coverage area, the total number of subscribers served by the base station will be smaller and hence, the traffic loading with respect to the smaller size cell is alleviated. With the same number of channels per cell, the capacity hotspot problems can be resolved. On the other hand, cell splitting requires additional investment on the new base stations and the associated real estate costs. Hence, in the growth stage, the cellular network will have cells of mixed sizes, ranging from macrocell for rural coverage to microcells for higher capacity in urban areas as illustrated in Figure 5.4

Finally, as the cells are split into smaller and smaller cells (such as microcells or picocells), there is a limit on how small a cell can be. Hence, when this limit is hit, the operator cannot further increase the capacity by cell splitting only. In this case, the network is said to reach the matured or saturated stage. The operator has to acquire more licensed spectrum (which is usually difficult) or upgrade the technology to utilize the limited spectrum more effectively.

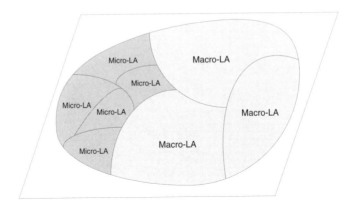

Figure 5.4 Illustration of macrocells and microcells during the growth stage of a cellular network.

5.2.1 Evolution from 1G to 2G

First generation cellular systems are also called analog cellular systems because the voice channels are provisioned through analog modulations. For example, in both the AMPS and TACS systems, voice signals are modulated based on FM with a transmission bandwidth of 30kHz and 25kHz respectively. Signals are separated by FDMA. The total number of channels available to an operator is limited. Furthermore, due to the high SIR requirement for FM transmissions (SIR = 18 dB), the frequency reuse cluster cannot be too small ($K = 7$). Hence, the efficiency of frequency reuse is limited. Due to these two factors, the capacity and the spectral efficiency of the 1G systems are quite limited. 1G operators quickly run out of capacity after cell splitting and reached the mature stage. Hence, there is a major driving force to upgrade the technology to enhance the spectral efficiency in order to meet the ever increasing demand for mobile voice communications.

There are several common second generation cellular standards in the world. For example, we have IS54 (Digital AMPS), IS95 (CDMA), GSM, as well as PDC (Personal Digital Cellular) systems [144] Second generation cellular systems are designed based on digital communication technologies. For instance, the speech signals are digitalized and compressed to 8kbps or 13.3kbps bit streams. The digital bit streams are protected by error correction coding and transmitted over the air using digital modulation methods as introduced in earlier chapters. The digital voice channels are partitioned from the spectrum based on hybrid FDMA/TDMA or hybrid FDMA/CDMA technologies. For example, in IS54 (digital AMPS), the same 30kHz bandwidth carrier is partitioned into three timeslots and hence, it can support 3 digital voice channels compared with one analog voice channel in AMPS system. Hence, due to the digital speech compression, the total number of voice channels available to the operator in D-AMPS is increased by three times compared to the AMPS system. Moreover, due to the digital modulation and error correction coding, the required SIR is reduced to less than 13 dB. Hence, together with sectorization, a more aggressive frequency reuse factor (e.g., $K = 4$) can be employed and this substantially increases the capacity scaling as more base stations are installed. Hence, these two factors contribute to the substantial increase in the overall network spectral efficiency of 2G cellular systems.

Another major driving force for the evolution of cellular systems from 1G systems to 2G systems is the ability to roam between different countries. There are various incompatible 1G systems such as the AMPS in the US and TACS in the UK. Many of the 1G systems are incompatible with each other and hence, mobile users of one system cannot roam in countries using other 1G systems. In the evolution from 1G to 2G, the US and the European countries took on different approaches. In the US, there is a widespread AMPS deployment over the country. Since US has a large domestic market and there is no issue of roaming across US because there is only one standard (AMPS) deployed, the US took an *evolutionary approach* in upgrading from 1G to 2G systems. Specifically, the 2G systems in the US are designed to be *backward compatible* with the AMPS systems. For example, IS54 or Digital AMPS system is designed to be backward compatible with AMPS systems. The channel bandwidth (30kHz) in D-AMPS is the same as that in AMPS system. In D-AMPS, the voice channels are further partitioned from the 30kHz carrier using TDMA. *Inter-system handoff* is supported when the D-AMPS mobile moves into AMPS coverage. Similar, IS95 (CDMA) system is also designed to be backward compatible with AMPS. Hence, in the US, the 2G digital cellular systems deployed co-exists with the 1G AMPS systems. In fact, a large percentage of rural areas in the US are still covered by the AMPS system.

On the other hand, the European countries took a different approach, namely the *revolutionary approach*, in upgrading from 1G to 2G systems. The reason behind is due to the incompatible standards in the 1G systems among European countries. Hence, during the upgrade to the 2G systems, the European countries have to scrapped the deployed 1G systems and replaced with a common standard 2G systems, namely the GSM systems. Hence, there is no such thing as a dual mode GSM / TACS handset. The GSM system also does not support inter-system handoff (say from GSM to TACS). In fact, GSM systems have around 474 networks in 190 countries worldwide and capture more than 70% of the cellular market. It has a rapid growth in subscribers reaching around 900 million subscribers in 2003. Lately, there are some GSM deployments in some hot spots of the US.

New spectrum has been allocated worldwide to support 2G cellular systems. There are *cellular spectrum* and *PCS (Personal Communication System) spectrum* allocated by the telecom regulators. In the US, the cellular spectrum is located at around 800MHz and the PCS spectrum is located at around 1.9GHz. In most of the other places (such as Europe, China, HK, Taiwan), the cellular spectrum is located at around at around 900MHz and the PCS spectrum is located at around 1.8GHz. Hence, a *tri-band* GSM mobile (covering the cellular spectrum 900MHz, the PCS spectrum

1.8GHz and the US PCS spectrum 1.9GHz) is required to support roaming from HK to the US. Note that the term *PCS* or *Cellular* refer to the spectrum service definitions and is usually technology independent. The operators are free to choose the appropriate technologies (such as GSM, CDMA, D-AMPS, PHS, etc.) to apply on the PCS or cellular spectrum respectively. The cellular spectrum was first allocated for the introduction of 2G systems. Later on, telecom regulators introduced the PCS spectrum to enhance competitions among the cellular industry. In fact, in some cases, the same operator has the license to operate in both the cellular spectrum and the PCS spectrum. To save cost, the operator may choose the same technology to deploy in these two spectrum. For example, we have GSM for cellular spectrum and GSM (also called DCS-1800) for the PCS spectrum.

In addition to the conventional voice service, the 2G cellular systems also support limited wireless data service such as low bit rate circuit switched service (9.6kbps or 14.4kbps transparent or non-transparent bearers) or fax service. Yet, the application of wireless data in 2G systems is limited due to the low bit rate and poor performance.

5.2.2   Evolution from 2G to 2.5G and 3G Systems

While the major driving force of moving from 1G to 2G systems is voice capacity and spectral efficiency, the major driving force for 2G to 3G is on the demand for high quality *wireless internet access*[62]. Similar to the 2G systems, the regulators worldwide has introduced new spectrum for the 3G systems. Figure 5.5 illustrates the worldwide spectrum allocations. Paired spectrum at 1.9GHz (Uplink) and 2.1GHz (Downlink) have been allocated for 3G applications.

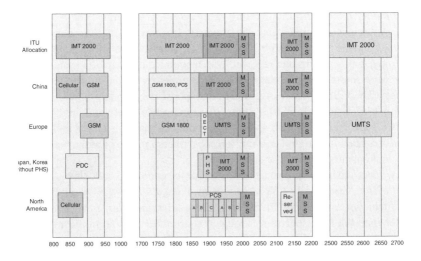

Figure 5.5   Illustration of spectrum allocation for 3G service.

As illustrated in the figure, the IMT2000 defined 3G spectrum is unused in most countries except the US. In the US, all the spectrum around the IMT2000 spectrum has been deployed for other services. Hence, around year 2000, people generally believe that US will be lagging behind the European countries in terms of 3G deployment. In fact, before the internet bubble burst around year 2000, many European operators have spent tens of billions of dollars to bid for the 3G spectrum. At that time, the business plans and financial plans of 3G operators worked fine to justify the expensive license fees. On the other hand, the US operators did not have the chance to bid for 3G spectrum because the spectrum is simply not available. Very soon after the internet bubble burst, the high technology industry collapsed and the financial forecasts / business plans of the 3G operators that sounded reasonable before became unrealistic after the burst. The demand for wireless data access and the *killer applications* became questionable. Hence, most of the 3G operators that spent a lot of money to bid for the license were in financial troubles. In the toughest times, some operators did not even have the budget to acquire 3G base stations and 3G infrastructure. Many operators have to form joint venture to share the equipment costs and the operating costs. Hence, the deployments of 3G systems were delayed for many years until around 2004. In the US, the situation is totally different. The PCS and cellular operators in the US somehow escaped the bidding of 3G licenses and hence, they are relatively *wealthy* after the internet bubble burst. They were ready to invest in the infrastructure to try out wireless internet access using *2.5G technologies* such as 3G1X systems. Hence, the deployment of the wireless data service in the US turned out to be earlier than the rest of the world.

In the new 3G spectrum allocated, there are new service definitions associated with it. One of the most important difference from the 2G systems is the introduction of packet switched connections. Conventionally, circuit switched connection is set up to support voice service. In circuit switched connection, the same dedicated radio resource is assigned to the user throughout the entire communication session. For voice service with *constant bit rate* (CBR) vocoder, the source bit rate is quite steady and it is very effective to support the CBR voice source using circuit switched connection. On the other hand, for data applications, the source bit rate is quite bursty and it is very ineffective to support the bursty application using circuit switched connection as illustrated in Figure 5.6 As illustrated in the figure, if the bit rate of the circuit switched connection is at the average source bit rate, then there will be overflow during the peak source rates. The overflow packets cannot be buffered due to the delay constraint and will be dropped at the transmitter. On the other hand, if the bit rate of the circuit switched connection is set at the peak rate, then most of the time, the connection will be under-utilized. In packet switched connection, the dedicated resource will be dynamically allocated according to the demand (buffer status). When the source demand drops, the dedicated radio resource will be released to the system. Hence, packet switched connection is more suitable to support bursty applications. This efficient use of scarce radio resources means that large numbers of users can potentially share the same bandwidth and be served from a single cell.

The actual number of users supported depends on the application being used and how much data is being transferred.

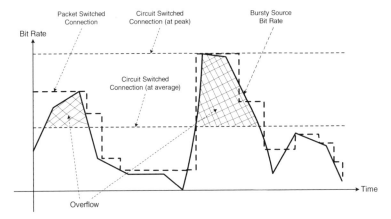

Figure 5.6 Illustration of circuit switched and packet switched connections to support bursty data applications.

The following briefly summarizes the new 3G service definitions.

Circuit Switched Voice Service: Similar to the conventional 2G systems, the 3G systems also support circuit switched voice services. Due to the circuit switched nature, the users are charged by *air time*.

High Speed Circuit Switched Data Service: High speed circuit switched data service up to 64kbps is supported in 3G system. Similarly, the users are charged by the *air time*.

High Speed Packet Switched Data Service: The 3G systems introduced a new data service, namely the *packet switched data service*. The radio resource assigned to support the packet switched data service is adaptive with respect to the traffic volume. Hence, when there is no active packets to transmit in the application, the dedicated radio resource will be released to the system to be shared by other active users. Hence, this mode of data service is very efficient with respect to the bursty source applications. The users will be charged by packet volumes instead of purely air time. Depending on the environment and the mobility of the users, the 3G systems support peak bit rates of 14.4kbps to 2Mbps. Table 5.1 illustrates the bit rates supported under different situations.

The 3G systems also support asymmetric traffic where the downlink bit rate can be different from the uplink bit rate. In 3G systems, there are four QoS classes defined, namely the *conversational class*, the *streaming class*, the *interactive class* and the *background class*. Table 5.2 illustrates typical configurations of the QoS class and the connection modes in various common applications.

Table 5.1   Illustration of peak bit rates of 3G systems in different scenarios.

Environment	Mobility	Peak Bit Rate (kbps)
Indoor	Pedestrian	2M
Outdoor	Vehicular	144
Outdoor	Pedestrian	384

Table 5.2   Illustration of services and applications mappings to 3G connections.

Type	Rate (kbps)	Connection Type	Mode	QoS Class
Voice	8–16	Circuit	Symmetric	N/A
Voice	8–16	Packet	Symmetric	Conversation
Web Browsing	128–2000	Packet	Asymmetric	Interactive
Video Streaming	64–384	Packet	Asymmetric	Streaming
Email, Fax, SMS	9.6–64	Packet	Asymmetric	Background

While 3G systems are designed with respect to the new 3G spectrum and the new 3G service definitions, there are some technology enhancements over the existing 2G systems that operates on the existing 2G PCS spectrum only. These new technologies are called *2.5G systems*. Examples are GPRS, EDGE and 3G1X systems. The 2.5G systems offer packet switched connections and offer an advantage for existing 2G operators to try out the market for wireless data access before fully upgrading the systems to 3G technologies. Figure 5.7 illustrates the evolution paths from 2G systems (GSM and IS95) to the 2.5G and 3G systems. According to Ovum, there will be 36.3 million users of EDGE technology at the end of 2005. Due to the very small incremental cost of including EDGE capability in GSM network deployment, virtually all new GSM infrastructure deployments are also EDGE capable and nearly all new mid- to high-level GSM devices also include EDGE radio technology.

5.3   TECHNICAL CHALLENGES TO REALIZE 3G SERVICES

In the previous sections, we have discussed the evolution of the cellular systems from 1G to 3G. In this section, we shall discuss briefly on the technical challenges involved in realizing the 3G services. In particular, there are three major aspects of technical challenge. They are briefly summarized in the following subsections. We shall outline using GPRS, EDGE and UMTS as examples on how these systems overcome the corresponding technical challenges.

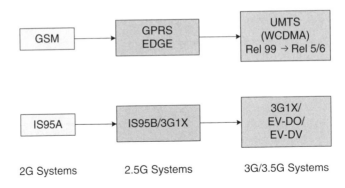

Figure 5.7 Evolution path of 2G systems (GSM, IS95) to 2.5G and 3G systems.

5.3.1 Transmission over the Air

In 3G systems, the physical bit rates of the radio connections are much higher than 2G systems. Hence, one important challenge is to design the radio interface to support such high bit rate requirement. As we have mentioned in earlier chapters, the radio interface is usually the bottleneck in the overall network capacity due to the limited radio spectrum. In other words, higher spectral efficiency is needed to support such higher bit rates given the limited bandwidth.

In GPRS, the air interface is very similar to GSM systems and higher bit rate is supported by *time slot bundling*. Specifically, a GPRS frame has eight time slots and by assigning multiple time slots to a mobile, the bit rate can be increased. In GPRS systems, the specification allows a maximum of assigning 8 time slots to a mobile. However, in practice, the number of time slot assigned to a mobile is limited by the capability of the GPRS mobile. In GPRS mobile, there is a specification of $N + 1$. This means that the GPRS mobile supports N downlink time slots and 1 uplink time slot. For example, most of the GPRS mobiles nowadays support $4 + 1$, meaning that it can support 4 time slots in the downlink. On each time slot, GPRS supports four frame formats (CS1, CS2, CS3 and CS4) with data rate ranging from 9.6kbps to about 21.5kbps. The CS1 format has a data bit rate of 9.6kbps and has the highest level of redundancy for error protection. The CS4 format has a data rate of around 21.5kbps and has no error protection at all. Hence, the theoretical maximum GPRS data rate is given by $8 \times 21.5kbps = 172kbps$. Achieving the theoretical maximum GPRS data transmission speed of 172 kbps would require a single user taking over all eight timeslots without any error protection. Clearly, it is unlikely that a network operator will allow all timeslots to be used by a single GPRS user. Additionally, the initial GPRS terminals are expected be severely limited- supporting only one, two, three or four timeslots. The bandwidth available to a GPRS user will therefore be severely limited. As such, the theoretical maximum GPRS speeds should be checked against the reality of constraints in the networks and terminals. Relatively high mobile data speeds may not be available to individual mobile users until Enhanced Data rates for GSM Evolution (EDGE) or Universal Mobile Telephone System (3GSM) are introduced.

Further enhancements to GSM networks are provided by Enhanced Data rates for GSM Evolution (EDGE) technology. EDGE provides up to three times the data capacity of GPRS. Basically, EDGE only introduce a new modulation technique (8PSK) and new channel coding that can be used to transmit both packet-switched and circuit-switched voice and data services. EDGE is therefore an add-on to the GPRS system and cannot work alone. Table 5.3 compares the air interface performance of GPRS and EDGE systems. Although the symbol duration of GPRS and EDGE systems are the same, the modulation bit rates are different. EDGE can transmit three times as many bits as GPRS during the same period of time and this is the main reason for the higher bit rates in EDGE system. Using EDGE, operators can handle three times more subscribers than GPRS; triple their data rate per subscriber, or add extra capacity to their voice communications. EDGE uses the same TDMA (Time Division Multiple Access) frame structure, logic channel and 200kHz carrier bandwidth as today's GSM networks, which allows it to be overlaid directly onto an existing GSM network.

Table 5.3 Comparison of GPRS and EDGE air interface specifications.

	GPRS	EDGE
Modulation	GMSK	8-PSK/GMSK
Symbol rate	270 ksym/s	270 ksym/s
Modulation bit rate	270 kb/s	810 kb/s
Radio data rate per time slot	22.8 kb/s	69.2 kb/s
User data rate per time slot	20 kb/s (CS4)	59.2 kb/s (MCS9)
User data rate (8 time slots)	160 kb/s (182.4 kbs/s)	473.6 kb/s (553.6 kb/s)

The EDGE system supports nine channel coding and modulation formats with bit rates ranging from 8.4kbps (MCS1) to 59.2kbps (MCS9). Both GPRS CS1 to CS4 and EDGE MCS1 to MCS4 use GMSK modulation (binary modulation) with slightly different throughput. Figure 5.8 illustrates the modulation and coding formats in GPRS and EDGE systems.

The spectral efficiency of the EDGE system is enhanced compared with GPRS system through *link adaptation*. For instance, the received SIR of mobile stations are time varying. For example, the average received SIR of a mobile maybe 15dB. However, occasionally, the instantaneous received SIR may be higher than the mean (say 20dB). At some other times, the instantaneous received SIR may be lower (say 10dB). Figure 5.9 illustrates the histogram of the received SIR of a mobile.

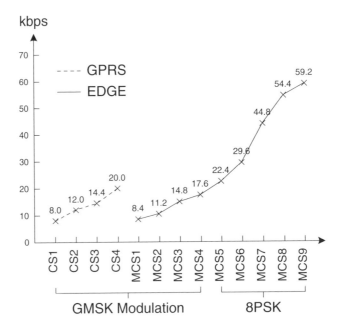

Figure 5.8 Illustration of GPRS and EDGE modulation and coding formats.

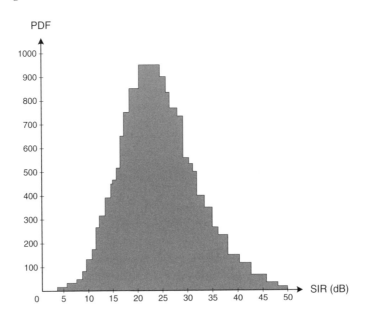

Figure 5.9 Histogram of the received SIR at the mobile.

As we have discussed in earlier chapters, 8PSK modulation offers a higher bit rate at the same bandwidth but a higher received SNR is required to maintain a similar error probability. Hence, by link adaptation, the transmitter (base station in the downlink or mobile station in the uplink) estimates the instantaneous rx SIR and selects the most appropriate modulation and coding formats. If the received SIR is high, the transmitter will selects higher MCS formats which results in higher instantaneous throughputs. On the other hand, if the received SIR is low, the transmitter will select more robust MCS formats for better error protection (resulting in lower throughputs). The modulation and coding formats can be changed for each radio blocks (4 bursts) but a change is usually triggered by new radio channel quality estimates. Hence, the practical link adaptation rate is determined by the measurement interval.

While GPRS and EDGE systems offer an average bit rate of 100kbps and 384kbps respectively, the UMTS system offers a peak bit rate of 2Mbps using wideband CDMA technology. The UMTS system achieves this high bit rate using *variable spreading factor* (VSF) and *multi-code transmissions*. In UMTS, the frequency bandwidth is 5MHz with a chip rate of 3.84Mcps and QPSK modulation. The spreading factor in the downlink changes from 512 (lowest bit rate)[1] to 4 (highest bit rate). The spreading factor in the uplink changes from 256 to 4. Hence, the bit rates of dedicated physical channels can be adjusted by adjusting the spreading factor of the channelization codes. One challenge of supporting variable spreading factor in UMTS is to maintain orthogonality among the codes. This is because for the same chip duration, channelization codes with different spreading factors will have different lengths. This is resolved in UMTS based on *Orthogonal Variable Spreading Factor* (OVSF) as illustrated in Figure 5.10 To maintain code orthogonality, all the codes belonging to the same sub-tree of the assigned code(s) should not be assigned to any users. For example, if $C_{2,1}$ has been assigned, the codes $C_{4,1}$ and $C_{4,2}$ cannot be assigned to any other users.

Due to the OVSF channelization codes, one issue in UMTS is the limitation of the number of orthogonal codes. For instance, if we assign spreading factor of 256 to all users in UMTS, we can support 256 orthogonal downlink channels using D-CDMA. However, if there is high bit rate data user assigned with code channel of spreading factor 4, this single data user will consume one quarter of the orthogonal codes space (with spreading factor 256). Hence, only 3×64 users (with SF=256) can be assigned together with this high bit rate data user (with SF=4). In other words, the UMTS system can easily become code limited if a few high bit rate users are assigned with low spreading factor channelization codes.

In UMTS uplink, a single code channel (with spreading factor 4) can support a bit rate of 400kpbs - 500kbps with coding. Similar to time slot bundling in GPRS, the bit rate of a user can be further increased by assigning multiple codes to a single user. In multicode transmission, the spreading factor of the codes is the same. In UMTS,

---

[1]In fact, spreading factor of 512 is seldom used in the downlink due to timing synchronization between Node-Bs involved in soft handover.

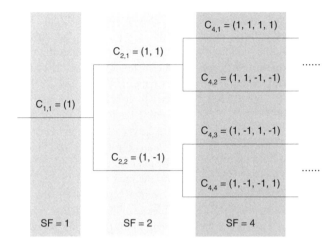

Figure 5.10 Illustration of Orthogonal Variable Spreading Factor (OVSF) in UMTS.

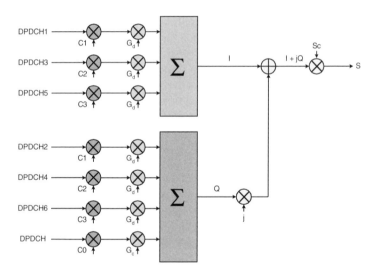

Figure 5.11 Illustration of multicode transmission in uplink of UMTS.

a maximum of six orthogonal codes can be assigned to a user in the uplink, with a total bit rate of around 2.4Mbps. Figure 5.11 and Table 5.4 illustrate, respectively, the multi-code structure as well as the supported data rates in the UMTS uplink.

Figure 5.12 and Table 5.5 illustrate, respectively, the multi-code structure and the supported data rates in the UMTS downlink. A maximum of 3 multi-codes (each

Table 5.4 Supported data rates in uplink of UMTS.

DPDCH Spreading Factor	DPDCH Channel Bit Rate (kbps)	Maximum User Data Rate with 1/2-Rate Coding (Approx., kbps)
256	15	7.5
128	30	15
64	60	30
32	120	60
16	240	120
8	480	240
4	960	480
4, with 6 parallel codes	5740	2.3M

with spreading factor 4) is allowed in downlink resulting in a data channel of bit rate 2.3Mbps.

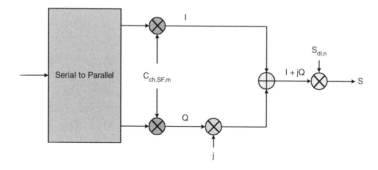

Figure 5.12   Illustration of multicode transmission in downlink of UMTS.

5.3.2   Quality-of-Service (QoS)

In 2G systems, the primary application is voice communications and is supported through circuit switched connections. Hence, there is no issue of maintaining Quality of Service (QoS) in the cellular systems[322]. However, to support wireless data applications in 3G systems, the QoS is an important issue. For instance, there are different types of data and multimedia applications with very different QoS requirement[135]. For example, email applications may be delay insensitive but error sensitive. On the other hand, video streaming may be delay sensitive but could tolerate a higher packet error rate. Hence, it is a challenging task to mix the data

Table 5.5   Supported data rates in downlink of UMTS.

Spreading Factor	Channel Symbol Rate (kbps)	Channel Bit Rate (kbps)	DPDCH Channel Bit Rate Range (kbps)	Maximum User Data Rate with 1/2-Rate Coding (Approx., kbps)
512	7.5	15	3–6	1–3
256	15	30	12–24	6–12
128	30	60	42–51	20–24
64	60	120	90	45
32	120	240	210	105
16	240	480	432	215
8	480	960	912	456
4	960	1920	1872	936
4, with 3 parallel codes	2880	5760	5616	2.3M

and multimedia applications with different QoS requirements efficiently in a cellular system (in particular the air interface QoS differentiation).

In 2.5G systems such as GPRS and EDGE, a descent bit rate is supported in the air interface. Moreover, packet switched data connection is supported for bursty applications. In fact, there were times when people questioned about the demand of 3G systems when GPRS or EDGE systems can deliver packet switched data connections with bit rates of 100kbps or 384kbps. One of the major difference between GPRS/EDGE and 3G systems (besides the physical bit rate) is in fact on the support for QoS differentiation in the air interface. In 2.5G systems, there is no QoS differentiation in the air interface and different users running different applications are treated with the same priority in the allocation radio resource (best effort delivery). This is probably sufficient for typical wireless data applications such as emails. However, for video streaming or interactive service, QoS differentiation is needed and therefore, in 3G systems, four QoS classes, namely the *conversational class*, *the interactive class*, the *streaming class* as well as the *background class*, have been defined. These classes are characterized by maximum bit rates, maximum packet size, transfer delay and traffic handling priority. Table 5.6 illustrates the typical parameter settings for the four QoS classes.

The end-to-end QoS support in UMTS is realized by *QoS provisioning* in both the radio interface as well as the fixed line infrastructure and *QoS signaling*[243]. QoS provisioning refers to the enforcement of QoS in the radio interface and vari-

Table 5.6   QoS parameters for the four QoS classes supported by UMTS Bearer (from 3GPP TS 23.107).

Traffic Class	Conversational Class	Streaming Class	Interactive Class	Background Class
Max. Bit Rate (kbps)	< 2048	< 2048	< 2048 – overhead	< 2048 – overhead
Delivery Order	Yes/No	Yes/No	Yes/No	Yes/No
Max. SDU Size (octets)	≤ 1500 or 1502	≤ 1500 or 1502	≤ 1500 or 1502	≤ 1500 or 1502
Erroneous SDUs Delivery	Yes/No/-	Yes/No/-	Yes/No/-	Yes/No/-
Residual BER	$5 \times 10^{-2}, 10^{-2},$ $5 \times 10^{-3}, 10^{-3},$ $10^{-4}, 10^{-5},$ 10^{-6}	$5 \times 10^{-2}, 10^{-2},$ $5 \times 10^{-3}, 10^{-3},$ $10^{-4}, 10^{-5},$ 10^{-6}	$4 \times 10^{-3},$ $10^{-5},$ 6×10^{-8}	$4 \times 10^{-3},$ $10^{-5},$ 6×10^{-8}
SDU Error Ratio	$10^{-2}, 7 \times 10^{-3},$ $10^{-3}, 10^{-4},$ 10^{-5}	$10^{-1}, 10^{-2},$ $7 \times 10^{-3}, 10^{-3},$ $10^{-4}, 10^{-5}$	$10^{-3}, 10^{-4},$ 10^{-6}	$10^{-3}, 10^{-4},$ 10^{-6}
Transfer Delay (ms)	100 – max. value	100 – max. value		
Guaranteed Bit Rate (kbps)	< 2048	< 2048		
Traffic Handling Priority			1, 2, 3	
Allocation/ Retention Priority	1, 2, 3	1, 2, 3	1, 2, 3	1, 2, 3

ous interfaces. QoS signaling refers to informing all the network entities about the QoS requirement of the UMTS bearer. Detail description will be elaborated in later chapters.

5.3.3 Infrastructure

For 2G networks, the primary application is circuit switched voice service. The target public network for circuit switched voice service is basically the *public switched telephone network* (PSTN), which is designed based on circuit switched architecture. Hence, the infrastructure (or core network) in 2G systems is designed based on circuit switched architecture and the interfacing between cellular core network and PSTN is straight-forward (via SS7 in the signalling path and via PCM-64 in the payload path).

However, in 3G networks, the mixed voice and data applications complicated the design of the core network. For instance, the target public network of circuit switched voice service is PSTN and this favors a circuit-switched core network in the cellular system. On the other hand, the target public network of wireless data network is the public internet, which is packet switched. Hence, this favors a packet switched architecture in the core network of the cellular systems. Hence, the challenge in designing the core network is to accommodate both PSTN and internet interfacing for both voice and wireless data applications.

In 2.5G systems such as GPRS and EDGE, the core network design is based on the *overlay approach*. Specifically, a new packet switched layer is added on top of the existing circuit switched layer in the core network as illustrated in Figure 5.13 The voice path involves the mobile station (MS), base station (BS), base station controller (BSC) and the mobile switching center (MSC), which is then interfaced to the PSTN via the gateway MSC. The packet switched data path involves the MS, the BS, the BSC and the SGSN, which is then interfaced to the internet via the GGSN. The GGSN and SGSN are similar to regular IP routers except with GPRS mobility management and GPRS session management capability.

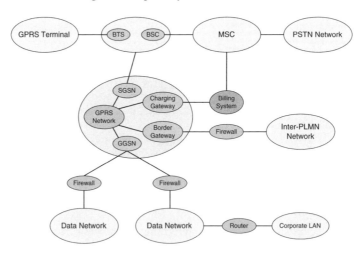

Figure 5.13   Overlay approach in GPRS core network to support circuit switched voice and packet switched data applications.

The advantage of the overlay approach is simple and it does not introduce any impact onto the existing circuit-switched core network (MSC). However, the overlay approach suffers from a disadvantage of higher operation and management cost because the operator has to maintain two core networks of very different nature (circuit switched core vs packet switched core).

In the earlier releases of 3G systems (such as UMTS Rel 99), the core network design is also based on the overlay approach with MSC nodes responsible for regular circuit switched voice service and SGSN/GGSN nodes responsible for the new packet switched data services. However, the latest trend (such as in UMTS Rel 6) is to integrate the core network into a single packet-switched architecture to support both PSTN and public internet[54]. The interface to the public internet is via GGSN which is pretty straight-forward as GGSN is also an IP router. On the other hand, the circuit switched voice is delivered to the core network in the form of VoIP packets. The interface to the PSTN is via VoIP gateway which converts the VoIP packets and SIP signalling into standard PCM-64 and SS7 signalling respectively.

5.4 SUMMARY

In this chapter, we have elaborated on the evolution of cellular systems from 1G to 3G systems. The primary driving force behind the evolution of 1G systems to 2G systems is voice capacity and international roaming. For instance, 1G systems are designed based on FM transmissions and have quite poor spectral efficiency. Hence, the voice capacity quickly saturated even after cell splitting is employed. Furthermore, the 1G systems are incompatible with each other and no inter-system roaming can be supported. In moving from 1G to 2G systems, the US and the Europe took different approaches. Evolutionary approach was taken in the US where 2G systems were designed to coexist with the existing AMPS operation. However, revolutionary approach was taken in the European countries where all the existing 1G systems were scrapped. GSM, IS54 (D-AMPS) and IS95 (CDMA) are typical examples of 2G systems where the radio interface is designed based on digital transmission technologies. The channels are partitioned based on hybrid FDMA/TDMA and hybrid FDMA/CDMA respectively. Due to the speech compression, more channels are available to the operator. Due to more robust digital transmission, more aggressive frequency reuse can be employed which increase the capacity scalability in multi-cell systems. Hence, the overall network spectral efficiency in 2G systems has been increased substantially compared to 1G systems.

In moving from 2G to 3G systems, the major driving force behind is no longer voice capacity but the demand for high speed wireless and multimedia data services. To support the bursty wireless data applications, packet switched connection is introduced. The dedicated radio resource associated with a data connection is dynamically adjusted according to the demand. Hence, the user can be logically *online* without having to pay for the air time because no dedicated radio resource has been attached to the logical connection. In 3G systems, the maximum peak bit rate supported is about 2Mbps.

Between 2G systems and 3G systems, a number of 2.5G technologies such as the GPRS and EDGE systems are introduced. GPRS increases the bit rate by time slot bundling whereas the EDGE system further increase the bit rate by higher modulation (8PSK) and link adaptation. In UMTS, higher bit rate is supported by variable spreading factor and multi-code transmissions.

Finally, we have elaborated on the three major technical challenges to realize the 3G services. For instance, we have the transmission challenge in the radio interface to support higher bit rate. We have the QoS challenge to support QoS differentiation and end-to-end QoS. We also have challenge of the infrastructure design to support both PSTN and internet interfacing for both circuit switched voice and packet switched data applications.

PROBLEMS

5.1   Second generation wireless systems cannot support digital data communications. [True/False]

5.2   UMTS system allows QoS differentiation in the air interface [True/False].

CHAPTER 6

CDMA (IS-95)

A secret may be sometimes best kept by keeping the secret of its being a secret.
—Henry Taylor: The Statement, 1836

6.1 INTRODUCTION

In this chapter, we shall elaborate the system architecture and the system operations of IS95 CDMA network as one of the 2G system example. We shall focus on the air interface of the IS95 and the circuit switched operations. Specifically, we shall first look at how physical channels are partitioned in the uplink and downlink based on *deterministic CDMA* and *random CDMA*. We shall then look at the call processing procedure in circuit-switched connections. We will look at the processing behind (such as cell search, physical synchronization) as soon as mobile station turns on the power. We will also look at the background process when the mobile station is in idle state (able to see the operator's logo). Next, we will explore the channel access procedure and explain why there is a "SEND" button in mobile stations (but not in fixed line telephones). Finally, we will elaborate the call origination and call termination processing between the mobile station and the infrastructure.

The next topic we will explore in this chapter is the power control. As explained in earlier chapters, power control is very important to CDMA systems due to near far problem. In fact, quite sophisticated power control procedures have been defined in the reverse link of IS95. The objective is much more than saving battery power but to maintain a reasonable system capacity. We will first look at the open loop power control and explain why the procedure is designed the way it is. We will then look at the inner loop power control and outer loop power control procedures. Finally, we will illustrate the operation with an example.

After power control, the next important topic in CDMA is soft handover. Soft handover is a unique feature in CDMA systems which is not found in GSM or IS54 systems. We shall first elaborate the reason how CDMA systems facilitate soft handover. We will then explore how the mobile station physically measure the pilot signal strength of the home cell and neighboring cell. Next, we will go over the mobile station processing for soft handover, followed by the corresponding infrastructure processing. Finally, we will illustrate the soft handover procedures using an example.

6.2 SYSTEM ARCHITECTURE OF IS95

The IS95 CDMA system can be viewed as a collection of network entities and the associated interfaces. Reference interface is defined as a reference point or interconnections between adjacent network elements. The reference interface is usually specified in terms of a set of procedures and associated signaling that define the operational responsibility of the connected network entities. Figure 6.1 illustrates the standard reference model or system architecture of IS95 system[127]. The IS95 system consists of three parts, namely the mobile station, the radio access network (RAN) as well as the core network (CN). The mobile station subsystem consists of CDMA mobiles (MS). The radio access network consists of base stations (BS) and base station controller (BSC). The core network consists of mobile switching center (MSC), interworking function (IWF), visiting location register (VLR) and home location register (HLR). Note that the IS95 interfaces are fully standardized or open interfaces, which allows the operators to deploy a network consisting of components from different manufacturers. For example, the Um interface is defined for the radio connection between the mobile and the base station (and is specified in the IS95 specification). The A-bis interface is defined for interconnection between base station and base station controller. The A interface[226] is defined for the interconnection between base station controller and the MSC (and is specified in the IS41 specification). In this chapter, we shall primarily focus on the IS95 radio interface.

The protocol architecture of the IS95 radio interface is illustrated in Figure 6.2 There is a user plane and control plane in the protocol stack. The user plane consists of user application data (such as circuit switched voice and circuit switched data).

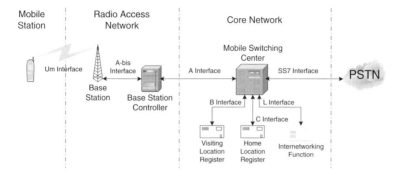

Figure 6.1   System architecture of IS95 CDMA systems.

The control plane supports a variety of functions such as the broadcasting of system information, radio resource management, mobility management and call processing. We will elaborate the functions of the control plane in later sections.

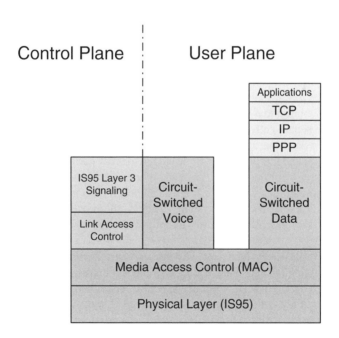

Figure 6.2   Protocol architecture of the IS95 radio interface.

6.3   PHYSICAL LAYER AND PHYSICAL CHANNELS

The physical layer of the IS95 radio interface represents the lowest layer of the IS95 protocol stack and defines the mobile station and base station interoperability procedure over the air interface. The physical layer specifies the physical channel baseband processing such as error correction encoding, interleaving, modulation and spreading (CDMA processing) at the transmitter side and the reverse operations (such as despreading, demodulation, deinterleaving and error correction decoding) at the receiver side. The physical layer also defines the modulation pulse shape, transmit power spectral mask and the radio frequency band over which IS95 can be operated. In this chapter, we shall primarily focus on the baseband processing (specifically CDMA processing) in the IS95 physical layer.

The mobile station and the base station communicates by means of physical channels, which are partitioned from a paired spectrum based on CDMA technology. By paired spectrum, we mean that a dedicated range of frequencies are defined for the uplink (mobile station to base station) and the downlink (base station to mobile station) respectively. The uplink and downlink are separated by *frequency division duplexing* (FDD). The corresponding frequency separation between the uplink channel and the downlink channel is 45MHz in the cellular spectrum (800MHz) and 80MHz in the PCS spectrum (1.9GHz). We shall elaborate on the channel partitioning and the channel structure in the uplink and downlink of the IS95 system.

6.3.1   Channelization in the Uplink

In the reverse link or uplink of IS95 (mobile to base station), the channels are partitioned based on random CDMA (R-CDMA) techniques. Specifically, all the uplink channels are separated by a spreading sequence called the *long code*. The long code is a unique pseudo-random (PN) sequence derived from the user equipment serial number (ESN) and international mobile station identity (IMSI). These long codes are designed not to be orthogonal but with small cross correlation. Hence, the system is not code limited (but interference limited) and each mobile user can have a unique long code in the entire CDMA system. In IS95, the processing gain is 64 and the modulation is BPSK. In other words, an information bit is chopped into 64 chips by the long code.

Figure 6.3 illustrates the reason why deterministic CDMA (D-CDMA) cannot be used in the uplink of IS95. Specifically, D-CDMA requires the spreading codes to be orthogonal to each other. However, the spreading codes can be orthogonal only if the timing offsets between the codes can be maintained at the receiver (synchronous or quasi-synchronous). In the uplink, it is a *many-to-one* configuration where multiple transmitters (at different locations) are communicating to the base station. Since the mobile stations are at random locations, the received signals from the mobile stations suffer from random propagation delays. Hence, the relative timing offsets between the spreading codes at the base station receiver are random and cannot be guaranteed to be completely synchronized. Hence, the orthogonality of the spreading codes cannot be guaranteed and R-CDMA has to be used to separate the channels.

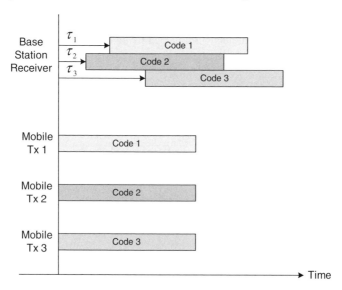

Figure 6.3   Illustration of the random timing offsets between the spreading codes in the uplink of IS95 systems.

While R-CDMA partitions the uplink spectrum into a number of uplink physical channels, the physical channels can be divided into common channels and dedicated channels. Common channels are also called overhead channels because they are shared by all mobiles in the base station. These common channels are always setup in the base station no matter how many active users are there in the base station. Dedicated channels, on the other hand, are assigned to one and only one mobile at a time. Hence, these dedicated channels are revenue generating because once assigned to a mobile station, the user has to start paying for the air time. In IS95, the common channel in the reverse link is called the *random access channel*. The dedicated channel in the reverse link is called the *dedicated traffic channels*. The random access channel is spread by a specific long code (which is defined in the specification and known to the system). The dedicated traffic channels are spread by the unique long codes derived from the mobile user ESN and MSN.

6.3.2 Channelization in the Downlink

In the downlink or forward link of IS95 (base station to mobile stations), the channels are partitioned based on hybrid deterministic CDMA (D-CDMA) and random CDMA (R-CDMA). Specifically, downlink channels are first spread by the *Walsh codes* with spreading factor $SF = 64$. Walsh codes are special spreading sequences which are mutually orthogonal (zero cross correlation) when they are synchronous (at zero offset). In the downlink, since it is a one-to-many configuration, the Walsh codes between the different downlink channels remain to be synchronous with each other at the mobile receiver as illustrated in Figure 6.4 Hence, the Walsh code spreading is used to separate the downlink channels within a cell by *deterministic CDMA* (D-CDMA).

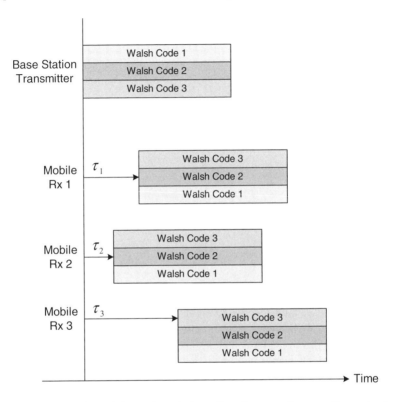

Figure 6.4 Illustration of the random timing offsets between the spreading codes in the downlink of IS95 systems.

In addition to the Walsh code spreading, IS95 downlink channels are further spread by a short PN sequence. The short PN sequence is a unique PN sequence per cell. All the downlink channels of the same cell are spread by the same PN sequence associated with the cell. The short PN sequence is used to mitigate inter-cell interference between different base stations in the system through random CDMA (R-CDMA). Hence, the downlink channels are separated by hybrid D-CDMA (same cell channels) and R-CDMA (multi-cell channels). In IS95, all the base stations are synchronous with each other (i.e., they have the same $t = 0$ time reference.). There is a common master short PN sequence defined in the system which is the same short PN sequence used in all the base stations. Each base station has a parameter called *PN Offset* (PNO). The actual short PN sequence transmitted from a base station is derived from the common master PN sequence except with an offset in the starting phase given by the PNO as illustrated in Figure 6.5 The advantage of this arrangement is simpler cell search procedure (as will be elaborated in the next section). However, the disadvantage is the requirement to accurately synchronize all the base stations in IS95 systems (up to the chip intervals). In practice, such accurate timing synchronization can be achieved using GPS on each base station. Yet, in some densely populated urban areas where the GPS reception is poor, the synchronization between base station will be quite challenging.

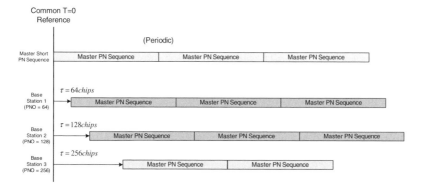

Figure 6.5   Illustration of the common time reference in all base stations and the PN offsets (PNO) in IS95 systems.

Similarly, downlink channels are partitioned into common channels and dedicated channels [190]. Common downlink channels in IS95 include the *pilot channel*, *SYNC channel* and *PAGING channel(s)*. The pilot channel is spread by Walsh code 0 and contains no data (all 1 sequence). It is used for timing acquisition, tracking, channel estimation and soft handover measurement. We shall elaborate on the usage of the pilot channels in later sections. The SYNC channel (spread by Walsh code 32) is a broadcast channel used to convey the *SYNC channel message*. The SYNC channel message contains various important system parameters such as the *system ID* (SID), *network ID* (NID), base station ID (PNO), system time, long code state, paging channel data rate, etc. The paging channel is spread by Walsh code 1 - 7 which carries the *system parameter message, access parameter message, neighbor list message, CDMA channel list message* or *page message*. We shall elaborate the important parameters contained in these system messages in later sections.

On the other hand, the dedicated downlink channels are traffic channels (spread by the remaining Walsh codes) to be assigned to dedicated users. Theoretically, in IS95 systems, there are at most 64 downlink channels per cell. However, in practice, since the dedicated traffic channels are assigned in pairs (uplink and downlink) and the uplink of the IS95 system is interference limited (supporting about 15-16 channels per sector). Hence, the actual number of voice channels is only around 15-16 per sector.

6.4   CALL PROCESSING

In this section, we shall focus on the call processing procedure in IS95 system. Call processing here refers to the procedure and handshaking between the mobile phone and the infrastructure (radio access network) from the time a mobile phone is turned on to the time that a user press the "SEND" button and establish a traffic channel.

A mobile phone (after power on) can be in one of the four substates, namely the *Initialization State*, the *Mobile Idle State*, the *Mobile Access State* as well as the *Traffic Channel State* [190]. This is illustrated in Figure 6.6 We shall elaborate the processing in each of the substates in the following subsections.

6.4.1   Initialization and Cell Search State

When mobile station is turned on, it will enter into the mobile initialization state. The mobile initialization state can be further broken into the following substates as illustrated in Figure 6.7

System Determination Substate: When the mobile station is powered on, it will first determine which system to use according to user selected preference. As we have discussed in earlier chapters, the IS95 system in the US is designed to co-exist with the AMPS system. Hence, the IS95 specification allows the mobile (dual mode) to acquire onto the AMPS system or the IS95 system. User can specify to acquire CDMA system (IS95) only or CDMA system (IS95) preferred.

In IS95, the channels are partitioned based on hybrid FDMA/CDMA where the cellular spectrum (or PCS spectrum) is first partitioned into a number of 1.5MHz carrier. For each 1.5MHz CDMA carrier, the physical channels are partitioned based on R-CDMA and R-CDMA/D-CDMA as we have explained in the previous section. While the number of CDMA carriers assigned to an operator may be limited, the IS95 mobile should be able to scan all the CDMA carriers in the cellular spectrum (or PCS spectrum).

The specific algorithm for system determination is not specified in the IS95 specification. However, most mobile phone vendors make use of a *preferred roaming list*. The roaming list is a list of CDMA carriers with priority that the mobile phone would attempt to acquire the pilot channel. The preferred roaming list is used in such a way that the mobile station, upon power up, tries to acquire the home system (the system which it is subscribed). Each entry of the roaming list contains the system identifier (SID) and the network identifier (NID) of the system to acquire, the block of CDMA carrier numbers, and the

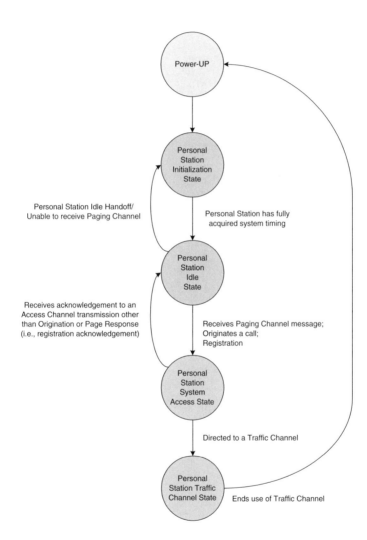

Figure 6.6 Illustration of the four system states when a mobile phone is turned on in IS95 systems.

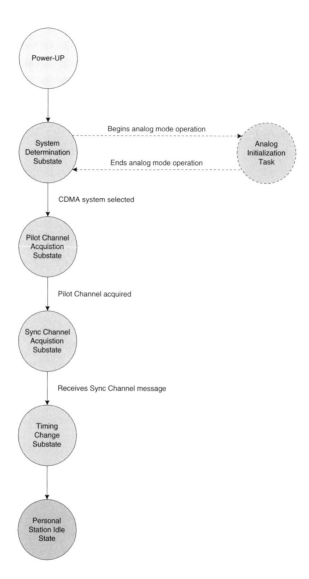

Figure 6.7 Illustration of the mobile initialization states when a mobile phone is turned on in IS95 systems.

corresponding roaming flag. The roaming flag, if turned on, makes the roaming indicator of the mobile station's display flash once service is provided on this entry. The algorithm also accounts for the previously visited systems to speed up acquisition. Some vendors may also maintain a dynamic table listing the CDMA carriers (channels) on which the service was most recently provided.

The system determination procedure builds an ordered scanning list. The top entries are from the most recently visited CDMA carriers, followed by those in the preferred roaming list. Once the scan list is built, the mobile station sequentially attempts to acquire each channel in the scan list using the procedures described below.

Pilot Channel Acquisition Substate: Say the mobile phone attempts to acquire to CDMA system. The mobile will enter into the *pilot channel acquisition substate* and it will attempt to acquire the downlink pilot channel on the selected CDMA carrier. The acquisition to the pilot channel is very important and is the first step before the mobile can decode any channels in the system. As we have discussed in Chapter 3, a CDMA receiver decode the information through the despreading process. The despreading process requires knowledge of the spreading sequences used to spread the information at the transmitter as well as the received timing (or received phase) of the spreading sequence. In IS95 downlink, channels are first spread by the Walsh code and followed by the short PN code. Both the Walsh code (Walsh code 0 for pilot channel) and the short PN code used in spreading the target downlink channel are known to the mobile station. Hence, the only unknown to the mobile is the starting phase (PNO) and the received chip timing. Hence, the objective of pilot acquisition is to estimate the unknown parameters.

Figure 6.8 illustrates the process of pilot acquisition. The mobile station has a local copy of the same short PN sequence and correlates with the received signal. Since the unknown parameter at the mobile is the timing offset and the mobile has no prior information about the timing offset, the mobile has to perform exhaustive search over all the possible offsets. When the current local PN sequence has a non-zero offset with the received PN phase, the result of the correlation will be quite small. However, when the local PN sequence is *in-phase* with the received PN sequence, the correlation value will be large. Hence, the estimated offset (or PN phase) is given by the delay position at which a large correlation value is observed. Once the mobile state estimated the correct PN phase offset, it can use the same PN phase value to despread the other downlink channels (with the correct Walsh codes) because all the downlink channels have the same PN phase offset. Note that due to multipath and signals from adjacent base stations, there might be more than one correlation peaks. The value of the correlation peak is equal to the received path energy from the corresponding base station. Since the PNO between base stations are separated by at least 64 chips, the peaks from one base station (due to multipath) will be clustered together relative to the peaks from different base stations. The mobile station will pick the peaks with the highest received energy and therefore, camping to the strongest base station.

In any case, if the mobile fails to acquire to any pilot channel in any CDMA carrier within 15 sec, it will return to the system determination substate to try to acquire the pilot channel on a different CDMA carrier.

Sync Channel Acquisition Substate: After the mobile has acquired to the pilot channel of a base station, the mobile station attempts to decode for the SYNC channel using Walsh code 32 and the PN code offset (obtained from the pilot acquisition search) to obtain the system timing information. For instance, while the mobile has searched on the appropriate local PN phase offset to acquire the pilot channel, the mobile station has no idea on where is the system $T = 0$ reference. By decoding the SYNC channel message, the mobile station obtains the current PNO of the base station. Hence, the mobile station can align the local $T = 0$ reference with the system $T = 0$ reference. Furthermore, in the SYNC message, the mobile station obtains the operator and system ID (SID and NID) and display the operator logo or roaming status. The mobile also synchronizes the local long code generator with the long code state of the base station.

Timing Change Substate: After the mobile station obtains the timing related information from the SYNC message, it will update the local timing and align with the system timing accordingly. From this point onwards, the mobile and the base stations in the IS95 system have a common $T = 0$ reference.

After the mobile station has completed the initialization process, the mobile station enters the idle state and is ready to originate or receive calls.

6.4.2 Idle State

During the idle state, the mobile station periodically monitors the messages in the paging channel as well as the pilot strength of the home cell and the neighboring cells. In particular, the mobile station monitors the paging channel for the *System Parameter Message*, the *Access Parameter Message*, the *Neighbor List Message*, the *CDMA Channel List Message* as well as the *Page Message*. All the messages (except the page message) are broadcast to all mobiles and contain important system parameters for the mobile to participate in call processing in the base station. For example, the system parameter message contains the search window for pilot, power control parameters as well as pilot thresholds in soft handover. The access parameter message contains the important parameters related to the random access procedure (such as the number of access channels, initial access power, access channel timing information and authentication parameters). The neighbor list message contains the

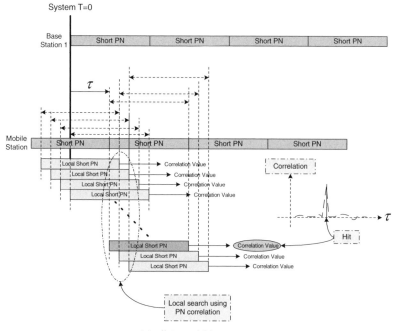

(a) pilot acquisition process

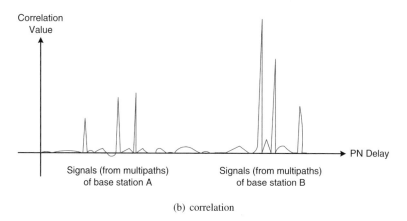

(b) correlation

Figure 6.8 Illustration of the pilot acquisition process in IS95 systems.

list of PNOs in the neighbor base stations and this is very important to speed up pilot strength measurement of neighbor cells during soft handoff. The CDMA channel list message contains a list of active 1.5MHz CDMA carriers in the current cell. These parameters will be elaborated in later sections when we discuss the corresponding procedures. Finally, the paging message is a dedicated message to a mobile user. It is used to notify a mobile station about call request (mobile terminated call).

When a mobile station is in idle state, it will continuously monitor the pilot strength from neighbor cells and perform idle handoff to the new (and stronger) base station when the home cell signal becomes too weak. Note that while IS95 supports soft handover, the idle handover is not soft. The handover decision during idle state is entirely made and initiated by the mobile and it is transparent to the infrastructure.

There are three possibilities that a mobile station will leave the idle state and enter the access state. They are namely, the *call origination process*, the *call termination process* and the *location update process*. When a user presses a SEND button, the call origination process will be triggered and the mobile will enter the access state to communicate this request to the infrastructure. When there is a mobile terminated call, the concerned base station will page the target mobile via the paging channel. The mobile station upon receiving the paging message will enter the access state and communicate the *page response message* to the infrastructure. Finally, when a mobile station moves to a new *location area* (a new cell with new location area identity (LAI)), the mobile station enters into the access state to send the location update message to the infrastructure.

6.4.3 Access State

In the system access state, the mobile station exchanges signaling message with the base station in response to messages requiring acknowledge (page message), to register with the infrastructure (location update message), or call origination. Suppose the user triggers a mobile originated call request by pressing the SEND button in the mobile phone. The mobile station attempts to send the "Call Origination Request" message to the infrastructure. However, at this stage, the only uplink channel available to the mobile station is the random access channel (there is no dedicated channel assigned to the mobile yet). Hence, the mobile station will leave the idle state and move to the channel access state, trying to utilize the random access channel (uplink) to convey the call origination request message to the infrastructure. In other words, a random access procedure is initiated by the mobile to send the call origination message. Figure 6.9 illustrates the random access procedure in IS95.

The entire random access attempts are grouped into *random access probe sequence*. Within each random access probe sequence, there are multiple *random access probes*. During the first random access attempt (the first random access probe in the first random access probe sequence), the mobile station transmit the random access message to the base station via the random access channel (spread by a long code assigned for access channel) at a transmit power given by *Initial Power*. The initial power is

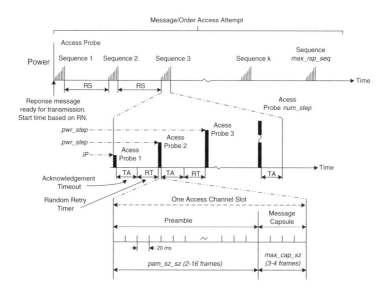

Figure 6.9   Illustration of the random access procedure in IS95 systems.

computed based on the following equation

$$P_i = nom_p wr + init_p wr - 76 - RSSI$$

where $RSSI$ denotes the received signal strength indicator of the downlink signals from the base station and $nom_p wr, init_p wr$ are access parameters defined in the access parameter message. If the mobile station does not receive an acknowledgement from the base station after a timeout period TA (which is also defined in the access parameter message), the mobile station will retransmit the access packet after a random waiting time RT. In the second attempt, the mobile station will transmit at a slight higher power defined by $pwr_s tep$ (which is another access parameter defined in the access parameter message). The process repeats until the mobile station receives the acknowledgement or the number of access attempts exceed $num_s tep$ (another access parameter). In the former case, the random access is successful and the mobile enters the traffic conversation state. In the latter case, the mobile station will backoff with a random time RS and starts another access probe sequence. The process repeats until the number of access probe sequence exceeds $max_r esp_s eq$ (another access parameter message). If so, the mobile station will move to system determination state and acquire the pilot channel on another CDMA carrier (channel) once more.

Observe that there are two possible reasons that the base station does not send back the acknowledgement to the mobile station. The first reason is because the mobile access probe is not transmitted at a sufficiently high power. Hence, the random access probes has an increasing power between repeated attempts. The setting of initial power is a tradeoff between the capacity loss due to random access[1] and the access delay because it takes a while for the transmit power of access probes to ramp up. The second reason that the base station may miss the access probe is *collision*. For instance, the random access channel is a common uplink channel shared by all the mobiles in the system. Hence, there are some cases when more than one mobile stations attempt to send access probes at similar time. In this case, the access probes may have been transmitted at sufficiently high power but due to collision, the access attempts between mobiles collide and nothing is received at the base station due to mutual interference. Hence, a random backoff timer (RT) between access probes and a random backoff timer (RS) between access probe sequence are introduced to randomize the chance of access collisions in the system. This also explains why there is a SEND button in mobile phones but not in fixed line telephone. In fixed line telephone, the local loop (telephone wire) is a dedicated channel not being shared with other fixed line telephones. However, in cellular systems such as IS95, the uplink access channel is a common channel shared by all the mobiles in the base station. Without a SEND button, the *channel occupancy time* of each access attempt (containing origination request with called telephone number) will be quite long and this increases the chance of collisions between access attempts of mobiles.

6.4.4 Traffic Conversation State

After the mobile station successfully transmit the signaling message (call origination or page response) to the base station via the uplink random access channel, the mobile will enter the traffic conversation state and try to setup dedicated traffic channel (both uplink and downlink)[2]. Figure 6.10 illustrates the three substates in the traffic conversation state.

Traffic Channel Initialization Substate: In this substate, the base station setup a downlink dedicated traffic channel (by assigning an unused Walsh code) and sends null frames at the eighth rate. The base station then notifies the mobile of the assigned Walsh code using a *Channel Assignment Message*. Note that the channel assignment message is sent to the mobile over the paging channel because at this time, there is no dedicated channels established between the base station and mobile station. The mobile station has to verify that it can properly demodulate and decode the downlink frames before starting transmission on the reverse traffic channel. In IS95, the mobile station has to receive two consecutive good frames (passed CRC check) on the dedicated traffic channel first before it can declare the downlink channel is of sufficient quality. The

---

[1]The background interference floor due to random access will be increased if mobiles transmit the access probe at higher than the required power.
[2]Note that if the mobile station enters the access state due to location update (or registration), it will return to idle state after successfully transmitted the location update message to the base station in the access state.

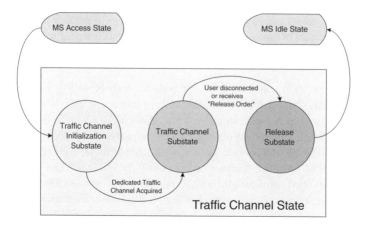

Figure 6.10 Mobile station substates in traffic channel state of IS95 systems.

mobile station will abort the call and re-enter the idle state if no two consecutive good frames are received within 1 second. Say the downlink traffic channel is good, the mobile station can start transmission on the uplink traffic channel. At this point, the mobile station will transmit uplink traffic channel preamble (with data part gated off). Note that the mobile station does not need inform the base station of the spreading code used to spread the uplink traffic channel because the uplink spreading code (long code) is derived from the ESN and IMSI of the mobile and the base station can derive the same long code from the ESN and IMSI of the mobile (which were communicated to the base station via the Call Origination Request Message or Page Response Message). Meanwhile, the base station attempts to acquire the reverse traffic channel preamble and send an acknowledgement message to the mobile station over the downlink traffic channel. This indicates to the mobile that it can start transmitting null data frames on the uplink traffic channel. At this point, both the uplink and downlink traffic channels have been setup properly and the mobile station enters the traffic channel substate.

Traffic Channel Substate: In this substate, the mobile station and the base station transmits signaling and data traffic on the forward and reverse dedicated traffic channel setup during the traffic channel initialization substate. Before the mobile and base station can exchange regular traffic frames, they have to negotiate the service options such as multiplex option, radio configuration and data rates. After the negotiation is completed, speech frames are exchanged between the mobile and the base station over the established dedicated traffic channel continuously once every 20ms. During this state, signaling messages are exchanged between the mobile and base station over the dedicated traffic channel using *dim-and-burst* or *blank and burst* signaling. Dim-and-burst signaling refers to carrying the signaling message in a 20ms traffic frame by gating off part of the speech data. Blank-and-burst signaling refers to carrying the signaling message in a 20ms traffic frame by gating off all of the speech data.

During the traffic channel substate, both the mobile station and base station perform supervision of the forward and reverse traffic channels, respectively, to detect loss of connection due to radio link failure. A radio link failure may occur if the mobile moves outside the coverage area of the access network. On the mobile side, the mobile station must disable its transmitter if it receives 12 consecutive traffic channel bad frames (CRC failed). The mobile station re-enables its transmitter only when receiving two consecutive good frames. In addition to disabling its transmission, the mobile station must decide when to release the connection based on *fade timer*. The timer is enabled when the mobile station first enables its transmitter upon entering the traffic channel state. The fade timer is reset for 5 seconds whenever the mobile station receives two consecutive good frames on the traffic channel. If the timer expires, the mobile station declares a loss of traffic channel and release the connection.

On the base station side, supervision is implementation dependent and is typically performed by monitoring the quality of the reverse traffic channel after frame selection (at the base station controller). The base station may also use a fade timer. For example, when the reverse traffic channel frames are in error consecutively for a period equal to the base station's fade timer, the base station declares loss of traffic channel and releases the connection. Usually, the fade timer at the base station is set to a value larger than that at the mobile station.

Upon the user disconnects the call or the mobile station receives *Release Order* from the base station, the mobile station enters the release substate and releases the dedicated traffic channel.

Release Substate: When a mobile station enters into the release substate, all the dedicated traffic channels are released and the radio resource is returned to the system. The mobile station moves to the idle state.

6.4.5   Example: Circuit Switched Connection for Voice

Figure 6.11 illustrates an example of call setup processing for mobile originated voice call. Steps (1) to (6) correspond to the procedures described in the *traffic channel initialization substate*. Steps (7) and (8) correspond to the procedures described in the *traffic channel state* where user is engaged in speech conversation. Step (9) corresponds to the *release substate* where the mobile receives a *release order* from the network and release the dedicated traffic channel. Note that the messages such as Call Origination Request and Release Order are terminated and generated from the core network. For mobile terminated calls, the procedures are similar except that the Call Origination Request message is replaced with Page Response message.

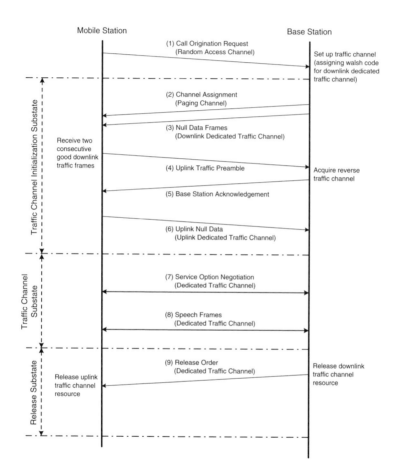

Figure 6.11 Call setup procedure for mobile originated call. In this example, the mobile initiated the call request and the network initiated the call release.

Figure 6.12 illustrates the end-to-end block diagram showing all the entities involved in the speech path for a mobile to fixed line call example. The voice signal is digitalized and compressed at the mobile station and the compressed speech frames are delivered to the base station over the air. The base station acts as a relay which delivers the compressed speech frames to the MSC where it terminates with a corresponding speech decoder. At the output of the speech decoder, the speech frames becomes standard PCM 64kbps speech data and interfaces directly to the PSTN via the gateway MSC. The speech decoder is usually located at the MSC because this could save the bandwidth occupied by a voice channel in the fixed line transmission systems between the base station and the MSC.

Figure 6.12   Illustration of all the network entities involved in the speech path for a mobile to fixed line call.

6.4.6   Example: Circuit Switched Connection for Data

The call setup procedure for circuit-switched data connection is similar to the procedure illustrated in Figure 6.11 except that the mobile has to indicate to the infrastructure that the circuit switched connection is for data service. This is because the infrastructure has to prepare different entity to terminate the data call. Figure 6.13 illustrates the end-to-end block diagram showing all the entities involved in the digital data path for a mobile to fixed line call example. The digital data (e.g., a file) from the laptop is transferred to the mobile via the infrared interface. The digital data is error protected and fit into the physical frames, which are delivered over the air to the base station with a bit rate of 9.6kbps or 14.4kbps. The base station acts as a relay which delivers the digital data to the InterWorking Function (IWF) (located at MSC). The IWF is a pool of modem gateways which interface the IS95 digital data with common modem standards (such as V.90). The V.90 side of the IWF handshakes directly with the V.90 modems in the ISP via the PSTN to establish the physical connection. Hence, the IWF can be regarded as a layer 1 gateway between the IS95 data frames and the V.90 modem standard. After the physical layers between the mobile and the IWF and between the IWF and the ISP modem are established, the PPP layer will handshake with the PPP layer in the ISP to activate the IP address in the mobile. Since then, the IP address in the mobile will be visible to the external internet and the applications (such as TCP) can work in the regular way as in the fixed line situation.

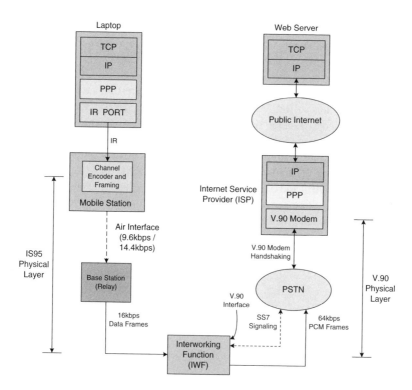

Figure 6.13 Illustration of all the network entities involved in the circuit switched data path for a mobile to fixed line data call.

6.5 POWER CONTROL

In this section, we shall focus on the power control in IS95 systems. In GSM or FDMA systems, power control is used primarily to save battery power of mobiles as well as reducing the interference between co-channel cells in the systems. The speed of the power control is therefore relatively slow because the channels for intracell users are orthogonal. However, in IS95 systems, the requirement for power control[131, 315] is more than saving battery power or reducing intercell interference between co-channel cells but rather to control the interference between intracell users as well. As we have explained in Chapter 4, the capacity of random CDMA systems are limited by the interference floor and hence, power control is very crucial to the normal functioning of CDMA systems. When the transmit power of the mobile station is too low, the voice quality will be degraded due to high FER. On the other hand, when the transmit power of the mobile is too high, there will be excessive interference to the other users (because the channels in R-CDMA are not orthogonal). Hence, the mobile station should be controlled to transmit only the minimum required power as any extra power will contribute to the rise of interference floor and consequently, reducing the system capacity. In IS95, there is sophisticated mechanism for power control in the reverse link (i.e., controlling the transmit power of mobiles)[368]. For instance, there are open loop power control and closed loop power control. Under the closed loop power control framework, there are fast inner-loop power control and outer-loop power control. The relationship of the power control algorithms in IS95 is illustrated in Figure 6.14

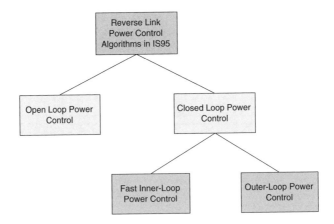

Figure 6.14   Relationship of power control algorithms in IS95 systems.

The open loop reverse link power control refers to the automatic adjustment of mobile station transmit power without any *feedback* from the base station. On the other hand, the closed loop power control mechanism refers to the adjustment of mobile station transmit power by explicit feedback of the base station. The open loop power control and the closed loop power control algorithms are designed to achieve an ideal operating condition of *equal received power at the base station from all the mobile stations*[86]. Obviously, this ideal operating condition can never be achieved perfectly. The errors in power control will be translated into *residual fluctuations* in the base station received power. The better the power control algorithms are, the smaller the dynamic range of the residual fluctuations of the base station received power are as illustrated in Figure 6.15 These power control mechanisms are elaborated in detail in the following subsections.

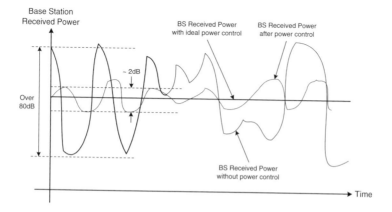

Figure 6.15 Performance of power control algorithms in CDMA systems.

6.5.1 Open Loop Power Control

The received power at the base station from the mobile stations fluctuates with time based on three factors, namely the path loss, shadowing and the microscopic fading. These three effects together account for over 80dB fluctuation of received signal strength from the mobiles as illustrated in Figure 6.15 In other words, the transmit power of the mobile station has to be adjusted over a dynamic range of more than 80dB in a relatively short time. This poses a great challenge to the closed loop power control design. Hence, in IS95 systems, there is an autonomous open-loop power control algorithm at the mobile station (without any explicit feedback from the base station) designed to reduce the dynamic range of the base station received power fluctuations from mobiles. Hence, after the open loop power control, the residual dynamic range requirement of the closed loop power control is substantially reduced. Note that the objective of the open loop power control is not on the accurate control of mobile transmit power but simply to reduce the dynamic range of the base station received power fluctuations so that the closed loop power control algorithms will have an easier job to fine tune the transmit power of the mobiles to the right target level.

We shall discuss how the open loop power control can reduce the dynamic range of the base station received power fluctuation without explicit feedbacks from the base station. The 80dB dynamic range is contributed by the path loss, shadowing and the microscopic fading. While the microscopic fading (due to multipath) may not be symmetrical between the uplink and the downlink, the first two factors (path loss and shadowing) are basically symmetrical between the uplink and downlink directions. In other words, when the path loss or shadowing from the base station to the mobile is large, it is very likely that the corresponding path loss or shadowing from the mobile station to the base station is large as well. Hence, the mobile station can have a first order estimate of the path loss and shadowing by measuring the received signal power from the base station. For instance, if the received power from the base station is large, it is likely that the mobile is close to the base station (path loss and shadowing are small) and the mobile station should transmit with a smaller power. On the other hand, if the received power from the base station is small, the mobile is likely to be far away from the base station and the mobile station should transmit at a larger power. Note that this logic works only if the transmit power from the base station is constant. In IS95, the pilot channel (Walsh channel 0) is always transmitted at constant power and could serve as a good reference for open loop power control. However, for implementation simplicity, the IS95 utilizes the total received signal strength (RSSI) from the base station for open loop power control. The transmit power of the mobile station is given by:

$$Pwr_{ms-tx}(dBm) = -76 - \overline{RSSI}(dBm)$$

The required response time for open loop power control is from 100ms to 300ms. Figure 6.16 illustrates the variation of mobile transmit power due to open loop power control.

Obviously, since the microscopic fading component is not symmetrical between the uplink and downlink, the open loop power control can never be perfect in adjusting the transmit power of the mobile. The fine tuning of the mobile station transmit power is the task for the closed loop power control.

6.5.2   Inner Loop Power Control

The inner loop power control algorithm in IS95 is the core of the power control algorithm. Basically, the base station measures the instantaneous received SIR and generates a 1-bit power control command (PCB) once per 1.25ms to control the mobile station transmit power. Hence, there are 800 PCBs per second transmitted from the base station to the mobile station and the speed of the inner loop power control is 800Hz. Figure 6.17 illustrates the block diagram of the inner loop power control in IS95 systems. There are two inputs, namely the *measured SIR*[86] and the *target SIR*, and one output, namely the *power control command*, in the inner loop power control

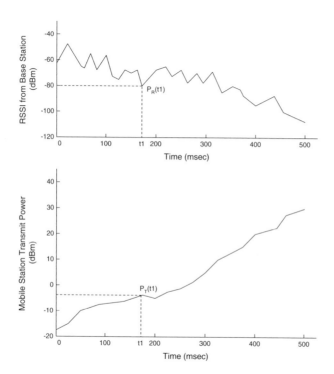

Figure 6.16   Example of open loop power control in IS95.

algorithm[359]. The base station measures the instantaneous received SIR from the uplink frames and compares with the target SIR. If the instantaneous SIR exceeds the target SIR, the mobile station is probably transmitting at excessive power and hence, a negative power control command (bit = 0) will be generated and the mobile station is asked to decrease the transmit power by a power control step size Δ. On the other hand, if the instantaneous SIR is less than the target SIR, a positive power control command (bit = 1) will be generated and the mobile station is asked to increase the transmit power by a power control step size Δ. Note that the 1-bit power control commands are sent to the mobile station in the downlink frame. Since a frame lasts for 20ms, there are 16 power control commands embedded in each downlink frame. To allow fast decoding of the power control commands in the mobiles, the power control commands are not protected by error correction codes. Furthermore, there is a dedicated power control loop per dedicated traffic channel.

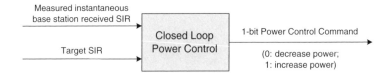

Figure 6.17   Block diagram of closed-loop power control in IS95.

Figure 6.18 illustrates an example of mobile transmit power adjustment as a result of inner loop power control in IS95 systems. During the first 20ms, the instantaneous base station received SIR exceeds the target SIR and hence, the mobile station transmit power is reduced by the power control step size Δ. When the instantaneous base station received SIR come close to the target SIR, the received SIR oscillates up and down about the target SIR and this is the *steady state* of the inner loop power control [189]. Note that due to the 1-bit power control action, the received SIR in steady state will not be a constant. This is a well-known characteristic of the *bang-bang control loop*. The target SIR may be adjusted to a new value once every 20ms frame. The power control commands are generated accordingly by comparing the instantaneous received SIR with the new target SIR. Note that choice of the power control step size Δ involves a tradeoff between the *steady state error* and the *response time*. For large Δ, the response time of the inner loop will be faster (i.e., the instantaneous received SIR will be brought to the steady state about the target SIR faster). However, since the received SIR fluctuates up and down about the target SIR in the steady state, large Δ will result in large steady state error of the received SIR, which is undesirable. In IS95, Δ is chosen to be 0.25dB–0.5dB. Furthermore, for a fixed Δ, a longer response time will be needed to bring the received SIR to the target level if the initial SIR offset from the target is large. This helps to justify the needs of open loop power control to reduce the dynamic range of the SIR offset.

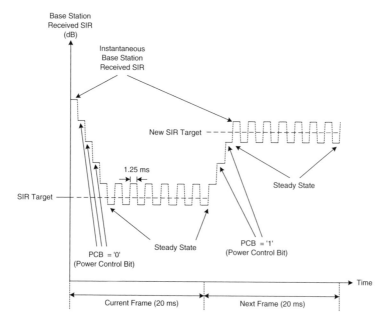

Figure 6.18 Example of mobile transmit power adjustment of inner-loop power control in IS95.

Note that after the inner loop power control, the base station received SIR will not be exactly at the target SIR value but will fluctuate with some residual variance as illustrated in Figure 6.15 The residual variance is contributed by the power control step size Δ as well as the delay in the power control commands and the errors in the power control commands. In particular, the residual variance of the received SIR at the base station increases with the mobile speed because the inner loop power control fails to track the microscopic fast fading at high mobility. To ensure small delay in the generation of power control commands, the inner loop power control algorithm resides at the base station.

6.5.3   Outer Loop Power Control

The inner loop power control algorithm is responsible to bring the received SIR at the base station to the target SIR level. However, the received SIR is not something that a user concerns. In fact, users only concern with the voice quality which is related directly to the frame error rate (FER). Hence, from the user's QoS perspective, the FER is the ultimate target rather than the SIR.

The translation from SIR to FER is usually obtained from the FER versus SIR curve of the physical layer. Given a target FER (e.g., 10^{-1}), we can read the target SIR from the physical layer curve. However, the mapping from the target FER to the target SIR is not always straight-forward and is time varying as well. For instance, different mobile speed will result in different coherence time and this will affect the FER curves of the physical layer. Hence, for the same target FER, the required target SIR is not a constant and changes with the mobility and the channel environment as the mobile moves around. In IS95, the outer-loop power control is designed to ensure an automatic adjustment of the target SIR in order to achieve the overall target FER. Figure 6.19 illustrates the conceptual block diagram of the outer-loop power control in IS95. The inputs and the output of the outer-loop algorithm are *target FER*, *instantaneous measured FER* and the *target SIR* respectively. The target SIR is updated once per 20ms. Basically, the target SIR is increased if the current FER exceeds the target FER and vice versa. However, the difficulty here lies in the ability to measure the FER with small delay. For example, to measure a FER of 10^{-2}, we need at least several hundred frames. Hence, brute force measurement of FER will incur unacceptable delay and will be useless. As a matter of fact, different vendors will have different implementations of the outer loop power control algorithm to get around this problem[185]. We shall elaborate one simple algorithm below.

Figure 6.20 illustrates an example of outerloop power control algorithm in IS95 systems. At the base station, the SIR target setpoint is reduced by a small amount ($down_{frr}$) for each consecutive frame until uplink frame error is detected. Note that frame error can be detected by CRC check in each frame. When uplink frame error is detected, the target SIR is increased by a large amount (up_{frr}). Within each 20ms where the target SIR setpoint remains constant, the inner loop power control will bring the actual instantaneous received SIR to the target SIR setpoint based on the

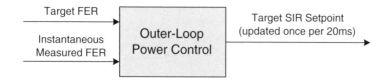

Figure 6.19 Conceptual block diagram of outer-loop power control in IS95.

1-bit power control commands. During steady state of the outer loop power control, there will be 2 uplink frame errors per cycle of $down_{frr}/up_{frr}$ (the factor of 2 is to account for the 1 frame latency in the outer loop). Hence, we have

$$FER_{\text{target}} = 2\frac{up_{frr}}{down_{frr}}$$

Since the response of the outerloop is just 50Hz and to incorporate the potential macrodiversity of soft-handover into the power control (will be elaborated in the next section), the outer loop power control algorithm resides in the base station controller.

6.5.4 Example

Figure 6.21 illustrates a summary of the reverse link power control in IS95, integrating the open loop and closed loop power control actions into a single example. The mobile station is initially in the clear with the base station and hence, receiving large RSSI from the base station. The corresponding mobile transmit power is small due to the action of open loop power control. At the same time, the inner loop power control reaches the steady state where the base station received SIR fluctuates up and down about the target SIR. When the mobile moves into a shadow, the received RSSI from the base station drops suddenly and this brings an immediate increase in the mobile transmit power. At the same time, the base station instantaneous received SIR drops below the target and the inner loop power control generates consecutive "+" commands to increase the mobile station transmit power. Furthermore, due to the frame error (when the mobile enters into the shadow), the target SIR setpoint is increased by up_{frr} due to the action of outer loop power control. As the mobile transmit power gradually increases due to the actions of inner loop power control, the base station instantaneous received SIR reaches the new target SIR level shortly and maintains at the steady state. Later on, when the mobile station leaves the shadow and enter into the clear again, the base station instantaneous received SIR suddenly exceeds the target SIR and the mobile station transmit power is decreased gradually through the actions of inner loop power control as well as the open loop power control. Shortly, the instantaneous base station received SIR reaches the steady state of the new target SIR again.

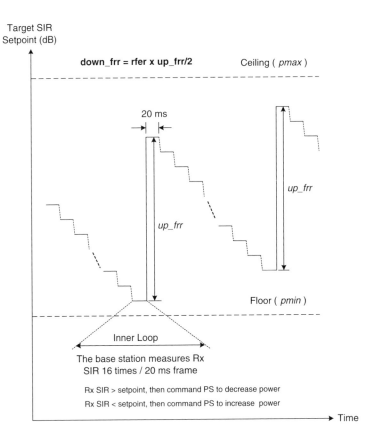

Figure 6.20 Example of outer-loop power control algorithm in IS95.

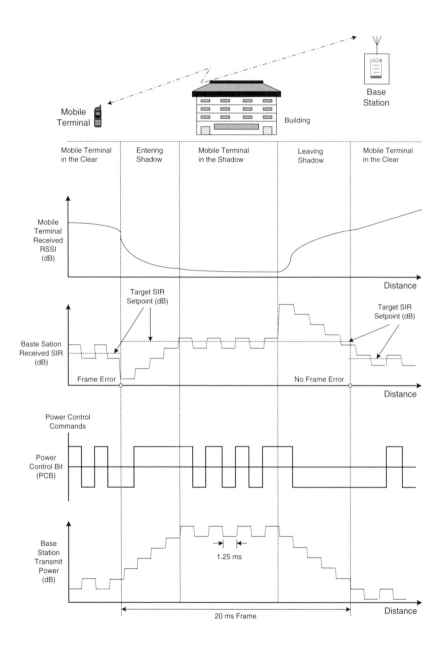

Figure 6.21   Integrated example of open loop and closed loop power control algorithms in IS95.

6.6   SOFT HANDOVER

In this section, we shall discuss the soft handover procedure in IS95. Handover is a generic process of transferring calls when a mobile moves out of the coverage of one cell and enters the coverage of another cell in the cellular system. In a handover process, the system must determines (i) the need for handover, (ii) the new cell that should continue to serve the call. To facilitate such knowledge, the mobile station as well as the cellular infrastructure have to coordinate the exchange of information. In IS95 systems, the mobile station assists the infrastructure by providing measurement information about the surrounding cells to the infrastructure. This refers to *mobile assisted handover*. In fact, mobile station plays a very important role to provide an accurate and timely picture of handover needs as well as handover direction to the cellular infrastructure.

After the infrastructure acquires the handover needs and handover direction information, the infrastructure have to coordinate with the new cells and mobile station to *execute* the handover decision. Note that while mobile station offers timely measurement data to the infrastructure, the handover decision is done at the infrastructure (base station controller). There are in general two common ways of handover execution, namely the *hard handover* and the *soft handover*. In hard handover, the existing connection (between the mobile and the home cell) is released before the new connection (between the mobile and the new cell) is established. In other words, we have a *break-before-make* situation. Examples of hard handover includes GSM and DAMPS systems. On the other hand, in soft-handover[110], the mobile station may establish the new connection (with the new cell) together with the existing connection (with the home cell) simultaneously during the transition. In other words, we have a *make-before-break* situation and therefore, a more reliable handover is possible with soft handover. Existing systems that can benefit from soft handover include IS95, CDMA2000 as well as UMTS [340]. Note that soft handover in CDMA systems refers to the dedicated traffic channel when the mobile is in conversational state. On the other hand, when the mobile is in idle state, no soft-handover will be supported in IS95[3]. In fact, the design of CDMA systems is well-suited for soft-handover and the detailed processing will be elaborated in the following subsections.

6.6.1   Pilot Strength Measurement and Soft Handover Combining

Before we elaborate on the soft handover algorithms in IS95 systems, we shall zoom into the physical layer and discuss how mobile station measures the signal strength of the environment. We shall also elaborate on why soft handover is well-suited for CDMA systems but not for other systems (such as GSM).

---

[3]Only hard handover will be supported when mobile is in idle state and this refers to *idle handover*. In idle handover, the decision is made entirely at the mobile station without any coordination from the infrastructure.

In IS95 systems, the mobile station has to measure the signal strength of the home cell as well as surrounding cells frequently so that the infrastructure will not miss any opportunities or needs for handover. However, how does the IS95 mobile physically measures the signal strength? Recall that every base station in IS95 system has a common downlink pilot channel (transmitted at constant power) spread by a master PN sequence at an offset PNO. Hence, the pilot strength measurements (measured at the mobile) from the home cell as well as from the surround cells are good indications for handover decision. Hence, it is important to understand how the mobile measures the pilot strength. When the mobile station powers up, it has to synchronize the local PN sequence with the pilot channel of the home cell. The delay position of the peaks in the autocorrelation measurement of the PN sequence at the mobile indicates the delay positions with synchronized PN timing. On the other hand, the magnitude of the peaks indicates the strength of the pilot and is proportional to the pilot strength. At the time of power up, the mobile station has no prior information about which is the home cell and hence, the mobile has to exhaustively search over all the delay offsets to synchronize with the strongest pilot. However, when the mobile is in conversation state, the pilot strength measurement is time critical as timely measurement is crucial to the appropriate execution of handover. To facilitate fast pilot strength measurement of the home cell as well as the surrounding cells, the base station periodically broadcasts a *neighbor list message* in paging channel. In the neighbor list message, there is a list of the PNO of the surrounding cells (with respect to the current home cell). Mobile station, upon decoding the neighbor list message, will have prior information about the target PNO of the pilot channel it is going to measure. For example, if the PNO of a neighbor cell B is 128 chips, the mobile station can shorten the time to measure the pilot strength of cell B by moving the local PN phase at the mobile to a delay position of $128 \pm \Delta$ where Δ is a measurement window to account for the delay uncertainty of the received pilot signal from base station B due to propagation delay and delay spread. Hence, instead of exhaustive search over all the possible PNO space, the mobile only needs to search over a small window Δ about the target PNO (e.g., PNO=128 chips) to measure the pilot strength from base station B. Note that the pilot strength is given by the sum of the peaks of the autocorrelation measurement at the mobile about the target PNO as illustrated in Figure 6.22 Typical delay in pilot strength measurements is of the order of 100ms in IS95 systems.

The next question is why soft handover is well-suited for CDMA systems but not say GSM systems. To understand this, we shall first discuss the general requirement of soft handover on the physical layer processing. During soft-handover, the mobile has to detect information signals from more than one base stations simultaneously. This poses a first requirement on the RF processing at the mobile. For example, in GSM systems, the mobile has to equip with two parallel RF units in order to support simultaneous detection of signals from two cells because the signals from two neighbor cells will be transmitted at different carriers (with different frequencies). On the other hand, for CDMA systems, only one RF unit is needed at the mobile to detect signals from more than one base stations simultaneously because the signals from neighbor

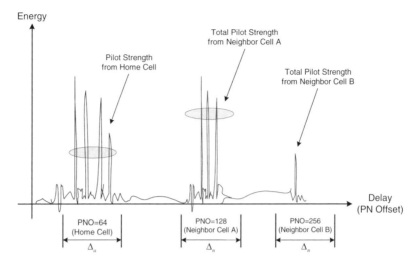

Energy

Pilot Strength
from Home Cell

Total Pilot Strength
from Neighbor Cell A

Total Pilot Strength
from Neighbor Cell B

Delay
(PN Offset)

PNO=64
(Home Cell)

Δ_a

PNO=128
(Neighbor Cell A)

Δ_n

PNO=256
(Neighbor Cell B)

Δ_n

Figure 6.22 An example of pilot strength measurement for a neighbor cell at the mobile.

cells are transmitted using the same carrier frequency (due to the interference suppression capability of CDMA). Hence, CDMA systems post a distinctive advantage
in terms of low cost consideration. A second requirement to support soft-handover
is the baseband processing involved to combine multiple signals (containing same
information) from different cells. For instance, consider a 2-way soft-handover scenario where a mobile is detecting information from cell A and cell B simultaneously.
The transmitted signals from cell A and cell B contains the same information and
hence, they serve as "independent observations" about the same information at the
mobile. However, in order to be able to detect the information as well as exploiting
the *macro-diversity* as a result of the two independent observations, the baseband has
to align and combine the two observations properly, which is not trivial. For GSM
systems, two baseband units will be required to achieve advantage of soft-handover,
which is costly. On the other hand, for CDMA systems, the RAKE receiver is well
suited for separating and combining these independent observations from cell A and
cell B. This is because there is virtually no difference in the processing complexity
for the RAKE receiver to track multipath signals from cell A only versus the RAKE
receiver to track signals from cell A and cell B simultaneously. For example, if there
are four RAKE fingers in the mobile, 5 significant multipaths from cell A and 3 significant multipaths from cell B, two of the RAKE fingers can be assigned to detect 2
strongest multipaths from cell A and 2 strongest multipaths from cell B during soft
handover. Hence, due to the complete frequency reuse and the RAKE receiver architecture, there is virtually no increase in mobile complexity to support soft-handover
in CDMA systems.

6.6.2 Mobile Soft Handover Processing

In IS95 systems, the soft handover procedures require actions from both the mobile station and infrastructure. For instance, mobile station has to measure the pilot strengths of the home cells as well as the neighbor cells frequently. When mobile spots any strong pilots from the neighbor, it has to report to the infrastructure at once in order not to miss any handover needs. The infrastructure will make a decision about whether to engage in soft handover or not based on the reported pilot strength as well as the target cell loading. Hence, whether the call can be maintained as the mobile station moves depends on how fast the mobile can measure the pilot strength from the neighbors. As we have explained in the above section, certain processing power (searching over the uncertainty window Δ) is needed for mobile to measure the strength of one pilot in the IS95 systems. Hence, the mobile station has to *prioritize* the measurement of pilots. This is elaborated below.

In mobile station, the pilots in the system are divided into the following four sets, namely the *active set*, the *candidate set*, the *neighbor set* and the *remaining set*. They are elaborated as follows.

Active Set: The active set contains all pilots (PNO) of the base stations that are directly engaged in carrying the payload traffic to and from the mobile. In other words, this is the set of all pilot offsets (PNO) corresponding to the active base stations with respect to the mobile. Note that when a mobile station is in conversational state, there must be at least one member in the active set. When there are more than one members in the active set, the mobile station is said to be engaged in *soft handover*. In IS95, the maximum entry of the active set is 6, meaning that there are at most 6 soft handover legs in any call.

Candidate Set: The candidate set contains pilots (PNO) that are not yet in active set but have sufficient signal strength to carry the mobile payload connection (successful detection of the payload frames). These are good candidates to be *promoted* into active sets and they are awaiting the decision from the infrastructure to be added into active set. In IS95, the maximum entry of the candidate set is 5.

Neighbor Set: The neighbor set contains pilots (PNO) that are not in the active set and candidate set but they are pilots from base stations that have potentials to be promoted to candidate set. In IS95, the maximum entry is 20.

Remaining Set: The remaining set contains some lower priority pilots (pilots that are not included in active set, neighbor set or candidate set).

The mobile station shall scan the pilot strength according to the priorities defined in the four pilot sets. Figure 6.23 illustrates a flow chart for the pilot measurement in the mobile station side. The mobile station first measures the signal strength of all pilots in the active set. Next, the mobile station measures the pilot strength in the neighbor set (trying to find strong pilots). If there is any pilot with strength greater than t_{add}, the pilot is regarded as *strong pilot* and is promoted to the candidate set by the mobile. At the same time, a signaling message called the *Pilot Strength Measurement Message* (PSMM) will be sent to the base station (via blank-and-burst or dim-and-burst over the dedicated traffic channel established). On the other hand, if none of the pilots in the neighbor set exceeds t_{add}, the mobile station then measures the pilot strength of the remaining set to look for strong pilots (pilot strength exceeds t_{add}). Similarly, if any strong pilot(s) is found, the pilot(s) will be promoted to candidate set and the corresponding PSMM message will be sent to the base station. This corresponds to *pilot addition event*. Note that one single PSMM message can contain pilot strength measurement of multiple strong candidate pilots to reduce the signalling overhead.

Next, the pilot strength of all the candidate pilots are compared with the active pilot strength. If any of them exceeds the weakest active pilot by t_{comp}, PSMM message will be sent to the base station so that the system may replace (or swap) the weakest active pilot with a stronger candidate. This refers to *pilot swapping event*. Finally, the mobile station compares the signal strength of all active pilots against a parameter t_{drop}. If there is any active pilot(s) with signal strength below t_{drop}, a timer is started at the mobile station. If the condition (the signal strength of that active pilot below t_{drop}) lasts for longer than t_{tdrop}, the mobile station sends the PSMM message indicating the weak active pilot. This corresponds to the *pilot drop event*.

Different mobile vendors may have different ways of implementing the pilot measurement and prioritization for pilots within a set. Note that a mobile station can decide on its own to move the pilots between the candidate set, neighbor set and remaining set. However, any movement of pilots into or outside the active set requires the approval of the infrastructure and is not decided by the mobile station alone.

6.6.3 Infrastructure Soft Handover Processing

The soft handover processing in the infrastructure resides at the base station controller. Hence, the base station just acts as a relay to deliver the handover signalling message to and from the base station controller and the mobile. At the base station controller, the handoff evaluation starts when the PSMM message from a mobile station is received. The base station controller forms a list of candidate pilots (from the PSMM message) that do not appear in the active list and rank the pilots by signal strength. The base station controller then checks the available resource for soft handover. For example, if the PSMM message contains a candidate pilot B and the resource at base station B is available to support the soft handover request, the base station controller will first inform the base station B to set up a full duplex (uplink and downlink) radio channel as well as a channel in the fixed line transmission between the base station controller and base station B. When the setups in base station B is completed, the base station controller will deliver a *Handover Direction Message* (HDM) to the mobile station via the home cell (e.g., base station A). In the HDM, the pilot (PNO) corresponds to the new active set configuration is delivered. The mobile station upon

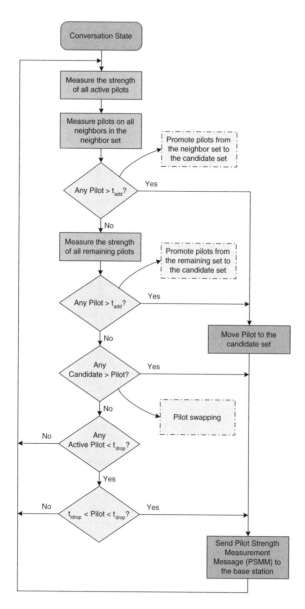

Figure 6.23   A flow chart of pilot measurement processing in the mobile station of IS95 systems.

receiving the HDM should reply with a *Handoff Completion Message* (HCM), which acts as an acknowledgement to complete the soft handoff transaction. Hence, one soft handover transaction starts with PSMM message, followed by HDM and HCM. There are three possible soft handoff transactions, namely the *pilot add transaction*, *pilot drop transaction* and the *pilot swap transaction*. Note that in the above example, if there is no resource available in base station B to support the soft handover, the base station will look at the second candidate in this list and so on. If none of the candidate base stations has the resource to support the soft handover, the PSMM message sent by the mobile will be ignored by the base station controller and the active set configuration of the mobile remains unchanged. Furthermore, note that the concept of active set, candidate set, neighbor set and remaining set applies to mobile station side only. At the base station controller, there are only the active list as well as the candidate list.

Figure 6.24 illustrates the data path of the uplink and downlink for mobile involved in soft handover. In the downlink path, the speech frames are duplicated at the base station controller and distributed to all the base station(s) involved in soft handover (all the base stations in the active set). The identical speech frames from the active base stations are delivered to the mobile over the radio interface. At the mobile, the signals from different active base stations are received using RAKE fingers as illustrated in Figure 6.22 and combined by one of the three diversity combining (selection diversity or equal gain combining or maximal ratio combining). In the uplink path, the speech frames delivered from the mobile are received respectively by the active base stations. The speech frame is detected at each of the base station independently. The detected frames from the active base stations are delivered to the base station controller where one out of N frames will be selected (where N is the number of active base stations). Hence, there is selection diversity only at the uplink.

6.6.4   Example

We shall complete the soft handoff description with an example. Figure 6.25 illustrates a hypothetical trace of pilot strength measurement by a mobile station. In this example, we have two base stations (A and B) in the system. At $t = 0$, the mobile station is in conversation state with pilot A in the active set (i.e., the home cell of the mobile at $t = 0$ is A.). The trace is the pilot strength of pilot from base station B. Hence, at $t = 0$, the mobile's active set has pilot A as member and the mobile's neighbor set has pilot B as the member. The trace in the figure denotes the pilot strength of B as measured by the mobile station. At $t = (1)$, the pilot B exceeds t_{add} and hence, a PSMM message is sent to the base station controller. At the same time, the pilot B is promoted from the neighbor set to the candidate set. At $t = (2)$, the base station controller confirms the addition of pilot B into active set by sending the HDM (with pilot A and pilot B). The mobile station acknowledges the HDM by sending HCM at $t = (3)$, completing the soft handover-pilot addition event. At the same time, pilot B is promoted into the active set (i.e., active set of the mobile consists of pilot A and

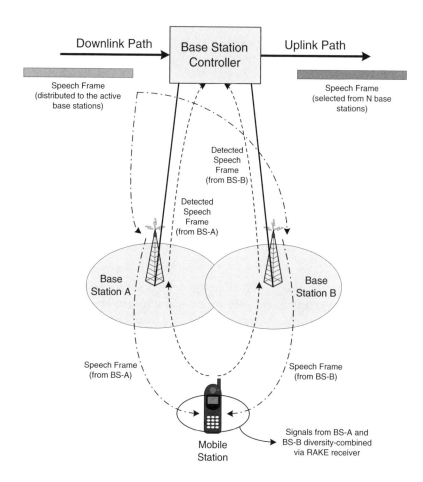

Figure 6.24 Uplink and downlink data path between mobile and base stations engaged in soft handover in IS95 systems.

pilot B). Since then, there are more than one members in the mobile's active set and the mobile is said to be engaged in *two-way soft handover* (from $t = (3)$ to $t = (7)$).

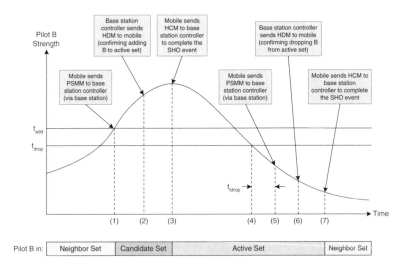

Figure 6.25 A flow chart of pilot measurement processing in the mobile station of IS95 systems.

At $t = (4)$, the pilot strength of B drops below t_{drop} and the mobile starts a timer. At this time, no message is sent to the base station controller and pilot B still remains in the active set. At $t = (5)$, the timer exceeds t_{tdrop} and hence, a PSMM message is sent to the base station controller, suggesting to drop off pilot B in the active set. Similarly, the base station controller sends the HDM at $t = (6)$ (with pilot A only) and the mobile station acknowledges with HCM at $t = (7)$. At the same time, pilot B is demoted from the active set to the neighbor set at the mobile and the active set contains pilot A only. After that, the mobile is not in soft handover anymore because there is only one member in the active set. This corresponds to a pilot drop event. Observe that the system is designed in such a way that pilot dropping is more careful compared with pilot addition. For example, to drop a pilot from an active set, the mobile has to make sure the signal strength is below t_{drop} for longer than t_{tdrop}. This is because the penalty of falsely dropping a pilot can be large (e.g., call drop).

Before we conclude this section, we shall elaborate the system consideration when soft handover and power control are considered together. In fact, power control or soft handover on its own perspective is relatively straight-forward. However, complications occur when we look at these two algorithms together. For instance, consider a 2-way soft handover example where a mobile station is engaged in 2-way soft handover with base station A and B respectively. There will be two closed-loop power control loops running between the base station A, base station B and the mobile. For instance, base station A will send power control commands to control the transmit power of the mobile based on the measured uplink SIR at A. Similarly, base station B will send power control commands to control the transmit power of the mobile based on the measured uplink SIR at B. Since the two base stations do not coordinate when generating the power control commands, there is a chance when base station A sends a power control command (0) and base station B sends a power control command (1) to the mobile. In IS95, the mobile will reduce the transmit power as long as there is a power control command of 0 from any of the active base stations. This is because after the frame selection at the base station controller, only one good frame is needed. Hence, the mobile station only needs to maintain the transmit power in such a way that at least one base station can detect the uplink frame.

6.7 SUMMARY

In this chapter, we have elaborated the system operation of IS95 systems. For instance, we have elaborated the multi-access configuration in the uplink and downlink of IS95 systems. In the uplink, mobile signals are separated by R-CDMA using a unique long code. In the downlink, intracell users are separated by D-CDMA using orthogonal Walsh codes. On the other hand, intercell users are separated by R-CDMA using short PN sequence (which is unique per base station). We have looked at the call processing procedure to establish a circuit-switched dedicated connection. There are four call processing states in a mobile station, namely the *initialization state, the idle state, the access state* and the *conversation state*. In the initialization state, the mobile station has to acquire the pilot channel by exhaustive search over the PNO space. In the idle state, the mobile station has to monitor the broadcast messages in various downlink common channels (such as SYNC channel and PAGING channel). When mobile originates a call request or when mobile receives a dedicated paging message or when mobile performs location updating, the mobile enters the access state and using the common uplink access channel, transmit the appropriate signaling messages to the base station based on the random access procedure. In the conversation state, dedicated traffic channels in the uplink and downlink are established to carry the payload.

Next, we have looked at the power control operation in IS95 systems. There is an open loop procedure and closed loop procedure to control the transmit power of mobile users. In the open loop procedure, mobile station adjusts the transmit power by measuring the RSSI from the base station downlink signals. The objective of the open loop power control is not to accurately adjust the mobile transmit power but to reduce the dynamic range of error power so that the closed loop procedure will have an easier job to bring the mobile transmit power to the target level. In the closed-loop power control procedure, there are inner loop power control and outer-loop power control. The inner loop procedure resides at the base station and transmit power control commands once per 1.25ms to control the mobile transmit power up or down by Δ dB so that the measured uplink SIR is similar to the target SIR. On the other hand, the outer-loop procedure adjusts the SIR target on a frame-by-frame basis based on a target FER.

Finally, we have elaborated on the soft handoff processing at both the mobile station and base stations in IS95 systems. Soft handoff refers to the handoff arrangement where the mobile establishes a new connection with the new base station before releasing the old connection (make-before-break). Soft handover is a unique feature in CDMA systems such as IS95, CDMA2000 and UMTS due to the low-cost mobile receiver to receive and combine signals from multiple base stations. We have elaborated the physical procedure to measure the pilot strength of the home cell and neighbor cells using CDMA structure. At the mobile station, pilots are partitioned into active set, candidate set, neighbor set and remaining set. Members of the active set are the base stations that are directly engaged in carrying payload of the user. When there are more than one members in the active set, the mobile station is engaged in soft handover. Pilot strength from the various sets are measured frequently at the mobile. If there are any strong pilots on the neighbor set or remaining set, the mobile station reports to the base station controller using PSMM . The base station controller replies with a handoff direction message (HDM) and the mobile station completes the transaction using HCM. There are three possible handoff transactions, namely the *pilot addition event*, the *pilot dropping event* and the *pilot swapping event*.

PROBLEMS

6.1   The outer loop power control is done at the base station of IS95 systems [True/False].

6.2   In IS95 system, the mobile station will combine messages (diversity combining) from the paging channels of the neighboring base stations to enhance the reliability of receiving the paging message [True/False].

CHAPTER 7

GSM

It takes all sorts to make a world.

—English Proverb, traced to the XVII century

7.1 INTRODUCTION

Global System for Mobile Communications (GSM) is a digital wireless network standard designed by standization committees of the European telecommunication operators and vendors [298, 239]. The objective is to design a high capacity cellular system to operate in the 900MHz cellular band with service portability, quality of service and security, high spectral efficiency as well as low cost design. The GSM system offers a common set of compatible circuit-switched services to all mobile users across European and the rest of the world. More than 70% of the mobile population in the world are GSM-based system. Hence, GSM represents a very important and successful deployment of digital cellular system and is one typical example of 2G cellular system. Around year 2000, the GSM system was evolved to GPRS system to

Wireless Internet and Mobile Computing. By Yu-Kwong Ricky Kwok and Vincent K. N. Lau
Copyright © 2007 John Wiley & Sons, Inc.

offer high speed packet switched data service. This refers to the so-called 2.5G system where packet-switched connection is the key new element. After GPRS, we have the EDGE system which further enhances the radio interface of the GPRS system to offer higher bit rate data services. Eventually, the GSM system will be evolved into UMTS system (today's 3G networks) to offer high quality wireless packet data access with quality of services. In this chapter, we shall provide an overview of the GSM and GPRS system. We first introduce the overall GSM/GPRS system architecture. Next, we shall look at the radio interface design. We will look at how logical channels are partitioned based on the hybrid FDMA/TDMA multi-access scheme. We will also look at the core network design in GSM/GPRS system and discuss some GSM-based services such as mobility management and international roaming. Moreover, we shall illustrate with practical illustration of GSM-based services such as the end-to-end call setup scenarios for circuit switched and packet switched connections. Finally, we shall conclude with a brief summary.

7.2   GSM SYSTEM ARCHITECTURE

We shall first elaborate on the GSM system architecture, followed by the GPRS evolution. Figure 7.1 illustrates the system architecture of a GSM network [324]. In this architecture, a mobile station (MS) communicates with a base station system (BSS) over the radio interface. The BSS is connected to the core network via a mobile switching center (MSC) using the *A-interface*. In the BSS, a number of base stations are connected to a base station controller (BSC). In the core network subsystem, the MSC connects with each other to switch calls between mobiles as well as interworking with external PSTN network.

7.2.1   Mobile Station

Unlike the IS95 system architecture in Chapter 6, the MS consists of two important parts [259]: the *subscriber identity module* (SIM) and the *mobile equipment* (ME). The SIM can be realized as a smart card or a smaller-sized "plug-in SIM". The SIM is protected by a personal identity number (PIN) between four to eight digits in length. The PIN is initially loaded by the network operator at the subscription time. Depending on the mobile implementation, this PIN can be deactivated or changed by the user. To use the mobile phone, the user may be asked to enter the PIN.

More importantly, the SIM contains very important subscription data (such as the service profile and the keys for encryption and authentication into the GSM network). The SIM also includes a list of abbreviated dialing, short messages received as well as the contact list. Parts of the SIM information can be modified by the subscriber either by using the keypad of the mobile or by a personal computer. However, some parts of the SIM information (such as the subscription data and the keys) cannot be modified by the user. In GSM, the SIM card can be updated over the air through

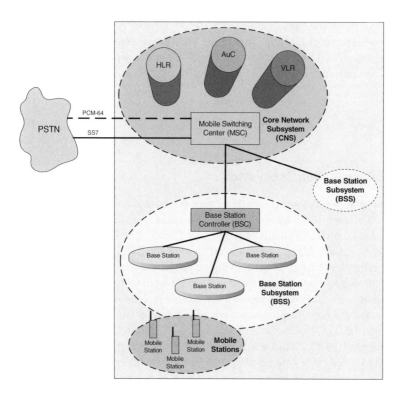

Figure 7.1   System architecture of GSM system.

the SIM toolkit as well. The network operator may remotely upgrade a mobile by sending codes through the SMS. These SMS messages are issued from a SimCard server and are received by mobiles equipped with SIM-toolkit capability.

There are two types of mobile station defined in GSM system, namely the *type 1* and *type 2* mobiles. Type 1 mobiles are not required to transmit and receive at the same time. Almost all GSM phones are of type 1 for simpler design. This means that although they operate in FDD mode with frequency offset of 45MHz between the uplink and downlink in the 900MHz cellular band or 95MHz in the 1900MHz PCS band, they are not operating in full duplex in strict sense. This is because they do not transmit or receive at the same time. On the other hand, type 2 mobiles are required to transmit and receive at the same time. Hence, they operate in full duplex mode and they usually require a duplexer in the RF circuit to separate the mobile transmit signal from jamming the mobile receive path.

7.2.2   Base Station Subsystem

The base station subsystem consists of multiple base stations connected with a base station controller (BSC) by fixed line transmission systems such as T1, E1 or ATM. The base station (BTS) interfaces the mobile station with the base station controller via the radio interface. On the radio interface side, the BTS contains radio transceivers as well as baseband digital channel elements specific to the radio interface of the GSM specification. On the fixed line interface, the BTS contains modems for T1/E1/ATM connections with the BSC. The fixed line interface is also called the *A-bis interface*. In fact, most of the signalling messages are terminated into the BSC or the core network rather than the BTS. Hence, the BTS simply acts as a physical relay for signals between the mobile and the BSC. An important part of the BTS is the transcoder and rate adapter unit (TRAU) that carries out GSM-specific speech encoding (downlink) and decoding (uplink) and rate adaptation in data transmission. However, in some implementations, the TRAU may be located in the MSC so that one voice channel occupies only 16kbps of the T1 time slot (instead of 64kbps) in the fixed line transmission link between the BTS, BSC and MSC.

On the other hand, the BSC is responsible for the switching functions and signaling protocol terminations (related to the radio resources) and is in turn connected to the MSC in the core network subsystem. The BSC supports radio channel allocations, handover management as well as other radio resource related functions (such as call admission). In GSM systems, multiple BTS can be connected to one BSC but one BSC can be connected to a BTS only. If there is only one BTS and one BSC in the base station subsystem, the BTS and the BSC may be integrated without the A-bis interface.

7.2.3   Core Network Subsystem

The core network subsystem [217] supports the call switching functions, subscriber profiles and mobility management (such as location update and paging). The basic switching function resides in MSC, which is a circuit-switched switching fabric similar to the PSTN switch. However, the software protocols in the MSC supports all GSM-based signalling protocols such as the GSM-MAP and mobility management protocols (MM). Multiple MSCs can be interconnected together in a hierarchical manner using fixed line transmission systems. There is a gateway MSC which interfaces with external public PSTN via PCM-64 in the speech payload path and SS7 in the signaling path. The MSC also communicates with other network entities in the core network subsystem such as the Visiting Location Register (VLR), Home Location Register (HLR) and the Authentication Center (AuC) using the GSM-MAP protocol (based on SS7). The VLR and HLR are very important elements (forming a distributed database) to keep track of the location of the mobile stations as well as supporting international roaming. When an MS moves from the home system to a visited system, its location is registered at the VLR of the visited system. The VLR informs the HLR of the mobile's home system of its current location. The AuC is used in the security data management for the authentication of the subscribers. The AuC may also be colocated with the HLR.

When a call is established between the mobile and the PSTN fixed line telephone, the entities involved in carrying the speech payload are the mobile, BTS, BSC, MSC and the gateway MSC. The entities involved in the signaling are the VLR, HLR, AuC (if outgoing call), MSC and gateway MSC. In GSM systems, since only the circuit-switched connection (for voice or data) is supported, the core network is designed based on circuit-switched architecture as well similar to the PSTN network.

7.3   GPRS SYSTEM ARCHITECTURE

In early 2000, there is strong demand for wireless data applications due to the bloom in the fixed line internet. While the existing GSM system supports wireless data connection, the bit rate and efficiency are very low. While 3G networks were still remote at that time, General Packet Radio Service (GPRS) serves as an interim solution to offer high quality, high data rate and low cost wireless data access based on the GSM network. This refers to the 2.5G solution that enables existing operators to *upgrade* the GSM network to offer wireless packet data access over the existing cellular and PCS spectrum. GPRS is designed based on an *overlay approach* that reuses the existing GSM architecture. For instance, in GPRS systems, both the traditional circuit-switched connection and the new packet-switched data connection are supported [325]. However, as explained in Chapter 5, the challenge in the infrastructure design lies in the interfacing between two different types of public networks (PSTN and internet). For instance, the PSTN is a circuit-switched network while the public internet is a packet-switched network. In GPRS system, the circuit-switched connection is provided through the core network subsystem (circuit-switched oriented). On the other hand, the packet-switched connection is provided through a new layer of network entity in GPRS (packet-switched oriented). This is illustrated in Figure 7.2

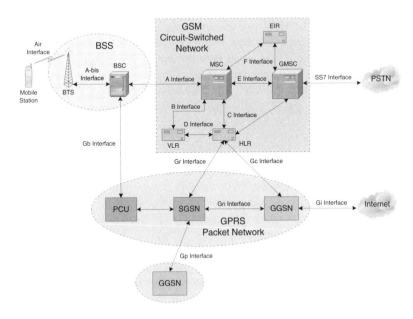

Figure 7.2   System architecture of GPRS system.

In GPRS system, the mobile, BSS, MSC, VLR and the HLR in the existing GSM system are modified. For example, the HLR is enhanced with GPRS subscriber information. In addition, a new packet switched layer is introduced for the core network in the GPRS system. The new packet-switched layer consists of two new entities, namely the serving GPRS support node (SGSN) and the gateway GPRS support node (GGSN). The SGSN and GGSN are IP routers which interconnects with each other via an IP-based GPRS backbone. The GGSN provides interworking with external public IP networks. The GGSN and SGSN are regular IP routers (in fixed line internet) except that they support mobility management protocols specific to the GPRS networks.

The BSS and the SGSN are connected via the Gb interface using ATM or frame relay. Within the same GPRS network, SGSNs and GGSNs are connected through the Gn interface. When SGSN and GGSN are in different GPRS networks, they are interconnected via the Gp interface. The GGSN connects to the public internet via the Gi interface. The HLR connects to the GGSN and the SGSN via the Gc and Gr interfaces respectively. Both the Gc and Gr interfaces follow the GSM-MAP protocol. The mobile, BTS, BSC, SGSN and GGSN are involved directly to carry the packet-switched data payload. The SGSN, GGSN, VLR and HLR are involved in the signaling for packet-switched connections.

7.3.1 GPRS Mobile

A GPRS mobile consists of a mobile terminal (MT) and terminal equipment (TE). An MT communicates with the BSS over the air based on GSM/GPRS radio interface specification. The MT is equipped with software for GPRS functionality in order to handshake and establish signaling and payload connections with the core network (SGSN/GGSN). We shall illustrate the protocol architecture of the GPRS system in later sections. A TE can be a computer attached to the MT. On the other hand, modern GPRS devices such as palm pilot or powerPC are equipped with GPRS modem. In that case, the MT and the TE are integrated into a single physical unit. Existing GSM mobiles do not support GPRS. For example, GPRS utilizes time slot bundling to increase the downlink bit rate. In GSM, mobile can only utilize a single time slot to deliver speech payload. Furthermore, there are four new encoding modes in GPRS which are not supported in regular GSM mobiles. Figure 7.3 illustrates the growth of GPRS enabled handsets over the past few years. As of 2002, 96 operators in 45 countries have GPRS systems in operation. GPRS enabled handset grows from a scant of 10 million in 2001 to over 280 million in 2005.

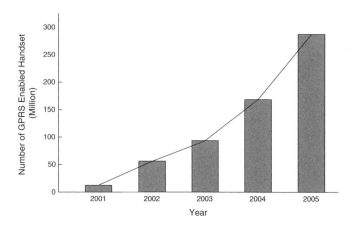

Figure 7.3   Growth of GPRS enabled handset.

Three operation modes were defined in GPRS, namely the *class A*, *class B* and *class C* mobiles. The class A mode of operation allows simultaneous circuit-switched and packet-switched services using different time slots within the GSM TDMA frame structure. Hence, class A mobile allows user to be in a GSM call or a GPRS data session and then monitor the applicable paging channel of the other service and react to it appropriately. Class B mobile supports GSM or GPRS sessions but not simultaneously. Once a user is in a GSM call or a GPRS data session, it is not required to monitor the paging channel for the other service. Hence, it is not required to respond to a page message if the mobile is paged for incoming calls from the network. Class C mobile supports only GPRS or GSM data services. These types of mobiles are typically modems. No provision is provided for the mobile to react to pages from either mode. By default, GSM circuit switched data mobiles are Mobile Station Class C mobiles.

The mobile stations that access GPRS services may or may not contain GPRS-aware subscriber identity modules (SIMs). An MS maintains mobility management (MM) and Packet Data Protocol (PDP) contexts to support GPRS mobility management. Some of the MM context fields stored in a GPRS aware SIM include international mobile subscriber identity (IMSI), which uniquely identifies the MS. In addition, the SIM stores the packet temporary mobile subscriber identity (P-TMSI), which is the GPRS equivalent of the TMSI in GSM. The SIM stores the routing area identity, which is a subset of the location area in GSM. If the SIM is not GPRS aware, these fields are stored in the mobile equipment. For data routing purposes, the MS maintains a PDP context which includes the PDP type (X.25, IP, PPP), PDP address, PDP state (ACTIVE/INACTIVE) as well as the QoS profile. We shall elaborate the handshaking and signaling involved in packet data session in later section.

Figure 7.4 illustrates the protocol stack in a GPRS mobile (packet data service only). We shall elaborate the details of the protocol layers and as well as the handshaking and signaling involved in packet data session in later section.

7.3.2 GPRS Base Station Subsystem

To accommodate GPRS, the base transceiver station (BTS) and the base station controller (BSC) in the BSS are modified. A new component, namely the packet control unit (PCU) is introduced. The BTS is modified to support new GPRS channel coding schemes. The BSC forwards circuit-switched calls to the MSC, and packet-switched data (through the PCU) to the SGSN. A BSC can connect to only one SGSN. A BSC can connect to only one SGSN. The Gb interface is implemented to accommodate functions such as paging and mobility management for GPRS. The BSS should also manage GPRS-related radio resources such as allocation of packet data traffic channels in cells.

The PCU can be viewed as an equivalent of the transcoder and rate adaptor unit (TRAU) for the packet switched service. The PCU is either colocated with the BTS or remotely located in the BSC or the SGSN. Most vendors follow the remote PCU option so that no hardware modifications to the BTS/BSC are required. In the remote option, the existing A-bis interface between the BTS and the BSC is reused and the GPRS data/signaling messages are transferred in modified TRAU frames with a fixed length of 320 bits (20ms). The PCU is responsible for medium access control (MAC) and radio link control layer functions, such as packet segmentation and reassembly,

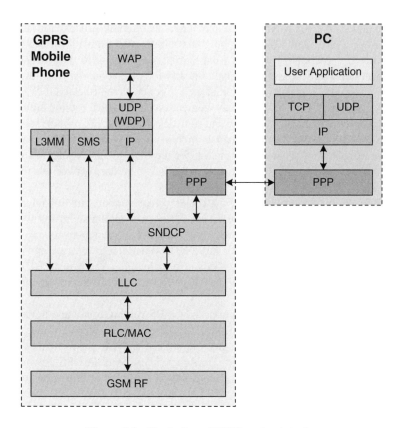

Figure 7.4   Illustration of GPRS protocol stack.

packet scheduling and ARQ as well as radio resource management (such as power control, congestion control and broadcast control information).

7.3.3 GPRS Support Nodes, HLR and VLR

There are two kinds of GPRS support nodes, namely the serving GPRS support node (SGSN) and the gateway GPRS support node (GGSN). The functionality of the SGSN and GGSN may be combined into a single physical node or distributed in separate nodes. A GSN is typically implemented using multi-processor architecture with hardware redundancy, hot swapping and robust software platform that support uninterrupted operation. The role of the SGSN is equivalent to the MSC/VLR in the current GSM network. It is involved in carry the packet data payload as well as signaling in a GPRS packet data session. SGSN connects the BSS to GGSN, which offers encryption, GPRS mobility management (such as GPRS routing area update and inter-network roaming), charging and statistics collection. To provide services to a GPRS mobile, the SGSN establishes an mobility management (MM) context that contains mobility and security information for the MS. At PDP context activation, the SGSN establishes a PDP context that is used to route data between the MS and the GGSN.

GGSN is primarily provisioned by a router, which supports traditional gateway functionality such as publishing subscriber addresses, mapping addresses as well as routing and tunneling packets. A GGSN may contain DNS functions to map routing area identifiers with serving SGSNs and DHCP functions to allocate dynamic IP address to the GPRS mobiles. The GGSN also maintains an activated PDP context for tunneling the incoming packets of the GPRS attached mobile to the corresponding SGSN.

To accommodate GPRS subscription and GPRS routing area, new fields in the GPRS MS record are introduced in the HLR. These new records are assessed by SGSN and GGSN using IMSI as the index key. In MSC/VLR, a new field called the SGSN number is added to indicate the SGSN currently serving the GPRS mobile. The MSC/VLR may contact SGSN to request location information or paging for voice calls.

7.4   RADIO INTERFACE

In GSM/GPRS system, the spectrum resource is partitioned into channels based on hybrid FDMA and TDMA. The uplink and downlink signals in GSM are separated in frequency (namely the frequency division duplexing (FDD)). For instance, the 900MHz cellular spectrum for the GSM downlink and uplink signals are 935–960 MHz and 890–915 MHz, respectively. The 1800 MHz PCS spectrum for the GSM (DCS-1800) downlink and uplink signals are 1805–1880 MHz and the 1710–1785 MHz respectively. In the cellular (900MHz) spectrum, the frequency band (both uplink and downlink) are first partitioned into 124 pairs of frequency duplex channels (carrier) with 200kHz carrier bandwidth. Each of the 200kHz carrier is further partitioned into 8 time slots based on TDMA. The time slot is the basic channel to be allocated to mobile users. Such a hybrid arrangement is to strike a balance between the number of radio transceivers needed at the base station and the equalization complexity of the mobile. This point has been explained in Chapter 3. In addition to the frequency separation of 45MHz for cellular spectrum (900MHz) and 95MHz for PCS spectrum (1800MHz), the uplink and downlink bursts of class 1 mobiles are further separated by 3 time slots. Hence, this removes the need for class 1 mobile to transmit and receive simultaneously and therfore, simplifies the design of the mobile. Figure 7.5 illustrates the timing schedule of uplink and downlink bursts in GSM for class 1 mobile. For instance, the mobile receives a downlink burst from the base station, re-tune to the uplink frequency and transmit the uplink burst 3 time slots later.

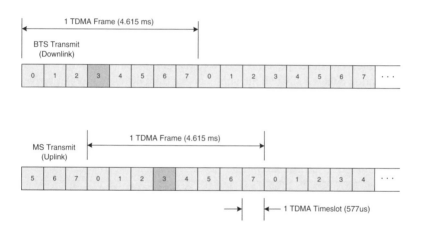

Figure 7.5   Timing schedule of uplink and downlink burst in GSM-FDD for class 1 mobiles.

Note that the propagation path loss of lower frequency band is in general smaller than that in higher frequency band. Hence, to save the mobile power, the uplink is usually assigned to the lower frequency band (890–915 MHz) while the downlink is assigned to the higher frequency band (935–960 MHz). To further save mobile battery power, GSM supports discontinuous transmission and discontinuous reception. Discontinuous transmission exploits the silence gaps between human conversation. The mobile station turns on the transmitter only when there is voice packet and turns off the transmitter during the silence gaps. In discontinuous reception, the mobile station can turn off the receiver during the time slots of the other users and the mobile only need to listen to its allocated time slot for paging. The 8 time slots per 200kHz carrier are the physical channels used to carry user traffic. However, to facilitate the call processing and controls, the time slot (physical channels) are first partitioned into a number of logical channels. We shall elaborate the important logical channels associated with GSM and GPRS systems in the following subsections.

7.4.1   GSM Radio Interface

As we have introduced, each GSM carrier (200kHz) is partitioned into 8 time slots based on TDMA and the data is transmitted in the form of *bursts* and the bursts are designed to fit within the time slots. In GSM, five different burst types [143] have been defined. The five burst types are namely the *normal burst* (NB), *frequency correction burst* (FB), *synchronization burst* (SB), *access burst* (AB) and the *dummy burst* (DB). Figure 7.6 illustrates the normal burst structure. The normal burst is the most commonly used bursts in GSM to carry user payload data. It contains 116 bits of information (user payload), 26 bit training sequence and a guard period of 8.25 bits. The training sequence is to allow the mobile station to estimate the channel fading response and to perform simple equalization (to alleviate the effect of frequency selective fading). For 200kHz carrier, the number of resolvable echos experienced by the carrier will be limited and hence, the equalization complexity is feasible in GSM mobile. The guard period is used as a margin to compensate for the variation of propagation delays so that all the uplink bursts from different mobiles arrive at the base station without *colliding* with each other. The length of the guard period will limit the maximum coverage of the GSM BTS.

Number of Bits:

Figure 7.6   Illustrate of GSM burst.

The frequency correction burst is used by the mobile to detect a special carrier which is transmitted by all the BTS in a GSM network. This carrier is called the broadcast channel (BCCH) carrier which delivers important system parameters related to the call processing. The synchronization burst (SB) is used by the mobile to synchronize to the BTS transmission timing to within a quarter-bit resolution based on the extended 64 bit training sequence. The access burst is used by the mobile to access the system (e.g., call request, page response) and it is the first uplink burst that a BTS will have to receive and demodulate from any particular mobile. Similar to the synchronization burst, the training sequence in the AB is extended to ease the demodulation process in the base station. Furthermore, the access burst is much shorter than the other burst and this results in a large guard period of 68.25 bits. This guard period is used to compensate for the propagation delay between the MS and the BTS. Such a large guard period is needed for AB because during access, there is no *timing advance mechanism* yet and hence, a large guard period is needed to cover the worst case propagation delay. For instance, 68.25 bit period is equivalent to 252us which allows the MS to be up to 38km from the BTS before the received uplink AB will spill into the next time slot. In other words, 38km is the maximum cell radius in GSM. Finally, the dummy burst is similar to the normal burst in that it has the same structure and uses the same training sequences. However, the information bits in the dummy burst are set to a predefined sequence in the DB. The DB is used to fill inactive time slots on the broadcast channel carrier, which is transmitted continuously at constant power.

The radio spectrum is partitioned into 200kHz carriers and then time slots. Hence, a combination of carrier number and time slot number identifies a *physical channel* uniquely in GSM system. However, not all physical channels can be assigned to users to carry user payload. The physical channels have to be mapped to a number of logical channels in the GSM system. There are two types of logical channels, namely the *traffic channels* (TCH) and the *control channels* (CCH). TCHs are intended to carry user information (e.g., speech packets or data packets) while CCHs are used to carry control or signaling information. The traffic and control channels are further elaborated below.

7.4.1.1 *Traffic Channels in GSM*   There are two kinds of TCHs defined in GSM, namely the *full-rate TCH* (TCH/F) and the *half rate TCH* (TCH/H). The TCH/F provides transmission speed of 13kbps for speech or 9.6kbps, 4.8kbps or 2.4kbps for data. A full rate TCH will occupy a complete physical channel (i.e., one time slot in each 4.615ms TDMA frame) on one uplink and one downlink carrier). The half rate TCH allows speech transmission at around 7kbps and data at primary user rates of 4.8 kbps and 2.4kbps. The half-rate channel uses one time slot in every other TDMA frame on average and hence, each physical channel (a time slot and a carrier) can support two half-rate TCH. The half rate channel is primarily intended to support the GSM half-rate speech coder, which produces digitalized speech at a reduced source rate of around 7kbps. TCH/F and TCH/H always use normal bursts.

Figure 7.7 illustrates the GSM frame and hyperframe structure. In GSM, one hyperframe consists of 2048 superframes. One superframe contains 51 *multi-frames* and each multi-frame contains 26 TDMA frames. The BTS and all the mobiles in the GSM system are time-synchronized from the hyper-frame level to the time slot level. In other words, a time slot in a carrier is *labeled* by a *time slot number*, a *TDMA frame number*, a *multi-frame number*, a *super-frame number* and a *hyper-frame number*. All the mobiles and the BTS have the same "notion" (or time reference) of the time slot number, TDMA frame number, Multi-frame number, super-frame number as well as the hyper-frame number. In fact, each of the 2715648 time slots in a hyperframe has a unique number and this is used in the ciphering (security) and frequency hopping algorithms.

Figure 7.8(a) illustrates the mapping of physical channels (physical time slots) into logical channel (TCH/F). The time slots in the first 12 TDMA frames (0-11) are used

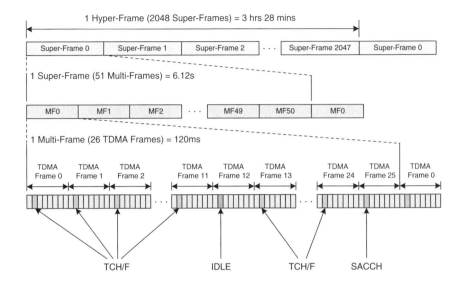

Figure 7.7 Illustration of GSM frame structure.

for TCH/F. The time slots in the 13-th TDMA frame (12) are not used for transmission and are labeled as IDLE. The next 12 TDMA frames (13-24) are used by the TCH/F. The time slots in the 26-th TDMA frame (25) are used by the slow associated control channel (SACCH), which is a control logical channel to be elaborated in the next subsection. The organization of the TCH/H is illustrated in Figure 7.8(b) with two half rate traffic channels, TCH0 and TCH1. The two half rate traffic channels share a time slot in every TDMA frame. Similarly, the time slots in TDMA frames 0-11 and TDMA frames 13 - 24 are used to carry TCH0 and TCH1. The time slots in TDMA frame 12 are used for SACCH corresponding to TCH0. The time slots in TDMA frame 25 are used for SACCH corresponding to TCH1.

7.4.1.2 *Control Channels in GSM* The control channels in GSM are intended to carry signaling information. Specifically, there are three types of CCHs defined in GSM, namely the *common control channels* (CCCH), the *dedicated control channels* (DCCH) and the *broadcast channels* (BCH). The control channels are needed to carry the signaling or control information between the mobiles and the BTS or to provide system information related to radio resource management and call processing.

The common control channels are shared by all mobiles in the BTS, which include the *paging channel* (PCH), the *access grant channel* (AGCH) and the *random access channel* (RACH). They are elaborated below.

Random Access Channel (RACH): The RACH is an uplink control channel used by the mobiles for initial access such as call origination, paging response or

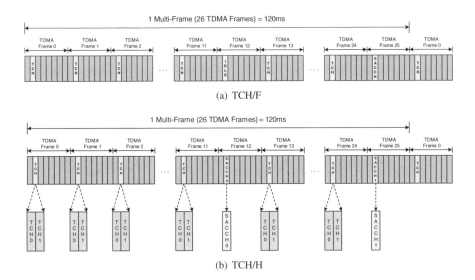

Figure 7.8 Illustration of the mapping from physical channel to logical traffic channels in GSM.

location update. Figure 7.9 illustrates the handshaking of a typical GSM call origination and call termination. The call request message is sent by the mobile via the RACH. Since all mobiles share the same RACH for initial access, there is potential collisions between access requests and hence, slotted Aloha (similar to that described in random access of IS95 in Chapter 6) is adopted in RACH to resolve potential access collision.

Access Grant Channel (AGCH): The AGCH is also a downlink control channel used by the network to indicate radio link allocation upon access requests from mobiles. For example, as illustrated in Figure 7.9(a), the BTS assigns a low-data rate SDCCH (introduced below) to carry the signaling messages and notify the mobile using the AGCH.

Paging Channel (PCH): The PCH is a downlink channel used by the network to page the target mobile in mobile-terminated calls. For example, as illustrated in Figure 7.9(b), the BTS sends a paging message to the target mobile via the PCH in call termination.

The dedicated control channels are control channels dedicated to a mobile in GSM. There are four types of dedicated control channels defined in GSM, namely the *stand-alone dedicated control channel* (SDCCH), the *slow associated dedicated control channel* (SACCH), the *fast associated dedicated control channel* (FACCH) and the

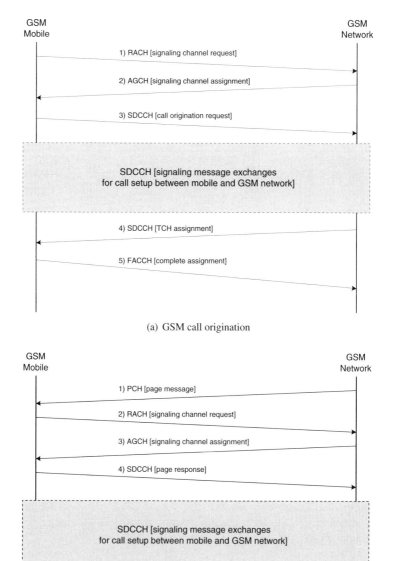

(a) GSM call origination

(b) GSM call termination

Figure 7.9 Illustration of the handshaking in GSM call origination and call termination.

cell broadcast channel (CBCH). The CBCH is used on the downlink only while the SDCCH, SACCH and FACCH are used in both downlink and uplink. They are elaborated as below.

Standalone Dedicated Control Channel (SDCCH): The SDCCH is a dedicated channel setup to carry signaling between a mobile and the BTS or for short message delivery. In fact, the SDCCH is similar to the TCH in the sense that they are both used to provide a dedicated connection between the mobile and the BTS. In some situations (such as location update and call setup), signaling information must flow between the mobile and the BTS. This may be accomplished by setting up the TCH directly. However, this may waste the radio resource because the data transfer requirement of the signaling message is far less than that of speech transmission. Hence, the SDCCH is defined to be a low-date rate dedicated control channel which has only 1/8 of the capacity of a TCH. For example, as illustrated in Figures 7.9(a) and 7.9(b), a SDCCH is first setup in both call origination and call termination scenarios to carry the call origination request message (or page response message) from the mobile to the BTS. The TCH assignment is also delivered from the BTS to the mobile via the SDCCH.

Slow Associated Dedicated Control Channel (SACCH): The SACCH (as illustrated in Figure 7.8) is associated with either a TCH or an SDCCH. The SACCH is used for non-urgent control messages such as the transmission of power control and time alignment information over the downlink, and measurement reports from the mobiles over the uplink. As a result, the power control in GSM (delivered in SACCH) is a slow closed-loop control (as compared with IS95 closed loop control) because the power control commands are delivered once per 120ms.

Fast Associated Dedicated Control Channel (FACCH): The FACCH is used for time-critical control messages such as call-establishment progress, authentication of mobiles, or handover signaling. The FACCH makes use of the TCH during a call and hence, there is a loss of user data because the FACCH steals the bandwidth of the TCH. This is similar to the blank-and-burst and dim-and-burst signaling in IS95 introduced in Chapter 6.

Cell Broadcast Channel (CBCH): Finally, the CBCH carries only the short message service cell broadcast messages, which use the same time slot as SDCCH.

The broadcast channels (BCH) are used by the BTS to broadcast system information to the mobiles in its coverage area. There are three types of BCHs, namely the *frequency correction channel* (FCCH), the *synchronization channel* (SCH) and the *broadcast control channel* (BCCH). The FCCH and the SCH carry information from the BTS to the mobile to allow the mobiles to acquire and achieve time synchronization (from time slot level to hyperframe level) with the GSM system. The BCCH provides system specific information such as the access parameters for the selected cell and the information related to the surrounding cells to support cell selection and location registration procedures in the mobiles.

7.4.2 GPRS Radio Interface

To support high speed packet data access, the TDMA structure in GSM is extended and new logical channels are introduced to support control signaling as well as packet traffic flow over the GPRS radio interface. In GPRS, mobile station can receive multiple downlink time slots as well as transmit on multiple uplink time slots. The burst structure and frame structure remain to be the same as GSM except that the multiframe has been extended to 52 frames. This is elaborated in Figure 7.10 The 52-multiframe consists of 52 consecutive assigned time slots with the same number (e.g., TS 2 as in Figure 7.10 Of the 52 time slots, 48 are used for sending the actual GPRS data (radio blocks). Of the remaining four, the two time slots marked 'X' (numbers 25 and 51) are used for neighbor cell identification, similar to the IDLE frame in the 26 multiframe in the GSM speech case. The mobile searches the SCH burst that holds the BSIC and timing information for the neighbor. During the 'X' time slot, the serving cell does not receive or transmit anything. Finally, the two time slots marked 'P' (numbers 12 and 38) form the *packet timing advance control channel* (PTCCH) channel used for timing advance regulation. We shall elaborate the new logical channels for the GPRS service in this section.

The physical channel dedicated to packet data traffic is called a *packet data channel* (PDCH). Different packet data logical channels can occur on the same PDCH. Similarly, there are two types of logical channels for packet data, namely the *packet data traffic channels* (PDTCH) and the *packet control channels* (PCCHs). The PDTCH is the logical channel setup to carry user dedicated information for packet switched sessions. The mapping of physical channel (PDCH) to PDTCH is illustrated in Figure 7.10 A user may simultaneously occupy multiple PDTCHs (multiple time slots) for high speed access. On the other hand, one PDTCH may be shared by multiple users for low data rate packet access.

Several new packet control channels are introduced to support GPRS packet data access. The packet control channels can be divided into the *packet common control channels* (PCCCHs), which are the logical control channels shared by all mobiles in the cell, and the *packet dedicated control channels* (PDCCHs), which are the logical control channels dedicated to mobile users. For PCCCHs, we have the *packet random access channel* (PRACH), the *packet paging channel* (PPCH), the *packet access grant channel* (PAGCH), the *packet notification channel* (PNCH) as well as the *packet broadcast control channel* (PBCCH). They are elaborated below.

Packet Random Access Channel (PRACH): The PRACH is the only uplink PCCH shared by all mobiles for initial access or transfer of bursty packet data. For example, Figure 7.11 illustrates the handshaking between the GPRS mo-

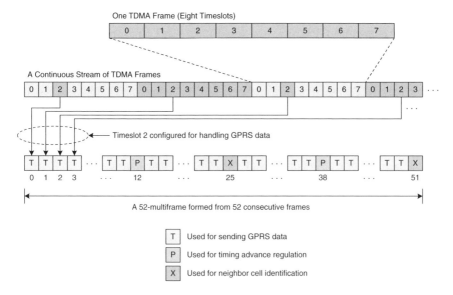

Figure 7.10   Illustration of GPRS frame structure and the mapping of physical channels to packet data traffic channels (PDTCH).

bile and the BTS for uplink packet data access. The GPRS transmits a packet channel request to the BTS via the PRACH. Alternatively, the GPRS mobile may use the RACH to send the access request.

Packet Access Grant Channel (PAGCH): This is the common logical control channel used by the BTS in the packet data transfer establishment phase to assign uplink packet channel (either PACCH [to be introduced later] or PDTCH [for delivering a small number of uplink packets]) in response of the PRACH access message sent by the mobile. For example, in Figure 7.11, the BTS may respond to a mobile packet access request using a *one-phase procedure* or a *two-phase procedure*. In the one-phase procedure, the BTS assigns an uplink packet data channel for a number of radio blocks to be transferred. One of more PDTCH may be assigned depending on the requested resource indicated in the access message in PRACH. The network informs the mobile of this assignment through the PAGCH. In the two-phase procedure, the network needs more information for resource reservation and it indicates the need for two-phase procedure in the PAGCH. Specifically, the BTS setup the uplink resource of PACCH (to be introduced later) and informs the mobile via the PAGCH. The mobile and the network exchange packet resource request and packet resource assignment signaling messages on the PACCH to complete the resource negotiation.

On the other hand, the mobile may use RACH to send the access request and in that case, the BTS can assign at most two PDTCH via the downlink AGCH for uplink packet data transfer.

Packet Paging Channel (PPCH): This is a common logical control channel used by the BTS to send paging message to mobiles for both circuit-switched and packet-switched services.

Packet Notification Channel (PNCH): This is a common logical control channel used by the BTS to send a *point-to-multipoint multicast* (PTM-M) notification to a group of mobiles prior to a PTM-M packet data transfer.

Packet Broadcast Control Channel (PBCCH): This common logical control channel is used by the BTS to broadcast system information specific for packet data. If PBCCH is not allocated in the physical channel, the packet data specific system information is broadcast on the existing GSM BCCH.

The packet dedicated control logical channels include *packet associated control channel* (PACCH) and the *packet timing advance control channel* (PTCCH). They are elaborated below.

Packet Associated Control Channel (PACCH): This is a dedicated control logical channel (dedicated to mobile) to convey mobile specific signaling messages such as power control, resource assignment and reassignment. The PACCH shares resources with PDTCHs. For example, in Figure 7.11, the packet resource request message (mobile to BTS) and the packet resource assignment message (BTS to mobile) are delivered in the PACCH. An mobile currently engaged in packet data transfer may also be paged for circuit switched services on PACCH.

Packet Timing Advance Control Channel (PTCCH): This is a full duplex dedicated control logical channel. The mobile uses the uplink PTCCH to transmit a random access burst. With this information, the BSS estimates the timing advance requirement. The timing advance information is delivered to the mobile using the downlink PTCCH.

7.5 CORE NETWORK INTERFACE AND SERVICES

The core network of the GSM/GPRS infrastructure supports the basic call switching functions as well as mobility management of mobiles. As illustrated in Figure 7.2, the GPRS/GSM core network adopts an overlay approach. The circuit-switched domain (MSC/IWF) interfaces with the PSTN via PCM-64 in the payload path and SS7 in the signaling path. The packet-switched domain (SGSN/GGSN) interfaces with the public internet via standard IP protocol [99]. The VLR/HLR distributed database

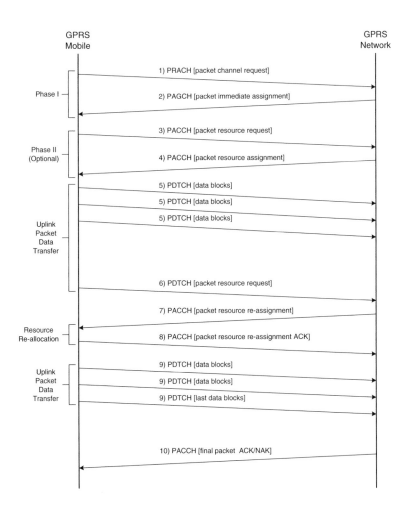

Figure 7.11 Illustration of GPRS uplink packet data access.

supports both the circuit-switched and the packet-switched core networks. We shall elaborate the circuit-switched and the packet-switched domains of the core network in the following subsections.

7.5.1   Circuit-Switched Domain—GSM Core Network

The critical components in the GSM core network includes the MSC, GMSC, IWF, VLR and HLR. The MSC and GMSC are directly involved in the call switching for circuit-switched voice connection. The IWF is involved for circuit-switched data connection. In both cases, the MSC/IWF, VLR and HLR are involved in the control signaling during call setup, call termination as well as mobility management. We shall elaborate the signaling procedures of these services with examples below.

Figure 7.12 illustrates the protocol architecture of the *control plane* of the circuit-switched domain. The BSC connects to the MSC through the A-interface. The MSC connects to the other MSC, GMSC, VLR and HLR through the GSM-MAP protocol, which is derived from the *Transaction Capabilities Application Part* (TCAP) of the SS7 framework. The GSM networks track the location of mobiles so that incoming calls can be routed to the target subscriber. To exercise the location tracking, a cluster of base stations is grouped to form a *location area* (LA). Mobile has to perform a *location update procedure* whenever it crosses a location area. The location update procedure is defined in the mobility management (MM) layer at the mobile. The corresponding peer of the MM layer is at the MSC. Above the MM layer is the *call control layer* (CC) which is responsible for the ISDN-like call setup and call termination procedures. The corresponding peer of the CC at the core network is the MSC.

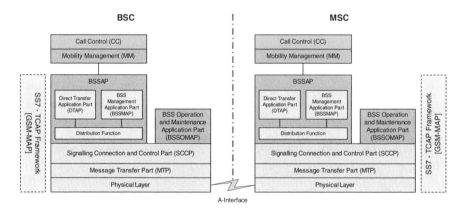

Figure 7.12   Protocol architecture of the control plane of the circuit-switched domain in GSM system.

The mobility management, call setup and call termination, as well as international roaming are all supported by the distributed database in the *visiting location register* (VLR) and *home location register* (HLR) [217]. One MSC is associated with one VLR in the core network and the entire core network is associated with one HLR. The structures of the VLR and HLR are elaborated below.

Home Location Register (HLR): The HLR maintains a permanent database record for every mobile in the GSM system. The search key used in HLR to identify a mobile record is the *IMSI*. A mobile record in the HLR includes the MSC number, the subscription profile as well as the address of the last VLR visited.

Visiting Location Register (VLR): The VLR maintains a dynamic database record for the mobiles under its location area. The search key used in VLR to identify a mobile record is the *TMSI*, which is a temporarily address generated by the VLR. The dynamic record in the VLR includes the IMSI, the address of the HLR, the address of the MSC currently serving the mobile (MSRN) and the current location area identity (LAI).

We shall elaborate the mobility management, call setup, call termination and international roaming procedures in the following subsections.

7.5.1.1 *Mobility Management: Location Update Procedure*

When a GSM network receives an incoming call, the GMSC has to decide where to route the call to and page the target mobile for the incoming call. When the network has no idea of the current location of the target mobile, all the BTS have to send paging message to look for the target mobile and hence, the paging overhead is enormous for each incoming call. This is obviously impractical. To reduce the paging overhead, the GSM core network is designed to keep track of the location of the mobile down to a *location area* (LA) resolution. A LA, identified by LAI, consists of a cluster of BTSs. Whenever the mobile (in idle state) crosses the LA boundary, the mobile has to initiate location update procedure. The size of the LA is a tradeoff of the uplink signaling overhead (location update overhead) and downlink paging overhead during idle state.

There are three cases for location update, namely the *inter-LAI movement within the same MSC*, the *inter-MSC movement* as well as the *inter-VLR movement*. Figure 7.13 illustrates the scenario of the first case. Say the mobile moves from LAI-1 to LAI-2 and both LAIs belong to the same MSC. The location update procedure is summarized into three steps below.

Step 1 (MS to MSC): The mobile station sends a *Location Update Request Message*, which is a signaling message generated at the MM layer of the mobile. The message includes the previous LAI, current LAI, previous MSC, previous VLR address and TMSI. The mobile (in steady state) identifies itself by the TMSI. However, if the VLR does not have record for the mobile (e.g., first time

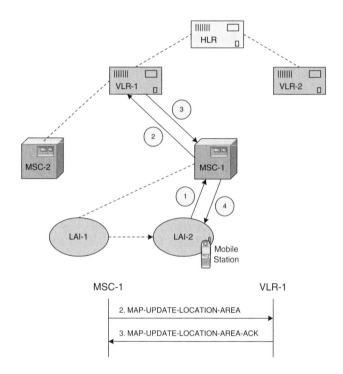

Figure 7.13   Case 1: Inter-LAI movement within the same MSC.

registration), the mobile will be asked to send the IMSI. The location update request message terminates at the MSC.

Step 2 (MSC to VLR): The MSC, upon receiving the location update message, forward the location update request to the VLR based on GSM-MAP dialogue. The location update transaction into the VLR includes the MSC address, TMSI, previous LAI and the current LAI. The VLR locates the mobile record based on the TMSI and update the LAI field.

Step 3 (Completion): Since the location update does not involve new MSC, the VLR sends an ACK to the MSC, completing the location update dialogue.

Figure 7.14 illustrates the scenario of the second case. Say the mobile moves from LAI-1 to LAI-2 which involves different MSCs. The location update procedure is summarized into four steps below.

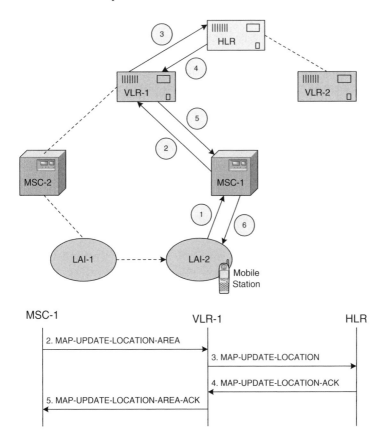

Figure 7.14 Case 2: Inter-LAI movement involving a change of MSC.

Step 1 (MS to MSC to VLR): The mobile station sends a location update message to the MSC with TMSI, previous LAI, current LAI and previous MSC address. The MSC forward the location update request to the VLR based on GSM-MAP dialogue.

Step 2 (VLR Processing): The VLR locates the mobile record based on TMSI as the key. VLR notices that LAI-1 and LAI-2 belongs to two different MSCs. Hence, the VLR updates the LAI and MSC address fields in the mobile record. In addition, it derives the HLR address from the IMSI (in VLR record) and sends a *MAP-UPDATE-LOCATION* to the HLR (containing IMSI, address of the new MSC as well as the address of the VLR).

Step 3 (HLR Processing): The HLR identifies the mobile record based on the IMSI and updates the MSC address in the record. HLR replies with an ACK to the VLR to complete the GSM-MAP dialogue.

Step 4 (Completion): The VLR, upon receiving the ACK from HLR, replies with an ACK to the MSC. This completes the location update procedure.

Finally, Figure 7.15 illustrates the scenario of the third case. Say the mobile moves from LAI-1 to LAI-2 which involves different MSCs as well as different VLRs. The location update procedure is summarized into four steps below.

Step 1 (MS to MSC to VLR): Similarly, the mobile sends a location update message to the MSC and the MSC forward the location update to the new VLR based on GSM-MAP dialogue.

Step 2 (VLR Processing): The new VLR does not have the record of the mobile based on the TMSI supplied from the location update request. Therefore, the new VLR sends a *MAP-SEND-IDENTIFICATION* message to the old VLR with the TMSI. The old VLR, based on the TMSI, retrieves the mobile record and delivers the IMSI of the mobile to the new VLR via the *MAP-SEND-IDENTIFICATION-ACK*. The new VLR creates a record for the mobile and generates a new TMSI for the record.

Step 3 (HLR Processing): The new VLR sends a registration message to the HLR (*MAP-LOCATION-UPDATE*) with IMSI, address of the new MSC as well as the address of the new VLR. The HLR locates the mobile record based on the IMSI and update the VLR address as well as the MSC address fields in the record. It completes the MAP dialogue with the MAP-LOCATION-UPDATE-ACK to the new VLR.

Step 4 (Completion): The new VLR sends the generated TMSI to the mobile via the new MSC. At the same time, the new VLR deletes the mobile record in the old VLR.

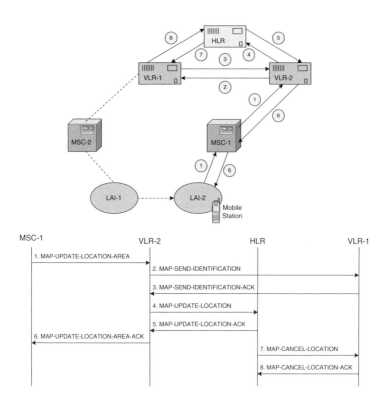

Figure 7.15   Case 3: Inter-LAI movement involving a change of MSC and VLR.

7.5.1.2 Call Control: Call Origination and Termination

Based on the location update procedures, the call setup procedures are described below. Figure 7.16 illustrates the call origination procedure at the core network level. In contrast to the call setup procedure described in earlier section, we shall focus on the end-to-end procedure involving the core network and the mobile.

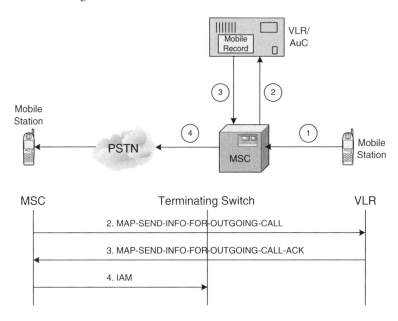

Figure 7.16 Core network procedure for the Call Origination in GSM.

Step 1 (MS to MSC): The mobile sends a *call origination message*, which is a signaling message at the *call control protocol layer*, to the MSC. Hence, the call origination message is transparent to the BTS and the BSC. In the call origination message, the mobile identifies itself using TMSI obtained during registration procedure.

Step 2 (MSC to VLR): The MSC forwards the call origination request to the VLR to locate the mobile record using *MAP-SEND-INFO-FOR-OUTGOING-CALL* GSM-MAP dialogue.

Step 3 (VLR Processing): The VLR retrieves the mobile record based on the TMSI. The VLR checks if the mobile is allowed to make outgoing calls and replies the MSC with *MAP-SEND-INFO-FOR-OUTGOING-CALL-ACK*, completing the GSM-MAP dialogue.

Step 4 (Call Setup): If the result is positive, the MSC shall setup voice trunks according to the standard PSTN procedures as well as the radio resource of the

BTS to carry the voice call. This triggers a number of radio interface hand-shaking to setup the voice channels as illustrated in Figure 7.9.

Figure 7.17 illustrates the end-to-end call termination procedure between the mobile and the core network. The detail procedure is outlined in the steps below.

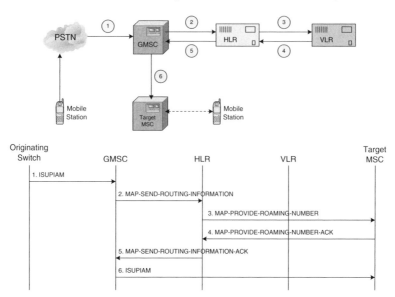

Figure 7.17   Core network procedure for the Call Termination in GSM.

Step 1: The PSTN phone initiates a fixed-line to mobile call and the call is routed to the GMSC from the PSTN via SS7 - IAM.

Step 2: The GMSC queries the HLR via *MAP-SEND-ROUTING-INFORMATION*, which contains the MSISDN (called number).

Step 3: The HLR identifies the mobile record based on the MSISDN and sends the *MAP-PROVIDE-ROAMING-NUMBER* to the target VLR to obtain the MSRN.

Step 4: The VLR creates the MSRN using the MSC address stored in the mobile record. The MSRN is sent back to the GMSC via the HLR, completing the GSM-MAP dialogue.

Step 5: The target MSC address can be derived from the MSRN at the GMSC. Hence, the voice trunk is setup from the GMSC to the target MSC using SS7 IAM. The mobile is paged from all the BTSs in the current LAI. Upon the mobile receives the page message and responds with page response message, the handshaking procedure over the radio interface defined in Figure 7.9 continues.

7.5.1.3 International Roaming   One of the important features supported in GSM network is the *international roaming*. In fact, international roaming can be regarded as a special case of call termination described in the above subsection, with an exception that the called mobile may be a visiting mobile. The international roaming is supported by the VLR/HLR distributed database architecture. Suppose a GSM mobile user (John) from Taiwan roams to Singapore. When the GSM mobile registers in the GSM network in Singapore, the VLR in Singapore network will update the HLR in Taiwan network based on the location update procedure in Case 3. Suppose a person in Taiwan calls John. The call will be processed according to the call termination procedure described in Figure 7.17 and hence, two local voice trucks (within Taiwan and within Singapore) and an international voice trunk (between Taiwan and Singapore) will be resulted. The caller is charged for a local GSM call but John is charged for an international call from Taiwan to Singapore. On the other hand, suppose a person from Hong Kong calls John. Following the call termination procedure in Figure 7.17, this will result in two international segments, from HK to Taiwan and from Taiwan to Singapore. Suppose now, a caller in Singapore calls John (who is roaming in Singapore), two international segments (Singapore to Taiwan and Taiwan to Singapore) will be resulted although both parties are in Singapore. This undesirable call setup is called the *Tromboming effect*. Figure 7.18 illustrates the call processing of the tromboming effect.

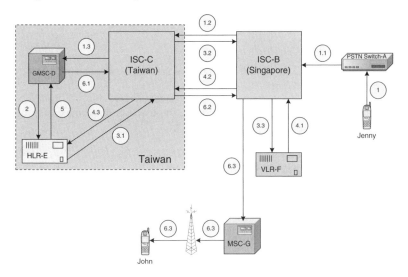

Figure 7.18   Illustration of Tromboming Effect in GSM International Roaming. A user in Singapore (Jenny) calls the roamer (John).

Step 1: Say Jenny calls John's number (Taiwan's GSM network). This is interpreted as an international call and hence, the SS7-IAM will be routed to an international switching center (ISC-B). The call will be routed to ISC-C in Taiwan, which will be routed to the GMSC-D in Taiwan's GSM network (John's home network).

Step 2: The GMSC-D in Taiwan queries the HLR-E in Taiwan for the MSRN. The HLR-E realizes that John is roaming in Singapore and queries the VLR-F in Singapore based on GSM-MAP dialogue. The MSRN is returned to GMSC-D.

Step 3: Based on the MSRN, the GMSC-D sets up the voice trunk to the MSC-G in the Singapore network, connecting to John. Hence, the voice path involves two international trunks, namely from Singapore to Taiwan (ISC-B to ISC-C) and from Taiwan to Singapore (GMSC-D to MSC-G), resulting in tromboning.

Obviously, the tromboning effect is highly undesirable. The key step is to stop the call being routed to the ISC-C as otherwise, the international segments cannot be eliminated in the voice path. We shall discuss several possible solutions below.

Roamer Location Cache (RLC): Figure 7.19 illustrates the architecture of the RLC-based solution. For instance, a RLC is colocated at the ISC-B in the visited system. The RLC acts as a HLR proxy for the visited system by intercepting the GSM-MAP message during the location update procedure. Specifically, during the location update procedure, the mobile (John) registers to the VLR in Singapore network. The VLR sends the *MAP-UPDATE-LOCATION* to the roamer's HLR (Taiwan). Since the HLR is in Taiwan, the message will be routed to ISC-B. The RLC in ISC-B duplicates the MAP message and creates a record with IMSI and VLR/MSC address for the mobile. RLC still does not have the MSISDN for the mobile. The RLC requests this information from the HLR (Taiwan) using "MAP-RESTORE-DATA". The MSISDN is returned from the HLR (Taiwan) to the RLC through the "MAP-INSERT-SUBSCRIBER-DATA".

During call processing, Jenny dials John's mobile number as before. Similarly, the call is routed to ISC-B. The ISC-B looks at the prefix of the called number and recognizes it as a potential roamer. Before routing the call to the ISC-C in Taiwan, the ISC-B searches the local RLC using the called MSISDN. If no entry is found, the call is routed to ISC-C as usual. If an entry is found for John's number, the call is routed to the MSC (Singapore). In this case, the RLC is acting as the HLR proxy. Hence, the international voice trunks are avoided in the speech path of the call. The advantage of the RLC solution is user transparency as the caller (Jenny) does not need to be aware of the special arrangement. On the other hand, the disadvantage of the RLC solution is the need to modify the ISC-B to understand GSM-MAP. Hence, agreements between the GSM operators and the international carriers are needed.

Roamer Location Cache within GSM (RLC-GSM): Figure 7.20 illustrates the architecture of the RLC solution within the GSM network. Unlike the previous solution, the RLC is now located within the GSM network so that no modification of the ISC-B is needed. Instead, the RLC is colocated at the operator-owned

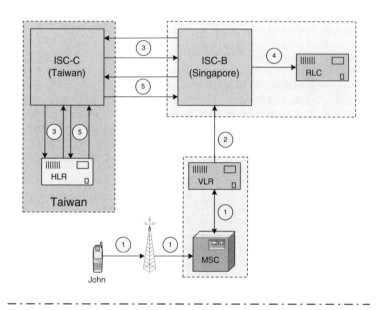

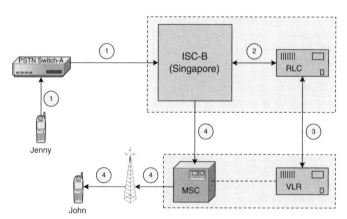

Figure 7.19   Illustration of the Roamer Location Cache (RLC) in GSM system.

switch (switch D). During the location update procedure, the VLR recognizes the location update is from a roaming user. Hence, the VLR sends the "MAP-UPDATE-LOCATION" to the RLC. The RLC creates a record for the mobile with IMSI, VLR and MSC address. The RLC relays the "MAP-UPDATE-LOCATION" to the HLR in Taiwan. After the location update procedure is completed, the RLC obtains the MSISDN similar to the RLC-based solution.

During call processing, Jenny calls a special access number (which is routed to switch D). After connecting to switch D, Jenny is asked for John's number. Switch D will search the RLC for John's record using the MSISDN. If there is no such entry, the call will be routed to ISC-B as usual. However, if an entry is found in RLC, the call is routed to the GMSC locally in Singapore network. Similar to the above solution, the RLC is acting as the HLR proxy. The advantage of this solution is that only modification within the GSM network is needed. However, extra modification at the VLR is needed and the dialing procedure for user (Jenny) is different from the regular procedure. Hence, the solution is not transparent to the general users.

Extractor: Figure 7.21 illustrates the roaming solution using *extractor*. The extractor-based solution is similar to the RLC-based solution except that no modification is needed in the GSM network. Instead, a new element called *extractor* is introduced into the GSM network. The extractor monitors (but does not modify) the GSM-MAP message between the VLR and the foreign HLR. It will capture the GSM-MAP message when there is location update dialogue between the VLR and the foreign HLR. RLC obtains the MSISDN similar to the previous solutions and acts as the HLR proxy during call processing. The call processing operation is exactly the same as the RLC within GSM solution. The advantage of the extractor-based solution is that it is transparent to the VLR and no modification of VLR is needed. On the other hand, the solution is not transparent to the end users.

International Roamer Access Code (IRAC): In fact, the key to avoid the tromboning effect is to route the call into the visited GSM network before it reaches the ISC. The GSM operators of the visited network reserves an IRAC. In order to make a call to John, Jenny dials the NDC1 + IRAC + John's number (MSISDN). Based on NDC1, the PSTN routes the call to the GMSC of the visited network. From IRAC, the GSM network recognizes that it is an international roaming call. From the MSISDN, the GMSC search for John's record in the HLR (Taiwan) to obtain the MSRN for normal call processing. The GMSC of the visited network will route the call to the local MSC and this results in local trunks to carry the voice call. The advantage of this approach is that no modification of the GSM network or ISC is needed. However, the solution is not transparent to the end users as special IRAC dialing rule is needed.

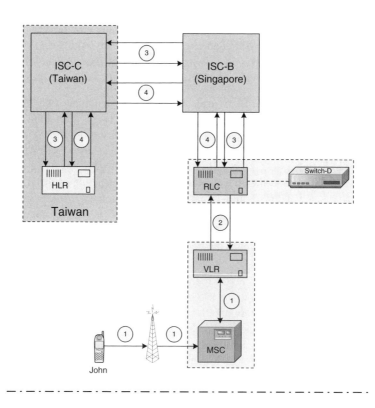

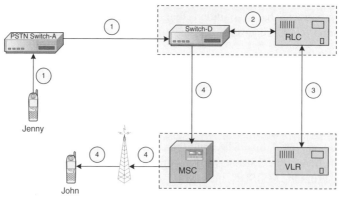

Figure 7.20 Illustration of the Roamer Location Cache (RLC) within GSM system.

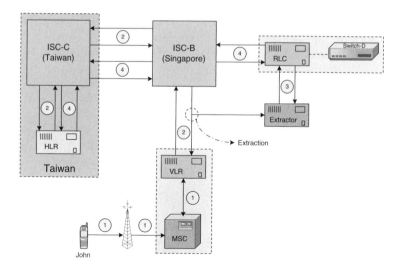

Figure 7.21   Illustration of the Extractor-based solution for International Roaming in GSM system.

7.5.2   Packet-Switched Domain—GPRS Core Network

Unlike the GSM core network, the GPRS core network (SGSN and GGSN) is packet-switched (IP) based. Figure 7.22(a) and 7.22(b) illustrate the protocol architecture of the user plane and control plane for GPRS system. The user plane illustrates the protocol stacks involved in delivering the user application across the GPRS network. The control plane illustrates the protocol stacks involved in signaling across the GPRS networks to establish packet-switched connections. Both the user plane and control plane are supported by the LLC/RLC layers. The RLC layer offers a reliable logical channel between the mobile and the BSC based on ARQ and retransmission. and BSC. The LLC layer defines a logical link control layer protocol between the mobile and the SGSN. The LLC spans from the mobile to the SGSN and is intended for use with both acknowledged and unacknowledged data transfer. Encryption is also provided in the LLC layer for secured data transmission.

On the user plane, the application data (such as TCP/IP) is delivered to the SGSN over the LLC logical link. The relay function in SGSN delivers the application data to the GGSN via the GPRS Tunneling Protocol (GTP).

On top of the LLC layer, there are two important protocol layers in the control plane. The main function of the GPRS Mobility Management (GMM) is to support mobility management functionality of a mobile such as GPRS attach, GPRS detach, security, routing area update, location update. The main function of the Session Management (GMM/SM) protocol supports mobility management functionality of a mobile such as GPRS attach, GPRS detach, security, routing area update, location update. The main function of the SM layer is to support PDP context handling of the user terminal. SM comprises of procedures for the PDP context activation, deactivation, and modification. These procedures will be elaborated later in this section.

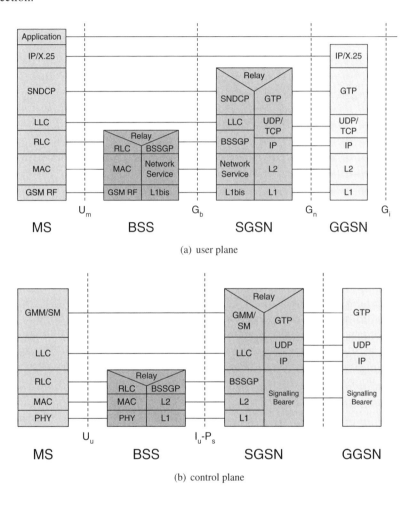

(a) user plane

(b) control plane

Figure 7.22   Protocol architecture of the user plane and control plane of the packet-switched domain in GPRS system.

Figure 7.23 illustrates the typical flow of events for a GPRS mobile. The GPRS mobile first physically attaches to the strongest cell and monitors the system broadcast messages and paging message. After that, the GPRS mobile registers to either the CS domain (IMSI attach) or PS domain (GPRS attach) to establish a mobility management logical connection. Say the GPRS mobile attaches to the PS domain. On top of the GMM connection, the GPRS mobile establishes a SM connection with the GGSN. Application data packets will be delivered via the logical channel established. During this state, the GPRS mobile may have to location area / routing area update as well as SGSN relocation. Finally, the GPRS mobile release the logical connection in the mobility management (GPRS detach or IMSI detach).

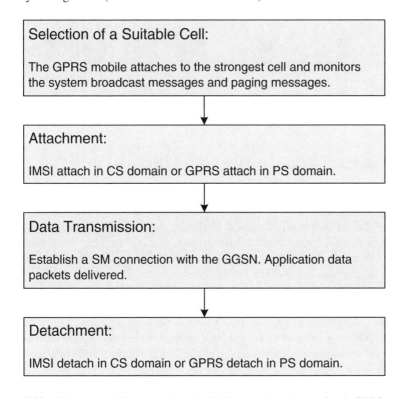

Figure 7.23 Illustration of the events involved during a packet data session in GPRS system.

7.5.2.1 GPRS Mobility Management In this section, we shall elaborate on the signaling procedures related to the GPRS mobility management functions. Specifically, we shall discuss the GPRS attach and detach procedures. Before a logical channel for the packet data flow can be established, the GPRS mobile has to establish a GMM connection with the SGSN. The signaling procedure for the GMM connection establishment refers to *GPRS attach procedure*. Similarly, the signaling procedure to release the GMM connection refers to *GPRS detach procedure*. Figure 7.24 illustrates the scope of the GMM connection in a GPRS network. In other words, the GPRS attach/detach refers to the registration/cancellation of the mobile to the SGSN in the packet-switched (PS) domain. On the other hand, the IMSI attach/detach refers to the registration/cancellation of the mobile to the MSC/VLR in the circuit-switched (CS) domain.

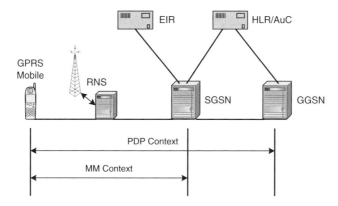

Figure 7.24   Illustration of the MM context and PDP context in GPRS systems.

When a GMM connection is established (GPRS attach), a mobility management context (MM Context) will be created for the mobile. The MM context at the SGSN includes the IMSI, TMSI, MSISDN, MM state, Routing Area Identity (RAI) and cell identity, the address of the VLR currently serving the MS, IP address of the new SGSN where buffered packets should be forwarded,.etc. The MM context is maintained throughout the entire GMM connection.

Figure 7.25 illustrates the signaling flow of a typical GPRS attach procedure. The GPRS attach is initiated by the GPRS mobile and is divided into three phases. In the first phase, a *GPRS Attach Request* message (GMM layer message) will be delivered to the SGSN from the mobile via the logical channel provided by the LLC layer. In the second phase, the SGSN authenticates the identity of the GPRS mobile user and the equipment. The data exchanged will be ciphered. During the third phase, the SGSN address will be registered in the mobile record in HLR (similar to the location update procedure in GSM network described in the previous section). Finally, the SGSN/VLR will generate a temporarily ID (P-TMSI) and notify the mobile with the *GPRS Attach Accept* message via the LLC connection.

Figure 7.26 illustrates the signaling flow of GPRS detach. The GPRS detach can be initiated by the mobile or the SGSN. The PDP context between the mobile and the GGSN will be released and the buffered user data in the SGSN will be deleted. The LLC connection will be released and the corresponding mobile record in the HLR will be purged.

7.5.2.2   GPRS Session Management   The GPRS session management (SM) establishes, modifies and releases packet data sessions with well-defined QoS between the mobile and the GGSN in GPRS system. Before the packet data session can start, the mobile and the GGSN has to establish a logical connection in the SM layer.

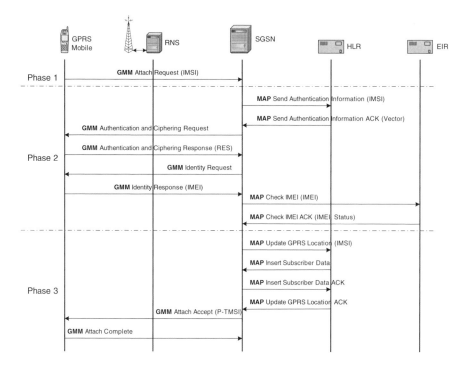

Figure 7.25 Illustration of the signaling flow in GPRS attach.

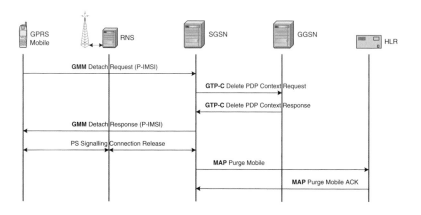

Figure 7.26 Illustration of the signaling flow in GPRS detach.

When the MM connection is established, a packet data protocol (PDP) context, which contains the important QoS attributes about the packet data session, will be created. PDP is a generic name for the user application protocol, which can be X.25 or IP. A mobile may have multiple PDP contexts simultaneously with each PDP context associated with a SM protocol entity. Figure 7.27 illustrates the information contained in the PDP context of the mobile, SGSN and GGSN.

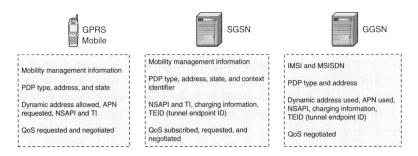

Figure 7.27   Illustration of the PDP context stored at the mobile, SGSN and GGSN.

There are three types of SM procedures defined for GPRS, namely the *PDP Context Activation*, *PDP Context Modification* and *PDP Context Deactivation*. All these procedures requires the GMM connection being established. Note that the PDP context can be activated by the mobile or the network. Figure 7.28 illustrates the signaling flow in a PDP context activation scenario (activated by the mobile). The mobile sends a *PDP Context Activation Request* to the SGSN over the GMM connection. The request contains the PDP type, PDP address (static/dynamic) as well as QoS requirement. The SGSN forwards the PDP activation request to the GGSN where a GTP tunnel is created between the PDP context in SGSN and the PDP context in GGSN. The GGSN obtains or activate the PDP address (e.g., IP address) from the external data network (e.g., ISP). If the GSGN reply (3.b) is positive, the SGSN activates the PDP context and is ready to forward / relay packets between the mobile and GGSN. In addition, the LLC connection (between SGSN and BSC) and the RLC connection (radio access bearer) between the BSC and the mobile will be established. Finally, the SGSN delivers the PDP Context Activation Accept message to the mobile, completing the PDP context activation procedure. From this point onward, the PDP address in the GPRS mobile is visible to the external network and PDP data can flow between the mobile and external network (e.g., internet). The GPRS mobile is said to be *online*. Note that due to the packet switched nature of the connection, the GPRS mobile can be online but at the same time, no dedicated radio resource is allocated at the BTS to support the logical connection.

Similarly, PDP context deactivation can be initiated by the mobile or the network. Figure 7.28 illustrates the signaling flow for mobile-initiated PDP context deactivation. The mobile sends the *Deactivate PDP context request* (with a transaction

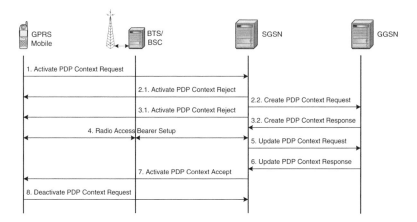

Figure 7.28 Illustration of the signaling flow in PDP context activation (activated by mobile) of GRPS network.

identifier TI to identify which PDP context to deactivate) to the SGSN via the LLC connection. The SGSN and GGSN exchange the signaling message to deactivate the PDP context. The GGSN deletes the PDP context and reclaims the dynamic PDP address into the system pool (if any). After all the concerned PDP contexts are released, the SGSN returns the accept message to the mobile. Note that if the mobile indicates "Tear Down Indicator" in the deactivation request message, all the PDP contexts of the current PDP address will be released.

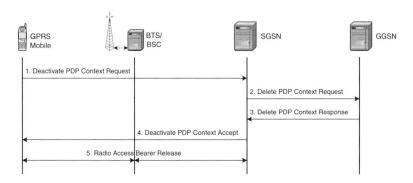

Figure 7.29 Illustration of the signaling flow in PDP context deactivation (initiated by mobile) of GRPS network.

Figure 7.30 illustrates the end-to-end wireless internet access based on GPRS network. The RLC layer between the GPRS/UMTS mobile and the BSC/RNC offers a logical channel (with ARQ protection). Above the RLC, there are user plane and control plane. The user plane consists of regular TCP/IP stack which terminates into the SGSN/GGSN. The control plane consists of RRC (UMTS), GMM and SM (GPRS/UMTS). The SM is responsible for handshaking with GGSN on the end-to-end QoS negotiation as well as IP address activation (static/dynamic IP). After the SM layer established the PDP context, the IP address is activated and visible to the external internet. The TCP layer establishes a socket with the peer TCP layer in the web server. Hence, the GPRS/UMTS network offers an IP-based wireless bearer (between the mobile and the GGSN). Applications such as web browsing and VPN can be run transparently on top of the IP-based wireless data bearer.

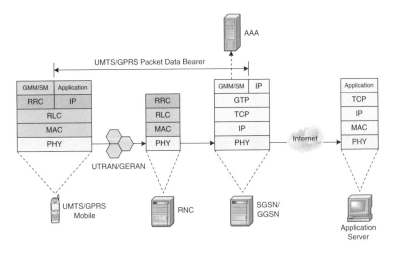

Figure 7.30   End-to-end protocol architecture of wireless internet access for GPRS/UMTS system.

7.6 SUMMARY

In this chapter, we have reviewed the system architecture of GSM and GPRS network. We started from the radio interface of the GSM and GPRS systems. Both radio interfaces are designed based on FDMA/TDMA where the spectrum is partitioned into 200kHz carrier by FDMA. Each 200kHz carrier is further partitioned into 8 time slots based on TDMA. In GSM system, each mobile can occupy one time slot only. The physical time slots are grouped into various logical channels. There are two types of logical channels defined in GSM systems, namely the *traffic channels* (TCH) and the *control channels* (CCH). TCHs are intended to carry user information (e.g., speech packets or data packets) while CCHs are used to carry control or signaling information. We have elaborated on the super-frame structure of the GSM system as well as the specific types and usage of the control logical channels and traffic logical channels. In GPRS system, each mobile is allowed to be allocated multiple downlink time slots to boost up the physical bit rate for high speed wireless data applications. Similarly, we have elaborated on the various types of new logical channels to support the GPRS operation.

Next, we have focused on the core network design. The GSM core network is designed based on circuit-switched architecture and therefore interfaces directly to the PSTN (which is also circuit-switched) via PCM-64 (payload path) and SS7 (signaling path). To support packet switched data service, the GPRS core network overlays another layer of packet-switched based core network (SGSN/GGSN). Hence, this new layer of GPRS core network interfaces naturally to the external public data network such as internet. In the GSM system, the protocol architecture for the control plane as well as the mobility management, call setup, and international roaming procedures are introduced. Examples are given to illustrate the signaling flow and interactions between the MSC, VLR and HLR. In the GPRS core network, we have elaborated the protocol architecture in both the user plane and control plane. We have elaborated the GPRS mobility management procedures (GPRS attach/detach) as well as the session management procedures (PDP context activation/modification/deactivation). A GPRS mobile can be online (or visible to the external network) only after GPRS attach and PDP context activation. Finally, an example is given to illustrate the protocol architecture of the end-to-end IP-based bearer offered by GPRS network.

PROBLEMS

7.1 To upgrade from GPRS to EDGE, the operator does not need to change any base station hardware [True/False].

7.2 Which of the following entities are involved in the transport of traffic payload information of circuit switched data connection in GSM system?

- Base Transceiver Station (BTS)

- Interworking Function (IWF)

- Visiting Location Register (VLR)

- Home Location Register (HLR)

7.3 Which of the following is involved in the packet data traffic flow of GPRS system?

- Interworking Function (IWF).

- Mobile Switching Center (MSC).

- GGSN

- SGSN

7.4   To avoid tromboning effect in international roaming, one has to avoid routing the call setup to the GMSC of the home network whenever possible [True/False].

7.5   Location update is used to balance out the paging signaling loading [True/False].

CHAPTER 8

WIDEBAND CDMA AND BEYOND

One half of the world must sweat and groan that the other half may dream.
—H. W. Longfellow: Hyperion, I, 1839

8.1 INTRODUCTION

Engineers and practitioners have always been trying hard to utilize the elegant results from wireless communication research to produce high quality systems that can meet the ever increasing demands from consumers. Third generation (3G) mobile communication systems are the most recent prominent examples. Specifically, in 3G systems, channel-adaptive technologies are used so as to further boost the bandwidth efficiency of the wireless spectra. As detailed in this chapter, high performance channel-adaptive scheduling techniques are implemented in 3G systems to enable high quality services to the heterogeneous demands of various diverse mobile applications such as video phone, multimedia messaging and traditional voice services. We shall focus on the Universal Mobile Telecommunications System (UMTS) as illustration.

Wireless Internet and Mobile Computing. By Yu-Kwong Ricky Kwok and Vincent K. N. Lau   **219**
Copyright © 2007 John Wiley & Sons, Inc.

UMTS, being a wideband CDMA (WCDMA) standard, can manifest in two different versions in pairing uplink and downlink: FDD (frequency division duplex) and TDD (time division duplex). In the FDD mode (the paired spectrum mode due to the need of two 5 MHz wide frequency bands for uplink and downlink.), each physical channel comprises of a unique code. In the TDD mode (the unpaired specturm mode due to the need of only one 5 MHz wide frequency band), each physical channel comprises of a unique code and unique time-slots in a frame. Notice that the chip rate in UMTS is 3.84 Mcps and the frame duration is 10 ms. The 10ms frame is further divided into 15 slots. Thus, we have a total of 2,560 chips per slot, leading to a symbol rate of 2,560 symbols per slot. With a spreading factor ranging from 4 to 256 for an FDD uplink and from 4 to 512 for an FDD downlink, the channel symbol rate ranges from 7,500 symbols per second to 960 Ksymbols per second. On the other hand, the spreading factor in TDD mode ranges from 1 to 16 and thus, the channel symbol rate ranges from 240 Ksymbols per second to 3.84 Msymbols per second.

UMTS can support both circuit switched connections (e.g., for conventional voice services) and packet switched connections. Indeed, packet switched connections are envisioned as the major driving force in the 3G market because such connections enable on-demand variable rate application services (e.g., multimedia messaging, video phone and location based services, etc.) and is well-suited for bursty sources.

The scheduling actions in the early UMTS standard—Rel 99—are located at the network (RNC). As will be elaborated in this chapter, the packet data scheduling in UMTS—Rel 99 is based on a *macroscopic time scale* where the resources allocated to a *data bearer* are dynamically adjusted based on the source buffer status. For example, if there are ten packet switched data users in the cell, there will be ten packet switched data bearers setup in which the radio resource assigned to these ten data bearers will be dynamically adaptive to the individual buffer status. On one hand when the buffer has plenty of packets, high data rate dedicated physical traffic channels will be setup for the data bearer. On the other hand, when the buffer is empty, the dedicated physical traffic channels associated with the data bearer will be released. As discussed in Chapter 3, this approach does not exploit the multi-user selection diversity and therefore, the bandwidth efficiency is low. Nevertheless, the performance of the packet data services in the UMTS Rel 99 is already much more superior than that in GPRS or EDGE systems due to the faster data rates in the dedicated physical channels.

To exploit the advantages of channel adaptation and multi-user selection diversity, the High-Speed Downlink Packet Access (HSDPA) scheme is incorporated in a more recent version of the UMTS standard—Rel 5. The HSDPA system adopts a *microscopic scheduling* approach on the radio resource (time-slots/codes) for its serving active users over very short duty cycles (up to 2ms). To facilitate the microscopic scheduling, the scheduling algorithm resides at the base station instead of the RNC to minimize the potential delay in the execution of scheduling. The base station obtains the instantaneous channel quality estimates from the mobiles (UEs) and selects one mobile (UE) to transmit on the high data rate shared traffic channel at the current time-slot based on a scheduling algorithm (which takes into consideration of system throughput, QoS requirements, fairness, or a combination of these). This is fundamentally different from the Rel 99 approach. For example, if there are ten packet switched data users in the cell, there will be one high data rate traffic channel shared dynamically between these ten users in the HSDPA systems. In fact, it is shown that the microscopic scheduling approach in the HDSPA systems is theoretically optimal due to the multi-user selection diversity.

In Section 8.2, we first briefly introduce the key features of the UMTS architecture. For the fine details and complete specifications, the reader is referred to the following excellent texts and research articles: [192, 24, 141, 151, 96, 343, 125, 260]. In Section 8.3, we shall discuss the design of the *macroscopic scheduling* mechanism in the UMTS Rel 99. In Section 8.4, we shall focus on the *microscopic scheduling* mechanism in the HDSPA systems.

8.2 UMTS ARCHITECTURE

A high level view of the UMTS architecture is shown in Figure 8.1 We can see that the UMTS architecture is divided into three main components: User Equipment (UE), UTRAN (UMTS Terrestial Radio Access Network), and Core Network (CN). Following the GSM architecture, UE is a mobile device consisting of a Subscriber Identification Module (SIM) and a wireless transceiver with the appropriate air interface equipment (WCDMA in this case). UTRAN is responsible for handling all radio-related functionalities. CN, adopted from the GSM architecture for easy migration of technology, is responsible for switching/routing calls/data connections to/from external networks such as ISDN, PSTN, and the Internet.

In the UTRAN, the radio resource management functions such as the overload control, admission control, code allocation, outer-loop power control, soft handover and dynamic bearer reconfigurations are primarily residing in the RNC. The Node B is responsible for the physical layer processing over the air such as channel coding, interleaving, rate adaptation and spreading. Manifested as a base station (BS), a Node B also handles inner loop power control and diversity combining for softer handover between different sectors of the Node B. In Rel 5, it is also responsible for "fast scheduling", as detailed in Section 8.4. Each RNC is designated to handle multiple Node Bs and there can be multiple RNCs in a UTRAN. The Node B and RNC are connected using fixed line transmission facilities such as ATM and the connection is defined as the Iub interface. The RNC may also be inter-connected with each other and the connection is defined as the Iur interface. Besides offering higher physical bit rate over the air, the UMTS system is differentiated from the 2.5G systems (such as GPRS) in the Quality of Service (QoS) dimension as well. For instance, UMTS offers four QoS classes over the wireless interface and the QoS differentiation is enforced at the RNC as well.

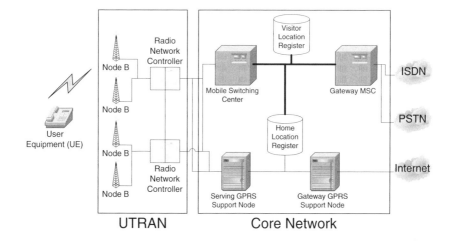

Figure 8.1   A high level view of the UMTS architecture, composed of the User Equipment (UE), Radio Network Controller (RNC), Node B, Home Location Register (HLR), Visitor Location Register (VLR), Mobile Switching Center (MSC), Gateway MSC, Serving GPRS Support Node (SGSN), and Gateway GPRS Support Node (GGSN).

In the CN, the primary role is to support the switching of circuit switched connections as well as packet switched connections. Unlike the GSM or IS-95, UMTS offers both circuit switched services and packet switched services. However, there is an implicit challenge in the design of the core network in UMTS. On one hand, the target public network of the circuit switched service is essentially the Public Switched Telephone Network (PSTN), which is based on circuit switching architecture. On the other hand, the target public network of the packet switched service is essentially the Internet, which is based on the packet switching architecture. Hence, a fundamental issue on the core network design of UMTS system is whether it should be circuit switched or packet switched. In Rel 99, an *overlay approach* is adopted. Basically, the core network has two layers, namely the circuit switched layer and the packet switched layer. In the circuit switched layer, it consists of Mobile Switching Center (MSC), which is the core element for mobility management, authentication, and switching of users' calls. The CN is connected to the PSTN via the Gateway MSC (GMSC) over the PCM-64 payload interface and SS7 signaling interface respectively. In the packet switched layer, it consists of Serving GPRS Support Node (SGSN) and Gateway GPRS Support Node (GGSN). Both the SGSN and GGSN are IP routers with additional functionality such as mobility management, authentication and data encryption for packet switched services. The CN is connected to the public internet via the GGSN which routes the IP packets between the UMTS CN and the internet. In addition to the switching elements in the CN, there are distributed database to support user mobility and supplementary services such as call forwarding. They are the Home Location Register (HLR) and the Visiting Location Register (VLR). HLR is a database for storing the master copy of a user's service profile which includes the user's service specifications and dynamic status information such as call forwarding details. VLR is a database for storing a visiting user's service profile.

UMTS is an evolving standard. The above-mentioned architecture is the basic structure as specified in UMTS standard Rel 99. More recent releases are Rel 4 and Rel 5 (as such, Rel 99 is retrospectively called Rel 3). The major addition in Rel 4 is the specification of using ATM for QoS (quality-of-service) control inside the CN. On the other hand, in Rel 5, an all IP based network is specified to be used inside the CN. In other words, a single IP-based core network is proposed in the UMTS-Rel 5 to interface with both the PSTN and public Internet. While the interface to the public internet is straight-forward, the interface of the circuit switched voice service to the PSTN is done via VoIP gateway. In addition, an IP Multimedia Sub-system (IMS) is added to interface with the GGSN in the CN. In the IMS, there are three components, namely the Media Resource Function (MRF), Call Session Control Function (CSCF), and Media Gateway Control Function (MGCF). MRF is responsible for controlling media stream resources and the multiplexing of different media streams. CSCF is the gateway interfacing the GGSN and the IMS. CSCF is also used as a firewall separating the CN from other operators' network. MCCF is for handling protocol conversions. We shall not elaborate further on the core network operation as this is beyond the scope of this chapter.

8.2.1   Radio Interface

The radio interface protocol architecture is shown in Figure 8.2 The protocol stack can be partitioned into two vertical planes, namely the control plane and the user plane. The control plane is responsible for the signaling, coordination and control functions in call setup (both circuit switched and packet switched), call release, radio resource management, session management as well as mobility management. The user plane is responsible for the user applications such as voice, video streaming as well as TCP/IP. In practice, the protocol layers above the physical layer are usually implemented in firmware and the protocol layers in both planes co-exist as concurrent threads or processes in the mobile phone.

The upper protocol layers in both the user plane and the control plane are all supported by a common suite of lower layers, namely the Radio Link Control (RLC) layer, the Medium Access Control (MAC) layer as well as the Physical layer (PHY). The RLC layer is responsible for offering *logical channels* to the upper layer applications for both planes. For example, packet retransmission, segmentation and re-assembly are the primary services offered by the RLC layer. The MAC layer is responsible for the multiplexing between the *transport channels* (offered by the physical layer) and the logical channels (offered to the RLC layer). The PHY layer is responsible for offering various *transport channels* to the MAC layer based on various physical channels. It is responsible for the modulation, coding, spreading, power control and

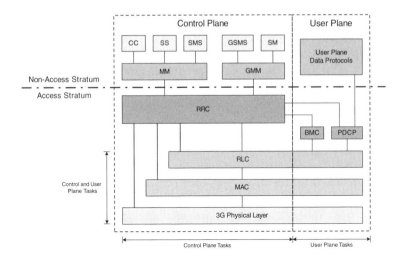

Figure 8.2   The radio interface protocol stack in UMTS.

the multiplexing of packets from various transport channels into physical channels. Note that the RLC layer and the MAC layer terminate at the RNC while the PHY layer terminates at the Node B.

In the user plane, user specific applications such as TCP/IP are interfaced to the logical channels offered by the RLC layer via the Packet Data Convergent Protocol (PDCP), which is mainly responsible for IP header compression. The PDCP layer terminates at the RNC and therefore, the IP header is re-generated at the RNC and delivered to the SGSN as a regular IP packet as illustrated in Figure 8.3 The IP layer in the user specific application terminates at the GGSN (transparent to the UTRAN) and the TCP layer is transparent to the UMTS network and the internet.

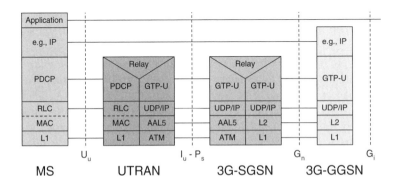

Figure 8.3   UMTS user plane protocol architecture.

In the control plane, there are various upper layer signalling protocols, namely the Radio Resource Manangement (RRC), the *Mobility Management* (MM) as well as the Session Management (SM). The RRC layer, which terminates at the RNC, is responsible for the radio dependent resource management such as outer-loop power control, soft handoff, admission control, overload control and dynamic bearer reconfiguration. The MM and SM layer terminate at the core network and are responsible for the location area / routing area update as well as QoS management of the established sessions. This is illustrated in Figure 8.4 We shall elaborate the functions of the PHY layer, MAC layer, RLC layer and RRC layer in the following sections.

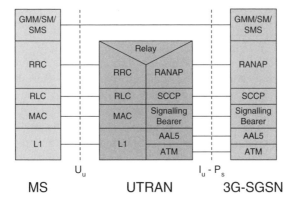

Figure 8.4   UMTS control plane protocol architecture.

8.2.2 PHY Layer

The PHY layer of UMTS is based on wideband CDMA technology. Specifically, the user separation is done based on hybrid *deterministic CDMA*[1] and *random CDMA* approaches. In the downlink, each physical channel is spread using a *channelization code* and a *scrambling code* as illustrated in Figure 8.5 The channelization code is used for the actual spreading operation (bandwidth expansion) and is targeted for the separation of intra-cell users whereas the scrambling code is used for the suppression of inter-cell interference. The downlink scrambling code is unique per cell. In the uplink, the physical channels from the mobiles are spread by both the channelization code and the scrambling code. However, the channelization code, which is used for the actual spreading operation (bandwidth expansion), is targeted to separate different physical channels from the same mobile whereas the scrambling code is used to suppress interference between different users (intracell as well as intercell).

---

[1] Deterministic CDMA refers to the case where the spreading codes between different users are completely orthogonal.

Unlike the downlink case, each mobile is assigned a unique scrambling code (by the RNC).

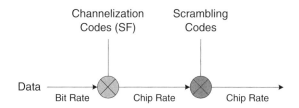

Figure 8.5 Channelization code and scrambling code in UMTS.

To support various bit rate in the physical channels, variable spreading factor is adopt in the design of channelization codes in UMTS. The spreading factor ranges from 4 to 512 and 4 to 256 in the FDD downlink and uplink systems respectively. The smaller the spreading factor of the channelization code, the higher the physical bit rate is. Unlike the IS-95 systems where the capacity bottleneck is usually in the uplink direction (interference limited), the UMTS system might be code limited in the downlink direction. This is because the orthogonal channelization codes are designed based on the *Orthogonal Variable Spreading Factor* (OVSF) codes as illustrated in Figure 8.6 On one hand, if all the users are low data rate users or voice users using channelization codes with spreading factor of 128, the downlink can support at most 128 orthogonal channels. However, if there is one high data rate user using a physical channel with spreading factor of 4 ($c_{4,1}$), all the other codes (of higher spreading factor) derived from the code $c_{4,1}$ cannot be used in order to maintain mutual orthogonality between code channels. Hence, a single high data rate user consumes $1/4$ of the voice capacity in the cell. If a UMTS cell has a few high data rate users, the number of voice channels will be quite limited due to the shortage of orthogonal codes. In other words, the system may be *code-limited*. This situation is aggregated by the way that packet switched data connections are serviced in UMTS Rel 99. For instance, if there are ten packet switched data users, there might be ten dedicated physical channels set up. One solution is to deploy the secondary scrambling code in the base station at the expense of increased mutual interference between code channels in the two scrambling code domains. In UMTS Rel 5 (HSDPA), the situation will be eased because only one high speed shared physical channel will be setup to service the various packet data users.

In UMTS systems, channels can be organized into three layers, namely the *physical channels*, the *transport channels* and the *logical channels*. A physical channel is characterized by a channelization code and a scrambling code based on WCDMA technology. A transport channel defines how and with which type of characteristics the data is transferred in the physical layer and it is the interface between the MAC layer and the PHY layer. A logical channel is a virtual connection for disseminating a

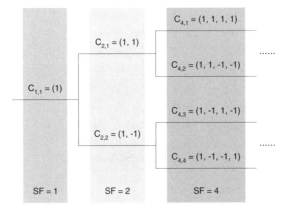

Figure 8.6 Orthogonal Variable Spreading Factor (OVSF) codes.

certain type of information (which can be control information from the control plane or data applications from the user plane) and it is the interface between the MAC layer and the RLC layer. Figure 8.7 shows the relationships among the three layers of channels in UMTS.

The UMTS logical channels are briefly described below:

- **Common Control Channel** (CCCH): a channel that is used for both the downlink and uplink for carrying control information between the UE and the RNC before any dedicated control channel is set up;

- **Dedicated Control Channel** (DCCH): a bidirectional dedicated channel carrying control information between a specific UE and the RNC;

- **Dedicated Traffic Channel** (DTCH): a bidirectional channel between a specific UE and the CN carrying traffic data (user plane);

- **Common Traffic Channel** (CTCH): a downlink channel carrying traffic data to a designated group of UEs;

- **Broadcast Control Channel** (BCCH): a common downlink channel carrying all the necessary system or cell specific information, such as the random access codes or time-slots, to all mobiles in the cell;

- **Paging Control Channel** (PCCH): a common downlink channel carrying paging or notification messages from the network to a specific UE, i.e., when the network wants to initiate communication with the terminal;

These logical channels are multiplexed or encapsulated by the transport channels in the MAC layer. Transport channels can be classified into three categories, namely

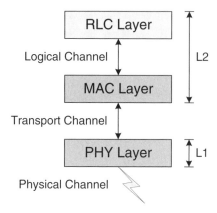

(a) three layers of channels

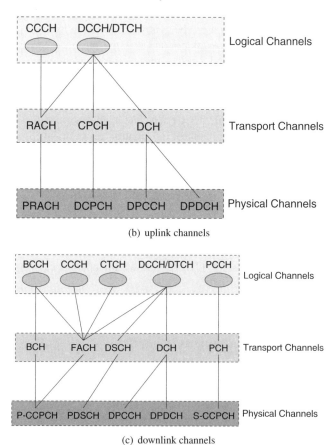

(b) uplink channels

(c) downlink channels

Figure 8.7   Three layers of channels in UMTS.

the *common transport channels*, the *dedicated transport channels* and the *shared transport channels*. Common transport channels are shared by all users in the cell and are sometimes referred to *overhead channels* because they usually do not contribute to the transmission of user data[2]. Shared transport channels are shared by a designated group of users only. Usually, there is no closed loop power control or soft handoff in both the common transport channels and the shared transport channels. Dedicated transport channels are set up for a specific user only and they support closed loop power control and soft handoff. Note that a dedicated logical channel is not neccessarily mapped to a dedicated transport channel in the MAC layer. This is further elaborated in Section 8.3. A list of transport channels are elaborated in the following.

- **Broadcast Channel** (BCH): a downlink common transport channel supporting the BCCH on the downlink;

- **Paging Channel** (PCH): a downlink common transport channel supporting the PCCH on the downlink;

- **Forward Access Channel** (FACH): a downlink common transport channel supporting several logical channels such as the CCCH, CTCH, DCCH, and DTCH. The FACH channel can be used to carry signaling information or a small amount of user data to a UE within a cell. A cell can have several FACH channels but the primary one must have low data rate in order to be received by all users. In-band signaling is used in the FACH to indicated for which user the data is intended;

- **Random Access Channel** (RACH): an uplink common transport channel supporting the CCCH and DCCH to carry control information or supporting DTCH to carry small amount of bursty data;

- **Common Packet Channel** (CPCH): an uplink common transport channel supporting the DCCH to carry control information or DTCH to carry small amount of bursty data;

- **Downlink Shared Channel** (DSCH): a downlink shared transport channel supporting DCCH or DTCH;

- **Dedicated Channel** (DCH): a dedicated downlink and uplink transport channel supporting DCCH and DTCH between the network and a specific UE;

The data carried by the transport channels are delivered by the MAC layer to the physical layer in the basic units of *transport blocks* once every *Transmission Time Interval* (TTI) (10ms, 20ms, 40ms or 80ms). The transport format of each transport block is identified by the Transport Format Indicator (TFI), which is used in the inter-layer communication between the MAC layer and the PHY layer. Several transport

---

[2]In fact, RACH and FACH can be used to carry short burst of user data.

channels can be multiplexed together into a single Code Composite Transport Channel (CCTrCh) in the physical layer as illustrated in Figure 8.8 Cyclic redundancy check (CRC) and channel encoding are performed on a per-transport channel basis. The encoded bit stream from a transport channel is zero-padded, interleaved (over the TTI) and partitioned into a number of radio frames. For example, a transport channel with TTI = 40ms is partitioned into four radio frames after channel encoding and CRC. Various transport channels, each may have different TTIs, channel coding rates as well as data size, are multiplexed together to form a single CCTrCh based on a *rate matching algorithm* on a 10ms frame-by-frame basis. The CCTrCh is further mapped into a number of physical channels and transmit over the air using CDMA.

In fact, UMTS protocol architecture allows QoS specification per transport channel because each transport channel can have different channel encoding, TTI, block size, FER target as well as rate matching attributes[3]. This is a unique feature that cannot be found in other competing standards such as 3G1X. Similar to the transport channels, there are also three types of physical channels, namely the *common channels*, the *dedicated channels* and the *shared channels*. They are elaborated below.

- **Common Pilot Channel (CPICH):** This is a downlink common channel with a constant spreading factor of 256 scrambled by the *cell-specific* primary scrambling code. This is used as a common reference for channel estimation and timing acquisition at the mobile.

- **Synchronization Channel (SCH):** This is a common physical channel used for cell searching by the mobiles. There are two sub-channels, namely the primary SCH and secondary SCH as illustrated in Figure 8.9 Both the P-SCH and S-SCH are sent only during the first 256 chips of each time-slot and they are transmitted in parallel. Please refer to [4, 5] for the details of the cell search operation.

- **Primary Common Control Physical Channel** (P-CCPCH): This is a common physical channel supporting the BCH transport channel, carrying cell broadcast information. It has a fixed data rate of 30 kbps and a fixed spreading factor of 256. Since the P-CCPCH is time-multiplexed with the SCH, the first 256 chips in every time-slot is not transmitted and hence, the effective data rate is reduced to 27kbps. There is no closed-loop power control in this channel and it needs to be demodulated by all terminals in the cell.

- **Secondary Common Control Physical Channel** (S-CCPCH): This is a common physical channel used to support the PCH and FACH (transport channels) in the downlink. A fixed spreading factor of 256 is used and variable data rates can be supported by Discontinuous Transmission (DTX) and rate matching. Similar to the P-CCPCH, there is no closed-loop power control in S-CCPCH.

---

[3]Rate matching attribute is a parameter defined per transport channel indicating the priority of bit puncturing or repetition in the CCTrCh multiplexing.

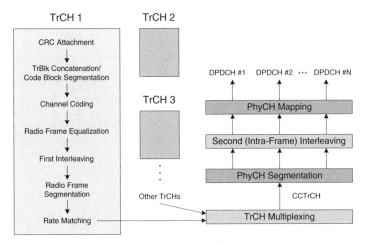

(a) uplink CCTrCh multiplexing

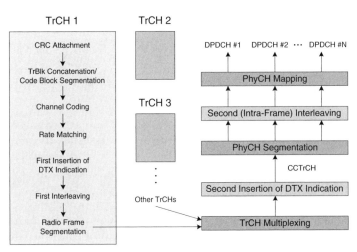

(b) downlink CCTrCh multiplexing

Figure 8.8 Multiplexing of several transport channels into a single CCTrCh in the physical layer.

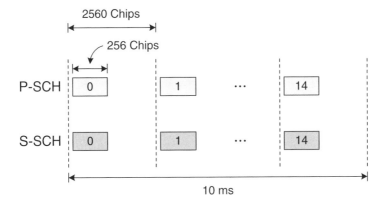

Figure 8.9   Structure of P-SCH and S-SCH for cell search.

- **Physical Random Access Channel** (PRACH): This is a common physical channel supporting the RACH transport channel. It is a contention based uplink channel which can be used to carry control information or small amount of user data from the mobile to the network. Similarly, no closed-loop power control is supported.

- **Physical Common Packet Channel** (PCPCH): This is a common physical channel supporting the CPCH transport channel for carrying bursty data traffic in the uplink. However, unlike the PRACH, channel can be reserved for several frames and fast power control is supported.

- **Physical Downlink Shared Channel** (PDSCH): This is a shared physical channel supporting the DSCH transport channel on the downlink. It is used to carry control information or bursty data and is shared by several designated users. It supports the use of fast power control but there is no soft handoff for this channel. In addition, it supports a variable spreading factor on a frame-by-frame basis so that the bit rate is adaptive to the physical channel condition. The PDSCH must always be associated with a downlink DPCH to indicate which terminal should decode the information in the PDSCH and the associated spreading code. Since the information of the PDSCH is associated with a downlink DPCH, the PDSCH frame is not started before three time-slots after the end of the associated downlink DPCH.

- **Dedicated Physical Channel** (DPCH): This is a dedicated physical channel in both uplink and downlink used to support the DCH between the RNC and a specific mobile terminal. There are two sub-channels that constitute the DPCH, namely the *Dedicated Physical Data Channel* (DPDCH) and the **Dedicated Physical Control Channel** (DPCCH). The DPDCH carries the data part of the

DCH whereas the DPCCH carries the physical layer control information such as the fast power control commands, dedicated pilots as well as the Transport Format Combination Indicator (TFCI)[4]. In the downlink, the DPDCH and the DPCCH are time multiplexed and use normal QPSK modulation as illustrated in Figure 8.10(a). The spreading factor of the DPCH does not vary on a frame-by-frame basis and the data rate is varied by the rate matching operation by bit puncturing, bit repetition or discontinuous transmission (DTX). In the uplink, the DPDCH and the DPCCH are I-Q multiplexed with different channelization codes c_D and c_C respectively as illustrated in figure 8.10(b). The DPCCH always has a fixed spreading factor of 256 whereas the DPDCH has a variable spreading factor on a frame-by-frame basis (4–256).

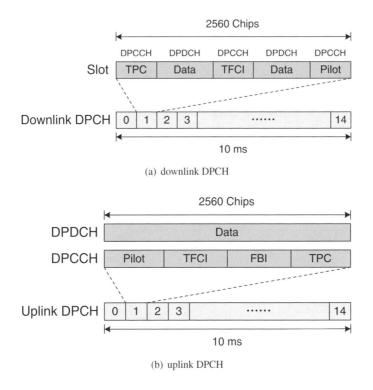

(a) downlink DPCH

(b) uplink DPCH

Figure 8.10 Multiplexing of DPDCH and DPCCH in downlink and uplink.

Table 8.1 illustrates the set of possible spreading factors and the associated bit rates of DPCH in the downlink and uplink.

---

[4]TFCI is used at the receiving side to demultiplex data from a single CCTrCh into a number of transport channels on a frame-by-frame basis.

Table 8.1 Spreading factors and bit rates in DPCH.

(a) downlink DPCH

Spreading Factor	Channel Symbol Rate (kbps)	Channel Bit Rate (kbps)	DPDCH Channel Bit Rate Range (kbps)	Maximum User Data Rate with 1/2-Rate Coding (Approx., kbps)
512	7.5	15	3–6	1–3
256	15	30	12–24	6–12
128	30	60	42–51	20–24
64	60	120	90	45
32	120	240	210	105
16	240	480	432	215
8	480	960	912	456
4	960	1920	1872	936
4, with 3 parallel codes	2880	5760	5616	2.3M

(b) uplink DPCH

DPDCH Spreading Factor	DPDCH Channel Bit Rate (kbps)	Maximum User Data Rate with 1/2-Rate Coding (Approx., kbps)
256	15	7.5
128	30	15
64	60	30
32	120	60
16	240	120
8	480	240
4	960	480
4, with 6 parallel codes	5740	2.3M

Table 8.2 summarizes the mapping of transport channels to physical channels in UMTS systems.

Table 8.2 A summary of transport channel mappings in UMTS.

Transport Channel	Physical Channel
(UL/DL) Dedicated channel DCH	Dedicated physical data channel DPDCH
	Dedicated physical control channel DPCCH
(UL) Random access channel RACH	Physical random access channel PRACH
(UL) Common packet channel CPCH	Physical common packet channel PCPCH
(DL) Broadcast channel BCH	Primary common control physical channel P-CCPCH
(DL) Forward access channel FACH (DL) Paging channel PCH	Secondary common control physical channel S-CCPCH
(DL) Downlink shared channel DSCH	Physical downlink shared channel PDSCH
Signaling physical channels	Synchronization channel SCH
	Common pilot channel CPICH
	Acquisition indication channel AICH
	Paging indication channel PICH
	CPCH status indication channel CSICH
	Collision detection/Channel assignment indicator channel CD/CA-ICH

8.2.3 MAC Layer

CPCH The MAC layer in the UMTS protocol stack is a very important component in that it serves many purposes:

- Mapping logical channels to transport channels;

- Choosing suitable transport format for each transport channel based on the instantaneous source data rate;

- Managing priority in a set of data traffic flows within each UE;

- Managing priority among different UEs using dynamic scheduling techniques;

- Identifying specific UEs on the common transport channels;

- Marshaling of data PDUs between transport block sets and the underlying physical layer dedicated channels;

- Measuring and monitoring of traffic volume;

- Ciphering for transparent mode RLC;

- Choosing access service class for RACH and CPCH sessions.

As shown in Figure 8.11, the MAC layer can be divided into four major components: MAC-d, MAC-c/sh, MAC-b and MAC-hs. MAC-d is responsible for handling dedicated channels and is residing at the Serving RNC (S-RNC)[5]. MAC-c/sh is responsible for handling the common and shared channels. MAC-b is responsible for handling the broadcast channel. Both MAC-c and MAC-b are residing at the Controlling RNC (C-RNC)[6]. MAC-hs is a high-speed extension of the MAC layer in HSDPA and is responsible for the high speed DSCH transport channel. Unlike the other MAC components, the MAC-hs is residing at the Node B to facilitate fast scheduling.

8.2.4   RLC Layer

The major role of the RLC layer is to provide segmentation and retransmission functions for the logical channels offered by the MAC layer and is residing at the S-RNC. There are three different operation modes, namely the transparent mode (Tr), unacknowledged mode (UM), and acknowledged mode (AM). The architecture of the RLC layer is shown in Figure 8.12 We can see that Tr and UM modes are characterized as unidirectional and AM mode is featured as bidirectional.

The Tr mode offers basic Service Data Unit (SDU) transfer, segmentation and re-assembly and SDU discard and it does not add any protocol overhead to higher layer SDUs. Specifically, corrupted SDUs are either discarded or marked as erroneous. Segmentation and re-assembly is an optional feature in the Tr mode. If segmentation is not configured in the RLC layer, the SDU from the upper layers is used as the Protocol Data Unit (PDU) to the MAC layer. On the other hand, if segmentation is configured, the packet length of the SDU received from the upper layers must be an integer multiple of the PDU length so that segmentation is done without any RLC overhead. This mode is useful for real-time applications, such as streaming or conversational applications, where low-overhead logical channel is needed.

The UM mode offers concatenation, padding, ciphering and sequence number checking in addition to the services offered by the Tr mode. It does not offer retransmission function and as such, data delivery is not guaranteed. Consequently, corrupted data PDUs are either marked or discarded depending on the system configuration. Unlike the Tr mode, RLC protocol overhead is added in the UM mode

---

[5]The S-RNC is the anchor point with respect to a UE when the call is setup. As the mobile moves to other Node Bs, the S-RNC will not be changed unless the S-RNC relocation procedure has been executed. With respect to a UE, the S-RNC is unique.
[6]Each Node B has a unique RNC and this is the C-RNC.

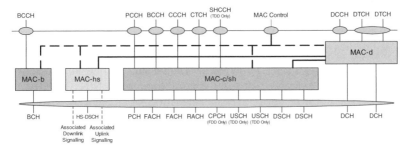

(a) MAC layer components in each UE

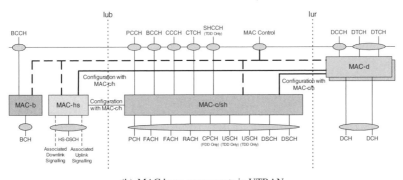

(b) MAC layer components in UTRAN

Figure 8.11   Three components of the MAC layer in the UE and the UTRAN.

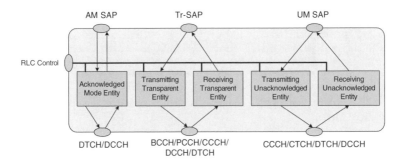

Figure 8.12   The architecture of the RLC layer.

to facilitate concatenation, padding, ciphering and sequence number checking. For instance, the UM mode is used for certain RRC signaling procedures whereas only a unidirectional logical channel is required and there are acknowledgment and retransmission procedures defined in the RRC signaling already.

In the AM mode, an ARQ (Automatic Repeat Request) mechanism is employed for error correction. Figure 8.13 shows the components and mechanisms involved in the AM mode. We can see that the transmitting party of the AM-RLC entity receives RLC SDUs from upper layers via the AM-SAP. These RLC SDUs are then segmented (in case of a large SDU) into fixed length AMD PDUs. Alternatively, the SDUs may be concatenated into an AMD PDU if the SDUs are smaller than the AMD PDU length. The AMD PDUs buffered in the retransmission buffer are either deleted or retransmitted. The MUX multiplexes the AMD PDUs from the Retransmission buffer that need to be retransmitted. Ciphering is performed (if configured) followed by sending the encrypted PDUs to either one or two DCCH or DTCH logical channels.

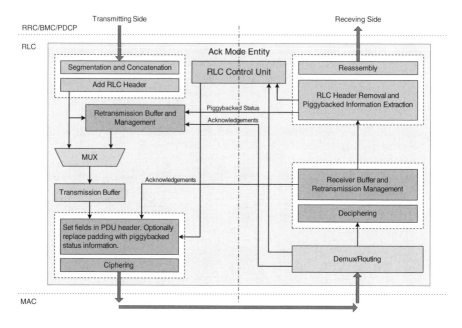

Figure 8.13   Illustration of the components and mechanisms involved in the RLC-Acknowledged Mode.

On the other hand, the receiving party of the AM-RLC entity receives AMD and control PDUs via the specific logical channels. Afterward, the AMD PDUs are routed to the Deciphering unit followed by delivery to the receiving buffer. Reassembly is performed in that the AMD PDUs are accumulated in the buffer until a complete RLC

SDU is received. The receiver then acknowledges successful reception (or requests for retransmission of missing AMD PDUs).

In summary, the RLC layer provides the following services:

- Segmentation and reassembly of higher-layer PDUs into/from smaller RLC payload units (RLC-Tr, RLC-UM, RLC-AM);

- Assembling by concatenation of RLC SDUs (RLC-UM, RLC-AM);

- Padding (RLC-UM, RLC-AM);

- Transfer of user data (RLC-Tr, RLC-UM, RLC-AM);

- Error correction (RLC-AM);

- In-order delivery of higher-layer PDUs (RLC-AM);

- Duplicate detection (RLC-AM);

- Flow control (RLC-AM);

- Sequence number check (RLC-UM);

- Protocol error-detection and recovery (RLC-AM);

- Ciphering (RLC-UM, RLC-AM);

8.2.5 RRC Layer

RRC layer is a very important layer in the control plane responsible for the radio resource management. It resides at the S-RNC and using RRC messages, the set-up, modification, and tear-down of layers 2 and 1 protocol entities can be accomplished. Mobility management functions in the UTRAN level such as measurements, handovers and cell updates with respect to a specific UE are also under the control of RRC layer. The RRC architecture is shown in Figure 8.14 We can see that there are four major functional entities:

- The Dedicated Control Function Entity (DCFE) is responsible for handling all functions and signaling for one particular UE. It relies mostly on AM mode RLC.

- The Paging and Notification control Function Entity (PNFE) is responsible for handling the paging of idel mode UEs. It mainly employs the PCCH logical channel.

- The Broadcast Control Function Entity (BCFE) is responsible for handling the system information broadcasting. It utilizes either BCCH or FACH logical channels.

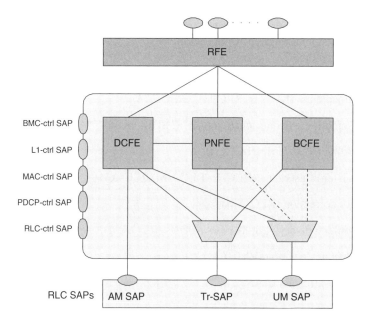

Figure 8.14 Protocol architecture of the RRC layer.

- The Routing Function Entity (RFE) is responsible for the routing of higher layer messages.

RRC interacts closely with each layer in the UMTS protocol stack so as to provide control information, and in turn receive measurement feedback from these layers. This is illustrated in Figure 8.15 Specifically, RRC provides the following functions:

- Cell Broadcast Service (CBS) control;

- Initial cell selection and cell re-selection;

- Paging;

- Broadcast of information;

- Establishment, maintenance, and release of an RRC connection between a UE and the UTRAN;

- Assignment, reconfiguration, and release of radio resources for the RRC connection;

- Control of requested QoS;

- UE measurement reporting and control of the reporting;

UTRAN   UE

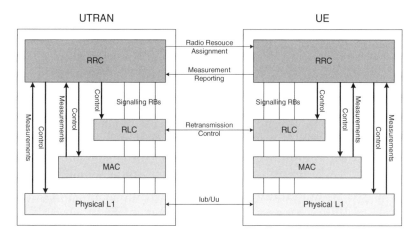

Figure 8.15   Interactions between the RRC layer and other protocol components.

- RRC message integrity protection;

- Arbitration of radio resources on uplink DCH;

- Slow dynamic channel allocation (DCA) in the TDD mode;

- Timing advance in the TDD mode;

- RRC connection mobility functions;

- Outer loop power control;

- Control of ciphering functions.

Figure 8.16 shows the transition mechanism of a UE's RRC states. After the UE is powered on, it chooses a suitable cell with the strongest pilot signal strength for service through monitoring the common downlink pilot channel. After the UE acquires the pilot channel through cell search procedure, the UE is able to receive system information and cell broadcast messages from the BCH channel. At this moment, the RRC layer at the UE is at the idle state. The UE stays in the idle mode until it sends a request to the UTRAN to establish a dedicated control channel (DCCH) with the RRC peer at the S-RNC. When this happens, the RRC layer is moved from the idle state to the *connected state*. Note that setting up a DCCH does not necessarily imply the setting up of dedicated transport channels and dedicated physical channels.

When a logical connection (DCCH) is set up between the UE and the S-RNC, both the RRC peers in the UE and the S-RNC will be at the connected state. However, there are four sub-states in the connected state, namely the *Cell-DCH* state, the *Cell-FACH* state, the *Cell-PCH* state as well as the *URA-PCH* state. In the *Cell-DCH* state, a

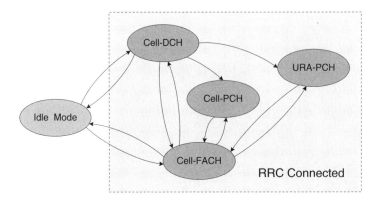

Figure 8.16 State transition diagram of RRC layer.

dedicated physical channel is allocated to the UE for the logical DCCH and DTCH (or radio bearer). In addition, the S-RNC recognizes the location of the UE down to the cell level. The UE carries out measurements and transmits measurement reports according to downlink measurement control information from the RNC. This state can be entered directly from the idle state during call setup or from the *Cell-FACH* state (as triggered by traffic volume change).

In the *Cell-FACH* state, the UE is not allocated any dedicated physical channel. The DCCH and DTCH logical channels are supported by the RACH, CPCH and FACH transport channels. In this state, the UE is also capable of listening to the BCH for acquiring system information as well as performing cell re-selections. Similar to the *Cell-DCH* state, the S-RNC recognizes the location of the UE down to the cell level. Hence, after a re-selection process is performed, the UE needs to send a *cell update* message to notify the S-RNC about its new location. This state can be entered directly from the idle state or from the *Cell-DCH* state (as triggered by traffic volume change).

In the *Cell-PCH* state, the UE is still recognized on a cell level but it can only be reached via the PCH. The major feature of this state is that the battery consumption is less than that in the *Cell-FACH* state because monitoring of the paging channel includes a discontinuous reception (DRX) capability. The UE also obtains system information on the BCH. If the UE performs a cell re-selection, it has to move to the *Cell-FACH* state first so as to carry out the *cell update* procedure.

Finally, in the *URA-PCH* state, it is very similar to the *Cell-PCH* state except that the UE needs to execute the *cell update* procedure only if the UE detects a change in the *UTRAN Registration Area* (URA). The UE obtains the URA of the current cell from the BCH and a cluster of cells will share the same URA. This arrangement is to reduce the uplink signaling loading to the system due to *cell updating*. The UE in this state is basically in a dormant mode and the DCCH cannot be used in this state.

The UE has to move to *Cell-FACH* state to re-activate the DCCH for any potential activities.

We shall illustrate the important role of the RRC states in the dynamic bearer reconfiguration of packet switched connections in the next section.

8.3 PACKET SWITCHED CONNECTIONS IN UMTS (REL 99)

As variable rate bursty data services are the major new incentives for using 3G wireless systems, great efforts have been exerted on designing efficient algorithms to enable such services. However, provisioning of packet data services efficiently is much more challenging than that of the constant bit rate voice services because of the inherent heterogeneity in the users' requirements and their associated resource conditions (e.g., signal-to-noise ratio, channel quality, data urgency, etc.).

Specifically, there are four basic types of traffic classes supported by UMTS:

- **Conversational Class**: real-time symmetric traffic, mostly voice services;

- **Streaming Class**: real-time asymmetric traffic, usually from the network to the user, e.g., video streaming;

- **Interactive Class**: non-real-time packet traffic, high integrity, e.g., Web surfing;

- **Background Class**: non-real-time packet traffic, delay insensitive, e.g., emails.

Judicious radio resource management is needed to satisfy users' QoS requirements while maximizing the system's performance and utilization. In UMTS Rel 99, the scheduler is residing at the S-RNC and is responsible for the dynamic allocation of the radio resources, in terms of time durations, codes, and power levels, in the radio bearers set up for the packet data users over a macroscopic time scale. As will be discussed below, the scheduler needs to decide on what types of channels to use for different UEs in a dynamic manner according to the system loading and the user buffer levels. In the followings, we shall focus on the packet switched data connection setup and dynamic reconfiguration incorporated in the Rel 99 standard of UMTS.

8.3.1 Radio Bearer for Packet Switched Users

In the UMTS systems, the *radio bearer* is defined as a logical channel between the UE and the RNC to carry data payload (from the user plane) or signaling data (from the control plane). In other words, the radio bearer is the logical channel offered by the RLC layer to the upper layers. If the radio bearer is used to carry signaling data, it is called the *signaling radio bearer*. Otherwise, it is called the *data radio bearer*.

While the data radio bearer is a logical channel for the user plane, it can be mapped into three types of transport channels, namely the *common transport channels* (CPCH,

RACH, FACH), the *dedicated transport channels* (DCH) and the *shared transport channels* (DSCH). Common transport channels can be used to carry user packet data. In a typical deployment scenario, there are usually only one or just a handful of RACH and FACH channels per sector. A distinctive merit of using common transport channels for user data is that the set-up time is minimal because these channels are already setup. However, there is no closed loop power control as well as soft handoff. Instead, the cell re-selection mechanism is employed when the UE moves to the next cell. Hence, this is most suitable for transmitting small and bursty user packets.

Dedicated transport channels are bidirectional in nature and support fast power control and soft handoff. Radio link level performance is thus greatly enhanced. However, the set-up time for dedicated channels is usually quite long and this introduces setup overhead. In addition, the dedicated transport channels will consume more code space in the downlink. Hence, it is most suitable for transmitting large volume of user packets.

The shared transport channel (DSCH) operates in a time-division scheduling manner in that a single orthogonal code is shared among many users in the time domain. Compared with the dedicated transport channels, the DSCH can save the code consumption in the downlink as well as reduce the associated setup time. Hence, it is suitable for transmitting large volume or bursty user packets. In fact, in HSDPA, a high speed version of the DSCH (HS-DSCH) is created to support fast scheduling of packet data users. A short-coming of the DSCH is that it does not support soft handoff.

Table 8.3 summarizes the possible mappings of various transport channels for supporting data radio bearer.

8.3.2 Setup of Packet Switched Connection

In order to set up a packet switched connection for a data user, the UE has to first establish a DCCH (either based on DCH or FACH/RACH). This is initiated by the UE sending a *RRC connection request* message to the S-RNC via the RACH channel (which is mapped to the CCCH). The corresponding RRC state change can be from the idle state to the Cell-DCH state or from the idle state to the Cell-FACH state. Once the RRC connection is successfully set up, the next step is to establish a data radio bearer. Specifically, a *radio bearer setup* message is sent from the S-RNC to the UE over the DCCH (already established RRC connection). Figure 8.17 illustrates an example of packet switched data connection setup with low traffic volume. The RRC-state of the UE is changed to the *Cell-FACH* after the DCCH is set up. Next, the data radio bearer (RB1) is also setup based on the FACH transport channel.

Unlike the circuit switched connection, the resource requirement of a packet switched connection has to be dynamically adjusted because of the bursty nature of the source. This is called the *dynamic bearer reconfiguration* or scheduling over macroscopic time scale and the decision is made by the RRC layer at the S-RNC. To facilitate the S-RNC to perform dynamic bearer reconfiguration, the RRC layer

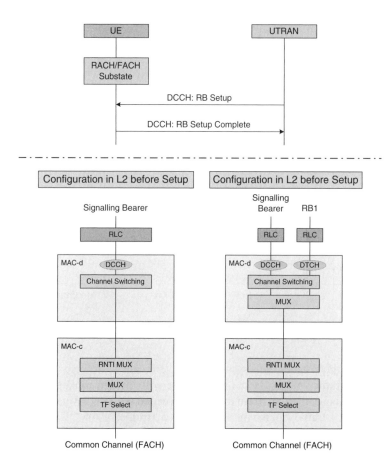

Figure 8.17   An example of packet switched data connection setup in UMTS with low traffic volume.

Table 8.3   A summary of transport channel mappings to support data radio bearer in UMTS.

TrCh	DCH	RACH	FACH	CPCH	DSCH
TrCH Type	Dedicated	Common	Common	Common	Shared
Applicable UE State	CELL_DCH	CELL_FACH	CELL_FACH	CELL_FACH	CELL_FACH
Direction	Both	Uplink	Downlink	Uplink	Downlink
Code Usage	Accordingly to max. bit rate	Fixed code allocations in a cell	Fixed code allocations in a cell	Fixed code allocations in a cell	Fixed code allocations in a cell
Power Control	Fast closed-loop	Open-loop	Open-loop	Fast closed-loop	Fast closed-loop
SHO Support	Yes	No	No	No	No
Target Data Traffic Volume	Medium or high	Small	Small	Small or medium	Medium or high
Suitability for Bursty Data	Poor	Good	Good	Good	Good
Setup Time	High	Low	Low	Low	Low
Relative Radio Performance	High	Low	Low	Medium	Medium or high

has to monitor the instantaneous system loading (from the MAC layer) as well as the instantaneous traffic volume (from the RLC layer). In the downlink, these measurement can be obtained directly from the MAC layer and the RLC layer of the S-RNC. In the uplink, the traffic volume measurement is performed at the UE RLC layer per instruction of the S-RNC[7]. This is illustrated in Figure 8.18

Suppose the downlink traffic volume increases such that the buffer level at the RLC rises. The S-RNC may trigger a *transport channel re-configuration* as illustrated in Figure 8.19 Dedicated transport channels (DCH1 and DCH2) are setup to support the existing DCCH and the radio bearer (RB1) in the reconfiguration in response to the rise in the downlink traffic volume. In this case, the RRC state changes from the *Cell-FACH* state to the *Cell-DCH* state because dedicated transport channels are setup after reconfiguration.

---

[7]To trigger uplink traffic volume measurement, the S-RNC can send a *traffic volume measurement* request to the UE and the UE responses by sending a *measurement report* to the S-RNC. The traffic volume measurement can also be configured as periodic or event driven.

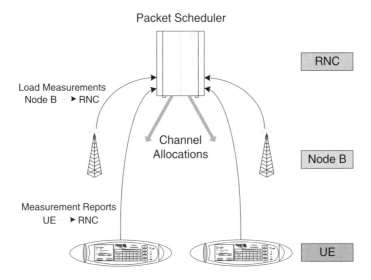

Figure 8.18   A schematic diagram illustrating the scheduling mechanism in UMTS Rel 99.

On the other hand, the dedicated transport channels may be released and replaced by the common transport channels to support the DCCH and RB1 in the example when the traffic volume drops. This transition is usually triggered by the *inactivity timer* which times the idle period when the buffer in the RLC layer is empty. When this happens, the RRC state will change from the *Cell-DCH* state back to *Cell-FACH* state. Figure 8.20 summarizes the RRC state changes in the lifetime of a packet switched data connection.

8.4 PACKET SCHEDULING IN HSDPA (REL 5)

The HSDPA concept can be seen as a continue evolution of the DSCH time division scheduling and a new transport channel, namely the *high speed DSCH* (HS-DSCH) is defined for more efficient scheduling of high bit rate packet data users. In addition, fast scheduling is made possible by introducing a new MAC layer called the *MAC-hs* residing at the Node B. The fast scheduling (over 2ms intervals) together with the HS-DSCH transport channel allow the HSDPA system to exploit multi-user selection diversity over the microscopic fading channels and therefore achieving higher peak data rate, spectral efficiency as well as QoS control for bursty and downlink asymmetric packet data users. In the following, we shall give an overview of HSDPA.

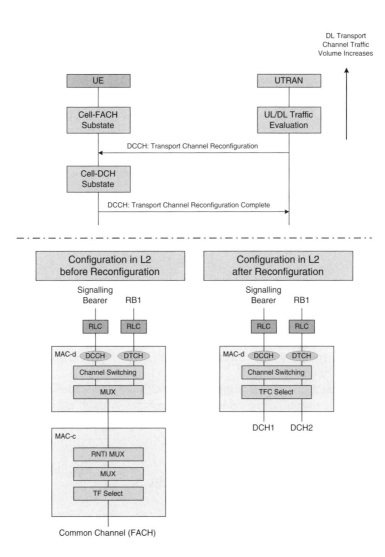

Figure 8.19   An example of transport channel re-configuration due to an increase of downlink traffic volume.

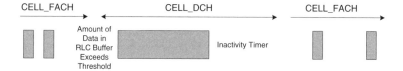

Figure 8.20 An example of RRC state changes in the lifetime of a packet switched data connection.

8.4.1 Key Enabling Technologies in HSDPA

There are two key enabling features in HSDPA, namely the adaptive modulation and channel coding, hybrid ARQ in the physical layer (HS-DSCH) as well as the fast scheduling at the MAC-hs layer in Node B. They are elaborated as follows.

- **Adaptive Modulation, Channel Coding and Multi-code transmissions:**

  On the HS-DSCH channel, two fundamental CDMA features, namely the variable spreading factor and the fast power control[8], have been deactivated and replaced by *short packet size, adaptive modulation and coding* (AMC), *multi-code operation* and *fast hybrid ARQ*. While being more complicated, the replacement of fast power control with AMC yields a power efficiency gain due to the elimination of the inherit power control overheads. In the HS-DSCH, the spreading factor is fixed to be 16 which gives good data rate resolution with reasonable complexity. In order to increase the efficiency of AMC to exploit the microscopic channel fading, the packet duration is reduced to 2ms.

  The means of adaptation in HDSPA are the channel coding rate, modulation level, the number of multi-code employed as well as the transmit power allocated per code. To be able to exploit the microscopic fading, the AMC and the multi-code of the HSDPA must be able to cover a wide dynamic range. In the modulation design, the HSDPA incorporated 16QAM in addition to the QPSK modulation in order to increase the peak data rate. The combination of rate 3/4 encoding and 16QAM gives a peak bit rate of 712 kbps per code channel (SF=16). On the other hand, higher robustness is available with rate 1/4 and QPSK modulation giving a peak bit rate of 119 kbps per code channel. Given very good channel condition, a single user can simultaneously receive up to 16 code channels with an aggregate bit rate of 10.8Mbps. Table 8.4 illustrates various combinations of channel encoding and modulation level as well as the corresponding data rate in HDSPA.

---

[8]Moreover, due to the existence of buffer for the bursty data sources and a relaxed delay constraint relative to the circuit switched voice users, the optimal transmission strategy should be *water-filling* instead of *equalizing* the fading channel. Hence, the removal of fast power control does not incur any fundamental performance penalty in information theoretical sense.

Table 8.4   Various data rates available in DS-DSCH of HSDPA using different modulation, channel coding and multi-code allocations.

TFRC	Data Rate (1 code)	Data Rate (5 codes)	Data Rate (15 codes)
QPSK, Rate 1/4	120 kbps	600 kbps	1.8 Mbps
QPSK, Rate 1/2	240 kbps	1.2 Mbps	3.6 Mbps
QPSK, Rate 3/4	360 kbps	1.8 Mbps	5.3 Mbps
16QAM, Rate 1/2	477 kbps	2.4 Mbps	7.2 Mbps
16QAM, Rate 3/4	712 kbps	3.6 Mbps	10.8 Mbps

The dynamic range of the AMC for single code and multi-code systems is illustrated in Figure 8.21 The curve includes the gain from fast hybrid ARQ based on chase combining which significantly improves the throughput at low E_s/η_0. The dynamic range of E_s/η_0 variation is usually over 20dB due to the Rayleigh fading. Observe that the multi-code operation allows a smaller granularity of the AMC and resulting in a smoother throughput curve. Furthermore, the multi-code operation allows the instantaneous throughput to vary over a wider dynamic range of E_s/η_0 (around 32 dB).

- **Fast Hybrid ARQ:**

  The hybrid ARQ (HARQ) protocol adopted by the HSDPA system is a stop-and-wait protocol. The transmitter persistently transmits the current packet until it has been successfully received by the UE (indicated through the acknowledgement (ACK)). In order to avoid resource idling (and therefore a waste) while waiting for the ACK, a maximum of 8 parallel HARQ processes may be setup for the UE so that different HARQ processes transmit in different TTI (2ms each). The control of HARQ is located at Node B over the new MAC-hs layer so that the storage of the unacknowledged data packets and the subsequent scheduling of retransmissions does not involve the RNC. Hence, Iub signaling (between Node B and RNC) is avoided and the resulting retransmission delay in HSDPA is much lower than the conventional retransmission in the RNC (RLC layer). Typical retransmission delay in HSDPA is around 8 – 12ms.

  The HSDPA concept supports two types of retransmission strategies, namely the *chase combining* (CC) and the *incremental redundancy* (IR). The basic idea of CC is to transmit an identical version of the erroneously detected data packet and then for the decoder to combine the soft information of the received copies weighted by the SNR prior to decoding. In the IR scheme, additional redundancy information is incrementally transmitted if the decoding fails in the first attempt. In general, the IR scheme requires a lower SNR relative to

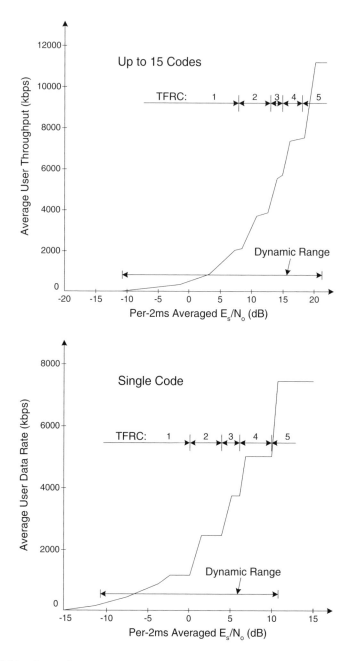

Figure 8.21   Dynamic range of AMC for single code and multi code HSDPA systems.

the CC scheme at the same target BLER but the disadvantage of the IR scheme is a higher memory requirement in the UE. Hence, the possibility of using IR scheme is defined by the UE capability class.

- **Fast Scheduling:** While the AMC and multi-code transmission in HS-DSCH allow a smooth link level adaptation over a large dynamic range of SNR variation, fast scheduling is introduced in the MAC-hs residing in the Node B to facilitate multi-user selection diversity over the Rayleigh fading channels. The scheduling decisions as well as the packet buffers are located at the Node B and no Iub signaling (between Node B and RNC) is needed in HSDPA. Hence, the HS-DSCH can be assigned to one packet data user in 2ms TTI resolution. This is shorter than the coherence time of most pedestrian users in the system and therefore, the microscopic fading remains quasi-static within a packet transmission.

In HSDPA, there are three new channels introduced in the physical layer as outlined below.

- **High-Speed Downlink Shared Channel (HS-DSCH)**: This is the channel carrying user data over the downlink and thus, it is the major resource for the scheduling algorithm to manipulate.

- **High-Speed Shared Control Channel (HS-SCCH)**: This is the channel carrying the physical layer control information over the downlink for several key purposes including decoding, combining, and retransmission.

- **High-Speed Dedicated Physical Control Channel (HS-DPCCH)**: This is an uplink channel carrying important control information such as the ARQ acknowledgments (ACK) and the downlink channel quality indicator (CQI).

Similar to the DSCH in UMTS Rel 99, there is always a low data rate downlink DCH associated with the HS-DSCH. While the HS-DSCH is shared by a number of UEs, the associated downlink DCH is dedicated per UE. Besides the downlink DCH, there is also an uplink HS-DPCCH (per UE) associated with the HS-DSCH in HSDPA. Figure 8.22 illustrates the multi-user fast scheduling concept in HSDPA. The Node B tracks the channel quality of each active packet data users in the downlink direction by monitoring the transmit power of the downlink DCH associated with the HS-DSCH. In addition, the UE can also be requested to report a CQI periodically over the uplink HS-DPCCH. The CQI is an indication of the adaptation mode and number of codes currently supported by the UE based on the current channel condition. The feedback cycle of the CQI can be set as a network parameter adjustable in 2ms steps. To facilitate HARQ, the UE is also required to send an ACK/NAK response on the HS-DPCCH.

In Node B, the CSI of a UE is estimated based on the power level of the downlink DCH, the ACK/NACK ratio as well as the CQI. Depending on the scheduling algorithm, the Node B then schedules data transmission to UE on the HS-DSCH. Prior to sending data to the HS-DSCH, the Node B sends a detail message to the active packet data users via the associated HS-SCCH. This message describes the employed AMC mode, the multi-code set as well as the HARQ process of the HS-DSCH transmission and it is transmitted 2 slots in advance of the HS-DSCH as illustrated in Figure 8.23

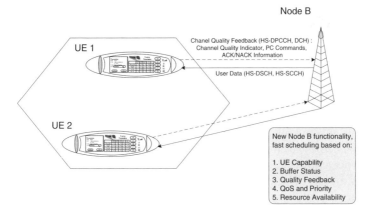

Figure 8.22   A schematic diagram illustrating the scheduling mechanism (HSDPA) in the UMTS Rel 5.

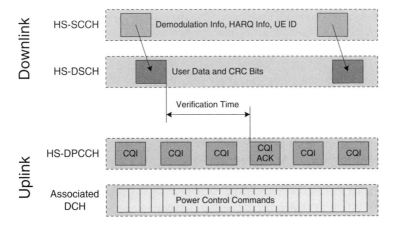

Figure 8.23   Inter-channel operation of HSDPA scheduling.

8.4.2 Continued Evolution

HSDPA provides a significant cell capacity gain for packet data traffic in UMTS and is thus an important part of the continuous 3G evolution. Since the HSDPA systems offers improved code efficiency and dynamic range in the user data rate, it can utilize the improvements in the detector performance such as channel equalizations, multi-user and multi-code interference cancellations as well as advanced MIMO techniques. In addition, the introduction of fast scheduling in HSDPA paves the way for cross-layer scheduling design incorporating the time, frequency and spatial dimensions on a multi-user perspective.

Naturally, the "high speed" idea can be extended to the uplink also. That is the so-called HSUPA (High Speed Uplink Packet Access). The idea is that except for scheduling, all the actions are reversed in HSUPA compared with HSDPA. That is, scheduling is still under the control of the Node B. But the measurements and combining actions are also moved to the Node B from the UEs. However, there is one major obstacle in realizing HSUPA—the power control problem. Over the downlink, the power source is centralized at the Node B. Thus, the power level can be dynamically adjusted solely at the Node B's discretion. By contrast, over the uplink, there are numerous heterogeneous power sources—the UEs. Thus, the interference induced is not easily controllable at the receiver located at the Node B. Indeed, uplink power control cannot be abandoned as in the downlink due to the near-far problem. Much more research is needed in this area.

8.5 SUMMARY

In this chapter, we have elaborated the application packet scheduling using UMTS as an example. We have reviewed a general architecture of UMTS. A UMTS network can be partitioned into a UTRAN and a CN. The UTRAN is handling radio dependent functions (such as radio resource control) whereas the CN is handling radio independent functions (such as switching of connections and packets, mobility management and session management). Inside the UTRAN, we have the RNC acting as a base station controller controlling a number of Node Bs (base stations).

In the UTRAN, channels are partitioned into three layers, namely the logical channels (or radio bearer), transport channels and the physical channels. The logical channels form a common interface for both the user plane and control plane protocols. Various logical channels are mapped to different combinations of transport channels in the MAC layer. Various transport channels are further multiplexed into different combinations of physical channels in the PHY layer.

In UMTS Rel 99, packet scheduling is residing at the S-RNC (RRC layer) and therefore, it fails to exploit the microscopic channel variation in the fading channels due to the inherit delay in the Iub signaling (over 500ms). Packet data users can be served on DCH based on code scheduling or DSCH based on time scheduling. In the former case, the physical radio resource assigned to the active UE are dynamically adjusted based on radio bearer reconfiguration as triggered by traffic volume changes in the RRC layer. It suffers from a larger channel setup overhead and is suitable only for large volume of bursty data. In the latter case, a number of packet data users are time sharing the DSCH and it incurs a lower setup overhead and delay. In both cases, the peak data rate of the packet user is around 2Mbps.

In UMTS Rel 5, HSDPA is introduced to significantly improve the spectral efficiency of packet data users. There are three enabling technologies behind HSDPA, namely the AMC, the fast HARQ and the fast scheduling. Three new channels (HS-DSCH, HS-SCCH, HS-DPCCH) are created to facilitate an adaptive link with large dynamic range over the supported bit rates. In addition, a new MAC layer (MAC-hs) is introduced at the Node B and fast scheduling can be done at the Node B without going through the Iub signaling. This substantially reduces the scheduling latency and the scheduling is based on time division over the shared HS-DSCH. The scheduling period is 2ms and this is sufficient to exploit the microscopic fading for pesdestrian users. As a result of the fast scheduling and the adaptive physical layer, multi-user selection diversity can be effectively exploited and the HSDPA could deliver a much higher spectral efficiency relative to UMTS Rel 99.

PROBLEMS

8.1 UMTS system allows QoS differentiations in the air interface [True/False].

8.2 In UMTS Rel 99, the system capacity of high bit rate packet data users mixed with regular voice users will be limited by the number of orthogonal codes [True/False].

8.3 The outerloop power control is done at the Node-B of UMTS systems [True/False].

8.4 In UMTS, one way to increase the physical bit rate is to assign multiple codes to the same user [True/False].

8.5 In UMTS Rel-99, the RNC is responsible for dynamically controlling the physical resources allocated to the data bearer [True/False].

8.6 Since the RLC layer in UMTS mobile has packet retransmission capability in case a packet is corrupted, there is no need for the TCP layer to do any TCP packet retransmission because all the potential error packets are taken care by the lower layer (RLC layer) in the protocol stack [True/False].

PART III

SHORT-RANGE WIRELESS TECHNOLOGIES

CHAPTER 9

IEEE 802.11X WLAN STANDARDS

The ether of space is nothing but nominative of the verb to undulate.

—Author unidentified

9.1 INTRODUCTION

Internet becomes truly pervasive when low-cost wireless technologies are widely available. Specifically, short-range wireless communications have by and large enabled people to get network access anytime anywhere. In this chapter we talk about one of the two most important short-range wireless technologies.

Interoperability:

Wireless LAN (WLAN) has been in widespread deployment for several years and the first IEEE 802.11 standard was introduced in 1997. The main reason for developing this standard is to provide interoperability in two dimensions: interoperability between wireless LANs and existing wired LANs, for example, IEEE 802.3 Ethernet, and also interoperability between wireless devices from different vendors.

Wireless Internet and Mobile Computing. By Yu-Kwong Ricky Kwok and Vincent K. N. Lau
Copyright © 2007 John Wiley & Sons, Inc.

In order to facilitate the cooperation between wired and wireless LANs, the IEEE 802.11 architecture has been developed with the concept of Basic Service Set (BSS) and how an IEEE 802.11 device communicates with other networking technologies through the Access Point (AP) and Distribution System (DS). By employing the DS concept, a WLAN not only can interconnect with other WLANs' BSS, but also can integrate with existing wired LANs. The IEEE 802.11 architecture is discussed in detail in Section 9.3 below.

On the other hand, before the IEEE 802.11 standard was created, different vendors developed their own wireless LAN products based on their own ideas, and this lead to a serious problem of incompatibility between wireless devices from different vendors. In order to facilitate the interoperability issue, the IEEE 802.11 working group has been set up to provide generally acceptable standards and specifications on Medium Access Control (MAC) mechanisms and Physical (PHY) links for wireless devices from different vendors, so as to share the wireless medium in a fair manner. These issues are discussed in detail in Section 9.4 and Section 9.5.

Performance:

In view of the demand for different performance requirements from the wireless LANs, different task groups have been developed including the famous ones like A, B, G, E. The standard provided by Task Group A, that is IEEE 802.11a, can provide theoretical capacity up to 54 Mbps with 24 orthogonal channels. This is a tempting capacity comparable to wired networks. However, it is not as popular as that of IEEE 802.11b. The main reasons are discussed in detail in Section 9.5. On the other hand, IEEE 802.11a/b/g has been developed to focus on data exchange in the wireless medium. However, only limited QoS requirements have been considered. In order to solve this problem, another task group IEEE 802.11e has been developed to highlight the issue of multimedia delivery in wireless LANs, based on existing IEEE 802.11a/b/g physical layers. This topic is discussed in Section 9.6.

9.2   DESIGN GOALS

There are several objectives in designing IEEE 802.11, which are summarized in Table 9.1 below.

Besides the objectives listed in Table 9.1, there may be some other goals which are under development, such as providing a larger throughput capacity, improving the coverage area and providing wireline comparable functions, etc. Indeed, as user applications' demands keep increasing, future design goals of WLAN extensions would be even more ambitious.

9.3   IEEE 802 ARCHITECTURE

The development of WLAN is assisted by the IEEE 802.11 working group which is responsible for providing specifications and standards on WLAN. The IEEE 802.11

Table 9.1   The design goals of WLAN.

Objective	Description
Seamless Wireline LAN Extension	Wireline LAN has been developed and used for quite a long time with many innovative inventions involved. People prefer new technology development to at least exploit some of those innovative ideas and provide backward compatibility. In order to do so, the design of IEEE 802.11 architecture has put into consideration of interoperability between different BSSs and wireline networks. At the same time, only minimum amount of changes are applied to WLAN's MAC and PHY layers, with upper layers unaffected.
Wireless and Mobile Capabilities	WLAN provides the stations with the capability to connect to an existing network through the wireless medium. This in return provides users with great convenience, as users can conveniently move with their wireless devices without the consideration of the limitation and location of the corresponding wired access point.
Low Cost	There are some considerations in achieving low cost: • The frequency bands used by IEEE 802.11x are all by and large unlicensed bands, so they can be used without charge as long as frequency band usage and power regulation has been followed, and the issue of collision with other wireless technologies is addressed in the design. • The popularity of WLAN devices in return leads to low cost mass production. Nowadays, WLAN devices can be obtained with reasonable and generally affordable price. • The interoperability with wireline LAN is achievable with the help of an access point to form BSS. No new device is required to integrate with existing infrastructure except an access point.

architecture has been suggested in the MAC/PHY specification. There are several components involved and some of the important components are listed in Table 9.2 [323, 6].

The basic building block of the architecture is Basic Service Set (BSS). A BSS may be composed of a single or multiple station(s). Since the wireless medium within a BBS is shared by all the stations, it is necessary for them to use the same MAC protocol and to be controlled by a single coordination function. There are two main types of coordination functions suggested by the IEEE 802.11 working group—Distributed Coordination Function (DCF) and Point Coordination Function (PCF). These functions are discussed in detail in Section 9.4.

Table 9.2   The components of the IEEE 802.11 architecture.

Architecture Component	Description
Station (STA)	A device that has implemented standard IEEE 802.11 Media Access Control (MAC) and Physical (PHY) layers.
Access Point (AP)	A device that has implemented station's functionalities and provided an interface for other associated wireless stations to access a Distribution System.
Basic Service Set (BSS)	A group of stations managed by a single coordination function.
Distribution System (DS)	A system which connects multiple Basic Service Sets and/or integrates with other networking technologies, for example, IEEE 802.x LANs to form an Extended Service Set.
Portal	A logical location where the Media Access Control Sevice Data Unit (MSDU) enters the Distribution System (DS).
Extended Service Set (ESS)	A group of one or more BSSs and integrated LANs that shown to be a single BBS to Link Logical Control (LLC) Layer of any station inside any associated BBS.
MAC Protocol Data Unit (MPDU)	The data unit exchanged by two peer MAC users using the service of the PHY layer.
MAC Service Data Unit (MSDU)	The information delivered by MAC protocol data unit.

Sometimes it is preferable to have a larger wireless network. However, due to physical limitation of the wireless medium and limitation on PHY standard, every station can only transmit data within a certain range. This causes a limitation on coverage area to a BSS. In view of this, the IEEE 802.11 working group suggested the concept of Distribution System (DS). DS is a backbone system in which interconnection between multiple BSSs are facilitated by connecting to it through Access Points (AP). An Access Point is a station which communicates with other stations within the same BBS via the wireless medium. At the same time, it communicates with other entities within the DS. In fact, though the IEEE 802.11 working group has defined the concept of DS, it has not limited the implementation of DS, so that DS can be a switch, another wireless or wired network.

In order to allow other networking technologies' network, for example, Ethernet, to integrate with the IEEE 802.11 architecture, a portal is used to connect to DS. Similar to AP, a portal can be a switch, bridge or hub which acts as an interface between the network and DS.

The Extended Service Set (ESS) consists of DS and BSSs and/or other networking technologies' networks. By definition, ESS refers to the integration of components listed Table 9.1 and appears to be a single BSS to Logical Link Control (LLC) Layer of devices within the network. In other words, it appears to be a single LAN. With this concept, a station's movement from one BSS to another BSS is possible to be transparent to LLC.

The concepts discussed above are illustrated in Figure 9.1. Here, each station is connected to a single BSS. While BSSs are connected to DS through AP, at the same time, an IEEE 802.x LAN is integrated with BSSs by connecting to DS through a portal. This is one of the simplest configurations.

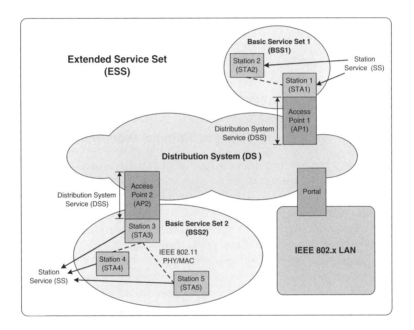

Figure 9.1   IEEE 802.11 architecture.

IEEE 802.11 Service

IEEE 802.11 defines nine services [323, 6]. Six of them are used to support MSDU movement between different stations within ESS and three of them are used for controlling the network access and privacy.

The nine services are:

1. Authentication

2. Association

3. Deauthentication

4. Deassociation

5. Distribution

6. Integration

7. Privacy

8. Reassociation

9. MSDU delivery

Obviously, from the list above, authentication, deauthentication and privacy are three services used to control network access and confidentiality. All other services are used to support MSDU delivery.

The services listed above can also be divided into two groups—Station Service (SS) and Distribution System Service (DSS). Their functions are summarized in Table 9.3

9.4 IEEE 802.11 MAC LAYER

9.4.1 Hidden Terminal and Exposed Terminal Problems

The transmission medium, no matter wired or wireless, is shared by all the users within the network, so it is necessary to have a mechanism to avoid simultaneous sending of data by the users. In a wired network, this is usually done by Carrier Sense Multiple Access/Collision Detection (CSMA/CD), which is briefly reviewed below.

In a wired network, for example, the Ethernet, the sender senses the medium to see if any signal exists. If some signal exists, before sending frames, it waits until the medium becomes "quiet". During sending out the frames, it continues auditing the medium to see if any collision occurs. If a collision occurs, the sender stops and send out a jamming signal. Figure 9.2 shows how CSMA/CD works in Ethernet. We can see that Computers A and C sent a frame to Computer B simultaneously since they both found the medium is free. Then a collision occurs near Computer B. The collision can then be detected by all the computers inside the network. The two computers then wait a certain amount of time before trying to send out their frame again.

As can be seen in Figure 9.2, the collision is detected by the sender instead of receiver. However, we are only interested in whether the receiver can receive a "clean" frame. In a wired network, this is not a problem at all as the signal level of a "clean" frame is similar everywhere within the network including the sender side. However, this is not the case in a wireless network. Figure 9.3 shows the problem of CSMA/CD in a wireless environment. Here, Stations A and C are within the transmission range of each other, but they are near the end of each other's transmission range. Stations A and C are sending a frame to Station B simultaneously. However, since they are sending a frame at that moment, the signal level of their own frame is much higher than that of their opponent. In this case, it is very difficult for them to detect whether there is a collision.

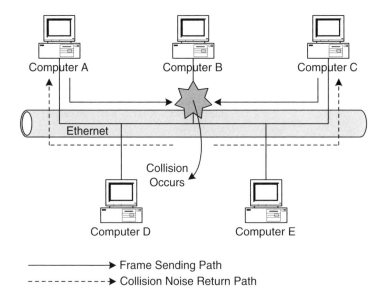

Figure 9.2   Wired network CSMA/CD illustration.

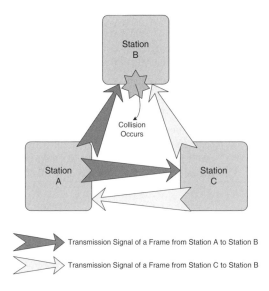

Figure 9.3   Wireless network CSMA/CD problem.

Table 9.3   IEEE 802.11 services.

Type	Service	Description
SS	Authentication	Unlike wired networks, the transmission medium used in wireless networks is open and shared. This service is required for stations to establish the identities of their communication peers within BSS. IEEE 802.11 does not fix the scheme used, but it requires mutual authentication to be carried out before association with AP.
	Deauthentication	This service is activated whenever any existing authentication is about to be terminated.
	Privacy	With reasons similar to those of authentication, this service is provided for preventing the message to be read by recipients other than the intended one.
	MSDU delivery	This service is used to deliver the data unit from one station to another.
DSS	Association	Before a station is allowed to send and receive frames, it has to associate with certain AP. In this case, either station and AP can determine the address and identity of each other. AP can then supply the associated station information to DS, so DS can route frames within ESS.
	Deassociation	Either a station or AP within the same BSS sends out a notification to notify others about the termination of existing association.
	Reassociation	Whenever a mobile station moves out of the transmission range of current AP, established association is then transferable with the help of this service.
	Distribution	This is the primary service used by stations to exchange frames through DS. That is, the frames are routed from one BSS to another BSS within ESS.
	Integration	Integration service allows the IEEE 802.x network to be connected to DS through a portal. It performs any necessary media and address translation when the messages are sent by station in IEEE 802.11 to that of IEEE 802.x or vice versa.

Other than the problem described previously, there are two main problems in the wireless environment—the hidden terminal and exposed terminal problems [313] that make schemes like CSMA/CD in a wired network fail and/or inefficient.

First of all, let us focus on the following scenario. In Figure 9.4, the transmission ranges of Stations A and C only cover up to Station B and so their signal cannot reach each other. This causes the so-called hidden terminal problem. Specifically, the problem is caused by the fact that Station A does not realize the existence of station C and vice versa. In this case, when Station A senses the environment, it finds out that the environment is free, so it sends out a frame to station B. At this moment, Station C also wants to send out a frame to station B. However, as it cannot hear the transmission from station A, it also thinks that the environment is free and sends out its frame. This leads to a collision at Station B. However, the collision is not realized by both Stations A and C, and so they continue sending their frame until they finish sending.

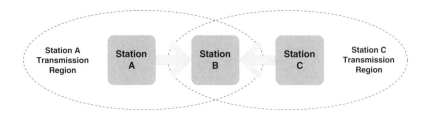

Figure 9.4 The hidden terminal problem.

Another issue is the exposed terminal problem. In order to explain this problem, let us focus on the scenario in Figure 9.5. We can see that Station B's transmission range covers both Stations A and C, but not Station D, while the transmission range of Station C covers both Stations B and D, but not Station A. In this case, Station B senses the environment and finds out that it is free, so it sends out the frame to Station A. However, at this moment, Station C also wants to send a frame to Station D. It senses that Station B is sending out a frame, so in order to avoid collision, it waits until Station B finishes sending. Indeed, we only care about the frame at receiver side instead of sender side. Thus, Station C, in fact, can also send out its frame simultaneously, without waiting for Station B to finish sending.

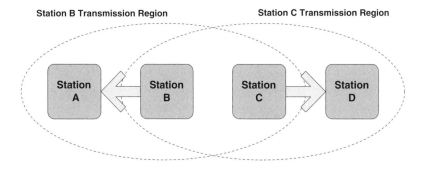

Figure 9.5 The exposed terminal problem.

9.4.2   Four-Frame Sequence

As explained in the previous section, there are hidden terminal problem and exposed terminal problem in using carrier sense mechanism in wireless environment. In order to solve the hidden terminal problem, a four-frame sequence is introduced. The sequence consists of Request-to-Send (RTS) frame, Clear-to-Send (CTS) frame, Data frame and Acknowledgement (ACK) frame.

Before sending out data frames, the sender sends out RTS after it finds that the medium is free. If the target recipient receives RTS, it sends out CTS in reply if it has not received any RTS/CTS from other stations before. When the sender receives CTS, it knows that the environment around the receiver is free to receive data frames, so it sends out data frames. While stations other than the sender receive RTS/CTS , they expect that there is a sender preparing to send data frames to a recipient within their transmission range, so they wait until the receiver receives the data frames and sends out ACK which is used as a signal of finishing sending, or until Network Allocation Vector (NAV) is deactivated. NAV is a flag which is used to indicate stations that the medium is busy, so that the stations activate it when they receive RTS/CTS. During the period when NAV is activated, the stations ignore other RTS/CTS and do not send out their own frames.

After explaining the four-frame sequence, the details on how this sequence can be used to solve the hidden terminal problem is explained in Figure 9.6. Here, Stations A and C are out of the transmission range of each other, i.e., Stations A and C are hidden from each other. Only Station B can receive signal from both stations. The whole sequence is explained as follows:

1. Station A senses that the medium is free and sends out RTS to Station B.

2. Station B receives RTS and sends out CTS as reply to Station A. Station C also receives CTS, so it knows that Station A is going to send data to Station B and so it sets the NAV (CTS) value.

3. Station A starts sending out data frames.

4. Station C wants to send data, but it has received CTS from Station B, so it waits until receiving ACK.

5. Station B receives data frames and sends out ACK, both Stations A and C receive it.

6. Station C sends out RTS now, as it knows that Station A has finished sending.

7. The process of RTS-CTS-DATA-ACK sequence repeats once again.

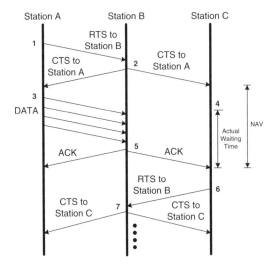

Figure 9.6 RTS/CTS solution to the hidden terminal problem.

9.4.3 DCF and PCF

To introduce the concepts of DCF (distributed coordination function) and PCF (point coordination function), let us consider the MAC architecture of IEEE 802.11. Figure 9.7 shows that there are two medium access coordination mechanisms in IEEE 802.11 MAC. They are DCF and PCF [323, 6]. DCF only provides contention service. By using DCF, stations are required to compete for the wireless medium. However, as shown in Figure 9.7, PCF is built on top of DCF, it uses the services of DCF in order to provide the functions of contention-free service for upper layers.

As discussed in the previous section, collision detection is difficult to be implemented in the wireless medium. In DCF, Carrier Sense Multiple Access with Collision Avoidance (CSMA/CA) is used instead. By using this algorithm, first of all, a station senses the medium if it wants to send out an MAC frame. It sends out the frame only if it discovers that the medium is free; otherwise, it waits until the medium becomes free again.

Even though CSMA/CA algorithm is used in DCF, there is still a high probability that multiple stations sense that the medium is free and send out their frame simultaneously, especially at the moment immediately after a sender finishes sending a frame and the medium becomes free again. In order to avoid this, various types of Interframe Space (IFS) are introduced together with a random backoff mechanism. IFS is the time between two frames, with variable length depending on the types. Different priority is then given to different services or stations. Generally speaking, the shorter the length of IFS, the higher the chance that a station can control the medium. Table 9.4 gives the description of various IFS.

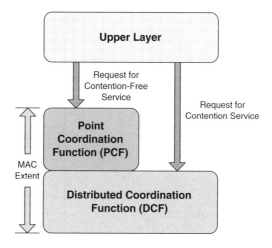

Figure 9.7   IEEE 802.11 MAC architecture.

Table 9.4   The various types of IFS.

Type	Description
Short IFS (SIFS)	This is the shortest length of IFS. It is the amount of time waited by stations in reply of RTS, CTS and data frame. Due to its short duration, using it essentially prevents other stations waiting for the medium access from using the medium, so the whole sequence of data frame transmission can be finished.
Point Coordination Function IFS (PIFS)	This is the second shortest length of IFS. In PCF, a centralized controller continuously polls the stations for traffics during contention free period. PIFS is the time waited before the start of contention free period and also the time wait for the response of polling.
Distributed Coordination Function IFS (DIFS)	This is the longest length of IFS. It is the minimum amount of time waited by stations after the medium becomes free in DCF.

Other than using various types of IFS, in order to avoid collisions, random backoff mechanism is also introduced in DCF. When the stations wait for DIFS after the medium becomes free, they enter backoff time. They continue waiting for a certain amount of time based on the slot randomly chosen. Initially, there are eight slots to choose from. However, if a collision occurs, the number of slots is then doubled (exponentially, and up to 256 slots) after each retransmission. This method is called binary exponential backoff. As long as one of the stations successfully sends out the frame during backoff time, the remaining waiting time slots chosen by all other stations are kept so that their backoff waiting process can be resumed in next contention window.

Figure 9.8 demonstrates a basic access method of DCF, illustrating the operation of various IFS types and random backoff mechanism.

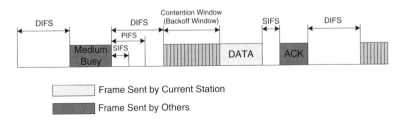

Figure 9.8   DCF basic access method.

Since DCF is contention-based in nature, it is only suitable for asynchronous traffic which does not mandate any QoS requirements. However, in order to support time-sensitive traffics, the scheme must provide some contention-free period dedicated for that kind of traffics. This leads to the concept of PCF. As indicated in the MAC architecture shown in Figure 9.7, PCF is built on top of DCF. It is possible for it to support both asynchronous and time-sensitive traffics simultaneously with the use of alternating contention-free period and contention period structure. This is indicated in Figure 9.9.

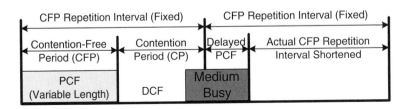

Figure 9.9   Alternating contention-free period (CFP) and contention period (CP).

As shown in Figure 9.9, there is a repeated fixed period of time for a CFP (contention-free period) to occur. This repeated fixed period of time is divided into two subperiods, they are CFP and CP. During CFP, the central point coordinator (usually the AP) polls each of the stations within the BSS. The polling structure of CFP is shown in Figure 9.10. We can see that before entering CFP, the point coordinator waits for PIFS and accesses the medium afterwards and this indicates the beginning of CFP. Afterwards, the point coordinator polls Station 1 as shown in the figure. If Station 1 has frames to send, it replies with a frame containing the DATA and ACK after waiting for SIFS. Then the point coordinator polls the next station, Station 2, as illustrated in Figure 9.10. Since Station 2 does not have any frame to send and thus does not reply to the polling messages. The coordinator then waits for PIFS and poll the next stations or send out CF-END frame which indicates the end of CFP.

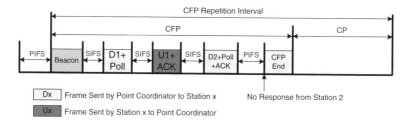

Figure 9.10   CFP structure.

After CFP, it comes to CP. During this period, each station acts just like that in DCF. They compete for the medium with the help of IFS and random backoff as described previously. However, as indicated in Figure 9.9, there may be chances that the medium is busy during the start of CFP. In this case, the point coordinator needs to wait until the medium becomes free and starts the CFP afterwards. In this case, the actual CFP repeat period for this time is shortened to meet the original schedule.

9.5   IEEE 802.11 PHYSICAL LAYERS

9.5.1   Physical Link Features

The transmission capabilities of the physical link of IEEE 802.11 keep on improving. With the help of techniques like Orthogonal Frequency Division Multiplexing (OFDM) and the release of more and more unlicensed band for usage, the bandwidth capacity has been greatly increased. Originally, the first specification was released in 1997, in which three physical link types were proposed—Direct Sequence Spread Spectrum (DSSS), Frequency Hopping Spread Spectrum (FHSS), and infrared [323, 128].

DSSS is a technology which divides the 2.4 GHz ISM frequency band into channels with spacing of 5 MHz in center frequency. The center frequency of channel 1 is located at 2.412GHz and so on. Each channel then takes up 22 MHz bandwidth. There are altogether 14 channels defined. However, due to spectrum regulation in different countries, there are at most 13 channels available. The system implementing DSSS divides a data bit into 11 chips according to Barker Sequence as shown in Figure 9.11. By doing so, the original narrow band signal is spread into a relatively wide band signal. The receiver side receiving the wideband signal can use the same sequence to decode the original narrow band signal due to good autocorrelation property of Barker Sequence.

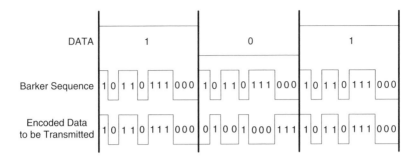

Figure 9.11 DSSS mechanism.

In the original IEEE 802.11, only two data rates are provided, they are 1 Mbps and 2 Mbps. They are using DBPSK and DQPSK as modulation scheme respectively. DBPSK is able to encode one bit in one symbol, while DQPSK can encode two bits in one symbol.

FHSS is a technology which divides the 2.4 GHz ISM frequency band into frequency slots of 1 MHz each. The center frequency of a frequency slot in return determines the channel number of the slot. Channel 1 corresponds to the central frequency of 2.401 GHz and so on. The total number of channels depends on countries, ranging from 23 channels in Japan up to 78 channels in the United States and most European countries. By dividing the frequency band into channels, the stations implementing FHSS keep on changing channels where they do not stay more than 0.4 second in each hop according to US regulations. Which channels are to be hopped next is determined by some predefined hopping sequence set with regulation on minimum hopping distance depending on countries. Figure 9.12 demonstrates a sample FHSS system with two stations, without consideration of hopping distance.

As mentioned before, the two data rates defined in IEEE 802.11 are 1 Mbps and 2 Mbps. In the FHSS case, with 1 MHz occupied frequency and two-level GFSK, 1 Mbps data rate can be obtained. By using two-level GFSK, only one bit is encoded into one symbol, so the data rate supported is 1 Mbps. Similar to the case of 1 Mbps, for the date rate of 2 Mbps, four-level GFSK modulation is used, with two bits encoded into one symbol.

Infrared can be divided into two main categories—line of sight infrared and diffused infrared. IEEE 802.11 defines diffused infrared in which a device is not required to be strictly within line of sight of each other. The infrared light is reflected by walls, ceiling, etc. IEEE 802.11 infrared defines two data rates—1 Mbps and 2 Mbps. For 1 Mbps, four data bits are mapped to a predefined group of 16-bit symbol. This modulation scheme is called 16 Pulse Position Modulation (PPM). For the case of 2 Mbps, two data bits are mapped to a predefined group of four bits symbol and it is called 4 PPM.

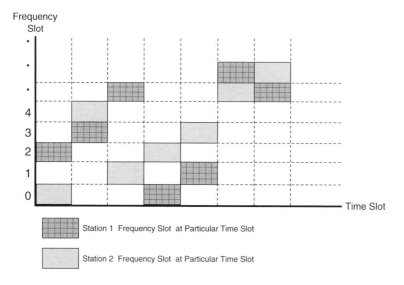

Figure 9.12   Orthogonal FHSS hopping sequence without consideration of hopping distance.

Orthogonal Frequency Division Multiplexing (OFDM) is not defined in original IEEE 802.11 PHY. However, it is proved to be effective in wireless medium usage and so it is included in IEEE 802.11a/g PHY. It is similar to Discrete Multi Tone (DMT) used in the Digital Subscriber Line (DSL). The frequency channels in the frequency band are divided into channels of 5 MHz spacing in center frequency, while OFDM occupies four channels with a total of 20 MHz. Within this 20 MHz, it is further divided into 52 smaller frequency slots with 0.3125 MHz apart and are called subcarriers. Within these subcarriers, only 48 are used to transmit data, while the remaining four are used for control. In this case, data bits from a single source is possible to be sent out through all of the subcarriers in parallel and more advanced modulation techniques like QAM, which can code more bits into a symbol, can be applied.

9.5.2   IEEE 802.11a

IEEE 802.11a specifies the usage of WLAN in 5GHz unlicensed band—Unlicensed National Information Infrastructure (UNII) [323, 7]. This band provides much larger amount of bandwidth for usage and is less crowded than that of 2.4 GHz ISM band. With the usage of OFDM, there are 13 (original) + 11 (currently new) orthogonal channels for usage according to United States Regulation. That is, 24 channels can be operated within the same region without interfering with each other. The original channels occupy the frequency from 5.15–5.35 GHz and 5.725–5.825 GHz, while the new channels occupy the frequency from 5.470–5.725 GHz according to new FCC rules in November 2003.

As we have discussed earlier, OFDM and wide bandwidth channels allow more complicated wireline modulation techniques to be used. They support modulations like BPSK, QPSK, 16-QAM, 64-QAM. This in return provides a higher throughput capacity. With the help of OFDM and different modulation schemes supported, data rates ranging from 6 Mbps to 54 Mbps are supported

9.5.3   IEEE 802.11b

IEEE 802.11b was proposed in 1999 together with IEEE 802.11a [323, 8]. It uses the 2.4 GHz ISM band. There are several design objectives of IEEE 802.11b that make it more popular than IEEE 802.11a. One of the main reasons is that IEEE 802.11b is using the 2.4 GHz ISM band which is the same as that defined in IEEE 802.11. Together with the support of low data rate DSSS (1–2 Mbps), backward compatibility is provided. This facilitates the interoperability of IEEE 802.11b and existing WLAN products on the market. At the same time, with relatively simpler modulation techniques and mass production, IEEE 802.11b devices are relatively cheaper.

IEEE 802.11b uses the 2.4 GHz ISM band and has three orthogonal channels. However, in order to provide data rates larger than 2 Mbps, another chip spreading method is used. In original DSSS, 11 bits Barker Sequence is used to spread the data bit into 11-bit transmitted symbol, which is then recovered at the receiver side. Consequently, 1M symbols are transmitted in a second. However, with the usage of Complementary Code Keying (CCK), an eight-bit sequence is used to encode four bits or eight bits data. In this case, spreading effect can be achieved together with multi-bits data encoding that results in a higher data rate. Thus, by using CCK, 1.375M symbols are transmitted in a second which results in a data rate of 5.5 Mbps with the use of 4-bit encoded symbol and even up to 11 Mbps with the use of 8-bit encoded symbol.

9.5.4   IEEE 802.11g

Although IEEE 802.11b is more popular than IEEE 802.11a, there is a gap in throughput capacity between these two standards. People enjoy better interoperability brought by IEEE 802.11b on one hand, while they would like to have a larger throughput capacity on the other hand. In order to get the best of both worlds, IEEE 802.11g is created to provide a higher throughput capacity in the 2.4 GHz ISM band [9].

To get a higher throughput capacity, OFDM is adopted in IEEE 802.11g. However, to provide backward compatibility to IEEE 802.11b devices, DSSS is also implemented. In order to use OFDM with the compatibility of IEEE 802.11b, another mode of operation is introduced in IEEE 802.11g, that is DSSS-OFDM mode. In using this mode of operation, the headers and permeable are sent by using DSSS, so that other IEEE 802.11b devices notice about the frame sent by IEEE 802.11g, while the data are sent by OFDM to achieve the highest possible data rate. With different mode of operations, IEEE 802.11g can support data rates ranging from 1 Mbps up to 54 Mbps with three coexisting orthogonal channels.

9.5.5 IEEE 802.11ag Dual Band

As indicated in the previous section, the main aim of IEEE 802.11g is to provide a throughput capacity higher than IEEE 802.11b by applying OFDM to the 2.4 GHz medium [71]. However, due to several limitations, using solely an IEEE 802.11g network may not be a good solution. It is because:

- IEEE 802.11g has three orthogonal channels only.

- The 2.4 GHz ISM band is crowded with many devices including microwave oven, cordless phone, Bluetooth, etc. Significant interferences are expected.

- IEEE 802.11g network's performance is severely affected in the presence of IEEE 802.11b stations due to the backward compatibility issue.

On the other hand, using solely an IEEE 802.11a network may not be a good solution either. Because:

- IEEE 802.11a does not provide backward compatibility to other IEEE 802.11x networks.

- Generally speaking, IEEE 802.11a devices are more expensive than IEEE 802.11b or IEEE 802.11g devices.

- IEEE 802.11a devices are less popular than that IEEE 802.11b.

Since both IEEE 802.11g and IEEE 802.11a have their own limitations and advantages, it is possible to combine both devices in order to get the best of both worlds. In view of this, dual-band IEEE 802.11/ag devices are implemented and are expected to dominate the WLAN market in near future. Generally speaking, since IEEE 802.11a has more orthogonal channels and aggregated bandwidth, and it is using 5 GHz UNII band which is less crowded, it is more suitable to be used to support QoS-based applications which generally require larger bandwidth and are more sensitive to delays. On the other hand, since IEEE 802.11g devices are more popular and are integrated into many devices which do not require or do not have strict QoS requirements, it is most suitable for some "best effort" applications, for example, normal data networking, some relatively low quality audio and video streaming etc.

Since WLAN becomes more and more popular, and the users of both QoS-based and "best effort" applications are expected to increase due to the integration of WLAN to new electronic devices, IEEE 802.11ag clearly becomes one of the best solutions to improve the performance of WLAN in future.

9.6   IEEE 802.11E FOR QOS PROVISIONING

The MAC and PHY of various IEEE 802.11 standards have been discussed in previous sections. In IEEE 802.11 MAC, only PCF has addressed some QoS-related issues. However, it does not provide a very clear picture on what and how QoS to be supported in WLAN. At the same time, there are two severe problems in PCF that make QoS support less effective. The two problems are unpredictable Beacon delay due to busy medium in CP and the unknown transmission time of the polled stations in CFP. In view of the problems, IEEE 802.11e standard is being developed based on the PHY of IEEE 802.11a/b/g. In the IEEE 802.11e draft standard, two main types of traffic classification are addressed—Traffic Category and Traffic Stream. These two traffic classification methods are described in Figure 9.13 [126].

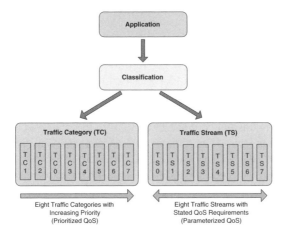

Figure 9.13   IEEE 802.11e traffic classification types.

In Traffic Category (TC), traffic flows are categorized into eight distinct classes with different priorities according to their traffic types predefined, from TC0–TC7. As shown in Table 9.5, a traffic priority is assigned based on the urgency to use the medium, the lowest priority is assigned to TC1 which refers to background traffic, while the highest priority is assigned to TC7 which refers to network control traffic. It is similar to the Differentiated Service suggested in wired networks.

For the case of Traffic Stream (TS), it is similar to that of Integrated Service in wired networks. In each station, eight traffic streams can be supported. Each stream is required to be characterized by the Traffic Specification (TSPEC). In TSPEC, information characterizing the traffic stream, including the packet size, arrival rate, maximum delay and delay variance should be specified.

In order to support these two traffic classifications, two new MAC modes of operation have been introduced in IEEE 802.11e. They are Enhanced Distributed Coordi-

Table 9.5 IEEE 802.11e traffic categories.

Traffic Category	Type	Priority
TC1	Background traffics	1 (lowest)
TC2	Spare traffics	2
TC0	Best effort data traffics	3
TC3	Excellent data traffics	4
TC4	Controlled load data traffics	5
TC5	Multimedia traffics with delay less than 100 msec	6
TC6	Multimedia traffics with delay less than 10 msec	7
TC7	Network control traffics	8 (highest)

nation Function (EDCF) and Hybrid Coordination Function (HCF). Similar to legacy IEEE 802.11 PCF, there are two periods of time within the CFP repetition interval. Similar to the role of DCF in PCF, EDCF is used within CP only and HCF is used in CFP. Unlike that of PCF, it is possible for HCF to be used within CP. A station that has implemented this function is called enhanced station. It may be possible for this station to act as centralized coordinator (similar to the role of PC in PCF) and this coordinator is called Hybrid Coordinator (HC). At the same time, the BSS with IEEE 802.11e supported HC and stations is called QBSS.

First of all, let us focus on EDCF. EDCF is mainly used to support traffic category (TC) that has been discussed previously [126]. It provides differentiated services for different priority queues waiting for accessing the shared wireless medium. It can support up to a maximum of eight priority queues. For differentiation purpose, first of all, the packets are classified into different queues, then the queues are prioritized by setting different Arbitration IFS (AIFS) values with AIFS length at least being equal to DIFS. The main reason for this is to provide support for non-QoS supported stations, so they can still compete for the wireless medium using the original DCF method. As shown in Figure 9.14, the higher the priority, the lower the AIFS value is set to that queue.

Similar to DCF, EDCF is using CSMA/CA. That is, before sending out a frame, the station is required to sense the medium to make sure that it is free, then each queue within the station waits for AIFS and make sure the medium is free. Then the queues can get into the contention window period similar to that of DCF. However, in EDCF, each queue can be associated with a persistence factor (PF) which controls the contention window size after each failure transmission of frame from that specific queue. In case that multiple backoff counters reach zero simultaneously (virtual collision), the medium access right is given to the highest priority queue, all other queues within the station undergo backoff again. The queue successfully gets the right to transmit within the station may have collision with other stations, however if collision does not occur, it can transmit their frames up to a maximum duration predefined in Beacon and this is called Transmission Opportunity (TXOP).

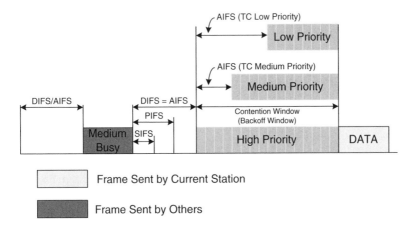

Figure 9.14 IEEE 802.11e EDCF.

HCF is mainly targeted for Traffic Stream (TS) [126] as we described earlier. The concept of HCF is very similar to PCF in the sense that there is a centralized coordinator which controls the access of medium by polling the stations. However, unlike that of PCF, stations are required to signal the HC with the QoS specifications (TSPEC) which state the requirements of the traffic stream of applications. In this case, HC grants certain TXOP to specific stations at certain period of time during CP or CFP.

As shown in Figure 9.15, during the CFP, the station polled by HC is allowed to transmit the frame within the TXOP specified in the CF-POLL frame. The CFP is terminated by the time specified in Beacon or specifically by a CF-END frame. During the CP, the QoS-enabled stations compete for the medium using EDCF as stated previously. However, it is also possible for HC to poll stations specially during CP. Since HC only waits for PIFS, while all other stations wait for AIFS (at least DIFS), HC has the priority access to the medium.

By using HCF and EDCF, the problems of delayed start of CFP and unknown duration of polled stations transmission can be solved, since the transmission duration is controlled by TXOP, QoS CF-Poll frame and Beacon frame. By doing so, QoS can be supported in WLAN environment.

9.7 ADVANCED DEVELOPMENTS

As we have discussed in Section 9.5, various Physical Link technologies have been proposed and eventually various IEEE 802.11 standards have been developed to suit

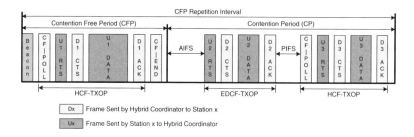

Figure 9.15 IEEE 802.11e HCF.

the needs of different users. However, users constantly request for better performances including higher throughput and larger coverage area. This comes up to some non-standardized ideas for increasing the throughput and extending the support range of standardized IEEE 802.11.

In order to increase the currently supported throughput capacity of the IEEE 802.11g, some WLAN vendors suggested and implemented some solutions like bursting, combined frames, multiple channels support etc. and these ideas are elaborated in detail as follows.

First of all, let us start with bursting. If a station knows that it has multiple frames to send out, after it takes control of the medium (after DIFS and random backoff), it continues controlling the medium by only waiting for SIFS which is the shortest possible IFS defined in IEEE 802.11 after each successive data frames are sent. In this case, a burst of frame can be sent out consecutively without the overhead of relatively large DIFS and contention windows. The whole sequence of actions is explained in Figure 9.16 [73].

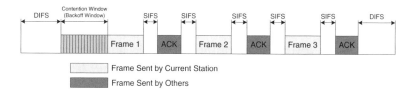

Figure 9.16 Bursting of MAC frames.

Other than bursting, another approach suggested in increasing the throughput is the combined frames approach. The scenario, in fact, is very similar to the case of bursting. When a station knows that it has multiple frames to send out, it can then make use of the opportunity that it has successfully taken control of the medium to send out a combined frames up to a maximum possible size. In this case, frames of different sizes are sent out instead. This approach is illustrated in Figure 9.17.

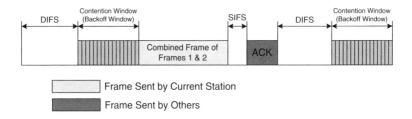

Figure 9.17 Combined MAC frames.

Furthermore, as IEEE 802.11b/g has three orthogonal channels, while IEEE 802.11a has 24 orthogonal channels, it is possible for an AP to choose more than one channel to transmit data if it finds that the bandwidth available is not enough to cope with the demands. In this case, the stations can make use of channels which are not used by other APs to increase the bandwidth significantly.

With the use of above idea, a network with more than 54 Mbps of data rate can be achieved. Other than higher throughput, users also expect longer supported range, larger coverage area and more reliable transmission of WLAN. In order to meet the demands of users, some vendors suggest the idea of enhancing the received sensitivity based on standard IEEE 802.11 . As shown in Figure 9.18 [72], when the station is within the original transmission range of IEEE 802.11, AP communicates with it using standard IEEE 802.11a/g with the transmission rate of 54 Mbps to 6 Mbps. However, once the station moves out of the original range, communication is carried out with smaller additional transmission rates from 3M to 0.25M. It is because a smaller transmission rate, in fact, has lower requirements in receive sensitivity and accuracy by the use of some simple modulations. In other words, signals with lower receive sensitivity can still be recognized and successfully decoded. Together with the improved MAC and PHY to support low Signal to Noise Ratio (SNR), the coverage area can be greatly increased.

Although the idea suggested results in a higher throughput and extended transmission range, there are some problems. The most serious problem is that the suggested idea is non-standardized. In other words, vendors can implement their own design and this leads to a serious problem in interoperability between WLANs from different vendors. At the same time, although the vendors suggesting these idea insist that they have backward compatibility to third party standard IEEE 802.11 design, their idea, in fact, leads to other problems that may break the design goals of IEEE 802.11. Just like the case of bursting and combined frames, these two methods lead to a higher throughput to one station, however, other stations implementing standard IEEE 802.11 may be required to wait for longer time before they can access the medium and this leads to the problem of fairness.

Finally, a recent development in further enhancing the capability of WLAN is to use the multiple-input-multiple-output (MIMO) physical layer technology. In simple

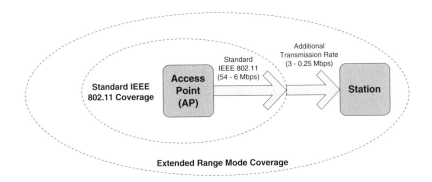

Figure 9.18   WLAN extended range.

terms, using multiple (say n) transmit antennae and multiple receive antennae, the data rate can be increased by n times. Indeed, it is estimated that the raw data rate of the MIMO based WLAN (called IEEE 802.11n) will be over 200Mbps.

9.8   PRACTICAL ILLUSTRATION: HOME NETWORK

WLAN is widely used in many places, for example, at home and at company, to provide networking facilities to the location where putting additional wires is impossible. Recently, there is a trend that WLAN is integrated into shopping malls and coffee shops to create hot spots for users to have temporary access to the Internet. In this section, we focus on a foreseen usage scenario of wireless network in near future—wireless home network [264].

In a home environment shown in Figure 9.19, coax cable is used to communicate with outside world. After the cable gets into the home, it is further split into three cables by cable splitter to support different locations within the home. There are three types of stations existing in the home environment. The first one is the station directly supported by the cable, while the second one is the station which communicates with the repeater which acts as a bridge between the cable and wireless medium, so as to extend the coverage of wireless network by transmitting the signal back to an access point. The third one is the station which communicates with an access point which implements both wireless station and coax station functionalities.

As shown in Figure 9.19, audio and video sources such as television programmes, movies, songs etc. is temporarily stored in the Set-Top-Box (STB) inside the home and transmitted through the coax cable. When Station 1 requested a certain audio or video programme, the programme is then moved to the thin Set-Top-Box and sent to Station 1 through the wireless medium. Since IEEE 802.11a/g has a bandwidth capacity of up to 54 Mbps, together with IEEE 802.11e QoS proposal, it is sufficient in supporting multimedia distribution. At the same time, when Station 2 wants to surf the Internet, it can send the request to the repeater which then forwards the request to the access point which is connected with a cable modem for Internet access. In this case, solely coax cable and IEEE 802.11x wireless access technologies are sufficient in providing entertainment like VOD and Internet access to all users in all locations at home.

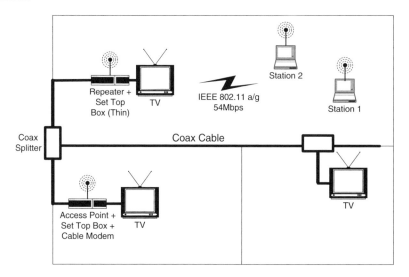

Figure 9.19   Wireless home network scenario.

In this scenario, solely cable may not provide the coverage area comparable with that of combination of both cable and wireless technologies, while solely wireless technologies may not provide enough bandwidth and coverage area, as some of the locations may be blocked by thick walls which cause great signal attenuation. Combination of both technologies can provide a highly satisfactory solution and it is expected to be more and more popular in near future.

9.9 SUMMARY

In this chapter, various IEEE 802.11x standards have been discussed. Within the standards, various MAC modes of operation, from the original DCF and PCF to the currently proposed QoS-supported HCF and EDCF have been discussed in detail. Moreover, various PHY link types including DSSS, FHSS, infrared and OFDM have also been mentioned together with their integration and various usages in IEEE 802.11x standards.

Since WLAN brings us many conveniences, more and more people and electronic devices vendors choose WLAN as a medium of communication. These lead to the constant demand for increasing the bandwidth and coverage area. In view of this, vendors have tried to develop their own improved add-on technologies based on the standard IEEE 802.11x in order to achieve the above-mentioned goals and to provide the backward compatibility to those third party devices. Besides, they have also started thinking about the possibility of combination and implementation of various standards on a single device.

Same as many people, the authors also expect the potential usages of WLAN in various environment. Just like the home network example given in Section 9.8, with the bandwidth comparable to that of wired networks and large coverage area supported without the need of extra wires, IEEE 802.11x technologies can be used in home in combination of coax to provide the features of home theater and Internet access everywhere. Innovative applications of the IEEE 802.11x technologies are only bounded by our imagination.

PROBLEMS

9.1   Describe the differences between different modes of operation in IEEE 802.11x MAC layer (DCF and PCF) and how HCF and EDCF can be used to solve the problem of PCF in supporting QoS in wireless network.

9.2   Describe the methods used by vendors to increase the throughput capacity and the coverage area of WLAN, and how these actions in return affect the development of WLAN.

9.3   Briefly describe OFDM. Describe the differences between IEEE 802.11a and IEEE 802.11g and explain why IEEE 802.11ag may be a good solution in solving the problem of lacking bandwidth capacity.

CHAPTER 10

BLUETOOTH WPAN

No one is born a master.

—Italian Proverb

10.1 INTRODUCTION

While IEEE 802.11x WLAN technologies discussed in the previous chapter are already of a low cost, they are still considered not viable for many low-end commodity devices such as keyboard, mouse, etc. Indeed, to enable "cable replacement" for such commodity devices, WLAN technologies are too heavy-weight. In this chapter, we discuss about a low cost but effective short-range wireless technology.

Interoperability:

Bluetooth was originally developed by Ericsson in 1994 and its development is now managed by Bluetooth Special Interest Group (SIG). Bluetooth was developed in view of the great necessity in short-range cable replacement solutions, while existing solutions are insufficient in satisfying the needs of users. During the development of Bluetooth, special care has been taken on interoperability in two dimensions—interoperability with existing "cable replacement" interfaces and interoperability with other Bluetooth devices.

In order to facilitate fast pace development of Bluetooth, Bluetooth SIG decided to build up the protocols in protocol stacks from scratch that are necessary for the functions of Bluetooth only. Examples include the Core Protocols which include both Bluetooth PHY and MAC. On the other hand, some other protocols in Bluetooth are modified and/or adopted from existing solutions. There are quite a number of such protocols. For example, OBEX is an existing protocol developed for object exchange in infrared, and WAP is also an existing protocol developed for carrying Internet materials for portable devices like mobile phone. These existing protocols have been adopted into Bluetooth. Consequently, applications originally designed for other cable replacement interfaces can be used in Bluetooth also. This not only increases the development pace of Bluetooth, but also provides interoperability with current existing cable replacement solutions. Together with Bluetooth protocol architecture, these issues are further discussed in Section 10.3.

On the other hand, Bluetooth also provides interoperability with other Bluetooth devices by standardizing the protocol stacks and profiles. Profiles, in fact, record the protocol requirements necessary to be implemented for supporting specific usage model. By standardizing usage model with profiles, any products claimed to support specific usage models and profiles should implement the same protocol stacks under the same requirements. In this manner, interoperability between different Bluetooth devices can be ensured. Together with Bluetooth profiles, these issues are further discussed in Section 10.3.

Performance:

Bluetooth, similar to IEEE 802.11 b/g, uses the 2.4 GHz ISM band. It provides a theoretical capacity up to 1 Mbps due to the design of its PHY and MAC. This is covered in detail in Section 10.4. On the other hand, Bluetooth supports both asynchronous traffics and synchronous traffics with the help of two physical channels—ACL and SCO. With the help of SCO, some time slots can be reserved for time sensitive applications like audio, which provides certain capability of QoS. This is further discussed in Section 10.4. Moreover, some current suggestions on the improvement of QoS support in Bluetooth are discussed in Section 10.5. Bluetooth can provide a theoretical capacity up to 1 Mbps. However, in actual scenarios, this theoretical capacity cannot be achieved. Together with some currently released analytical and simulated results, this issue is further elaborated in Section 10.6.

10.2 DESIGN GOALS

IEEE 802.11 has been developed for years. There was a great controversy on the necessity of the development of a new wireless technology in complementary to the existing one. Bluetooth has been developed with several objectives that defy the controversy. These objectives can be summarized in Table 10.1.

Table 10.1   Bluetooth design goals.

Objective	Description
Low Cost	Bluetooth is a one-chip solution with the goal of US$5 per chip in the long run due to the expected mass production. With this low cost, vendors would be able to integrate Bluetooth into their products without great increase in their production costs and users would be able to afford the products.
Cable Replacement	The development objective of Bluetooth is not aggressive, it mainly targets at functioning as a short-range cable replacement for devices like headset of mobile phone, PDA, etc. Together with its low power consumption and low cost, more and more users use Bluetooth devices as a replacement for lengthy, inconvenient cables.
Ad Hoc Networking	In the market, many devices like remote control, PDA, mobile phone, etc. provide infrared as wireless point-to-point communication technologies. This has been considered appropriate because most of the connections between these devices are only set up temporarily (i.e., ad hoc in manner). In view of this, with the interoperability brought by Profiles specification, different products can communicate with each other and set up a temporarily connection as long as they have implemented the same profiles.

Currently, more and more usage of Bluetooth has been discovered and implemented. New design objectives keep emerging to support new usage models. For example, there is a proposal of increasing the theoretical capacity of Bluetooth from the original 1 Mbps to 2 Mbps or more.

10.3   BLUETOOTH PROTOCOL STACK

The development of Bluetooth technologies and Bluetooth protocol specifications are carried out by Bluetooth Special Interest Group (SIG). Originally, Bluetooth specifications are divided into two parts—Core specification and Profiles specification. The Core specification covers a layered protocol stacks architecture, in which protocols including the Bluetooth Physical layer up to Bluetooth specific Logical Link Control layer are defined. At the same time, some implementation details including maximum and minimum power dissipation, and the maximum supporting range, are also included. On the other hand, Profile Specification mainly focuses on specific usage models of Bluetooth. The specification v1.1 covers 13 Profiles and more extended profiles are released currently. In each profile, the specification defines clearly on what Bluetooth Protocol Stacks are required to implement and what other actions could be implemented as an options. By doing so, devices from different vendors which claim to support specific profiles would be able to interoperate [1, 2, 323].

First of all, let us consider the Core specification in some more details. In Core specification, a layered Bluetooth Protocol Stack are defined as shown in Figure 10.1. We can see that within the protocol stack, four group of protocols are involved. They are Core Protocols, Cable Replacement Protocol, Telephony Control Protocol and Adopted Protocols. Their components and corresponding functions are summarized from Table 10.2 to Table 10.5.

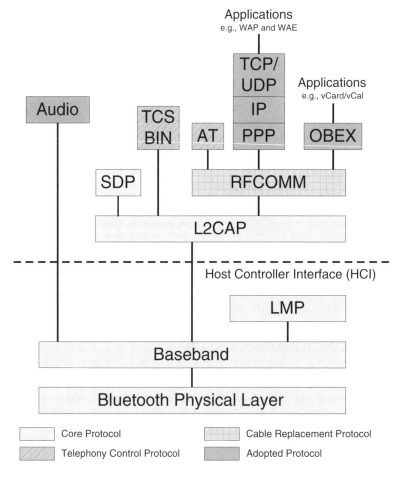

Figure 10.1 Bluetooth protocol stack.

After discussing the protocol stack in Core specification, We now proceed to describe the Profiles specification. Figure 10.2 shows Profiles specification v1.1

Table 10.2   Members of Core Protocols and their functions.

Member	Description
Bluetooth Physical Layer	This layer specifies the details about the radio characteristics of Bluetooth, including the frequency, frequency hopping usage, power and modulation.
Baseband	This layer specifies connection establishment in piconet including the physical link establishment, clock synchronization, etc.
Link Manager Protocol (LMP)	This layer manages the link setup and ongoing link management. On the other hand, it is also responsible for security issues including authentication and encryption.
Logical Link Control and Adaptation Protocol (L2CAP)	This layer acts as an interface between the upper protocol layer and baseband which multiplex traffic flows that are connectionless and connection-oriented. It also provides packet segmentation and reassembly for better management of the link usages.
Service Discovery Protocol (SDP)	This layer is mainly responsible for discovering other Bluetooth devices in nearby physical locations, and their corresponding characteristics including the services provided and some basic information, for example, the profiles, the link types and the frame types supported, etc.

Table 10.3   Member of Cable Replacement Protocol and its function.

Member	Description
Radio Frequency Communications Protocol (RFCOMM)	This layer is used to emulate a serial port over L2CAP layer based on European Telecommunications Standards Institute (ETSI)'s Technical Standard (TS) 07.10 that is originally designed for Global System for Mobile (GSM) communications. This protocol emulates RS232 serial port (or commonly named as COM port)'s control and data signal over L2CAP protocol. By doing so, upper layer protocols generally used for wired networks (for example, PPP over modem) can be used here.

Table 10.4 Member of Telephony Control Protocol and its function.

Member	Description
Telephony Control Specification Binary (TCS BIN)	This layer is a bit-oriented protocol defined for call control signaling which is used to build up a data and speech call between Bluetooth devices based on the ITU-T Q.931. At the same time, mobility management procedures are also provided to maintain group of Bluetooth devices.

Table 10.5 Members of Adopted Protocol and their functions.

Member	Description
Point-to-point Protocol (PPP)	This is a protocol defined by Internet Engineering Task Force (IETF) for carrying IP datagrams traffic over serial point to point link (for example, RS232 link).
Internet Protocol (IP)	Connectionless network protocol commonly used in Internet.
Transmission Control Protocol (TCP) and User Datagram Protocol (UDP)	TCP and UDP are respectively connection-oriented and connection-less transport layer protocol used in Internet.
Object Exchange Protocol (OBEX)	This protocol is originally developed by Infrared Data Association (IrDA) to support object exchange between devices. This protocol is subsequently adopted into Bluetooth for object exchange between devices.
Wireless Application Protocol (WAP) and Wireless Application Environment (WAE)	A protocol which provides sending and receiving Internet materials on small wireless devices like mobile phone. This is discussed in detail in Chapter 15.

and their corresponding relationship. As we have discussed earlier in this section, Bluetooth Profiles specification v1.1 has included 13 profiles and there are constantly some more profiles released as extended profiles due to the discovery of more usage models of Bluetooth. Examples include Personal Area Networking (PAN) for building up a network, Human Interface Device Profile (HID) for Bluetooth keyboard, mouse or gamepad, Basic Printing Profile (BPP) for Bluetooth enabled printer, etc.

In fact, the main function of profiles is to define which specific components of protocol stacks should be implemented in supporting a certain usage model. Consequently, interoperability is provided between products from different vendors that claim to support specific profiles. The most basic one is Generic Access Profile which specifies most core protocols. As a result, most profiles depend on Generic Access Profile, as shown in Figure 10.2. On top of it are Telephony Cordless Binary Based Profiles which include TCS BIN layer, and Serial Port Profile which includes RF-COMM layer. With the help of RFCOMM, many other profiles can be implemented, such as General Object Exchange Profile which includes OBEX layer, LAN Access Profile (LAP)[1] which includes PPP layer and optional TCP/IP layer, Dial-Up Networking Profile which includes AT commands layer for modem control and optional PPP layer, etc. By building profiles on top of others, adopted/additional functions can be implemented easily and this allows Bluetooth to interoperate with programs using protocols originally designed for other technologies. OBEX originally designed for infrared for object exchange is one of the best examples.

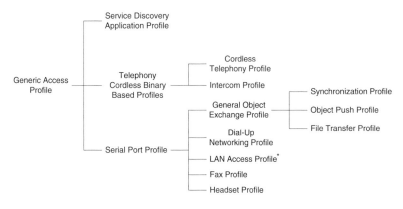

* deprecated in specification v1.2

Figure 10.2   Bluetooth profiles specification v1.1.

[1]LAP is deprecated in v1.2 [2] due to the release of PAN, which is specially designed for supporting networking applications. Some profiles, for example, WAP, which rely on LAP are also moved above PAN in the new release.

In fact, due to the development of more and more usage models on Bluetooth, more and more profiles are expected and developed. As a result, Bluetooth protocol stacks keep on expanding with some newly developed layers specifically for Bluetooth, for example Bluetooth Network Encapsulation Protocol (BNEP) in PAN profile, and some newly adopted layers.

10.4   BLUETOOTH PHYSICAL AND MAC LAYERS

10.4.1   Bluetooth Physical Layer

Bluetooth uses the 2.4 GHz ISM frequency band, similar to that of IEEE 802.11 discussed in Chapter 9. This frequency band is unlicensed and can be used by devices as long as specific frequency and power regulations are followed. Accordingly, this frequency band is crowded and so any technology using this band should have mechanism in withstanding collisions with other technologies using the same frequency band.

In view of this, Bluetooth adopted Frequency Hopping Spread Spectrum (FHSS) technique described in Section 9.5 in its Physical (PHY) layer with 79 channels[2] and bandwidth of 1 MHz each. The frequency band details are described in Table 10.6 [1, 2, 53, 323]. Within this frequency band, Bluetooth has a relatively high hopping rate (1600 hops per second) so as to get a better resistance to interference from devices using the same band. Together with GFSK as the modulation scheme, it can provide up to a maximum throughput capacity of 1 Mbps.

Table 10.6   The frequency band usage of Bluetooth.

Frequency Band Range	2.400 - 2.4835 GHz
RF Channel	$f = 2.402$ GHz $+ n$ MHz, $n = 0, ..., 78$
Lower Guard Band	2 MHz
Upper Guard Band	3.5 MHz

Other than frequency bands and channels requirements, radio specification in Bluetooth also specifies three power classes which in return determine the transmission range of Bluetooth devices. The classes and their corresponding power output are summarized in Table 10.7.

10.4.2   Bluetooth MAC Layer

In this section, we describe Bluetooth MAC layer in detail [1, 2, 53, 323]. Bluetooth Baseband, in fact, plays a crucial role in the Bluetooth MAC layer, as it supports the functions of MAC and helps in frequency hopping, clock synchronization, etc.

---

[2]In specification v1.1, both 79 and 23 hopping sequence channels are defined, but the 23 hopping sequence channels are deprecated in specification v1.2.

Table 10.7   The three power classes of Bluetooth.

Class	Maximum Power Output	Minimum Power Output	Remarks
1	100 mW (20 dBm)	1 mW (0 dBm)	Longest transmission range with power control over +4 dBm. Power control is optionally below +4 dBm.
2	2.5 mW (4 dBm)	0.25 mW (-6 dBm)	Optional power control.
3	1 mW (0 dBm)	N/A	Optional power control.

The linkage between Bluetooth devices is based on a master-slave relationship, in which one device becomes the master and other devices become slaves. A master can control not more than seven active slaves. This structure is called *piconet* and is further discussed in Section 10.5. Within a piconet, the master device and slave devices communicate in time division duplex (TDD) manner in which the data transmissions use up the whole frequency slot (1 MHz) with downlink (from the master to slaves) and uplink (from slaves to master) transmission divided into different time slots. This is illustrated in Figure 10.3. Similar to the case of PCF in IEEE 802.11, the master has the function similar to that of point coordinator in the sense that the master polls each slave by sending them a downlink frame. If a slave has backlogged data to send, it then replies with a uplink frame in next time slot. Unlike that of PCF, each time slot in Bluetooth is fixed to be $625\mu s$ which corresponds to 1600 hops in a second, while PCF does not have the concept of time slot instead. As indicated in Figure 10.3, within each time slot, the frame is transmitted using a specific frequency $f(k)$. That is, the master sends the frame using frequency $f(k)$ in the first time slot, then the slave replies with the frame using another frequency $f(k+1)$ in the next time slot.

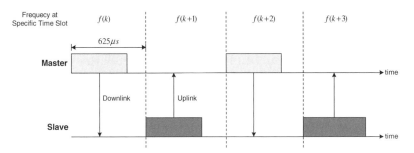

Figure 10.3   Time division duplex in Bluetooth MAC.

The frequency that is used in next time slot is specific to each piconet and this is called Frequency Hopping Sequence. This sequence is derived from the master BD_ADDR and clock. BD_ADDR is extended from IEEE 802 architecture, and because of this, BD_ADDR is unique for each Bluetooth device. Even though frequency hopping sequence is used, there are still chances that the frequency used by different piconets within the same time slot is the same and this leads to collisions. However, with the use of frequency hopping techniques, the chances of colliding within specific time slot becomes smaller and consecutive collisions become unlikely. With the use of error detection and correction techniques like Forward Error Correction (FEC), which is elaborated below, the effect of collision can be practically handled.

Figure 10.3 shows alternating transmission of frame making use of 1-slot time only. However, it is possible for either side to send out frames which take up multi-slots. There are three types of frames—1, 3, 5 slots frames and they are shown in Figure 10.4. Within the transmission of the multi-slot frames, the frequency is unchanged. However, after the transmission of the frame, the next frequency used is the frequency that should be used after multiple time slots, as shown in the Figure 10.4.

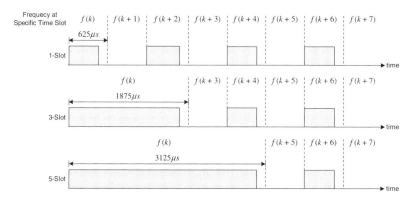

Figure 10.4   Multi-slot frames in Bluetooth.

There are several types of packets defined in Bluetooth. They are control packets, Asynchronous Connectionless (ACL) packets, Synchronous Connection-Oriented (SCO) packet, and extended Synchronous Connection-Oriented (eSCO). Control packets include Identity (ID), NULL, POLL and Frequency Hop Synchronization (FHS) packets. Table 10.8 shows ACL packet types, whereas Table 10.9 shows SCO and eSCO types of packets. Different types of packets are used for supporting different physical links defined in Bluetooth, which are based on different error concealment techniques. There are two main types of physical links that can be established between Bluetooth devices—ACL link and SCO link. The characteristics of these two links are discussed in detail below.

ACL refers to the point-to-multipoint connections between a master and its associated slaves within a piconet. The master exchanges packets with its slaves in a round-robin fashion (although other ordering methods may also be possible), only the slaves with ACL connection addressed by the master is allowed to transmit at that specific time slots. ACL connection is still allowed even though a SCO connec-

Table 10.8 Bluetooth ACL packet types.

Type	Slot	Max Symmetric Rate (kbps)	Max Asymmetric Rate (kbps)	Max Payload (Bytes)	Description
DM1	1	108.8	108.8 | 108.8	17	With CRC and 2/3 FEC.
DH1	1	172.8	172.8 | 172.8	27	With CRC, but no FEC.
DM3	3	258.1	387.2 | 54.4	121	With CRC and 2/3 FEC.
DH3	3	390.4	585.6 | 86.4	183	With CRC, but no FEC.
DM5	5	286.7	477.8 | 36.3	224	With CRC and 2/3 FEC.
DH5	5	433.9	723.2 | 57.6	339	With CRC, but no FEC.
AUX1	1	185.6	185.6 | 185.6	29	No CRC and FEC.

Table 10.9 Bluetooth SCO and eSCO packet types.

Type	Slot	Max Symmetric Rate (kbps)	Max Payload (Bytes)	Description
HV1	1	64.0	10	No CRC, but 1/3 FEC.
HV2	1	64.0	20	No CRC, but 2/3 FEC.
HV3	1	64.0	30	No CRC and FEC.
DV	1	64.0 + 57.6 (Data)	19 (10 + 9 (Data))	No CRC and FEC (voice); with CRC and 2/3 FEC (data).
EV3	1	96	30	With CRC, but no FEC.
EV4	3	182	120	With CRC and 2/3 FEC.
EV5	3	288	180	With CRC and FEC.

tion is established between two Bluetooth devices and there must only be one ACL connection between two Bluetooth devices.

ACL is mainly used to support asynchronous type of traffic, for example, file data, and as such, no reservation is allowed (i.e., packet switching connection). As shown in Table 10.8, there are two main types of packets—DMx and DHx, where x is a number corresponding to the number of slots occupied by that packet type. The main difference between these two types of packets is that 2/3 FEC with (15,10) shortened Hamming code is used in DMx packet type. That is the reason why the maximum throughput capacity achieved by using DMx traffic is around 2/3 that of DHx traffic. Apart from FEC, Automatic Retransmission Request (ARQ) is used in ACL traffics to give the packet further protection.

SCO refers to the point-to-point connection between a master and a single slave with a fixed bandwidth allocated to support this connection. It is mainly used to support synchronous type of traffic, for example, voice (circuit switching connection). Due to this reason, the master reserves a pair of time slots at regular interval for the corresponding slave. Within the reserved time slots, the slave can send out a packet in next time slot no matter whether the packet from master is addressed to it. There are maximum of three simultaneous SCO connections.

Other than DM1, there are four basic types of packets allowed in SCO. They are HVx and DV. As shown in Table 10.9, the main difference between various HVx packets is on the types of FEC used: HV1 uses 1/3 FEC, HV2 uses 2/3 FEC and HV3 does not use FEC. They all provide a fixed transmission rate of 64 kbps which is a standard bandwidth requirement for 8-bit digitized voice. While for the case of DV, both voice and data are be encoded into it with the voice portion providing up to 64 kbps and data portion providing up to 57.6 kbps. For this type of packets, ARQ is generally not required. However, it is adopted in data portion of DV. On the other hand, EVx is the packet types defined in specification v1.2 in eSCO [2]. eSCO is also considered SCO type because it also reserves slots between a master and specific slave. However, the main difference between them is that eSCO allows retransmission in all its packet and allows both symmetric and asymmetric with up to three slots to be used.

10.5 PICONETS AND SCATTERNETS

10.5.1 Overview

There are three topologies defined in the specifications—point-to-point connection, single point-to-multipoint connection (piconet) and multiple point-to-multipoint connection (scatternet). Their basic structures are shown in Figure 10.5 [1, 2, 53]. In the figure, point-to-point connection is shown on the right single connection between a master and a slave. As shown in the figure, there are three different piconets. The masters of Piconet 1 and Piconet 2 connect to three slaves, respectively, while the master of Piconet 3 connects to two slaves. Together, these three piconets are connected together to form a single scatternet. As can be seen in Figure 10.5, there are two methods of linking two piconets together. The two methods are shown in the connection between Piconets 1 and 2, which makes use of master-slave bridge, and the connection between Piconet 1 and 3, which makes use of slave-slave bridge.

10.5.2 Piconet Formation

Before the scatternet formation details are discussed, let us focus on the piconet formation details. In order to set up a piconet, there are two main procedures required—inquiry and paging [1, 2, 53].

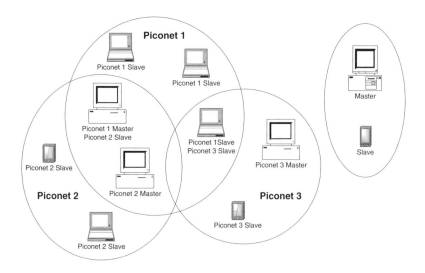

Figure 10.5   Bluetooth topologies.

Inquiry procedure is mainly used by a potential master device to find other sur-rounding devices which are willing to form a piconet. Initially, all the devices are either in Standby or Connection state as shown in Figure 10.6. Firstly, the potential master starts by entering the inquiry state and sends out Identity (ID) packet with inquiry access code (IAC). The potential master makes use of common frequency channels (known as wake up carriers) to send out the ID packet. The wake up fre-quency channels include 32 out of total 79 Bluetooth channels. The IAC ID packet is sent out over each of these wake up channels.

On the other hand, a potential slave enters the inquiry scan state periodically to look for IAC ID packets. In receiving IAC ID packet, the potential slave enters inquiry response substate and generate a Frequency Hop Synchronization (FHS) packet in re-sponse which includes the slave's BD_ADDR, clock and other information. However, there may be chances that multiple potential slaves receive IAC ID packet simulta-neously. In order to avoid collision in replying, potential slaves follow the random backoff mechanism stated in Section 9.4 before sending out the response. After re-sponse has been sent, the potential slaves then enter page scan state and wait for paging from the potential master. If no paging is received after a specific amount of time, the slaves switch back to inquiry scan state due to a possible collision of FHS packets in the previous state. On the other hand, the potential master may not want to reply the FHS packets from slaves as soon as possible. It is because it would like to wait for as many replies as possible from the surrounding. As such, it stays in inquiry state for some time. The duration of this procedure may last for 10.24s in order to provide enough time to get replies from all surrounding potential slaves.

After receiving enough responses, the potential master then enters page state as shown in Figure 10.6. The main purpose of this procedure is allow the potential master to establish connection with potential slaves in piconet and establish a synchronous hopping sequence between all devices for communication in the piconet. At the beginning, the master calculates a page hopping sequence based on specific slave's BD_ADDR and clock information. This specific frequency hopping can then be used to page specific potential slave in this procedure. The master sends an ID packet with device address code (DAC) which is the lower part of a slave's address using specific page frequency hopping. In receiving DAC ID packet, the slave echoes the same DAC ID packet to show that it has successfully received. The master then responds with a FHS packet indicating its own address and clock information. In receiving this packet, the potential slave then sends the DAC ID packet to show that it has received the packet successfully. At this time, the slave transits from page response substate to connection state and it then uses the connection hopping sequence derived from master's clock and address to communicate later. On the other hand, the master continues to page other potential slaves and enters connection state after all paging operations have been done.

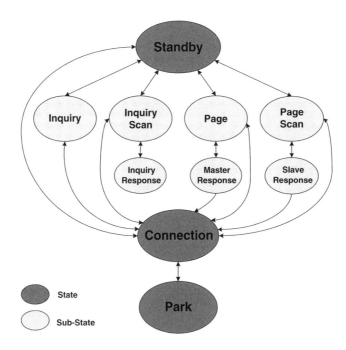

Figure 10.6 Bluetooth transition states.

At this point, a piconet is formed with a master and multiple slaves in connection state with a synchronized clock and hopping sequence. However, there are several modes that a slave can participate in the piconet. The modes are shown in Figure 10.7, while their functions are summarized in Table 10.10

Table 10.10 Bluetooth connection state modes.

Mode/State	Description
Active	The slave gets a 3-bit AM_ADDR code and actively participate in the activities of piconet. Since there are only 3-bit AM_ADDR, there are at most seven active, sniff and hold mode slaves in piconet.
Sniff	The slave also gets AM_ADDR, but only interests in the messages in specified time-slot to which master has been informed before joining the piconet. Sniff mode slave can enter reduced power mode in-between wakeup time.
Hold	The slave also gets AM_ADDR. In this mode of operation, ACL packet exchange is not supported, but SCO packet exchange can continue. The slave can entered reduced power mode in specific piconet. This mode is mainly used by bridge node in which actively participate in more than one piconet simultaneously is impossible, so the activities in particular piconet can be temporarily held and the node can switched between different piconets by using this mode and still be considered as connected member.
Park[a]	In this state, a slave is given a 8-bit parking member address (PM_ADDR) and is given up its own AM_ADDR. In this case, it can not actively participate in the piconet, but it is still considered as a part of it. With this state, the slaves can freely get into lowered power mode and the piconet is allowed to have more than seven members with 8-bit PM_ADDR in this state.

[a]Originally Park state is also considered as a mode in Connection State, it becomes a new state in specification v1.2 with similar functionalities as specification v1.1.

10.5.3 Scatternet Formation

As discussed in the previous section, a piconet cannot have more than seven active members. In order to increase the size of the network and also increase the coverage area, Bluetooth devices can form a scatternet, which is constructed by connecting multiple piconets together. Since there is still no conclusion on how scatternets should be formed, in this section, a number of methods currently suggested are discussed [53].

In scatternet formation, two main categories can be concluded—centralized approach and distributed approach. In the centralized approach, a leader is elected from all the Bluetooth devices within the transmission range of each other. Then the leader makes decision on the final topology of the scatternet. The final structure may be

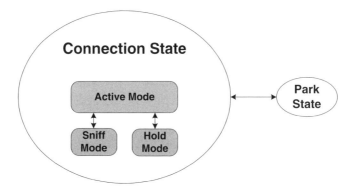

Figure 10.7   Bluetooth connection states.

a tree, a star, a ring, etc. In the distributed approach, discovery of neighbors, connections and other topology control actions are done locally. Consequently, different number of localized clusters are formed. These localized clusters are then combined to form a scatternet.

First of all, some centralized scatternet formation approaches are briefly discussed:

1. **Bluetooth Topology Construction Protocol (BTCP).** It is a three-phase scatternet formation protocol. The first phase involves an election of a leader by Voting process, in which each device is assumed to have one VOTE at the beginning and alternately performs inquiry and inquiry scan. Once connected, the one with a larger VOTE wins and get the VOTE of loser. The winner continues inquiring for more nodes, while the loser enters page scan state. At the end of first phase, a centralized controller with global information is elected. The second phase involves the topology construction decision making by the leader. The decisions on which nodes are designated as the master, slave and bridge are forwarded to specific devices. Then the actual connection is made in the third phase. The final topology may be a mesh, tree, etc.

2. **Bluetree Scatternet Formation.** This formation protocol is similar to BTCP where a leader is elected at the beginning to become a root, so called Blueroot. This root with complete information then initiates a spanning tree formation process with itself being the root of the spanning tree. As a result, a spanning tree structure is formed.

We now consider some semi-centralized or distributed scatternet formation approaches:

1. **Bluenet Scatternet Formation.** Bluenet Scatternet Formation is a three-phase scatternet formation protocol. In the first phase of this protocol, similar to centralized scatternet formation algorithms, alternating inquiry and inquiry scan

are done by each independent node to acquire the neighborhood information and exchange information with others. Afterwards, piconets are formed by several nodes nearby to form small clusters across a distributed area. In the second phase, any remaining independent nodes in phase one should enter page scan mode to wait for paging. In the last phase the slaves in a piconet then actively search for other slaves in other piconets to form a scatternet. The final topology may be a tree or some other sparsely connected structure.

2. **BlueConstellation Scatternet Formation.** This algorithm is another three-phase protocol for scatternet formation. Similar to the above protocols, the first phase is to exchange information among neighbouring nodes. In the second phase, Bluestars (piconets) is formed by electing the master with the largest weight of clustered nodes around. In the final phase, the Bluestars are connected to form BlueConstellation with the property that each master is within three hops' distance of each other. By doing so, the maximum number of hops of a node in one piconet to another node in another piconet can be determined and fixed.

3. **Randomized Scatternet Formation.** In this scatternet formation algorithm, all the nodes are the leaders at the beginning. Each leader randomly decides a probability within 1/3 to 2/3 on whether it performs a SEEK operation, which is an inquiry action, or stay on performing a SCAN operation, which is an inquiry scan action. When a match occurs, the two components are merged together with one of the leader retired, which will then restructure to maintain two constraints predefined (i.e., each leader has either no slave or one unshared slave, and each leader should have no more than k slaves). After a number of iterative leader retiring operations, there is only one leader left and finally a scatternet is formed.

In the above two categories of scatternet formation, there are several points in common. Firstly, an initial setup and maintenance is required to maintain the structure. Secondly, the topology formed is closely related to how a packet can be routed from one node in one piconet to another node in another piconet. However, most of the scatternet formation algorithms stated above do not consider the second point. The following scatternet formation algorithms have taken this issue into account and provide feature of "self-routing":

1. **Scatternet Formation Algorithm based on Search Tree.** This algorithm is very similar to the Bluetree Scatternet Formation. However, it is different in that the root is chosen based on the BD_ADDR with the node having the largest BD_ADDR to be the root. Then it makes decision on the tree topology construction based on a tree-sort like procedure. In this procedure, an invariant in a node A of the tree should be maintained with the property: $max(c_i) < min(c_j)$ for all $i < j$, where c_i and c_j are children of node A, while $max(c_i)$ and $min(c_j)$ are the largest BD_ADDR and smallest BD_ADDR in c_i and c_j,

respectively. By using this algorithm, a sorted tree structure will be formed, so that routing in such structure can be easily done.

2. **BlueRing Scatternet Formation.** Similar to most of the algorithms, the assumption that all nodes should be within the transmission range of each other is maintained. After the information exchange step, the one with the largest BD_ADDR is elected as the leader. The leader then assigns some other nodes in the network to be masters to form a ring with one slave destined as slave-slave bridge in between. With this structure, routing can be done by passing through ring of masters.

Some of the self-routing scatternet formation algorithms have been discussed above. However, the efforts of forming a scatternet may not be worthwhile in the case where there are low traffic and frequent topology changes. Due to these reasons, some researchers suggested the usage of traditional ad hoc routing techniques in carrying packets from a node to another without the necessity of forming a structure beforehand. There are several such routing algorithms suggested currently in Bluetooth, for example, Routing Vector Method, on-demand scatternet routing, zone routing protocol, which are adopted from traditional ad hoc routing protocols like Ad Hoc On Demand (AODV), Dynamic Source Routing (DSR) and Zone Routing, which are discussed in detail in Chapter 19.

10.5.4   QoS Consideration in Piconet and Scatternet

In the previous section, piconet and scatternet formation approaches have been discussed in detail. However, in the piconet and scatternet formation, only limited QoS is considered [53].

In the case of piconet, the master polls its slaves in simple round robin fashion. However, this method may not be an efficient one since some slaves may have more traffics than the others. Bluetooth uses TDD in MAC, so in order to improve QoS, the possible methods include using better scheduling algorithms in TDD and/or improve the Segmentation and Reassembly process (SAR) in the frame. In view of this, suggestions on improvement have been proposed and some of them are summarized below:

1. **Piconet Queue-State-Dependent Scheduling.** An example is head-of-line priority policy (HOL-PP). In HOL-PP, the master-slave communications are classified into three classes. The first class is the one with 100% utilization of channels (1-1,1-3,3-1). The second class is the one with 75% utilization of channels (3-0, 0-3). The third class is the one with only 50% of utilization (1-0, 0-1). The allocation of time-slots and polling frequency for specific master-slave pair are then be adjusted according to the classification.

2. **Piconet SAR and Scheduling.** Examples are SAR-BF and SAR-OSU. SAR-BF aims at optimizing the channels utilization by segmenting the upper layer

packet into the slot frames that best utilize the channels. On the other hand, SAR-OSU aims at reducing the overall delays. Consequently, it may prefer the usage of the slot frame with size larger than the upper layer packet size, so as to send out the frames as soon as possible.

On the other hand, focus has been given on scatternet formation. However, little has been suggested on the QoS issue of scatternets. In a scatternet, it relies on the bridge nodes to relay packet from one piconet to another. As such, in order to consider the issue of QoS in scatternets, better time scheduling has to be done on the bridge nodes. Some of the suggestions on improving the performance of scatternets are summarized in the following:

1. **Scatternet Nodes Randomized Rendezvous Scheduling.** The bridge nodes in a scatternet have neighbors (links) from different piconets, however, they cannot meet with more than one of them at the same time. In this case, Pseudo Random Coordinated Scatternet Scheduling (PCSS) has been suggested with the idea of randomly choosing the rendezvous time by using the master's clock and slaves' addresses.

2. **Locally Coordinated Scheduling (LCS) in Scatternets.** This approach also addresses the rendezvous issue of bridge nodes. However, instead of randomly choosing a point, this algorithm calculates the start time of next rendezvous by observing the data rate trend. At the same time, it also finds out the duration of next rendezvous by using the queue size and past transmission history, so that the duration is about right to transmit all the backlogged traffics.

In summary, some of the QoS considerations in piconets and scatternets have been suggested together with the general ideas on how to improve QoS. However, there is no consensus on which one is the best and so this topic is still under development.

10.6 PERFORMANCE ISSUES

As it has been discussed previously, Bluetooth is using the 2.4 GHz ISM frequency band, which is crowded with devices of other technologies. However, even though this issue is neglected, multiple Bluetooth piconets formation with overlapping coverage area also brings up great interferences since they are using the same frequency band independently. Unfortunately, this case happens frequently nowadays when Bluetooth becomes more and more popular with users forming their own Bluetooth personal area network (PAN) using multiple electronics devices such as PC, laptop, PDA, mobile phone, etc., which are generally not shared with other users. In this section, we describe a study on the performance issue [74].

In order to determine the performance of Bluetooth in a more realistic case, several issues are required to be mentioned beforehand. Firstly, the power received by a receiver depends on the distance between the transmitter and receiver, the transmitted power and path loss exponent used, etc. Secondly, the occurrence of interferences depends on the ratio between power received from the intended transmitter and all other surrounding transmitters. As long as the ratio is higher than certain threshold value, the received signal is still valid. This in turn determines the probability of a successful transmission of a packet in a piconet.

With the above information and the assumption that seven piconets are located within the region where a large portion of each piconet is interfered by the others, the following analytical and/or simulated results are shown:

1. The analytical results show that the average throughput of Bluetooth with interference is better than that of FDMA in which the whole bandwidth has to be divided into seven non-overlapping frequency bands for seven independent piconets.

2. Bluetooth performance varies with the number of obstacles and interferences with a given receive threshold value, so different average throughput can be expected in different environment.

3. The shorter the distance between the transmitter and receiver within the piconets, the higher the probability that it will receive packets correctly, which in turn increases the throughput.

4. By using the model for simulation, it has been shown that for the case in ACL DH1, DH3, DH5 packets, the simulated performance is 3% to 4% lower than the ideal case when interference is neglected, however, the performance drops by 14% to 30% when interference occurs.

From the results, it can be concluded that Bluetooth, in fact, has relatively good performance when compared to some access technologies like FDMA. However, the average throughput is varied greatly according to the environment with different obstacles and interferences.

10.7   PRACTICAL ILLUSTRATION: SENSOR NETWORK

Bluetooth has become a publicly accepted standard for short-range communication and has been included in many electronic devices as a complementary communication technology to infrared. More and more devices are integrated with Bluetooth technology, for example, PDA, laptop, mobile phone, printer, etc. In this section, we present a practical security system in which Bluetooth is used as a communication technology for assisting in building up a sensor network, which has attracted much researchers' attention.

There are three types of nodes involved in the system—sensor nodes, relay nodes and control node. A sensor node may be an embedded system which has the functions of detecting movement from outside environment and controlling the Bluetooth module to transmit the movement information out. In fact, relay nodes and control node are also sensor nodes that carry extra functions. Relay nodes are used to route information from the sensor nodes, while the control node acts as an interface between the sensor network and local security processing system. The nodes in the Bluetooth sensor network build up a tree-based hierarchical structure with the control node as a root of the tree and relay nodes as intermediate nodes, while sensor nodes as leaves. By using this simple configuration, simple routing can be achieved by constantly pushing the information to the root. Because of this, intermediate nodes failure recovery and route maintenance can be done easily.

The whole scenario can be summarized in Figure 10.8 [66], and the details are discussed as below:

1. Sensor Node A detects movement.

2. Sensor Node A converts the motion event into digital data and controls the Bluetooth module to transmit the corresponding information out.

3. Sensor Node B receives the data from Sensor Node A and helps in routing the data to next hop (Sensor Node C) towards the control node. So in this case, Sensor Node B acts as a relay node.

4. Similar to Sensor Node B, Sensor Node C also acts as a relay node and forwards the information to Control Node D through Bluetooth.

5. Sensor Node D is a control node which connects to a local security processing system (e.g., a workstation). In receiving the movement information from Sensor Node A routed through Sensor Nodes B and C, security guards located inside the security room monitoring the local security processing system receive a warning message about the intrusion into the building.

6. The intrusion information may also be routed to remote/centralized security processing system through Internet. In this case, multiple geographically locations can be monitored and actions can be made in a centralized manner.

In the above case, a Bluetooth sensor network is used in a security system. In fact, this configuration can also be used in many scenarios. A good example is remote site rescue—a temporary rescue center is set up outside the rescue site and Bluetooth sensor nodes are distributed in the rescue site to provide information inside to the control node outside with a local processing system, which will then be able to carry the information to the centralized rescue office in remote location.

10.8   SUMMARY

In this chapter, Bluetooth has been discussed in detail. First of all, we discuss the design goals of Bluetooth—a short-range cable replacement with low cost and low power

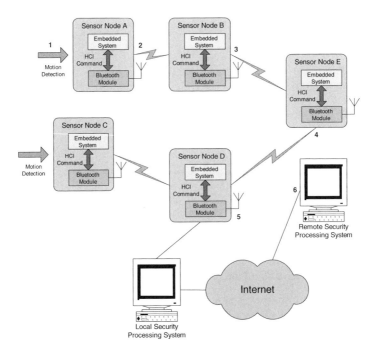

Figure 10.8 Bluetooth sensor network security system.

consumption. Then Bluetooth's protocol stack and profiles which aim at providing interoperability between devices from different vendors have been gone through in detail. Afterwards, Bluetooth PHY and MAC are discussed.

With the structural constraints of Bluetooth, Bluetooth single point-to-multipoint (piconets) and multiple point-to-multipoint (scatternets) have been discussed in detail with the focus on their formation, together with the possibility of QoS support. After that, we come to a performance study on Bluetooth MAC throughput which shows that Bluetooth, in fact, gives a good performance, but its performance may be unstable subject to environment conditions. Lastly a simple Bluetooth sensor network security system scenario is shown to demonstrate the practical usage of Bluetooth.

All in all, as for its original target, Bluetooth has been successful in providing a short-range (10m) cable replacement solution. People start thinking about other potential usage models on Bluetooth, which brings out more and more extended profiles recently. It is expected that Bluetooth's future development will also follow IEEE 802.11x in the sense that they will be capable of carrying a larger bandwidth capacity and have a longer transmission range. However, with the development of UWB (Ultra Wideband), which provides bandwidths comparable to wired networks and also targets on WPAN, the future of Bluetooth is difficult to be predicted at this moment.

PROBLEMS

10.1   Bluetooth uses the same frequency band and provides similar functionalities as Wireless LAN, e.g., IEEE 802.11b/g. Briefly describe the design goals of each technology and explain why the products of these two technologies can co-exist in the market.

10.2   Briefly describe the differences between Wireless LAN MAC and Bluetooth MAC, and how they can be used to support QoS applications.

10.3   What are the differences between piconets and scatternets? Briefly describe different methods of formation of scatternets.

CHAPTER 11

COEXISTENCE ISSUES

The art of life lies in a constant readjustment to our surroundings.

—Kakuzo Okakura

11.1 INTRODUCTION

Low cost short range wireless communications bring us great convenience. But such convenience could be haunted by the ubiquity of such devices. In this chapter, we discuss about the coexistence problem involving Bluetooth and WLAN devices.

Interoperability:

As we have learnt from previous chapters, ISM band is mainly provided for the usage of short range wireless communications. Ideally, devices using this band should be interoperable with each other. As detailed in Section 11.2, the major issue is to withstand interference from other devices. However, the spread spectrum techniques adopted by current popular technologies, e.g., Bluetooth that uses FHSS and WLAN that uses DSSS, are designed based on specific conditions without regard to the existence of devices using different short-range technologies. Indeed, these different ISM band short-range technologies operating in close proximity result in a serious packet collision problem. Section 11.3 will focus on this issue.

The approaches suggested in handling coexistence issues are described in Section 11.4. These solutions can be divided into two main categories: collaborative mechanisms and non-collaborative mechanisms. The collaborative solutions suggested provide interoperability among main interference technologies by colocating multiple (usually two) air interface solutions within the same physical unit. On the other hand, two main methods (AFH and MDMS) are suggested in the non-collaborative category. AFH approach involves changes in frequency hopping sequence. As a result, interoperability can only be supported among the devices that use AFH. On the other hand, the MDMS approach is based on skipping the usage of certain frequencies in an adaptive manner. Consequently, this approach only requires changes made on the master devices. Indeed, all other slaves devices can still interoperate without any modifications. AFH approach is discussed in detail in Section 11.4.2.3, Section 11.5 and Section 11.6. MDMS approach is discussed in detail in Section 11.7.

Performance:

From the simulation results reported in some studies, two main AFH approaches—TG2-AFH and ISOAFH—outperform traditional pseudo-random frequency hopping technique used in handling interference issue. Specifically, ISOAFH works well in resource and power limited environment. However, this is not the case for TG2-AFH which does not take into account resource and power constraints. On the other hand, in a performance study that involved TG2-AFH, ISOAFH and ISOMDMS, it is found that ISOAFH greatly outperforms other two mechanisms in handling interferences in all scenarios. Interestingly, ISOMDMS is able to produce better performance than TG2-AFH in some cases. This indicates that AFH may not be the only suitable technology in combating interference. We will further elaborate on these issues in Section 11.8.

11.2   THE ISM BAND SPECTRUM

In order to facilitate short range transmission technologies development, government has provided some unlicensed spectra specifically for this purpose. Unlicensed does not mean free from any regulations. In fact, the government has regulated the usage of these spectrum by specifying the maximum power transmitted, frequency band range, etc. The most notable example is ISM bands shown in Figure 11.1 [61].

ISM bands are mainly provided for wireless communication devices for commodity purposes. The devices using these bands do not need to pay license fee. Nevertheless, they must follow the power regulation and the usage of spread spectrum techniques in order to provide acceptable interference to others and survive from interference of others. Currently, there are three ISM bands. They are located at 900

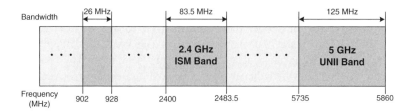

Figure 11.1   ISM bands in United States.

MHz, 2.4 GHz, and 5 GHz frequency regions as shown in Figure 11.1. The most notable one is 2.4 GHz ISM band which has been discussed briefly in Chapter 9 and Chapter 10. Indeed, 2.4 GHz ISM band is the frequency band used by many of the devices nowadays, including IEEE 802.11b/g, Bluetooth, Microwave Oven, cordless phone, etc. On the other hand, 5 GHz ISM band is less crowded currently, mainly used by IEEE 802.11a, HiperLAN2, etc.

The low cost and open usage of the ISM band inevitably leads to severe interference between different products integrated with different technologies using the same frequency band. This is further elaborated in Section 11.3 below.

11.3 PACKET COLLISION

In the previous section, several unlicensed bands have been introduced. Because any device can use unlicensed ISM bands free of charge as long as regulations defined by the government of specific countries have been followed, large number of products produced using these frequency bands lead to severe interference and packet collision. In this section, the general concept of packet collision is further elaborated [61].

The 2.4 GHz ISM band is one of the most commonly used unlicensed band nowadays. It has been shared by two popular kinds of devices—Bluetooth and IEEE 802.11b/g WLANs. As a result, these two types of device are two main sources of collision. First of all, let us briefly review some general characteristics of these two technologies here.

- IEEE 802.11b is a kind of short range wireless technology using 2.4 GHz frequency band. It divides the whole 2.4 GHz ISM band into channels with center frequency spacing of 5 MHz. Each channel takes up 22 MHz bandwidth in total. Under United States FCC regulations, there are 11 channels available in which only three of them can be used simultaneously. It uses Direct Sequence Spread Spectrum (DSSS) in physical layer. Consequently, its frequency band occupancy is relatively static and wide. Further details can be found in Section 9.5.

- Bluetooth is another kind of short range wireless technology using 2.4 GHz frequency band. It divides the whole 2.4 GHz ISM band in channels of 1 MHz. Thus, there are 79 channels in total. It uses Frequency Hopping Spread Spectrum (FHSS) in physical layer. A single piconet is formed by a master and a group of slaves in which they use the same hopping sequence to communicate. The frequencies used in the TDD time slots are predefined by this hopping sequence and specific to that piconet. Since master-slave communication is carried out using different frequency every time (hopping), its frequency band is relatively narrow and dynamic. The readers are referred to Section 10.4 for further details.

Generally speaking, packet transmission collision occurs only when both technologies use overlapping frequency in overlapping period of time. This concept illustrated in Figure 11.2. As can be seen, both Bluetooth and WLAN IEEE 802.11b devices are located within transmission range of others. We can also see that there are two packets collision occurred, namely the second and third WLAN packet transmissions, corresponding to third and seventh Bluetooth packet transmissions.

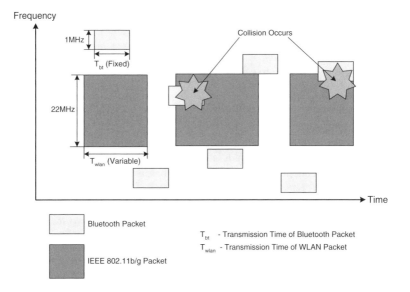

Figure 11.2   Packet collision example.

It should be noted that the severity of this collision, in fact, depends on many factors including the separation of devices, signal strength etc. Figure 11.2 shows that Bluetooth packets hop across different frequencies over different time slots. As a result, packet collision is only temporary. By using this technique, narrowband and fast-changing interferences can be coped with successfully. However, collision occurs frequently as WLAN frequency band is relatively static and wide. On the other hand, WLAN uses DSSS to cope with collision. As long as the signal to interference ratio (SIR) is lower than certain threshold value, the interference problem can be ignored. This technique is best for handling background noise. However, this may not be a good solution for narrowband and fast-changing interferences.

As a result, both Bluetooth and IEEE 802.11b WLAN use their own methods in solving the problem of interference. However, the methods adopted may not be an effective solution as these schemes are not targeted at the main interference source. Furthermore, with increasing popularity of both technologies nowadays, the interference problem becomes more and more severe. Together with serving as a reference for less crowded 5 GHz ISM band which is expected to be frequently used sooner or later, many possible solutions are proposed and they are introduced in Section 11.4.

11.4 POSSIBLE SOLUTIONS

In Section 11.3, the concept of packet collision has been introduced. Specifically, packet collision is assumed to occur only when more than one short-range wireless technologies use overlapping frequency during the same period of time. In view of this, we can easily think of two possible methods to cope with this problem, namely, obtaining frequency orthogonality, and/or time orthogonality among different kinds of technologies.

There are two main groups of solutions using the methods suggested above. They are collaborative mechanisms and non-collaborative mechanisms. The classification of coexistence mechanisms is shown in Figure 11.3 [61]. Most of the mechanisms suggested involve modifications in Bluetooth WPAN side instead of IEEE 802.11b WLAN side.

11.4.1 Collaborative Mechanisms

Collaborative mechanism mainly requires the cooperation of the affected wireless technologies to minimize the mutual interferences. It can be further divided into collocated and non-collocated [61].

11.4.1.1 Collocated Mechanism Collocated mechanism refers to the cooperation of the affected wireless technologies which are implemented into a single physical unit. Since they are implemented into the same physical unit, within such a short separation distance, the power transmitted by either Bluetooth and IEEE 802.11 b/g WLAN will cause interference to each other once they occupy adjacent frequency in overlapping period of time. Moreover, the transmission power by either side may cause interference even if non-overlapping frequency is used. Consequently, only time orthogonality method can be adopted to prevent interference. Time division

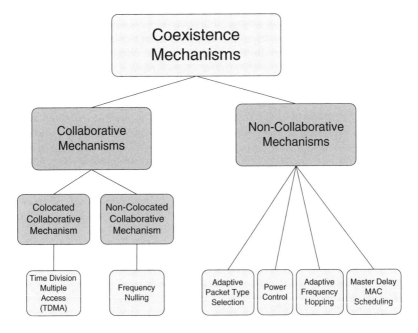

Figure 11.3   Various coexistence mechanisms.

multiple access (TDMA) technique is generally applied in this case. It is a kind of MAC layer solution which solve the problem by alternatively interleaving the transmission of Bluetooth and WLAN packets. In this case, either kind of packets will be transmitted at certain period of time and so coexistence can be achieved. The general concept of TDMA of multiple packets is shown in Figure 11.4.

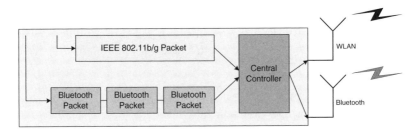

Figure 11.4   Collocated TDMA system conceptual diagram.

Some of the proposed TDMA approaches in collocated mechanism are summarized here.

- **Alternating Wireless Medium Access (AWMA) scheme.** This scheme simply uses a time division approach which divides a IEEE 802.11b Beacon interval into two sub-interval. They are Bluetooth sub-interval and IEEE 802.11 sub-interval. By doing so, alternating exclusive periods are allocated to both technologies to transmit their own packets.

- **MEHTA scheme.** This scheme requires the MAC layer information of both technologies. Whenever Bluetooth or IEEE 802.11b would like to transmit a packet, the transmission request is forwarded to a central controller (MEHTA) for approval. They can transmit their packets only when permission is granted. Permission is granted based on certain criteria including signal strength of the technology, the differences between the center frequencies of both technologies used, etc.

11.4.1.2 Non-collocated Meachanism   By definition, non-collocated mechanism refers to the mechanism that both technologies are implemented into different physical units, but they are required to cooperate to achieve coexistence. This approach is not commonly used. A notable example is Frequency Nulling.

This mechanism requires the IEEE 802.11b device to be integrated with a Bluetooth receiver which is mainly used to snoop Bluetooth packets around. By doing so, the frequency hopping sequence of Bluetooth piconet around is then known. In this case, a 1 MHz wide frequency null can be inserted into the appropriate region of 22 MHz of IEEE 802.11 channel whenever a collision on frequency domain is expected.

11.4.2   Non-Collaborative Mechanisms

As indicated in Figure 11.3 [61], there are four main types of solutions in this category. They are adaptive packet selection, power control, adaptive frequency hopping, and master delay MAC scheduling. Non-collaborative mechanism refers to the co-existence mechanism which does not involve any information exchange between possibly interference sources (i.e., the nearby WPAN and WLAN). Since information exchange is not required, it is more applicable in real life environment. Indeed, non-collaborative mechanism is generally considered as a more important approach. Because explicit information exchange is not carried out, these approaches generally involve two stages—channel classification and adaptive control actions. This is further elaborated below.

11.4.2.1 adaptive packet selection mechanism   Firstly, the adaptive packet selection mechanism is briefly described. In Section 10.4, different types of Bluetooth MAC layer frames are introduced. Specifically, they are frames occupying one slot, three slots and five slots and frames with different error detection and correction mechanisms—no FEC or 2/3 FEC. DH5 frame, which has no FEC and occupies five slots, can carry the largest amount of information with the least overhead. However, DH5 frame is more vulnerable to interferences as it occupies the channels with longer time and is not equipped with any error correction mechanism. On the other hand, DM1 frame carries the least information with largest overhead.

In view of this, an idea is that Bluetooth should select appropriate packet types for transmission at specific channels. At the beginning, a Bluetooth device has to store statistics on channels that encounter interference. Adaptive control mechanisms can then be employed. Some example control mechanisms are presented below.

- A longer frame (with more slots) is used in good channel which will then cover the slots that are originally assigned to bad channels immediately after the good channels.

- The successful packet transmission probability is dynamically calculated. Shorter frame is used if the successful transmission probability is low. This mechanism is known as IBLUES.

11.4.2.2 Power Control mechanism
In this mechanism, transmission power of Bluetooth can be limited at minimum level. At this level, only required performance and data integrity are maintained. By controlling the transmission power, all the interference sources nearby will experience reduced level of interference.

Thus, mechanism has been proposed for the channel detection process. The transmission power is maintained at minimum level with a margin. Accordingly, when data lost caused by propagation loss is detected, the transmission power is increased. By using such heuristic power adjustment, maximum data rate can be achieved by using minimum amount of power.

11.4.2.3 Adaptive Frequency Hopping
Adaptive Frequency Hopping (AFH) is a non-collaborative mechanism generally accepted as a solution for solving coexistence problem in ISM band. Indeed, AFH has already been adopted in Bluetooth specification v1.2. Originally the frequency hopping sequence is generated randomly and the devices within the piconet use that generated sequence to communicate with each other, without regard to the fact that some channels in the sequence may be overlapping with other interference sources. However, in the case of Adaptive Frequency Hopping, whenever the frequency hops generated are occupied by some interference sources, the affected hops are swapped to some other frequency hops that are non-overlapping with others. The general principle of AFH is illustrated in Figure 11.5. Here, channel 10 of Bluetooth is selected to be used for communication within the piconet. However, this channel is also occupied by a interference source nearby. As a result, channel 10 is swapped to channel 40, which is non-overlapping with others during that time slot.

In general, AFH involves several processes. They are illustrated in Figure 11.6 and their corresponding functions are summarized below:

1. Device identification is mainly used to make sure that the devices involved support AFH mode. In other words, it is mainly added for backward compatibility.

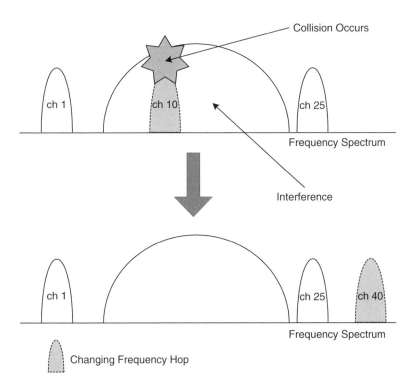

Figure 11.5 Adaptive Frequency Hopping: a conceptual illustration.

2. Channel Classification is mainly used to classify the quality of the channels.

3. Classification Information Exchange is a process for exchanging measurement information between master and slaves within the piconet.

4. Adaptive Frequency Hopping mechanism is the actual process which make changes to the frequency hops generated dynamically based on the information of the channels.

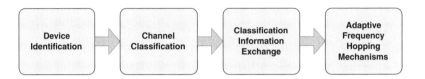

Figure 11.6   Adaptive Frequency Hopping processes.

Due to its demonstrated efficacy and practicality, AFH is generally accepted as a promising solution in avoiding packet collision. Indeed, AFH is adopted by IEEE 802.15 TG2 as a mechanism to tackle the issue. Variants of the AFH mechanism are further discussed in Section 11.5. In Section 11.6, we also describe an idea which further improves AFH solution suggested in IEEE 802.15 TG2.

11.4.2.4 Master Delay MAC Scheduling   Master Delay MAC Scheduling (MDMS) is another non-collaborative mechanism in solving the problem of packet collision. The main idea behind is simple—whenever the frequency hop generated is found to be occupied by other sources. Transmission at that time slot is skipped. This idea is further elaborated in Section 11.7.

11.5   IEEE 802.15 TG2

The interference between WLAN (e.g., IEEE 802.11 b/g) and WPAN (e.g., Bluetooth) has caught significant attention. Bluetooth SIG and IEEE 802.15 then organized a task group to work on this issue specifically. The task group is known as coexistence Task Group 2 (TG2)[1]. TG2 is mainly responsible for quantifying the mutual interference between WLAN and WPAN. It is also responsible for developing a set of practices to facilitate combating the coexistence issue.

Specifically, IEEE 802.15 TG2 recommended AFH mechanism to be implemented in Bluetooth to avoid packet collision between WLAN and WPAN. This involves the appending an adaptive frequency hops selection scheme after the original Bluetooth frequency hops selection scheme as shown in Figure 11.7. Originally, the frequency

---

[1]The task group is now in hibernation until further notice.

hopping sequence is generated randomly with the use of parts of master BT_ADDR and its clock information. With the addition of new scheme, the sequence generated by the original one is then passed through a frequency remapping module which aims at modifying the hops inside the sequence that are classified as "bad" channels and maintaining the pseudo-random nature inside the sequence. The frequency remapping function makes use of the information from sequence generation module which stores the information of channels and decides when and which frequencies are to be used. After passing through remapping module, an adaptive frequency hopping sequence is formed.

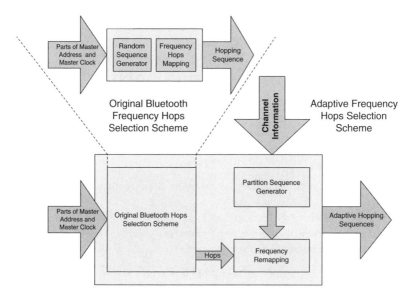

Figure 11.7   Structure of TG2 adopted AFH.

The two additional modules in adaptive frequency hops selection scheme performs two main functions in most of the non-collaborative mechanisms—channel classification and adaptive control actions.

In the channel classification process, TG2 mandatorily requires both the master and slaves to keep track of the channels conditions based on some channel assessment methods such as Bit Error Rate (BER), Packet Error Rate (PER), Packet Loss Rate (PLR) and Received Signal Strength Indicator (RSSI), etc. Generally speaking, it is common to use BER, PER and PLR for channel condition assessment.

After the measurements, the slaves then transmits the results back to master through an LMP message. Channel status is generated on channel-by-channel basis. After gathering the information from slaves, a list of "good" and "bad" channels are generated, each of which is then used to modify the frequency hops in adaptive control actions.

Let us consider the adaptive control actions on ACL links. In channel classification process, channels have been classified into "good" channels set and "bad" channels set. The frequency hops generated by using original Bluetooth frequency hops selection scheme are then compared against the elements in these two lists. Specifically, the frequency hop classified as a "bad" channel is substituted with a "good" channel by the remapping module. However, there is a regulation on the minimum number of frequency hops N_{min}. According to FCC regulation in United States, N_{min} is 15. Thus, if the number of available "good" channels is larger than N_{min}, some "bad" channels will be modified to one of the "good" channels in this process (Mode L operation); otherwise "bad" channel will continue to be used in order to comply with the regulation (Mode H operation). After remapping, the final sequence generated is called an adaptive frequency hopping sequence.

11.6   INTERFERENCE SOURCE ORIENTED ADAPTIVE FREQUENCY HOPPING

In Section 11.5, IEEE 802.15 TG2 adopted AFH mechanism in handling coexistence issue between WLAN and WPAN has been discussed in detail. However, this method still has room for improvement due to the following reasons.

1. TG2 adopted AFH mechanism assesses the channels on a channel-by-channel basis. However, it does not take into account the observation that if the interference source is WLAN, then the adjacent channels of "bad" channels actually have high probability of subjecting to interference also. This is because WLAN uses a wide frequency band statically.

2. Since channel assessment in TG2 AFH is done on a channel-by-channel basis, this generally results in a relatively longer response time from the presence of interference to the detection of interference and the response of appropriate actions. In other words, there is still room for improvement in channel classification mechanism.

According to above explanations, the improvements of TG2 AFH are usually focused on its deficiency in channel classification process. A recent attempt is an approach called Interference Source Oriented AFH (ISOAFH) [61]. This AFH mechanism mainly focuses on improving the channel classification process by identifying the interference source and analyzing its corresponding characteristics. Firstly, Table 11.1 gives a detailed comparison between IEEE 802.11b channel frequency allocation and that of Bluetooth.

From Table 11.1, we can see that if the interference source is IEEE 802.11b WLAN which makes use of channel 4, then the corresponding Bluetooth channels (channels 14–36) will be highly affected. With this observation, ISOAFH tries to locate the interference source (WLAN) channel(s) and then prevents Bluetooth frequency hops selection scheme to select the group of frequencies affected.

Table 11.1 Comparison of IEEE 802.11b WLAN and Bluetooth WPAN channel frequency allocation.

IEEE 802.11b channel Number (US)	Occupied frequency range (MHz)	Bluetooth channel Number
channel 1	2401 — 2423	channel 0 — 21
channel 2	2406 — 2428	channel 4 — 26
channel 3	2411 — 2433	channel 9 — 31
channel 4	2416 — 2438	channel 14 — 36
channel 5	2421 — 2443	channel 19 — 41
channel 6	2426 — 2448	channel 24 — 46
channel 7	2431 — 2453	channel 29 — 51
channel 8	2436 — 2458	channel 34 — 56
channel 9	2441 — 2463	channel 39 — 61
channel 10	2446 — 2468	channel 44 — 66
channel 11	2451 — 2473	channel 49 — 71

The channel classification process in ISOAFH first groups the Bluetooth channels according to IEEE 802.11b WLAN channels listed in Table 11.1. This results in 11 groups of Bluetooth channels. Instead of keeping track of every Bluetooth channel status, only the aggregate PER of these 11 groups is stored. These 11 groups include some overlapping Bluetooth channels, e.g., Bluetooth channel 10 is located in Group 1, 2 and 3. If a packet error is detected in Bluetooth channel 10, then Group 1, 2 and 3 PER will also increase. By doing so, the group corresponding to the interference source will have the highest PER and so the interference source channel can also be located. In this case, "good" groups set and "bad" groups set are formed.

In the associated Adaptive Control Actions process, the master informs its slaves in the piconet about the channel groups status by using modes of revocation that are numbered according to groups. In other words, if group 6 is marked as "bad" group, mode of revocation number 6 will be sent to each slave. In receiving the LMP message from master with revocation number, all slaves remap frequency hops in group 6 to other channels in "good" groups.

A critical system parameter, N_{min}, which indicates the minimum number of hops involved in a sequence has been briefly discussed in Section 11.5. According to FCC regulation in United States, the number should be at least 15. On the other hand, the maximum number of orthogonal channels in IEEE 802.11b WLAN is three which occupies channel 1, 6 and 11 according to US regulations. With the interferences of these three 802.11b channels, the remaining unaffected Bluetooth channels include channels 21–23, 47–49, and 73–78. Consequently, this minimum requirement can always be fulfilled.

11.7 INTERFERENCE SOURCE ORIENTED MASTER DELAY MAC SCHEDULING

We have discussed various AFH approaches in handling coexistence issue. AFH is an aggressive approach in that it aims at searching for another "good" channel to replace the "bad" channel. However, this approach has certain limitations. The main limitation is backward compatibility. In using AFH, it is necessary to make changes to all devices, including master and slaves. However, thousands or even millions of Bluetooth devices which does not support AFH have already been produced and are being used in the market. In view of this, MDMS approach is considered as a more practicable solution [61].

Let us first review the principles of MDMS. Specifically, MDMS is a non-collaborative coexistence mechanism which can also be divided into two main steps—channel classification and adaptive control actions. In channel classification process, it uses the method adopted by normal non-collaborative mechanism (same as the one shown in TG2-AFH). In adaptive control actions process, the master first obtains the "good" channels set and "bad" channels set. Afterwards, the master will check the sets against the existing hopping sequence in pair as master has to guarantee both uplink and downlink transmission in ACL connection. Whenever a "bad" channel is encountered, the master simply delays the transmission of the packet to specific slave until next "good" channel hop. The whole process is illustrated in Figure 11.8.

Since the whole process does not include any modification of the original hopping sequence and the master has the complete control on the usage of time slots, changes on slaves are unnecessary. Only master has to be changed with simple firmware upgrade to support MDMS. As a result, backward compatibility is supported by using this approach.

In view of the deficiency in TG2-AFH approach, the channel classification process in MDMS is also enhanced. The modification procedures are exactly the same as that described in Section 11.6. By using this method, the interference sources characteristics will be considered by grouping channels into 11 groups. The readers are referred to Section 11.6 for the details of the classification procedures. After classification of "good" groups of channels and "bad" groups of channels, the master applies the same techniques discussed above to skip the "bad" channels.

11.8 PERFORMANCE ISSUES

Non-collaborative approaches are more popular in tackling the problems of coexistence issues as they can be used readily in realistic application scenarios. Recently, a performance study of two main AFH approaches and traditional pseudo-random

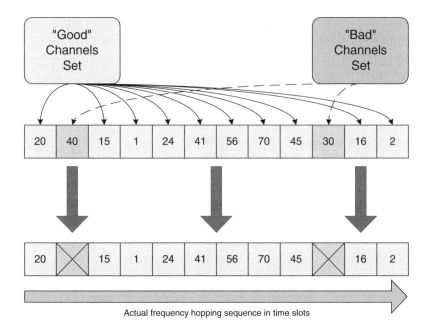

Figure 11.8 An illustration of the MDMS mechanism.

frequency hopping approach under limited resource and power environment has been carried out [61]. The study involved the three main non-collaborative approaches: TG2-AFH, ISOAFH and ISOMDMS.

Conclusions derived from the performance study [61] are summarized below.

- Both TG2-AFH and ISOAFH perform better than the traditional Pseudo-Random Frequency Hopping in the present of interference sources. The results are especially significant if the interference source is IEEE 802.11b WLAN. The main reason behind is that both AFH approaches obtain extra information in selecting frequencies for transmission.

- The performance of TG2-AFH mechanism is severely deteriorated in the presence of power and resource constraints while ISOAFH is less sensitive to these constraints. The main reason behind is that less information is concerned and stored in the case of ISOAFH (11 groups) than that of TG2-AFH (79 channels).

- AFH based techniques exhibit similar performance compared to that of pseudo-random frequency hopping techniques if the interference source is pure Bluetooth (i.e., pure "inter-piconet" interference) with varying number of piconets. It is because the narrow and fast-changing characteristics of interfering piconets affect the accuracy of channel and this greatly affects the ability of AFH based techniques to classify "good" and "bad" channels.

Moreover, in performance comparisons involving TG2-AFH, ISOAFH and ISOMDMS, the following observations were made.

- ISOAFH approach performs the best in terms of throughput and access delay among the three approaches considered.

- AFH mechanisms do not always perform better than MDMS mechanism from the perspective of the channel classification process. For low interference environment (less than 60% of IEEE 802.11b system loading), ISOMDMS performs better than that of TG2-AFH. It is because the channel classification mechanism of ISOMDMS allows it to response faster than that of TG2-AFH by skipping "bad" frequency channels in case of interference. However, when the interference becomes severe, ISOMDMS starts consecutively skipping "bad" frequency channels while TG2-AFH will swap "bad" frequency channels into "good" one. As a result, TG2-AFH performs better than ISOMDMS in high interference environment.

From the summary listed above, an overall conclusion is that AFH approaches generally obtain better performance than that of MDMS approaches. Indeed, despite AFH's implementation problems in existing devices, AFH should be the approach adopted for solving the problem of interference in long run.

11.9   PRACTICAL ILLUSTRATION: BLUETOOTH AND WI-FI COLOCATE TRANSMISSIONS

It is very common nowadays for owners to deploy IEEE 802.11b/g WLAN access point or Bluetooth access point in their coffee shops, shopping malls, etc., to act as internet access hotspot. Indeed, such facilities add values to the core services (e.g., coffee) purchased by the clients. However, due to interference problems, simultaneous deployment of both kinds of access points inevitably results in poor performance. Recently, a practical performance study [365] modeling a scenario using a collocated collaborative network access point has been carried out.

As discussed above, packet collision between Bluetooth and IEEE 802.11 b/g is a serious problem due to the popularity of these two technologies nowadays. Some access point manufacturers observe that it is actually possible to implement both Bluetooth and IEEE 802.11 b/g access point into a single device as shown in Figure 11.9. Here, both Bluetooth and IEEE 802.11 b/g WLAN client devices are served by the same network access point, which is equipped with dual air interface—Bluetooth and WLAN interfaces. As a result, Internet access service can be provided to both Bluetooth and WLAN clients.

In order to support dual air interfaces which are competing for frequency resources in nature, a collocated collaborative design and implementation detail are shown in Figure 11.10. This prototype system [365] is a PC that contains dual air interfaces

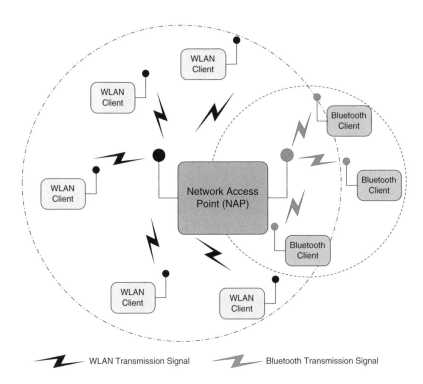

Figure 11.9   Dual air interface network access point scenario.

and operates in Linux environment. With the use of Linux operating system, each air interface will be associated with a software interface, i.e., bnep0 for Bluetooth and wlan0 for WLAN. This software interface can be used to control the packet traffics by assigning some rules in ingress, forward and egress points in this interface. As shown in the figure, the packets from both bnep0 and wlan0 are directed to a virtual interface imq0. Within imq0, the packets are queued up and packet scheduling algorithms can be applied to it. After applying the scheduling algorithm, the packet scheduled can then be transmitted by using the corresponding interface.

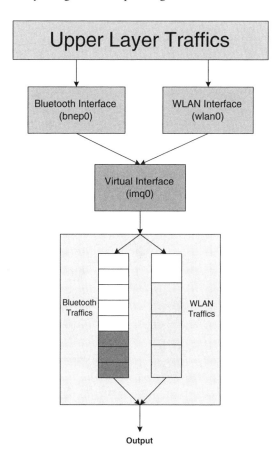

Figure 11.10   Bluetooth-WLAN network access point design and implementation.

Other than some common scheduling algorithms, the administrator of Bluetooth-WLAN NAP may also implement their own scheduling algorthms based on certain criteria such as Bandwidth Balancing, Round Robin Balancing, Interference Consideration, etc. The NAP keeps track of moving averages of the instantaneous packet transmission rates of Bluetooth and WLAN in order to classify the conditions of the corresponding wireless channels. Specifically, there are three cases: increasing, decreasing and steady. For instance, a persistence decreasing of the transmission rate gives a strong indication of interference. In this case, the virtual interface will skip dequeuing of a packet from either interfaces. By doing so, further interference can be avoided.

In this section, we have described a Bluetooth-WLAN NAP scenario which can be used to support both Bluetooth and WLAN clients simultaneously. However, we have to emphasize that this is only one possible implementation of collocated collaborative coexistence mechanism. Indeed, vendors keep on developing their own standards and many products have been sold in the market already. Because of this, standardization becomes a pressing issue for interoperability among these different NAPs.

11.10   SUMMARY

In this chapter, we first describe in detail the most commonly used ISM bands. Then, packet collision problem in ISM bands has been discussed in detail, especially for the case of Bluetooth and IEEE 802.11b/g WLAN in 2.4 GHz ISM band. Since collision is mainly a result of overlapping frequency band usage in overlapping period of time, obtaining orthogonality in either frequency domain or time domain can effectively solve the problem.

Indeed, there are many proposed solutions in tackling the coexistence problem. Specifically, in view of the severity of the problem, IEEE 802.15 has setup TG2 to focus on this issue and proposed some recommended solutions. AFH based solution has been suggested and proposed by TG2 and this solution is also recently incorporated in Bluetooth specification v1.2.

However, researchers have also found htat the solution proposed by IEEE 802.15 TG2 may not be a good solution since it does not exploit the knowledge about the type of interference source. Due to this reason, some researchers have proposed modified versions of AFH, one of which is known as ISOAFH. Generally, these improved versions include identification of interference source and channel classification specifically designed for identifying the channel occupied by interference source.

Although AFH approaches generally obtain good performance in tackling interference problem, all the devices involved are required to cooperate in order to carry out the same AFH mechanism. In other words, legacy devices are unable to enjoy the advantage brought about by AFH mechanisms. In view of this, MDMS mechanism which focus on time domain based orthogonality has also been proposed. The main advantage of this mechanism is that only the master device is required to make changes by updating the firmware, and the slaves devices are unaffected. A modified version called ISOMDMS which includes channel classification modifications has also been presented in this chapter. Finally, a practical scenario of collocated collaborative network access point has been covered and discussed in detail.

PROBLEMS

11.1   Briefly describe the original Bluetooth frequency hops selection scheme and how adaptive frequency hops selection scheme can be extended from it to solve the problem of coexistence interference.

11.2   Briefly describe the differences between ISOAFH and TG2-AFH. In particular, explain why ISOAFH could obtain better performance than TG2-AFH.

11.3   Briefly describe the differences between AFH and MDMS. Suggest some conditions that MDMS may obtain better performance.

CHAPTER 12

COMPETING TECHNOLOGIES

It takes all sorts to make a world.

—English Proverb, traced by Smith to 1620

12.1 INTRODUCTION

As short-range wireless communication is a lucrative market, there are many other similar technologies competing with Bluetooth and IEEE 802.11x WLAN. In this chapter we focus on three such competing technologies: IrDA, HomeRF, and Hiper-LAN.

Interoperability:

Interoperability is a main problem in the competing technologies for short-range wireless communications. One major competing technology that dominated the short-range realm is infrared. For using infrared as physical layer, there are standards from IEEE 802.11 and IrDA. Of the two technologies, IrDA is the one more commonly used. IrDA actually comprises several standards: IrDA-data, IrDA-control and the most recently proposed AIr (Advanced Infrared). For data communications purpose, IrDA-data is the one that is widely used. Multiple standards in irfrared greatly hinder its interoperability. From an implementation perspective, different manufacturers tend to have their considerations on the transmission range due to reasons like potential hazards posed to the human eye due to reception of infrared with higher power. As a result, some manufacturers implement IrDA with a transmission range shorter than the suggested one meter. This issue also greatly hinders the interoperability. Details about the infrared technologies are further discussed in Section 12.2.

HomeRF is another major competing technology in the short-range wireless domain. HomeRF is a major competitor for IEEE 802.11 WLAN as they both use similar frequency band and provide similar services. Most notably, HomeRF adopts the idea of IFS in IEEE 802.11 WLAN but its IFS scheme is incompatible with that of IEEE 802.11 WLAN. This hinders the wide-spread acceptance of the HomeRF technology. Indeed, with greater support from manufacturers and better specifications with multiple task groups, IEEE 802.11 is expected to overshadow HomeRF. HomeRF becomes fading out in the market. The details of HomeRF are further discussed in Section 12.3.

Last but not the least, HiperLAN/2 standards developed in Europe becomes a main competitor for IEEE 802.11a. HiperLAN/2 is widely touted to have a better performance than that of IEEE 802.11a. One special feature is that it includes the consideration of integration with other wired network. However, these idea greatly increase the complexity of implementing the HiperLAN/2 technologies. Consequently, this leads to serious interoperability and implementation problems due to complex frame structure and tight time constraint. More about HiperLAN/2 is further discussed in Section 12.4.

Performance:

IrDA offers wireless communication range for at least 1 m with variable throughput capacity up to 16 Mbps by using different kinds of modulation schemes to support different transmission rates. Indeed, IrDA's throughput capacity is much higher than that achievable by its current main competitor—Bluetooth. The details about different modulation schemes used in order to attain these different transmission rates are elaborated in Section 12.2.4.

On the other hand, HomeRF achieves the wireless communication range for at least 50 m with variable throughput capacity up to 800 Kbps for voice communication, 1.6 Mbps for voice and data communication, and 10 Mbps for pure data communication. We provide a detailed discussion about this in Section 12.3.4. Specifically, we discuss about the different MAC schemes used for different traffic flows.

Finally, HiperLAN/2 achieves wireless communication range for around 150 m with variable throughput capacity similar to that of IEEE 802.11a. The throughput ranges from 6 Mbps up to 54 Mbps. This is further elaborated in Section 12.4.4.

12.2 IRDA

12.2.1 Design Goal

Infrared has been developed for more than ten years. Similar to the case of Bluetooth, it was developed in view of the great market demand for short range cable replacement solutions for portable devices. As a result, it is designed based on similar objectives as that of Bluetooth—Low Cost, Cable Replacement and Ad Hoc connectivity. The readers are referred to Section 10.2 for the design goals of Bluetooth. The design goals of infrared are summarized in Table 12.1 below.

Table 12.1 Design goals of infrared.

Low Cost	Infrared transmitter and receiver can be produced with the cost as low as US$2—3 due to its simple hardware architecture (one LED for transmitter and one photodiode for receiver). As a result, many portable devices can integrate this solution with negligible cost.
Cable Replacement	One of the main objectives of infrared is to provide short range cable replacement solution to portable devices—PDA, mobile phone, portable computers, etc. As a wide range of portable devices are already equipped with IrDA for communication, its objective as cable replacement solution is considered as successful.
Ad Hoc Connectivity	Infrared's design expects the users to to set up temporary connections only. As a result, only some simple applications like exchange of business card, exchange of picture, etc. are suggested in its original design.
Secure	Infrared does not need to adapt highly complex security measure since the communication between devices has limited range and requires line-of-sight.

12.2.2 Transmission techniques

There are three basic transmission techniques using infrared light—directed beam infrared, omnidirectional infrared and diffused infrared [323].

Directed beam infrared's transmission signal is focused. It can be used to create a point-to-point link between two infrared devices. Its range of transmission depends on many factors such as transmission power and degree of focusing. One of the example usages of this type of transmission technique is to build a token ring LAN as shown in Figure 12.1. Here, each IR transceiver is used to support a computer or a group of computers connected together by hubs. As a result, a token ring network is built.

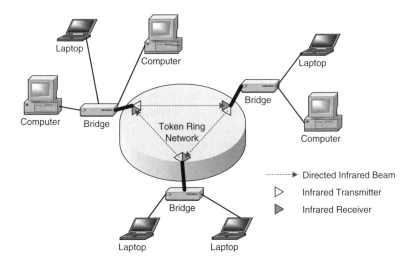

Figure 12.1   Directed beam infrared example.

Omnidirectional infrared generally involves a single access point located at the ceiling so as to provide coverage area large enough to cover the whole room. The access point is connected to an IR transceiver in which transmission signal is radiated omnidirectionally, so that receivers of stations within the coverage area are possible to receive the signals. On the other hand, the transmitters of other stations transmit directed signal pointed to the receiver of access point. The whole configuration is illustrated in Figure 12.2.

Figure 12.3 illustrates the operation of diffused infrared. Here, the transceiver attached to each station focuses at a point which is possible to reflect the IR signal. For example, the lightly painted ceiling in the room is such a target point. By doing so, the directed beam is reflected omnidirectionally by the ceiling and all the receivers within the area can receive the signal.

In the market nowadays, most devices are equipped with IrDA compatible infrared transceiver for data communications. As a result, most IR devices employ directed beam infrared type with point to point link transmission only. On the other hand, IEEE 802.11 PHY specifications also include infrared as a possible physical medium. Specifically, diffused infrared is specified. The readers are referred to Section 9.5 for more details about IEEE 802.11 infrared.

12.2.3   IrDA Protocol Stack

IrDA (Infrared Data Association), founded in June 1993, aims at providing an open standard for short range wireless infrared communication. IrDA was established in view of the fact that there are many manufacturers implementing their own standards

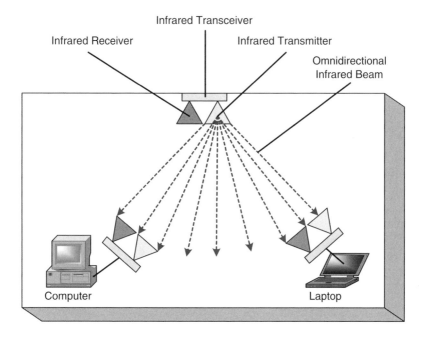

Figure 12.2 Omnidirectional infrared example.

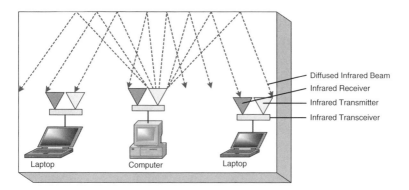

Figure 12.3 Diffused infrared example.

on infrared technology at that time. This leads to incompatibility among different infrared devices from different products. Indeed, even for IrDA standards on infrared, there are more than one standard, including IrDA-Data, IrDA-Control and AIr. In this section and the next, IrDA-Data communication protocol stack, its corresponding physical layer, and datalink layer are discussed in detail.

A simplified IrDA protocol stack is shown in Figure 12.4. Details about each protocol components are summarized below [349].

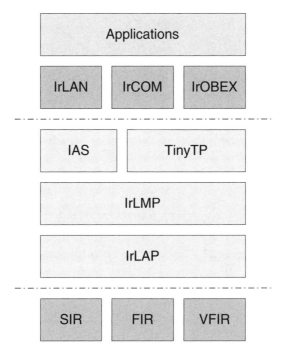

Figure 12.4   Simplified IrDA protocol stack.

- Physical layer involves SIR (Serial Infrared), FIR (Fast Infrared) and VFIR (Very Fast Infrared)[1]. SIR refers to the physical transmission rate less than 1.152 Mbps, FIR refers to the physical transmission rate of 4 Mbps, and VFIR refers to the physical transmission rate of 16 Mbps. These three types of physical layers involved different modulation and/or coding techniques. These will be further elaborated in the next section.

- IrLAP (Inrared Link Access Protocol) is defined based on a modified HDLC protocol. In general, it is responsible for device discovery, negotiation and

---

[1] SIR, FIR, VFIR are obsolete terms in specifying Infrared physical layer. However, it is used in this chapter for presentation purpose.

establishment of a physical connection. IrLAP allows establishment of a single reliable connection between exactly two devices with the description on its establishment and termination including the supported connection rates of both side. In IrLAP connections, there are one primary device and multiple secondary devices. Despite that in theory it is possible to have more than one IrLAP connections in a device (point to multi-point connection), practically most devices only support one single IrLAP connection (point to point connection).

- IrLMP (Infrared Link Management Protocol) is mainly used to multiplex multiple infrared applications/services into a IrLAP connection. As a result, multiple applications listening for incoming connections without interfering with others are allowed.

- TinyTP (Tiny Transport Protocol) is mainly used to provide two basic functions—flow control, segmentation and reassembly. Specifically, flow control can be provided on virtual service channels similar to that of TCP. On the other hand, a large packet can be divided into smaller packets which will then be reassembled at the receiver side.

- IAS (Information Access Service) is mainly used by the user application to advertise their services and query the services provided by the opposing hosts. Its function is similar to that of SDP in Bluetooth protocol stack. At the beginning, only the IAS service is present in the service record. When an application starts up, it registers itself to IAS. Then, a specific selector—LSAP (Link Service Access Point) Selector is given to record this application so as to be accessed by other devices.

- IrCOMM (Infrared Communication)'s functions are similar to that of RF-COMM described in Bluetooth protocol stack in Section 10.3. This protocol is mainly used to emulate RS-232 serial port via infrared. As a result, legacy serial port applications can be used between IrDA devices.

- IrOBEX (Infrared Object Exchange Protocol) is a simple protocol in which PUT and GET commands generally used in HTTP protocol, are defined for object exchange. By using this protocol, binary data object can be transferred between devices supporting this protocol.

12.2.4 IrDA Physical layer and Data Link layer

IrDA operates at around 850–950nm wavelength. As it has been discussed in previous section, IrDA involves multiple physical layer transmission rates with different modulation and/or coding techniques. Common to all transmission rates, IrDA standard requires the devices to work at least 1 meter with 15 to 30 degree half angle deflection (in cone shape) for both transmitter and receiver under daylight environment. There are three main types of physical layers—SIR, FIR and VFIR [3, 349].

SIR refers to the transmission rate up to 1.152 Mbps. It is further divided into lower rate with throughput up to 115.2 kbps and higher rate up to 1.152 Mbps. In lower transmission rate type, pulse modulation—RZI (Return to Zero Inverted), with 3/16 length of the original duration of a bit (i.e., 3/16 mark-to-space ratio) or a fixed length of 1.63 μs pulse is used. The data format of this type of transmission is inherited from UART frame with a start bit and stop bit between data bits. With the help of start bit, asynchronous transmission word can be used.

On the other hand, for high-speed transmissions, the pulse modulation used is RZI with 1/4 mark-to-space ratio. The packet format is also a little bit different from that of lower rate in that two start words are transmitted at the beginning, followed by address word to indicate the target device, data, 16-bit CRC and a stop word. This type of transmission is synchronous in nature.

Figure 12.5 illustrates the pulse modulation schemes that is mentioned above. Here, RZI is compared with NRZ (Non-Return to Zero), which is analogous to data bits. NRZ is a pulse modulation scheme which uses a pulse period to represent data bit 1 and non-pulse period to represent data bit 0. At the same time, during the bit pulse period, the pulse does not return to zero position. RZI uses the opposite concept that a pulse represents data bit 0 and the pulse will return zero position during the bit period as shown in the figure.

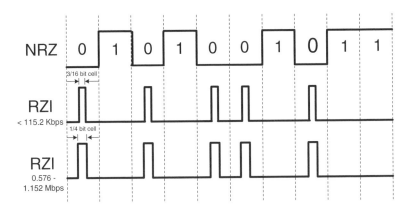

Figure 12.5 Pulse modulation of different SIR schemes.

Now let us consider FIR, which represents the infrared transmission rate of 4 Mbps. In order to achieve this transmission rate, 4 PPM modulation with 1/4 mark-to-space ratio is used. Figure 12.6 illustrates the idea of 4 PPM. As can be seen, the original data bits are paired up. Two bits of data are encoded into one of the four position pulses as shown in the figure. That is, 00 → 1000, 01 → 0100, 10 → 0010 and 11 → 0001. In case of data bits pair 00, the LED flashes at the first position (a chip period) and then there will be nothing for the three remaining position. Unlike the mechanism in SIR, FIR mechanism requires only one flash to represent two bits. This leads to two time faster speed for data transmission.

The packet format of FIR is a little bit different than that of SIR. In a packet, there is a permeable field existed at the beginning for receiver to establish phase lock. Then a start flag and information field are followed. Stop flag is the last. However, unlike that of SIR, FIR packet uses 32-bit CRC.

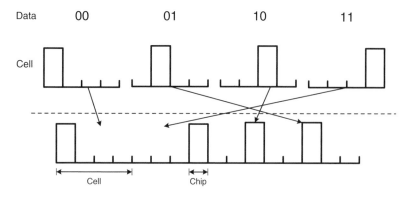

Figure 12.6   Four pulse position modulation scheme.

Finally, VFIR refers to the infrared transmission rate of 16 Mbps. It incorporates a new modulation code HHH (1, 13). The specified data rate is achieved by a low duty cycle, rate 2/3 and (1, 13) run-length limited (RLL) code. With (1, 13) RLL code, it guarantees that there is at least one space between each pulse and at most 13 spaces between each pulse. The chip clock is 24 Mchips/s. With the rate of 2/3, the IR transmission speed achieved is 16 Mbps.

The data encoding techniques of HHH(1, 13) are quite complicated. We provide a brief overview here. Firstly, there are 3-bit representing present state inside and 6-bit representing internal inputs. The state is first initialized to (1, 0, 0) and the internal inputs bits depend on initial input data bits pair. Depending on the input data bits pair, internal input bits and present state, next state is then determined with 3-bits output for transmission every time.

Let us move on to discuss the data link layer of infrared. As discussed earlier, the data link layer functionality of infrared can be regarded as provided by IrLAP protocol, which has the functions of device discovery and maintaining physical connection. Accordingly, it has two modes of operation—contention mode and connection mode. Contention mode is used in device discovery in which the transmission speed is fixed at 9600 bps. In this mode of operation, the device senses the media for at least 500 ms before sending out any query. This mode is mainly used to discover the devices within the visible region associated with their nicknames and MAC address. A MAC address in infrared is a 32-bit address randomly chosen by the device. If the device senses that there is an address conflict, an address conflict resolution procedures is carried out in this mode of operation; otherwise, a connection will be established with the target device with a negotiated rate. IrLAP then changes to connection mode. Communication can then be carried out between two infrared devices with that rate.

12.3 HOMERF

12.3.1 Design Goal

The main objective of HomeRF is to provide an interoperable wireless networking environment with both local and remote (Internet and/or PSTN) contents including voice, data and, audio and/or video streaming, etc. In other words, it targets at integrating all the communication services at home with both wired and wireless environment by using HomeRF.

The development of HomeRF was coordinated by a working group[2] formed by companies including Siemens, Xilinx, AT&T, Motorola, etc.

12.3.2 Structure of HomeRF

There are two main types of network structure in HomeRF—Ad Hoc Networks and Managed Networks [126]. Before these two structures are discussed in detail, three types of node involved are introduced below. It should be noted that it is possible for a device to implement more than one service. As a result, some devices may become nodes of multiple types.

- A-node (Asynchronous node). This type of node is mainly used to transfer asynchronous data services including file transfer. A number of machines are regarded as this type of node such as PCs, PDAs, laptop computers, etc.

- S-node (Streaming node). This type of node is mainly used to receive or transfer some streaming data (asynchronous in nature) from or to other devices. Since this type of node requires a moderately strict QoS support, it should receive priority service than normal A-node traffic. Devices having streaming of audio and/or video applications such as audio headset, PC streaming of radio programme, etc., are regarded as this type.

- I-node (Isochronous node). This type of node is mainly used to support isochronous data service with strict QoS requirements such as a telephone call. A cordless phone is a typical example.

The idea of Ad Hoc Network topology in HomeRF is very similar to that of IEEE 802.11 WLAN. In this structure, there is no centralized control on the medium access. Peer communication is allowed by competing for the shared wireless medium. This

---

[2]This workgroup was disbanded in January 2003.

type of network only allows A-nodes. Figure 12.7 illustrates an example of this kind of network.

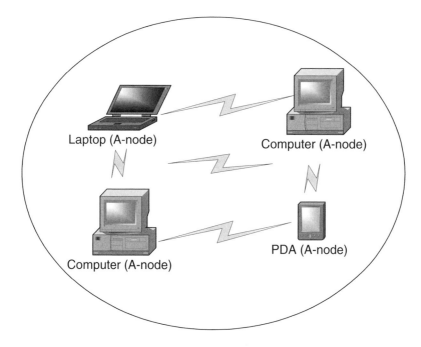

Figure 12.7   HomeRF ad hoc network example.

On the other hand, for a managed network, there will be a centralized control point, which manages the resources of the whole network by allowing A-nodes to exchange data among themselves, other wired networks, or the Internet. At the same time, it manages the S-nodes by assigning priority channel access to them through session setup process. It also manages the I-nodes by connecting them to PSTN telephone network and allocating dedicated time and bandwidth to maintain the conversation with specified QoS requirements. Figure 12.8 illustrates a managed network example.

12.3.3   Protocol Stack

Similar to all other technologies that have been mentioned, e.g., Bluetooth, IrDA, etc., HomeRF also has its own specialized protocol stack as shown in Figure 12.9. It supports three types of traffics: Asynchronous, Streaming and Isosynchronous [126]. In order to support different requirements of different traffic types, different MAC types are used. CSMA/CA is used for normal asynchronous traffic, CSMA/CA with priority access is used for streaming traffic as this such traffic requires a certain degree of QoS supports. Last but not the least, TDMA is used for isochronous traffic in order

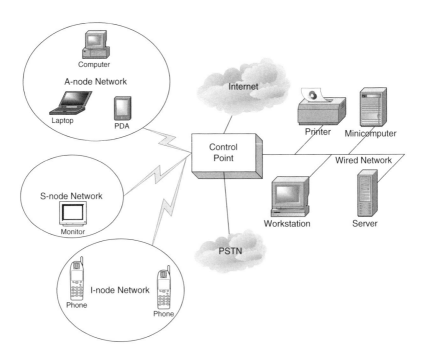

Figure 12.8   HomeRF managed network example.

to provide dedicated time and bandwidth for voice traffic. HomeRF MAC is further discussed in next section.

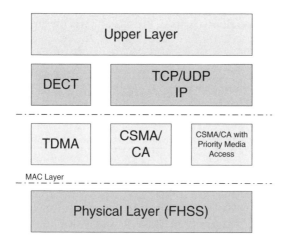

Figure 12.9   HomeRF protocol stack.

From Figure 12.9, we can see that TCP or UDP is used over IP for asynchronous and streaming traffic. However, the DECT (Digital Enhanced Cordless Telecommunications) which is a cordless phone standards in Europe is used for isochronous traffic. Due to the usage of DECT, the I-nodes in the network will be able to connect to normal PSTN network with the normal call setup.

12.3.4   HomeRF PHY and MAC

HomeRF operates at 2.4 GHz frequency band [126]. It chooses FHSS which has been discussed in detail in Bluetooth PHY in Section 10.4. Since it aims at supporting both data and voice communications, the channels division mechanism also depends on the communication types. For data and voice communication, frequency bands are divided into 75 channels with 1 MHz center frequency spacing. This results in 1.6 Mbps throughput rate. On the other hand, for pure data communication, 5 and 10 Mbps throughput rates are supported by dividing the frequency band into 15 channels with center spacing of 5 MHz each.

Meanwhile, certain adaptation mechanism is used in HomeRF to detect whether interference occurs. If interference is detected, it examines the hopping sequence so as to swap those hops within the interference region. The main objective of this mechanism is to prevent two consecutive hops from affecting by the interference source.

We can see that the PHY of HomeRF is very similar to that of Bluetooth. Now, let us consider the HomeRF MAC. Figure 12.10 illustrates the superframe of HomeRF

MAC. We can see that here are two main types of structures. The first one includes only contention period, while the second one includes both contention period and contention free period. A superframe lasts for a fixed period of 20 ms. Both types of superframe start with HOP (Hopping Period) which is used to change the hopping frequency used.

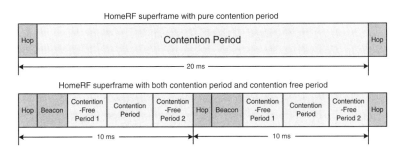

Figure 12.10   HomeRF superframe.

The contention period in both types of superframe are mainly used to transfer asynchronous traffics and streaming traffics. The standards define that the channel access method used in this period is CSMA/CA with priority access and time reservation. As a result, this period is mainly used by A-node and S-node for transferring data or streaming voice. The idea of contention resolution is very similar to that of DCF in IEEE 802.11 with the use of IFS and random backoff mechanism. Figure 12.11 illustrates the contention period with both asynchronous and streaming traffics. There are at most eight simultaneously streaming traffics exists. The CP assigns smaller stream slot number to the traffic which requires stricter QoS requirements. Recall IFS explanation in Section 9.4.3, the shorter the IFS, the higher the chance that the traffic can get control of the medium. In this case, the CF assigns the stricter QoS requirements traffic with shorter IFS (DIFS). As shown in Figure 12.11, the DIFS[3] of these eight streaming sessions is shorter than that of asynchronous traffics. Thus, priority access is given to these eight streaming traffics.

Now let us consider the contention free period. In order to provide a contention free period for isochronous traffics which are sensitive to delay and jitters, the CP has to provide a dedicated time slot and bandwidth. In view of this, HomeRF uses TDMA for this kind of traffic. Similar to that of streaming traffics, it allows eight simultaneous isochronous streams. Figure 12.12 illustrates an example of isochronous transmission in contention free period. We can see that there are two frames with each 10 ms. Within each frame, there are two contention free periods. The second contention period is mainly used to transmit the isochronous traffics as shown. On the other hand, the first contention period in the frame immediately after the previous

---

[3]DIFS in HomeRF is a little bit different from that of IEEE 802.11. The DIFS in HomeRF includes the contention windows time, while IEEE 802.11 does not include contention window time.

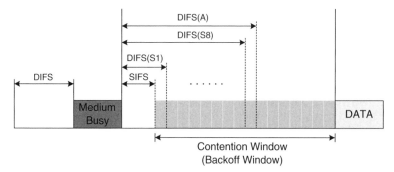

A - Asynchronous node

Sx - Streaming node with priority number x up to
    eight simultaneous streams

Figure 12.11   CSMA/CA with priority access.

frame is mainly used for retransmission of the lost traffics in previous contention free
period. In other words, this contention free period may not occur if there is no loss in
previous transmission. This approach supports high quality voice transmissions such
as DECT, which requires strict delay and jitter bounds.

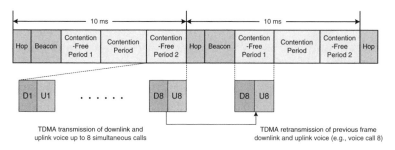

Figure 12.12   TDMA of HomeRF contention free period.

12.4   HIPERLAN

12.4.1   Design Goal

The main objective of HiperLAN is to provide a high performance WLAN in Europe which supports access to external network architecture such as Ethernet, ATM
network, 3G network, etc. On the other hand, it also targets at both traditional data
traffic together with multimedia traffics with certain QoS requirements.

In order to facilitate the development of HiperLAN[4], ETSI (European Telecommunications Standards Institute) Project BRAN (Broadband Radio Access Networks) was developed. ESTI works together with more than 50 companies to come up with the standards.

12.4.2 Structure of HiperLAN

Similar to that of HomeRF, HiperLAN is designed to support two main structures—Business Network and Residential Network [126, 284]. Let us first focus on the business network, which involves an access point acting as centralized control of the medium. All the traffics are routed through the access point. Figure 12.13 illustrates an example of business network model. This is also known as centralized mode of operation in HiperLAN.

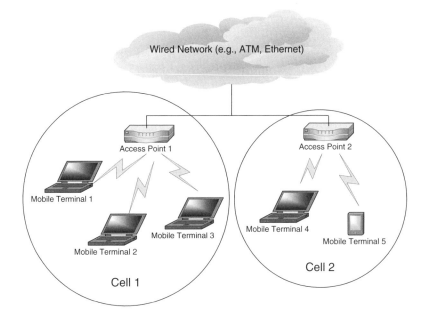

Figure 12.13 HiperLAN business network example.

On the other hand, there is another structure known as residential network. It is, in fact, an Ad Hoc network model. However, in this configuration, a CC (central controller) is needed to coordinate the medium usage among all the devices within the subnet. The CC can be dynamically selected from any Mobile Terminal. This is also known as direct mode of operation. Unlike centralized mode of operation, traffics

---

[4]Current version of HiperLAN is HiperLAN version 2 or HiperLAN/2. Throughout this chapter, HiperLAN refers to HiperLAN/2.

are sent in a peer-to-peer manner among mobile terminals. Figure 12.14 illustrates an example of this kind of structure.

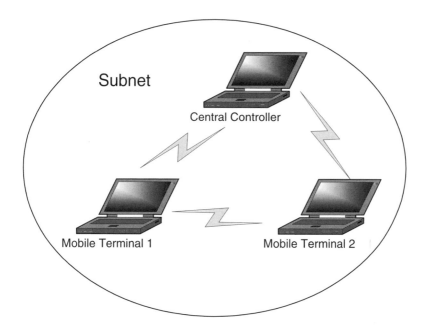

Figure 12.14   HiperLAN residential network example.

12.4.3   Protocol Stack

HiperLAN/2's protocol stack is shown in Figure 12.15 [108, 126, 284]. Here, it can be seen that HiperLAN/2 protocol stack actually composes of three layers—CL (Convergence Layer), DLC (Data Link Control) Layer including MAC and RLC, etc. and PHY (Physical) layer. The functionalities of PHY and DLC layer are elaborated in detail in the next section.

The convergence layer is a major component that enables HiperLAN/2 to provide interface between wireless mobile terminal and wired network or core network such as Ethernet, ATM, 3G network, etc. Convergence layer acts as an interface which performs packet transformation from fixed or variable size packet from upper layer to fixed size packet of proper priority class transmission in HiperLAN/2 network and vice versa. Essentially, convergence layer performs the functions of segmentation and reassembly (SAR) for packets.

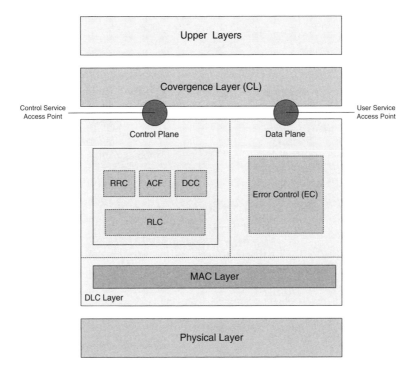

Figure 12.15 HiperLAN/2 protocol stack.

12.4.4   HiperLAN/2 PHY and DLC

In this section, we discuss about the detailed features of HiperLAN/2 PHY and DLC layer [108, 126, 284]. Let us consider the PHY layer first. The PHY layer in Hiper-LAN/2 is very similar to that of IEEE 802.11a WLAN. It operates at 5 GHz ISM band with OFDM. Since OFDM is adopted, it is required to operate into channels of 20 MHz center frequency spacing. Modulation schemes like BPSK, QPSK, 16-QAM and 64-QAM are also adopted into OFDM in order to provide transmission rate of 6 Mbps up to theoretical maximum of 54 Mbps.

Unlike that of IEEE 802.11a, dynamic frequency selection (DFS) scheme is adopted into HiperLAN/2. This scheme allows AP to choose the frequency based on the consideration of interference source. At the same time, P/CC can adjust their transmission rates by changing the modulation scheme used to adopt the channel conditions.

DLC in HiperLAN/2 is quite complicated. Indeed, the DLC structure is one of the main features of HiperLAN/2 that differentiates it from other technologies. As shown earlier in Figure 12.15, in the protocol stack, there are three main DLC components: RLC (Radio Link Control), EC (Error Control), and MAC layer. RLC and EC functions are summarized below.

- RLC implements the control plane of DLC and is mainly used to manage the use of wireless radio networks and exchange the control data between nodes. Multiple instances are present in AP with each of them represents a MT. However, only one instance exists in a MT. It consists of three main functions as shown in Figure 12.15.

    - DCC (DLC connection control) function is mainly used to manage the establishment, maintenance and termination of a connection between MT and AP. The establishment is managed by the set-up procedures which is usually initiated by a MT with the QoS parameters provided if necessary. Connection release procedures are also provided for termination of the connection.

    - ACF (Association Control Function) is mainly used to provide any necessary functions in association including, authentication, encryption, association, dissociation, link capabilities information exchanges, etc. As it has been suggested in Section 12.4.2, the wireless medium is under centralized control by AP/CC no matter in direct mode of operation or centralized mode of operation. As a result, association is a necessary function for every MT accessing HiperLAN/2 network.

    - RRC (Radio Resource Control) includes four main functions including dynamic frequency selection (DFS), handover, station availability check, power saving and control. These functions are necessary for efficient and effective use of the shared wireless medium.

- EC (Error Control) is implemented into the data plane of DLC for detecting and correcting (if possible) the error of the data frames sent and received. There are three modes of operation.

  - Acknowledged mode uses retransmission techniques to recover the unacknowledged packets. The Automatic Retransmission Request (ARQ) scheme used is selective repeat with variable window size. By doing so, reliable transmission is provided.

  - Repetition mode transmits arbitrary number of times of the same packets in consecutive order. It is unnecessary to inform the transmitter with acknowledgement. By multiple transmission of the same packet, reliable transmission can be provided with high probability.

  - Unacknowledged mode does not use any mechanism to retransmit the packets in case of error. As a result, the transmission is unreliable.

MAC layer [108, 126, 284] in HiperLAN/2 is quite complicated compared to that of IEEE 802.11 WLAN. It adopts Time Division Duplex (TDD) with Time Division Multiple Access mechanisms (TDMA). Thus, different time slots are assigned for downlink, direct link and uplink. The wireless medium is controlled by AP/CC. Each frame has a duration of 2 ms. Figure 12.16 illustrates an example of HiperLAN/2 MAC frame. As can be seen, some of the transport channel components of the MAC frame have also been shown. These includes Broadcast Channel (BCH), Frame Channel (FCH), Access Feedback Channel (ACH), Random Access Channel (RCH) and also the location of the three phrases—downlink, directlink and uplink phrases in the frame.

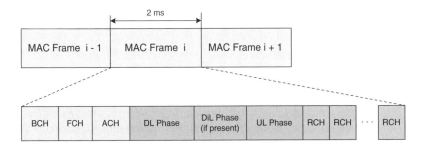

Figure 12.16 HiperLAN/2 MAC frame example.

In the MAC frame construction process, transport channel represents the message format in carrying the information from the logical channel. On the other hand, logical channel is mainly used to represent the types of message that a particular channel carry. In order words, the message (either control message or data message) is first mapped to an appropriate logical channel which will then be mapped to different transport channels in the MAC frame for transmission. There are altogether ten logical channels and six transport channels. Their corresponding mappings in downlink (from the AP/CC to MT), direct link (peer to peer transmission between MTs) and uplink (from MT to AP/CC) are shown in Figure 12.17.

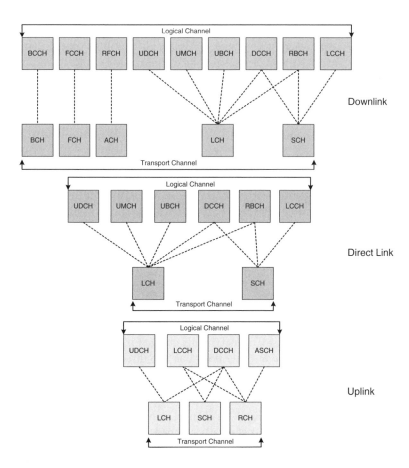

Figure 12.17   HiperLAN/2 logical and transport channels mapping in downlink, direct link and uplink phrases.

The corresponding functions of these altogether 16 channels are summarized below. First of all, let us start with the 10 logical channels.

- BCCH (Broadcast Control Channel). It carries downlink broadcast control message related to the whole radio network to all nodes in the network. It is mapped to Broadcast Channel (BCH) in transport channel.

- FCCH (Frame Control Channel). This is also a kind of downlink message to the MTs in the network. This channel carries the information related to the frame structure. This channel is also known as resource grants. It carries information about how many LCHs and SCHs a particular MT can get for transmitting and receiving data. This channel is mapped to Frame Channel (FCH) in transport channel.

- RFCH (Random Access Feedback Channel). It is also a downlink message which carries the reports of transmission attempt of MTs in Random Channel (RCH) in previous MAC frame. This channel is mapped to Access Feedback channel (ACH) in transport channel.

- RBCH (RLC Broadcast Channel). This channel is used to broadcast control message related to the whole radio network cell. This channel is used in downlink and direct link. Information like, MAC ID of the AP/CC, RLC broadcast message, some encryption related information and convergence layer ID information can use this channel. This channel is mapped to LCH and/or SCH in transport channel.

- DCCH (Dedicated Control Channel). This channel is mainly used to carry RLC messages in either downlink, uplink and direct link in LCH, SCH or RCH transport channel.

- UBCH (User Broadcast Channel). This channel is mainly used to carry the user broadcast message from the Convergence Layer (CL) in either downlink and direct link. It is mapped into LCH transport channel.

- UMCH (User Multicast Channel). This channel is mainly used to carry the user multicast message in either downlink and direct link. It is also mapped into LCH transport channel. Together with UBCH, this channel can either use repetition or unacknowledged mode for error control.

- UDCH (User Data Channel). This channel is mainly used to carry user data in all the links mentioned. It is generally mapped to LCH transport channel. This channel is granted for a connection described in FCCH.

- LCCH (Link Control Channel). This channel is mainly used to carry error control information such as ARQ results in all the link types mentioned. However, it is specifically used to carry the resource request (RR) message in uplink direction. As a result, this channel is mapped into SCH or RCH transport channel.

- ASCH (Association Control Channel). This channel is mainly used by MT to transmit association or handover message to an AP/CC in case it is not associated to any. This message is mainly transport through RCH transport channel.

The six transport channels are introduced below.

- BCH (Broadcast Channel). This channel is mainly used to carry information related to the whole radio cell including information like AP ID, power transmitted, etc. It transmits message in downlink direction.

- FCH (Frame Channel). This channel is mainly used to carry information related to the MAC frame structure. In other words, it transmits information on the allocation of the resources. Similar to BCH, this channel is also transmitted in downlink direction.

- ACH (Access Feedback channel). This channel is mainly used to carry information and results related to the access attempts in RCH of previous frame. Similar to the previous two channels, this one is also transmitted in downlink direction.

- LCH (Long Transport Channel). This channel is mainly used to carry user data and control information. This channel can be used in both direction.

- SCH (Short Transport Channel). This channel is mainly used to carry relatively short control information. This channel can be used in both direction.

- RCH (Random Access Channel). This channel is mainly used to give MTs which are not allocated for any SCH an opportunity to transmit their control information to AP/CC. This information can be resource request (RR), association message, etc. Consequently, this channel is mainly used in uplink direction.

12.4.5   QoS consideration in HiperLAN/2

The complicate DLC layer of HiperLAN/2 has been mentioned in previous section. In fact, this complicated structure allows HiperLAN/2 to support per-flow QoS [126, 284]. HiperLAN/2 adopts a centralized mode of operation in its design with an AP/CC assigned to perform the functions of allocating resources. Bandwidth allocation is performed by signaling: resource request and resource grants. Firstly, MTs send out the resource request indicating the characteristics of the traffic flows to AP/CC. AP/CC will then grant MTs with the numbers of packets which is allowed to send based on bandwidth resources availability, traffic characteristics, etc., through resource grant message. This centralized mode of operation effectively supports per-flow QoS in HiperLAN/2 with the help of signaling and admission control.

12.5 PRACTICAL ILLUSTRATIONS—IMPLEMENTATION OF HIPERLAN

DLC layer in HiperLAN/2 allows it to achieve high performance with QoS considerations. However, in the market, IEEE 802.11a products are widely accepted. One of the main reason is that HiperLAN/2 design is highly complicated due to the reason of complex DLC and restricted time constraint. In this section, we describe a practical implementation of HiperLAN/2 on a FPGA board to demonstrate the difficulties in the implementation of HiperLAN/2 [337].

Figure 12.18 illustrates HiperLAN/2 MAC protocol together with the implementation idea and division of software and hardware module. Such a division is necessary, as the restricted time constraint does not allow all the protocol processing to be done in software. In particular, hardware speed is needed for some of the time critical functions like parsing of broadcast message, CRC error checking, etc. In other words, hardware module is adopted to reduce the time of performing some time critical tasks.

As shown in Figure 12.18, the implementation of DLC involves a software scheduler for management of the available resources. The scheduler on AP involves generation of allocation map of a frame with information from traffic table which stores information related to QoS agreement with connected MTs. The execution time of the resource grants in the frame is used to sort the mappings. Afterwards, broadcast control frames are formed and stored. All these will be done by software parts in AP. On the other hand, the scheduler in MT will be required to map the resource grants carried in the broadcast control message in the traffic table. After the scheduler has finished its jobs, the software part generates a memory address list which contains the memory addresses of the connection list that is going to be sent out in current frame. These are the main functions done by the software modules in AP and MT HiperLAN/2 implementation.

As shown in Figure 12.18, the remaining parts of DLC and PHY layer are implemented in hardware logics in a FPGA board. The host controller interface is implemented by PCI controller which acts as a communication interface between the software module and hardware module. The most difficult part of hardware module design and implementation involves the implementation of frame processor in AP and MT. The frame processor needs to carry out many functions including the management of the air interface, the generation of acknowledgement map for the RCH field of previous frame, terminals synchronization, etc.

Since the time for executing certain events need to be precise, time descriptor which contains information about when and what events needed to be carried out are proposed. The generation of this description is done by parsing the broadcast message which involves information about the whole frame. The slot number suggested inside the time descriptor is used to match that of MAC frame slot. In order to have a precise matching which results in synchronization of AP and MT, a 40 MHz clock with slots exactly equal to the MAC frame slot is used after BCH is sensed.

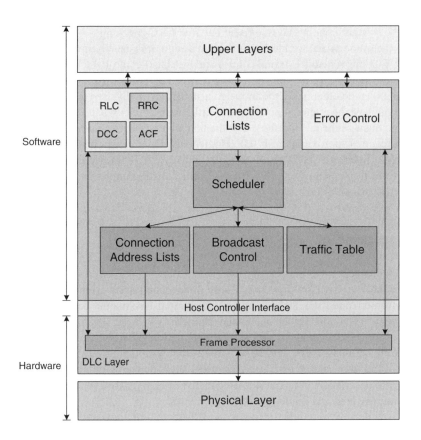

Figure 12.18 HiperLAN/2 MAC protocol architecture.

Moreover, ACH field in the current frame which is used to gives acknowledgment to the RCH in previous frame. As the RCH requires very fast response time, it is also implemented in frame processor. This is mainly done by handling the CRC checking of the received packets in the frame processor. As a result, an error occurs in RCH results in a CRC error which is detected by hardware module. The hardware module can then produce appropriate acknowledgment mappings accordingly. On the other hand, there is latency introduced by PCI bus and baseband modules. Thus, different queues are used to store the process data for either uplink or downlink. The data can then be brought to upper layers or sent out after CRC checking.

In the frame processor hardware module design, it is usually involves three main parts: PCI bus scheduler, frame control and baseband control. PCI bus scheduler is mainly used to handle the transfer of data between the software module and hardware module. On the other hand, the frame control part is usually used to handle the functions that have been mentioned above. Furthermore, the baseband control is mainly used to manage the air interface. For the hardware implementation of HiperLAN/2 AP, it requires approximately 170,000 logic units. On the other hand, the implementation of HiperLAN/2 MT requires approximately 150,000 logic units. These parts have to be implemented in a FPGA board, indicating the implementation difficulties of the HiperLAN/2 functions.

12.6 SUMMARY

In this chapter, three short-range wireless technologies—Infrared, HomeRF and HiperLAN are discussed in detail. These three technologies are major competitors for Bluetooth and IEEE 802.11x WLAN. Infrared is also mainly designed as a cable replacement solution. On the other hand, HomeRF also uses 2.4 GHz frequency band. We have discussed about its aim at providing an integrated wireless home network including both data and voice. HiperLAN/2 employs the 5 GHz frequency band and is designed to provide a high speed wireless network with QoS consideration. We have also described a practical implementation of HiperLAN/2 with FPGA board to illustrate the practical implementation difficulties.

It is widely expected that Infrared and Bluetooth will coexist in the market. It is because they are competing yet complementary in nature. The main advantages of IrDA are: it is easy to use, it comes with low implementation cost, it does not need complicated security. On the other hand, HomeRF, which occupies 2.4 GHz band, has very similar MAC layer with that of IEEE 802.11. Since this standard does not really provide many advantages over that of IEEE 802.11 and they are competing in many aspects, it seems that HomeRF does not have a competitive edge. Indeed, HomeRF working group was disbanded in January 2003. For the case of HiperLAN/2, its initial ideas of supporting multiple network architectures and incorporating QoS consideration are good. However, as illustrated in the practical illustration section, it requires both hardware and software support for precise time synchronization, framing support in order to build such a complex system. As a result, it is believed that IEEE 802.11 would drive out HiperLAN/2 due to its distributed approach and relatively less complicated design.

PROBLEMS

12.1   Briefly discuss the similarities and the differences between IrDA and Bluetooth in their protocol stack and some current usage models. Explain why these two technologies simultaneously exists in market with both competing and complementary effects.

12.2   Briefly discuss the PHY and MAC of HomeRF. Explain why this technology fades out in the market.

12.3   HiperLAN/2 is expected to be a competing technology of IEEE 802.11a WLAN. Briefly discuss the advantages and disadvantages of HiperLAN. Explain why people expect that IEEE 802.11 would drive out HiperLAN/2.

PART IV

PROTOCOL ADAPTATIONS FOR WIRELESS NETWORKING

CHAPTER 13

MOBILE IP

Home is home, be it never so homely.

—English Proverb, traced by Apperson to c. 1300

13.1 INTRODUCTION

The Internet extends its reach into the wireless realm with the help of various protocol adaptation mechanisms. For one thing, the location and addressing of hosts
are among the most important issues. In this chapter, we discuss about a mature
mobility management adaptation framework for making wired IP interoperable with
its wireless counterpart.

Interoperability:

Due to the advancement of wireless technology, there is a proliferation of mobile
devices which are equipped with wireless interface such as WLAN or Bluetooth.
Because these devices are mobile, it is very likely that the users may carry them from
one Internet access point (typically wireless access point) to another Internet access

Wireless Internet and Mobile Computing. By Yu-Kwong Ricky Kwok and Vincent K. N. Lau
Copyright © 2007 John Wiley & Sons, Inc.

point. Thus, it is beneficial to keep the connection when they move across different domains, as if they were connected to a wireline connection all the time. Mobile IP is designed to fulfill this goal. In Mobile IP, a mobile host uses some advertisement mechanisms to discover its home agents and foreign agents. When it moves from its original home network to a foreign network, it gets a care-of address from the foreign agent and registers the address to it home agent. After that, all packets destined to the mobile host will be intercepted by the home agent and then are sent through a tunnel to foreign agent for delivery to the mobile host. This makes the mobile host get connected wherever it goes. Interoperability is manifested here as a transparent connection migration with the help of different agents across multiple networks.

Performance:

Using Mobile IP, when a mobile host roams across different domains, it can get uninterrupted Internet services. However, this comes with a cost—there is overhead induced in the discovery and registration process. For example, if the mobility of the device is high, the registration process may cause excessive signalling traffic and a long service delay may be incurred. On the other hand, when an Internet host sends packets to the mobile host, since it is not aware of the location of the mobile host, all packets will be sent to the home agent first, regardless of whether the mobile host is close to it. All these request for efficient schemes for enhancing the performance of Mobile IP. Indeed, from a performance perspective, much work is still being done in optimizing the connection migration process.

13.2 ADVERTISEMENT MECHANISMS

Advertisement mechanisms comprise several protocols and procedures for a mobile host to discover its home agent when it is at its home network and foreign agent when it is away from home. Thus, effectively, an agent discovery process is carried out. Due to its similarity to the router advertisement process defined in Internet Control Message Protocol (ICMP), the agent discovery process in Mobile IP makes use of ICMP router advertisement messages by attaching special extensions to the standard ICMP messages.

As shown in Figure 13.1, the home agent and the foreign agent periodically broadcast advertisement messages. Specifically, at the home network, there might also be a foreign agent for serving other visiting mobile hosts (similarly, there could be a home agent at the foreign network). Thus, when a mobile host receives an agent advertisement message, it must compare the network portion of its home network address with the network portion of the advertising agent's IP address. When these portions match, the mobile host knows that it is at the home network, and it can receive IP datagrams with forwarding by the home agent to the foreign agent; otherwise, the mobile host is at the foreign network, and it needs to get a care-of address so as to receive IP datagrams forwarded by the home agent.

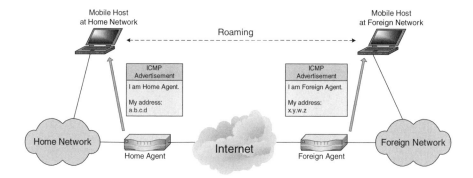

Figure 13.1   Agent advertisement.

From the illustration shown in Figure 13.1, we can see that the agent advertising process provides a mechanism for the mobile host to determine its point of attachment. When the mobile host is roaming, handoff from one network to another one occurs at the physical layer, without notification to the IP network layer. Consequently, the agent advertising process must be a continuous process, where the home and foreign agents need to periodically broadcast advertisement messages.

As mentioned before, the mobile host needs to get a care-of address when it discovers that it is at a foreign network. Now, let us see how it acquires such an address by utilizing the advertisement message received from the foreign agent. Figure 13.2 shows the ICMP router advertisement with the agent advertisement extension, and Table 13.1 summarizes the description of the various fields in the agent advertisement extension.

In Figure 13.2, it also shows an optional prefix-length extension succeeding the agent advertisement extension. The objective of this extension is to indicate the number of bits in the router's address that define the network number, allowing the mobile host to compare the network portion of its home IP address with that of the router's IP address. The value of the type field of the extension is 19. The prefix length field is the number of leading bits that define the network number of the corresponding router address listed in the ICMP router advertisement portion. Finally, the length field is the number of such prefix lengths.

From the agent advertisement message received, if the mobile host recognizes that it is on a foreign network, it will acquire a care-of address advertised in the agent advertisement message. After that, it will carry out the registration process to tell its home agent its care-of address. But before we go on to discuss the registration process, there are few points that are worth discussing:

Agent Solicitation: If a mobile host needs the information of the agents on its network immediately, it is allowed to send an ICMP router solicitation mes-

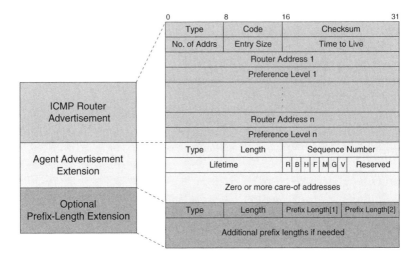

Figure 13.2 Agent advertisement message.

sage so as to elicit an agent advertisement message. Any agent receiving the solicitation message will issue an agent advertisement message.

Collocated Care-Of Address: Besides obtaining a care-of address form its foreign agent, a mobile host can alternatively acquire a care-of address from other independent services such as DHCP. This situation happens in some cases where there is no foreign agent available in the visiting network or all the foreign agents are busy. Specifically, the mobile host can obtain an IP address for its network interface via DHCP, and then use the address as the care-of address, which is known as a collocated care-of address.

Movement Detection: As mentioned earlier, by comparing the network portion of its home address and that of the advertising agent's address, a mobile host can determine whether it is on the home network or it has moved to a foreign network. In the other aspect, the mobile host can detect movement by using the lifetime field of the agent advertisement message. Specifically, when an agent advertisement message is received from an agent, the mobile host records the lifetime as a timer. Later on, if the timer expires before it can receive another advertisement message from the same agent, it can assume that contact has been lost with the agent.

Table 13.1 Description of the fields in the agent advertisement extension.

Field Name	Description
Type	Value 16, indicating that this is an agent advertisement.
Length	The length of the extension, which depends on how many care-of addresses are being advertised.
Sequence Number	The count of agent advertisement messages sent since the agent was initialized.
Lifetime	The maximum lifetime, in seconds, that this agent will accept a registration request from a mobile host.
R	Registration with this foreign agent (or another foreign agent on this network) is required. Mobile nodes that have already acquired a care-of address from this foreign agent must re-register.
B	The foreign agent is busy and will not accept any more registration. This bit is set to prevent existing customers from thinking that the foreign agent had crashed, and moving away unnecessarily.
H	The agent is a home agent.
F	The agent is a foreign agent. Note that bits F and H are not mutually exclusive, and that B cannot be set unless F is also set.
M	This agent supports minimal encapsulation (RFC 2004).
G	This agent supports GRE encapsulation (RFC 1701).
V	This agent supports Van Jacobson header compression (RFC 1144).
Care-Of Address	The care-of address(es) supported by this agent. There must be at least one care-of address if the bit is set.

13.3 REGISTRATION

After a mobile host recognizes that it is on a foreign network and then acquires a care-of address, it needs to inform its home agent about its care-of address to request the home agent to forward its IP datagrams to the care-of address. In addition, it also needs to tell the home agent how long it wants to use the care-of address and indicate special features that are available from the foreign agent. All these are done by carrying out a registration process. Two types of messages are defined in the registration process, they are registration request message and registration reply message, both sent using UDP. Figure 13.3 depicts the registration process, and the details are described as follows:

1. The mobile host sends the registration request message to the foreign agent to ask it to pass the message to the home agent.

2. The foreign agent relays the request message to the home agent.

3. The home agent decides whether to accept or reject the request from the mobile host, and sends the decision as a registration reply message to the foreign agent.

4. The foreign agent passes the reply message to the mobile host.

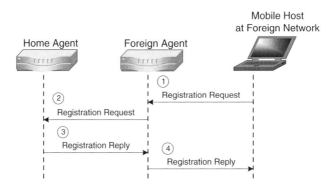

Figure 13.3 The registration process.

Note that during the registration process the foreign agent is considered a passive agent which agrees to pass the mobile host's request to the home agent and then pass back the reply from the home agent to the mobile host. On the other hand, if the mobile node is using a collocated care-of address, it will directly register with its home agent, without the intervention of a foreign agent.

Now, let us investigate the two types of registration messages, which are carried in UDP segments. The packet formats of the registration request message and the registration reply message are shown respectively in Figures 13.4 and 13.5, while Tables 13.2 and 13.3 describe the fields of the corresponding messages.

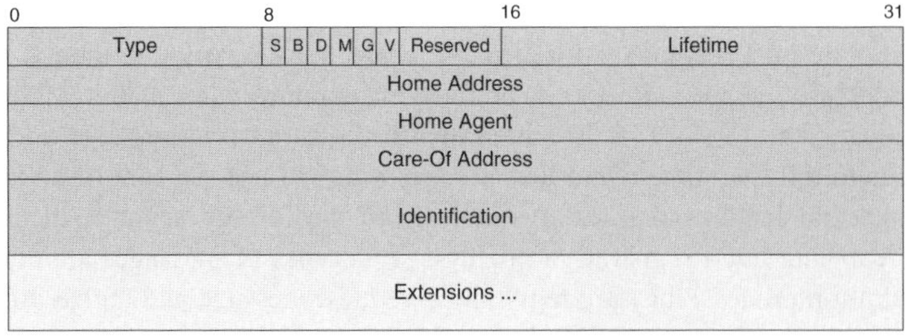

0	8						16	31
Type	S	B	D	M	G	V	Reserved	Lifetime
Home Address								
Home Agent								
Care-Of Address								
Identification								
Extensions ...								

Figure 13.4 Mobile IP registration request message.

Security Measures in the Registration Procedure

The mobile host relies on the registration messages to inform the home agent of its care-of address so as to receive IP datagrams forwarded by the home agent. This

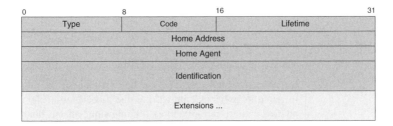

Figure 13.5   Mobile IP registration reply message.

makes it critical to to detect and reject fraudulent registration messages to prevent a malicious user from disrupting the communication between the home agent and the mobile host in the following cases:

1. A malicious user replays old registration request messages to the home agent to cut existing connection between the home agent and the mobile host.

2. A malicious user pretends to be a foreign user to send a registration request message (with the malicious use's IP address being the care-of address) to the home agent in order to intercept traffic destined to the mobile host.

In Mobile IP, the measures applied to protest against the above interruptions are based on the use of the identification field in the registration request and reply messages, and the use of authentication extensions. The identification field is basically a timestamp or newly generated random number (a nonce, a **n**umber used **once**), which makes each registration message different. Specifically, the home agent and the mobile host have to agree on reasonable values for the timestamp or nonce chosen, and resynchronization should be allowed by using the registration reply code 133 as shown in Table 13.4 By using the identification field, a registration reply can be matched to a registration request, and therefore suspicious registration messages (e.g., replayed messages) can be rejected.

The authentication extensions, which are appended to the registration request and reply messages, are used to authenticate the sender and receiver in the message sending between the home agent and the mobile host. Specifically, they are used to protect the fields of a registration message, particularly the important identification field for fighting against replay attacks. In the registration process, despite that the foreign agent acts as a relay to pass messages between the home agent and the mobile host, authentication extensions can also be used for authentication between the home agent and the foreign agent, and between the foreign agent and the mobile host. Thus, there are totally three types of authentication extensions defined in Mobile IP.

Figure 13.6 describes the format of an authentication extension. The type field indicates the type of authentication extension, while the length field indicates the size of the remainder of the authentication extension. The security parameter index (SPI)

Table 13.2 Description of the fields in the registration request messages.

Field Name	Description
Type	Value 1, indicating that this is a registration request message.
S	Simultaneous bindings. The mobile host requests the home agent to retain its previous binding, i.e., the association of the its home address with its care-of address. Using simultaneous bindings, the home agent will forward a datagram to each of the care-of addresses that are registered. In effect, multiple copies of a datagram will be sent to the mobile host to increase reliability in situations like wireless handoff.
B	Broadcast datagrams. It is used to tell the home agent to encapsulate broadcast datagrams from the home network to care-of address.
D	Decapsulation by mobile host. The mobile host is using a collocated care-of address and thus will decapsulate by itself the tunnelled datagrams.
M	It tells the home agent to use minimal encapsulation (RFC 2004).
G	It tells the home agent to use GRE encapsulation (RFC 1701).
V	It tells the home agent to use Van Jacobson header compression (RFC 1144).
Lifetime	The number of seconds that the mobile host requests the home agent to honor the registration. A value of zero is used for de-registration.
Home Address	The home address of the mobile host. This is for telling the home agent to receive datagrams destined to this home address and then forward to the care-of address.
Home Agent	The address of the mobile host's home agent. This tells the foreign agent to which this registration request message should send to.
Care-Of Address	The care-of address of the mobile host. It tells the home agent to which the datagrams of the mobile host should forward.
Identification	A 64-bit number used for replay protection, discussed below.
Extensions	Additional extensions that are required, such as the authentication extension, discussed below.

is an index which identifies a security context between the sender and the receiver. Specifically, it selects the authentication algorithm, mode, and secret (a shared key or public/prviate key pair) used to compute the authenticator in the authenticator field. The authenticator is a code generated by the sender to authenticate a registration message, and then used by the receiver to check that the message has not been altered.

In Mobile IP, the default authentication algorithm uses keyed-MD5 to in prefix+suffix mode to produce a one-way 128-bit message digest as the authenticator.

Table 13.3 Description of the fields in the registration reply messages.

Field Name	Description
Type	Value 3, indicating that this is a registration reply message.
Code	The result of registration request, detailed in Table 13.4
Lifetime	If the registration is accepted, it is the number of seconds that the registration will be honored by the home agent; a value of zero indicates that the mobile host has be de-registered.
Home Address	The home address of the mobile host. This is for telling the home agent to receive datagrams destined to this home address and then forward to the care-of address.
Home Agent	The address of the mobile host's home agent.
Identification	A 64-bit number used for replay protection. It is used for matching a registration reply to a registration request, discussed below.
Extensions	Additional extensions that are required, such as the authentication extension, discussed below.

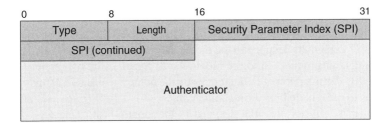

Figure 13.6 Mobile IP authentication extension.

The detailed operations are depicted in Figure 13.7. As shown in the figure, the 128-bit authenticator is computed over the shared key, and then the protected fields of the registration request/reply message, and finally the shared key again. With the authenticator inserted, the registration message is transmitted. After the receiver receives the registration message, the receiver computes its own 128-bit authenticator based on the shared key and the protected fields of the received registration message. Then a comparison is done between the computed authenticator and the authenticator in the received registration message. If the two are equal, the receiver can conclude that the registration message has not been altered.

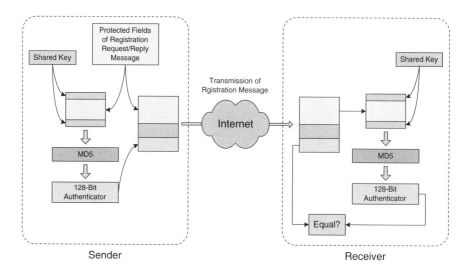

Figure 13.7   The keyed-MD5 authentication algorithm in Mobile IP.

13.4   TUNNELING APPROACHES

After the mobile host registers its care-of address at the home agent, the home agent intercepts IP datagrams destined to the mobile host and then tunnel the IP datagrams to the foreign agent and then to the mobile host. Specifically, two main processes are involved—interception of IP datagrams by the home agent and transmission of IP datagrams to the mobile host through tunneling.

There are two types of traffic destined to the mobile host. One type of traffic is originated from a node in the mobile host's home network. The other type is originated from outside the home network and is transmitted across the Internet (or from other topology of routers and links) and going through the home agent, which serves as a gateway for the home network. In the case of traffic going through the home agent, the interception of IP datagrams by home agent can be accomplished directly without any additional effort. However, if the traffic is generated from within the home networking, the home agent needs to perform a gratuitous ARP (address resolution protocol, a protocol for the translation between network-layer addresses and link-layer addresses) broadcasting beforehand, telling the other nodes in the home network to send IP datagrams destined to the mobile host to the home agent (at the link level).

The reason for performing a gratuitous ARP broadcasting by the home agent is that, while the mobile host is at home, the other nodes on the same network are likely to have ARP cache entries for the mobile host, which will become stale when the mobile host is away from home. Thus, it needs to invalidate those stale ARP cache entries in order for the mobile host to correctly receive its IP datagrams. For the same reason, when the mobile host returns home, it needs perform a gratuitous ARP broadcasting so that its home address is associated to its own link-layer address by the other nodes within the home network.

After a mobile host's IP datagrams are captured, the datagrams are tunneled to the foreign agent for delivery to the mobile host. The tunneling can be done by one of the following encapsulation algorithms:

1. **IP-within-IP Encapsulation.** Defined in RFC 2003. It is the simplest approach and must always be supported (default algorithm).

2. **Minimal Encapsulation.** Defined in RFC 2004. It involves fewer fields in an IP datagram.

3. **Generic Routing Encapsulation (GRE).** Defined in RFC 1701. It is a generic encapsulation algorithm developed before the development of Mobile IP.

In the IP-within-IP encapsulation approach, the entire IP datagram sent by the Internet host is inserted in a new IP datagram as the payload. The header in the original IP datagram is unchanged except that the TTL field is decreased by one. On the other hand, for the header in the new IP datagram, the protocol field is given a value of 4, indicating the presence of the encapsulated IP datagram. In addition, the IP address of the home agent is used as source address while the care-of address of the mobile host is used as the destination address; the values of other fields are selected from the header of the original IP datagram. At the exit of the tunnel, the original IP datagram is restored and delivered to the mobile host.

Minimal encapsulation can be used if the home agent, mobile host and foreign agent all agree to use. It results in lower overhead when compared to the IP-within-IP encapsulation approach. However, it does not support fragmentation to deal with tunnels with smaller path maximum transmission units (MTUs). Figure 13.8 shows the operation of minimal encapsulation. As can be seen in the figure, a value of 55 in the protocol field indicates the presence of minimal encapsulation. Furthermore, a new header called minimal encapsulation header is inserted between the header and payload of the original IP datagram, while the fields in the original header are modified accordingly to reflect the change, such as the length, checksum, source and destination addresses, etc. The fields in the minimal encapsulation header are detailed in Table 13.5.

At the exit of the tunnel, the fields in the minimal encapsulation header are restored to the original IP header and then the minimal encapsulation header is removed. The length and checksum fields are re-calculated to reflect the change.

13.5 ROUTE OPTIMIZATIONS

From the above-mentioned discussion, we know that IP datagrams from an Internet host and destined to a mobile host need to be routed to its home agent first and then

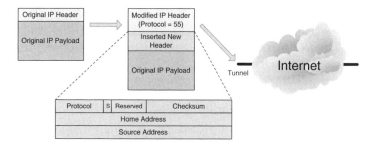

Figure 13.8   Minimal encapsulation.

tunneled to the foreign agent of the mobile host. On the other hand, IP datagrams from the mobile host can be directly routed to the Internet host. This kind of asymmetric routing, the so-called triangular routing [276] as illustrated in Figure 13.9, is far from optimal. The situation is even worse when the Internet host and the mobile host are close to each other, but datagrams from the Internet host to the mobile host are routed to and back from the distant home agent first.

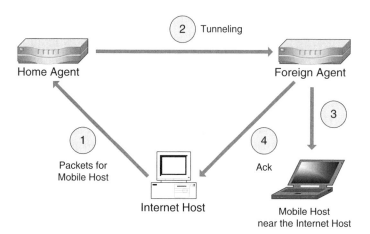

Figure 13.9   Triangular routing.

To solve the triangular routing problem, a route optimization protocol has been introduced. Basically, the protocol defines several messages so as to inform the correspondent host of an up-to-date mobility binding of the mobile host (i.e., the association of the mobile host's home address with its care-of address). With this binding, the correspondent host can directly send encapsulated datagrams to the care-of address of the mobile host. However, to achieve all these, it also means that changes are required in the correspondent cost so that it can understand the rout optimization protocol.

There are two main types of message in the route optimization protocol, namely binding request and binding update messages. A corresponding node can issue a binding request message to the home agent to query the current care-of address of the mobile host. The home agent will then reply with a binding update message. Besides sending binding update messages upon request, the home agent can also send binding update messages to corresponding nodes when it detects that it needs to tunnel datagrams received from the corresponding nodes.

Similar to the security issues in the registration process, a malicious user may send a forged binding update message to the correspondent host so as to eavesdrop datagrams of the mobile host. To secure the the binding update, similar measures employed in the registration process can be applied, such as the use of the identification field and authentications.

Another problem associated with route optimization is the handoff problem. When a mobile host transits from a network to another, it needs to find a new foreign agent and undergo the registration process to tell the home agent its new care-of address. Subsequently the home agent tell the correspondent host the new binding of the mobile host. However, between the time the mobile host leaves the previous foreign network and the time the correspondent host is notified of the new binding, a considerable amount of datagrams might have been sent to the previous foreign agent. The previous foreign agent will forward the datagrams back to the home agent to send to the mobile host, after it discovers that the mobile host has left. The effect of delayed transfer of datagrams will then be magnified in upper layers like TCP, which has adopted some traffic control mechanisms.

In order to solve the above handoff problem, once the mobile host finds a new foreign agent, it can ask the new foreign agent to send a binding update message to the previous foreign agent. Then before the registration process is completed, the previous foreign agent can directly send the mobile host's datagrams to the new foreign agent, without the involvement of the home agent. However, there is authentication issue here. Specifically, we need a way to persuade the foreign agent that the binding update by the new foreign agent is sent on behalf of the mobile host. There are many ways in achieving this kind of authentication and it is left as a problem at the end of this chapter.

In addition to route optimization, there are some other techniques that can improve the performance in Mobile IP, for example:

1. **Simultaneous bindings.** A mobile can perform multiple registrations at the same to register more than one care-of address. Doing so, the mobile host will receive multiple copies of a datagram, which helps to increase the reliability of data transfer via the error-prone wireless media.

2. **QoS management.** To provide differentiated services to a mobile host, at the home network, the mobile host can negotiate a service level specification (SLS)

with the bandwidth broker (BB) to get a certain level of service [12]. When the mobile host moves away from home to a foreign network, in order to maintain differentiated services to the mobile host, the SLS used in the home network can be transformed and transmitted to the foreign network via the BBs in both networks.

13.6   PRACTICAL ILLUSTRATIONS—HIERARCHICAL MOBILITY MANAGEMENT

In Mobile IP applications, most of the time, the mobile host roams across multiple subnets within a single network of domain, that is, the user's mobility is local to a single administrative domain. A typical example is that while you are walking around in the campus with a PDA equipped with a WLAN adapter, the campus network provides you with uninterrupted Internet access using Mobile IP, wherever you go (e.g., library, cafe, bookstore, etc.).

With reference to the above application scenario, since each time a mobile host comes to a new subnet it gets a new care-of address, when the mobility of the mobile host increases, it needs to register its new care-of addresses more often with its home agent. As a result, with frequent location update, the traffic induced in the registration process can be excessive. The situation is even worse when the population of mobile users is large in the home network. It is because the home agent is likely to be overwhelmed by the large volume of registration messages from the mobile users with high mobility, which in turn causes the registration process to take a longer time to finish. Subsequently, the performance of data transfer in the mobile most suffers.

To solve the above problems, intuitively, we need to make the home agent unaware of the mobile host's mobility within the foreign domain. Specifically, we need to make the home agent see a single unchanged care-of address so as to reduce registration messages. To accomplish this, we can adopt a hierarchy of foreign agents [230].

As illustrated in Figure 13.10, originally the mobile host enters the foreign domain at $subnet_1$ and gets a care-of address CoA_1 from the foreign agent FA_1 in that subnet. Then the mobile host (MH) registers CoA_1 with the home agent (HA). Later on, the mobile host moves to $subnet_2$ and gets a new care-of address CoA_2 from the foreign agent FA_2. Instead of registering the new care-of address CoA_2 with the home agent, the mobile host asks FA_2 to inform FA_1 of its new care-of address. This way, the home agent continues to tunnel the mobile host's IP datagrams to CoA_1 and then FA_1 re-tunnels the datagrams to FA_2 for delivery to the mobile host. However, it should be emphasized that when the level of hierarchy increases, the latency of datagram transfer from FA_1 to the mobile host will also increase. Whenever a threshold level of hierarchy is reached (the threshold can be dynamically adjusted based on the mobility of the mobile host and the traffic load), the mobile host needs to register its current care-of address with the home agent to set up a new foreign agent hierarchy.

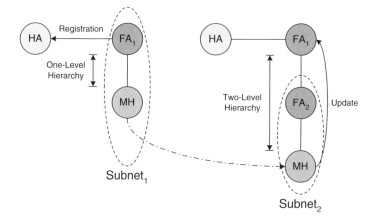

Figure 13.10   Dynamic hierarchical mobility management.

13.7 SUMMARY

In this chapter, we have discussed the three key design features of Mobile IP, namely advertisement mechanisms, registration process and tunneling approaches. The triangular routing problem is discussed and the corresponding route optimization approaches are presented. Finally, as a practical illustrative example, we talk about a hierarchical mobility management approach for reducing signaling traffic in the registration process.

PROBLEMS

13.1   How can a mobile host know whether it is at the home network or foreign network?

13.2   What is the triangular routing problem? How to solve it?

13.3   What will be the problem when a mobile host roams across different domains frequently? Describe a scheme to mitigate the problem.

Table 13.4   Code values for a registration reply message.

Code Value	Description
	Registration Successful
0	Registration accepted.
1	Registration accepted but simultaneous mobility bindings not supported.
	Registration Rejected by the Home Agent
128	Reason unspecified.
129	Administratively prohibited.
130	Insufficient resources.
131	Mobile host failed authentication.
132	Foreign agent failed authentication.
133	Registration identification mismatch.
134	Poorly formed request.
135	Too many simultaneous mobility bindings.
136	Unknown home agent address.
	Registration Rejected by the Foreign Agent
64	Reason unspecified.
65	Administratively prohibited.
66	Insufficient resources.
67	Mobile host failed authentication.
68	Home agent failed authentication.
69	Requested lifetime too long.
70	Poorly formed request.
71	Poorly formed reply.
72	Requested encapsulation unavailable.
73	Requested Van Jacobson header compression unavailable.
80	Home network unreachable (ICMP error received).
81	Home agent host unreachable (ICMP error received).
82	Home agent port unreachable (ICMP error received).
83	Home agent unreachable (other ICMP error received).

Table 13.5   Description of the fields in the inserted header of minimal encapsulation.

Field Name	Description
Protocol	It is copied from the protocol field of the original IP header for identifying the protocol type of the original IP payload.
S	It is 1 if original source address is present; otherwise, it is 0.
Checksum	Checksum of the header.
Home Address	It is copied from the destination address in the original IP header.
Source Address	It is present when the value of S is 1, and is copied from the source address of the original IP header.

CHAPTER 14

IPV6

Riches are often abused, never refused.

—Danish Proverb

14.1 INTRODUCTION

Driven by the shortage of IP addresses, an improved version of the Internet Protocol has been in pressing need for many years.

Interoperability:

IPv6 is designed, at the very beginning, to interoperate with IPv4 because a "flag day" for the transition from IPv4 networks to IPv6 networks is simply infeasible. Therefore, IPv4 networks and IPv6 networks will coexist for a certain period of time. Consequently, some transition mechanisms are needed. Currently, the main three types of transition mechanism are dual-stack, tunneling, and translation. On the other hand, a number of advanced features are incorporated in IPv6 in order to increase efficiency when different network nodes interoperate with each other. Features like

Wireless Internet and Mobile Computing. By Yu-Kwong Ricky Kwok and Vincent K. N. Lau
Copyright © 2007 John Wiley & Sons, Inc.

simplified header format and the use of extension headers provide enhanced routing efficiency when compared to IPv4. Mobile IPv6, which takes experience from Mobile IPv4 and utilizes advanced features in IPv6, provides mobile hosts with enhanced mobility service so as to facilitate efficient roaming across different networks.

Performance:

Since IPv6 was released as a draft standard in 1995, network device manufacturers and operating system providers have started their efforts to support IPv6. Since then, IPv6 implementations have matured, and the deployment of IPv6 networks worldwide has been transiting from experimental networks to operational networks. Because of this transition trend, it becomes a concern to investigate the performance of IPv6 on user applications.

Zeadally *et al.* have conducted an empirical performance comparison of IPv4 and IPv6 protocol stack implementations of commodity operating systems [369]. They investigated performance metrics such as throughput (TCP and UDP), round-trip time (RTT), socket creation time, and TCP connection time. The experiment was conducted on two identical workstations connected by a point-to-point link. Their performance results obtained from a Red Hat Linux operating system indicate that there is no significant difference on the performance between IPv4 and IPv6. Specifically, in the throughput experiments (TCP and UDP), there is notable decrease in IPv6 implementation for messages smaller than 512 bytes, which is due to the overhead of the larger header size of IPv6. On the other hand, when compared to IPv4, the round-trip latencies for IPv6 are lower. As for TCP connection time, IPv6 results in a slightly increased connection time due to the increased IPv6 header size, while there is more or less no difference in socket creation time.

On the other hand, Wang *et al.* have conducted a performance comparison of IPv6 and IPv4 in networks of a larger scale [347]. They collected packets from IPv6/IPv4 dual-stack Web servers located in 44 countries to study performance metrics like connectivity, packet loss rate, and RTT. Their measurement results indicate that IPv6 connections tend to have lower RTTs than their IPv4 counterparts. However, at the same time, IPv6 connections suffer higher packet loss rate. Finally, it is found that tunneling does not show notable performance degradation in IPv6.

14.2 DESIGN GOALS

The Internet Protocol version 4 (IPv4) was developed in the early 1970s to facilitate information exchange between educational and government entities in the United States. At the beginning, it was intended for use in a small sized and closed system of limited number of access points, and therefore the original IP protocol is rather simple and the developers did not foresee requirements such as security. However, as more and more machines are interconnected, it was realized that the features provided by the original IP protocol is insufficient to meet the needs in a large network. The situation is even worse when the Internet became world wide public and the emergence of World

Wide Web (WWW) in the early 1990s. Since then, a number of Internet applications have been developed and the problems of IPv4 have become more apparent. In order to mitigate the shortcomings of IPv4 found in practical use, many additional protocols and standards have been developed to work collaboratively with IPv4, such as Internet Control Message Protocol (ICMP) and Network Address Translation (NAT). This makes the current IPv4 protocol very complex.

Since the proliferation of web pages, emails, and instant messengers, the number of Internet users has been increased unprecedentedly. Besides, people are becoming more dependent on data and information services provided by the Internet. This trend continues to grow with the advancement of mobile communication technology. Consequently, there has been an explosive growth in IP-capable devices, since it is advantageous to utilize existing Internet infrastructure. Handheld game consoles, such as Sony Playstation Portable (PSP) and Nintendo DS (NDS), are equipped with Wi-Fi devices and can access to the Internet. Many PDAs and mobile phones are also Wi-Fi-enabled. In additional, IP-phones and IPTV are very common nowadays. Thus, IP-enabled devices are penetrating into every aspect of our life. With this explosive growth in IP-capable devices, one apparent and most serious problem of IPv4 is the shortage of IP addresses. To mitigate this address space problem, NAT has been suggested to map public IP addresses with private addresses such that a few number of IPv4 addresses are used to enable a high number of users with private addresses to connect to the Internet. However, NAT prevents end-to-end communication as it modifies end node addresses during address mapping. Consequently, some applications are not able to work with NAT, such as peer-to-peer applications. Therefore, inevitably, the problem of running out IPv4 addresses need to be solved radically.

In the early 1990s, the Internet Engineering Task Force (IETF) started to develop a successor protocol to IPv4. The Internet Protocol Next Generation (IPng) working group was started in 1993 to investigate different proposals and recommendations. In 1995, the RFC 1883 (Internet Protocol, Version 6 (IPv6) Specification) [82] was published, which was then obsoleted by RFC 2460 [83] in 1998.

Although solving the address space problem in IPv4 is the main driving force for the introduction of IPv6, IPv6 is not merely about extending the address space of IPv4. Based on the experience that we have from the development and usage of IPv4, IPv6 is also designed to augment many other aspects of IPv4 and many additional features are integrated in IPv6. Specifically, proven functions that make IPv4 successful have been retained and modified to extend the scalability and flexibility, limitations of IPv4 have been removed, additional functions that are originally designed to work collaboratively with IPv4 are now integrated in IPv6 to enhance the efficiency and security, and finally new features are added to ease network management, to provide mobility support to mobile nodes, etc.

The following gives an overview of the main features of IPv6:

Expanded Address Space: The length of IP address field is extended from 32 bits to 128 bits. In practice, it supports virtually infinite number of IP addresses.

In addition, the address space can be structured in a hierarchical way to allow optimized global routing of packets.

Simplified Header Format: The header of IPv6 has a fixed length of 40 bytes. Since 32 bytes are used for source and destination addresses, only 8 bytes are left for incorporating other information, and some fields of the IPv4 header have been either eliminated or made optional. This simplified header format of IPv6 allows for faster processing of IP packets.

Extension Headers: In IPv4, optional fields are integrated in the base header. However, in IPv6, optional fields are put in extension headers after the based header. Extension headers are added only when needed and may not be processed by intermediate nodes between the source and destination so as to allow fast processing of packets.

Autoconfiguration: IPv6 supports stateless and stateful autoconfiguration of IP addresses to reduce the network management costs. In the stateful mode, the operation is similar to the use of DHCP for IPv4, where a node requests for an IP address dynamically from a DHCPv6 server. On the other hand, the stateless mode allows a mobile node to be connected to a network in an easy and efficient way. For example, the mobile node can autoconfigure for a global IP address by using its MAC identifier, in addition to getting a network prefix from an IPv6 router.

Security: IPsec is mandatory in IPv6. The Authentication Header (AH) and the Encapsulating Security Payload Header (ESP) are included as extension headers.

Mobility Support: Mobility support provided by IPv6 removes some of the limitations found in IPv4 implementations. Mobile IPv6 aims at providing mobile users with an efficient way to maintain connections when they move between networks. For example, route optimization is supported in Mobile IPv6 and foreign agents are no longer needed.

Quality of Service (QoS) Support: To support QoS in IPv4, various additional protocols have be developed to work with IPv4. On the contrary, IPv6 integrates QoS support by using flow label. Specifically, packets belonging to different traffic flows can be labeled such that the sender can request special handling of the packets.

Unlike the switch from using Network Control Protocol (NCP) to IP in one day in 1983, there will be no such flag day for IPv6. Thus, another key feature of IPv6 is that IPv6 is designed to be interoperable with IPv4. Specifically, IPv4 and IPv6 coexisting schemes have been developed, such as dual IP layer approach and tunneling. This allows incremental upgrade from IPv4 to IPv6, since an IPv4 host can be upgraded anytime independently of other hosts and routers.

IPv6 Addressing

As IPv6 addresses are four times the length of 32-bit IPv4 addresses, the dotted decimal notation as adopted in IPv4 is inappropriate for IPv6 addresses. The IPv6 addressing architecture is defined in RFC 3513 [149].

Figure 14.1 shows an example IPv6 address taken from [81]. As can be seen, an IPv6 address typically consists of eight blocks of four hexadecimal digits (i.e., each block contains 16 bits), and each block is separated by a colon. In each block, if there are leading zeroes, then the zeroes can be omitted. Moreover, if the entire 16-bit block is zero, then the block can be reduced to two consecutive colons. If there are consecutive blocks of zeroes, the blocks can also be reduced to two consecutive colons. However, to avoid confusion, this kind of double-colon notation can only be used once in an IPv6 address.

Similar to IPv4 addresses, an IPv6 address can be divided into two portions, the network portion and the host portion. These two portions can be explicitly defined by using a prefix length (i.e., the number of bits used to represent the network portion), as illustrated in Figure 14.1

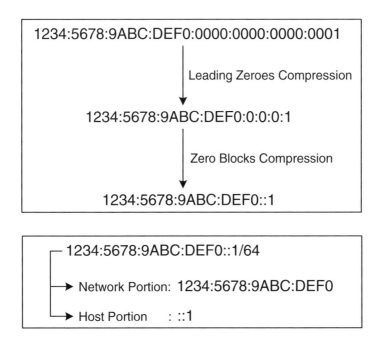

Figure 14.1 IPv6 addressing example.

In IPv4, there are three classes of address, namely unicast, multicast, and broadcast. In IPv6, broadcast is considered a special case of multicast, and a new class of address called anycast is introduced. Anycast addresses can be used when we want to assign a single global unicast address to more than one host.

IPv6 Header Format

Figure 14.2 shows a comparison between the IPv4 header format and the IPv6 header format. Table 14.1 describes he fields in the IPv6 header. By comparing the header of IPv4 with that of IPv6, we can find that several fields in the IPv4 header have been removed, meaning that some operations are revised in IPv6:

IHL (Internet Header Length): As the IPv6 header has a fixed length of 40 bytes, a length indicator is no longer needed.

Identification, Flags, Fragment Offset: These fields have been moved to an extension header, called the Fragment Header. In IPv6, fragmentation and reassembly can only be performed at the source node and destination node, respectively. This means that intermediate routers are not allowed to fragment and reassemble packets during routing of packets. During packet transmission, the Fragment Header is used to send a packet of size larger than the path MTU (maximum transmission unit) to a destination node, and the source node will divide the original packet into several smaller packets. If a router receives a packet that is too large to be forwarded, the packet will be discarded and a "Packet Too Big" ICMPv6 error message will be sent to the source node to inform the use of smaller packet size. Since fragmentation and reassembly are time-consumption operations, removing them at intermediate routers can help speed up the routing task significantly.

Header Checksum: Since other layers in the Internet protocol stack also perform error checking (e.g., TCP and UDP in the transport layer), error checking at the network layer seems to be redundant and can be removed completely. Another reason is that point-to-point connections nowadays have very low error rate. Again, removing error checking aims at allowing fast packet processing in intermediate routers.

Options: The Options field is no longer a part of the IPv6 header and has been moved to extension headers. The removal of this variable-length field results in the fixed 40-byte IPv6 header.

Transition Mechanisms

As the number of Internet-connected machines is significantly large, a "flag day" for upgrading from IPv4 to IPv6 is practically infeasible. Also, because of the upgrading cost, many systems may not upgrade from IPv4 to IPv6 in the immediate future. Despite that IPv6-capable systems can be made backward compatible with IPv4 systems, existing IPv4 systems are merely not able to handle IPv6 packets. Therefore, inevitably, IPv4 and IPv6 coexisting mechanisms are needed to allow the gradual transition from IPv4 networks to IPv6 networks, and more importantly to

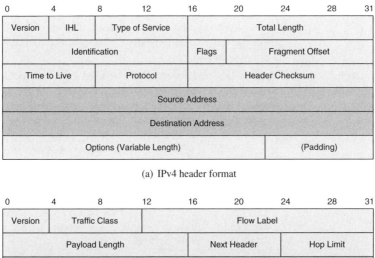

(a) IPv4 header format

(b) IPv6 header format

Figure 14.2   Header formats of IPv4 and IPv6.

Table 14.1 Description of the fields in the IPv6 header.

Field Name	Description
Version	Value 6, indicating that this is an IPv6 header.
Traffic Class	The usage of this field is similar to the TOS (type of service) field in IPv4. It is used to specify different types of IP packets so as to support differentiated service/quality of service. Intermediate routers use this field to distinguish different classes or priorities of packets.
Flow Label	A flow is a sequence of packets between the source and destination nodes. The 20-bit flow label field is used by the source node to label the sequence of packets such that special handling of the packets by intermediate routers can be requested (e.g., best effort or real-time service). A flow is identified by the combination of the source address and a non-zero flow label value. Hosts or routers that do not understand the flow label field should set the field to zero, and a packet with the flow label field set to zero does not belong to any flow.
Payload Length	This field specifies the length of the packet excluding the 40-byte header. This means that extension headers are also considered as part of the payload in IPv6.
Next Header	This field identifies the header immediately following the IPv6 header. It is used to specify the next protocol (e.g., TCP, UDP, etc.) in the data field of the packet. The field values adopted are the same as the protocol field in the IPv4 header.
Hop Limit	The 8-bit hop limit field replaces the TTL (time to live) field in the IPv4 header. It specifies how many hops a packet is allowed to be forwarded, and thus the value of this field is decremented by one by each router that forwards the packet. When the value of this field becomes zero, the packet is discarded.
Source Address	A 128-bit value specifying the originator of the packet.
Destination Address	A 128-bit value specifying the recipient of the packet.

allow the interoperability between the two networks. Existing transition mechanisms can be classified into three categories, dual-stack, tunneling, and translation.

The simplest way of allowing coexistence of both IPv4 and IPv6 protocol stacks is the dual-stack approach. In this approach, the network interface of a node is configured with an IPv4 address as well as an IPv6 address. Therefore, the node can run both protocols in parallel, and hence the node is called a IPv6/IPv4 node. Depending on whether the incoming packet is an IPv4 or IPv6 packet, appropriate protocol stack is used to handle the packet. Similarly, the IPv6/IPv4 node needs to determine whether the next node supports IPv6 or IPv4 only. This can be achieved by using DNS, which can return the IPv6 and/or IPv4 address(es) of the next node.

In the dual-stack approach, if both the source and destination nodes are IPv6-capable but there is intermediate node that can speak only IPv4, then it might turn out that the source and destination nodes are communicating using IPv4 packets finally. As Figure 14.3 [194] illustrates, node A is IPv6-capable and wants to send a packet to another IPv6-capable node, node E. Since the next node of node A, i.e., node B, is an IPv6/IPv4 node, node A can send an IPv6 packet to node B. Afterwards, node B needs to forward the IPv6 packet to node C. However, since node C can only understand IPv4 protocol, node B needs to convert the IPv6 packet to an IPv4 packet, by performing appropriate address mapping and dropping of some IPv6-specific fields. Consequently, when node D receives the IPv4 packet from node C and constructs an IPv6 packet for transmission to node E, some fields in the original IPv6 packet sent from node A are lost.

While the dual-stack approach is simple and flexible, it has the disadvantage of consuming more processor power and memory. The reason is that the IPv6/IPv4 node needs to keep all type of tables (e.g., routing table) twice for each protocol. Nevertheless, the dual-stack approach, as we will see later, forms the basis for other transition mechanisms.

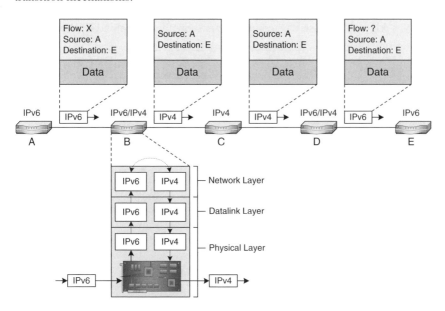

Figure 14.3 Dual-stack approach [194].

As mentioned earlier, in the dual-stack approach, if there is intermediate node that can handle only IPv4 packets, then some IPv6-specific fields will be lost during packet transmission between two IPv6-capable nodes. Tunneling approach is another transition mechanism that can be used to solve this problem. Specifically, by using tunneling, information/data from one protocol (e.g., an IPv6 packet) can be encapsulated inside the packet of another protocol (e.g., an IPv4 packet). Effectively, the latter acts as a tunnel for transmission of original information/data in the former without any augmentation.

Figure 14.4 [194] illustrates the idea of the tunneling approach. At node B, it discovers that node C can only handle IPv4 packets. Then node B encapsulates the IPv6 packet received from node A in an IPv4 packet and send the IPv4 packet to node C, who is situated inside an IPv4 tunnel. At this point, we say that node B is the tunnel entry point, and the IPv4 packet will be transmitted via node C to the tunnel exit point, i.e., node D. At node D, the received IPv4 packet will be decapsulated, and the original IPv6 packet sent from node A will be delivered to node E, without any modification of the IPv6-specific fields.

In the tunneling approach, it is required that the tunnel endpoint (e.g., the tunnel exit point at node D in Figure 14.4) should have a public IPv4 address. However, if the tunnel endpoint is located behind one or more IPv4 NATs, other mechanism is needed. Teredo is designed to allow IPv6 packets to traverse IPv4 NATs by using IPv4-based User Datagram Protocol (UDP) messages [267]. Teredo consists of three main components—Teredo server, Teredo client, and Teredo relay. Teredo, however, does not work with the case where both tunnel endpoints are behind IPv4 NATs.

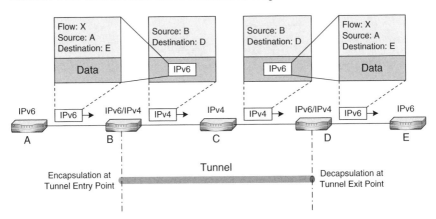

Figure 14.4   Tunneling approach [194].

The discussion above focuses on providing connectivity between IPv6-capable nodes via an IPv4 network. In order to provide connectivity between an IPv6 node and an IPv4 node, translation mechanisms can be applied, for example, the Network Address Translation-Protocol Translation (NAT-PT). NAP-PT is applied in a router that supports both IPv4 and IPv6. NAT-PT uses a pool of IPv4 addresses for dynamic assignment to IPv6 nodes when a session is initiated between an IPv6 node and an

IPv4 node. Then the NAP-PT router performs addressing mapping in the IPv4 and IPv6 headers. In addition, if an application program embeds IP addresses within the data of a packet, then address translation within the data should also be performed in order to match the address mapping in the headers. To perform this kind of application level translation, an Application Level Gateway (ALG) can be employed within the NAT-PT router.

Figure 14.5 illustrates how NAT-PT works. In the example, node A, whose IPv6 address is 1234:5678:9ABC::DEF0, wants to communicate with node B, whose IPv4 address is 238.12.234.56. Node A sends a packet destinated for node B with a destination address 2228:38::238.12.234.56. The prefix 2228:38::/96 is advertised by the NAT-PT router such that all packets sent from an IPv6 node will be routed through the NAT-PT router. When the NAT-PT router receives the IPv6 packet from node A, corresponding mapping of IP address and port number is carried out as depicted in Figure 14.5 For inbound traffic, the mapping is performed in a similar way.

There are several disadvantages of using NAT-PT. First of all, end-to-end security cannot be provided, which is a common pitfall of using any form of NAT. Secondly, there might be information lost when translating from one protocol to another protocol. Finally, the NAT-PT router is the performance bottleneck, since all packets from the same session need to go through the same NAT-PT router.

14.3 MOBILITY SUPPORT

As a result of the advanced mobile communication technology, mobile devices such as smart phones are penetrating into our lives. By using a mobile device, we can access Internet services anytime and anywhere. In addition, to leverage existing Internet infrastructures, IP services are commonly implemented on mobile devices. For mobile devices, mobility is a key concern. Specifically, it is advantageous if existing connections can be maintained when a mobile host roams across different wireless networks. In Chapter 13, we have discussed how Mobile IPv4 supports mobility to mobile hosts.

Using Mobile IP, when a mobile host is attached to a new network, its network prefix changes, meaning that a new care-of address (CoA) is assigned to it. Because of this reason, when a considerable number of mobile hosts are attached to a foreign network (due to the proliferation of mobile devices), limitation of address space in IPv4 may become an issue in Mobile IPv4. On the other hand, Mobile IPv4 has other limitations such as lack of support of route optimization. In view of these problems, Mobile IPv6 is defined in RFC 3775 [172]. Since Mobile IPv6 is designed based on the experience from Mobile IPv4 and advanced features of IPv6, the mobility service provided by Mobile IPv6 is considered as one of the killer applications in IPv6 [250].

When compared with Mobile IPv4, the key changes in Mobile IPv6 are as follows:

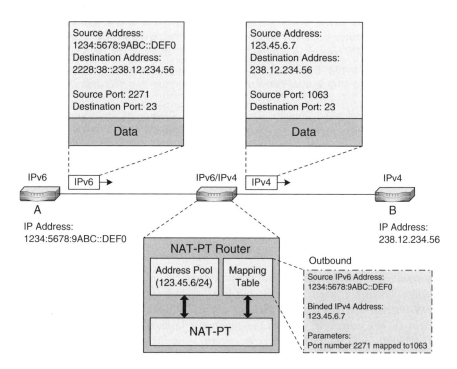

Figure 14.5   Translation approach—Network Address Translation-Protocol Translation (NAT-PT).

Foreign Agents: In Mobile IPv6, foreign agents are not necessary. Mobile hosts can obtain a CoA by using stateless or stateful autoconfiguration. In the stateful approach, a CoA can obtained from a DHCPv6 server. In the stateless approach, a mobile host can use the network prefix advertised by the router of the foreign network to form the upper 64 bits of the CoA, and to use its MAC address to form the lower 64 bits of the CoA.

Route Optimization: Route optimization is supported in Mobile IPv6. In route optimization, it is required that the mobile host registers its CoA with the correspondent host. As for the security measure required for this correspondent registration, a return routability procedure is defined for authenticating the mobile host to the correspondent host. Then, to send packets from the correspondent host to the mobile host, a special routing header (type 2) is used by the correspondent host.

Use of Neighbor Discovery Protocol: In Mobile IPv4, ARP (address resolution protocol) is used. However, the IPv6 neighbor discovery protocol is used in Mobile IPv6.

Home Agents Discovery: Mobile IPv6 supports multiple home agents. A mechanism called Dynamic Home Agent Address Discovery is defined in Mobile IPv6 in order to enable a mobile host to learn about the reconfiguration of the home network (e.g., a change of IP address of the home agent).

In Mobile IPv6, an extension header called Mobility Header (MH) is defined. This extension header has a next header value 135. Using the header, 11 mobility messages are defined. Mobility messages are used by the mobile host, correspondent host and home agent to carry out the return routability procedure, binding update procedure, and fast handover.

14.4 HOME AGENTS DISCOVERY

When a mobile host moves to a new foreign network, it needs to send a binding update message to its home agent. However, the mobile host needs to determine the address of its home agent first. Dynamic Home Agent Address Discovery is a mechanism in Mobile IPv6 for serving this purpose. It involves the use of ICMPv6 messages, a home agents list and anycast addresses. This mechanism is useful for a mobile host to find new home agents dynamically when its original home agent goes down.

As illustrated in Figure 14.6, when the mobile host needs to find the address of its home agent, firstly it sends an ICMPv6 Home Agent Address Discovery Request message to the home agent anycast address. Then in the home network, the home agents who are configured with the home agent anycast address will respond with a Home Agent Address Discovery Reply message. In the reply message, it consists of a Home Address field storing one or more home agent addresses. Finally, on getting a home agent address, the mobile host can send a binding update message to its home agent.

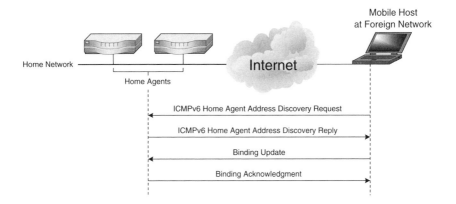

Figure 14.6   Dynamic Home Agent Address Discovery.

In order to reply a mobile host's Home Agent Address Discovery Request messages, each home agent is required to maintain a home agents list, which contains all the home agents on the home network link. Each home agent maintains the list by monitoring router advertisements on the network, since a router will advertise itself as a home agent by setting the H-bit in the router advertisement packet. Specifically, the list contains the following information:

- Link-local addresses of the home agents on the home network link.

- Remaining lifetimes of the entries for the home agents. The entry for the home agent whose lifetime has expired must be deleted.

- One or more global IPv6 unicast addresses for the home agents.

- The preference value for each home agent. This value can be found in router advertisements. A higher value means a higher preference. This preference value is used for sorting home agents when replying Home Agent Address Discovery Request messages.

14.5   PRACTICAL ILLUSTRATIONS—IPV6 BASED VTHD NETWORK

Since it is inevitable to use IPv6 in the long term, governments or organizations will benefit from earlier introduction of IPv6 services to their existing IPv4 networks. With this consideration in mind, in the VTHD project carried out in France, which is partially funded by the French government, IPv6 services have been introduced in the very high broadband IP/wavelength-division multiplexing (WDM) network (VTHD network) for new-generation Internet applications [11]. In VTHD, both IPv4 and IPv6 services are supported, and it has a public IPv6 subnet, which is connected to the global IPv6 network via Open Transit (France Telecom's international IP service for ISPs).

Figure 14.7 [11] shows the VTHD architecture. The VTHD backbone network has 10 transit routers interconnected by 2.5 Gb/s optical channels, with 12 additional edge routers connected to the backbone for aggregating network traffic from campuses or organizations. The VTHD network is assigned with a /42 prefix, namely 2001:688:1F80::/42, which allows 64 /48 prefixes ranging from 2001:688:1F80::/48 to 2001:688:1FBF::/48. Out of the 10 transit routers in the backbone network, seven of them are connected with external sites. Therefore, address space 2001:688:1F80::/42 can be divided into eight blocks, each having eight /48 prefixes, as illustrated in Figure 14.7 (the first block, 2001:688:1F80—2001:688:1F87, is used by the backbone). In each block, only the first quartet of the eight /48 prefixes is allocated while the second quartet is reserved for use in the case of network increase. When a partner node is connected to a transit router node, it will obtain a /48 prefix from the first quartet of the node. If all the /48 prefixes in the first quartet have been used, a /48 prefix can be obtained in the second quartet.

As the VTHD network supports both IPv4 and IPv6 services, some transition mechanisms are implemented in the VTHD network. Specifically, the dual-stack approach is mainly implemented in the IPv6 backbone nodes. In addition to the dual-stack approach, VTHD also implements some tunneling transition mechanisms, namely IPv6-in-IPv4 tunneling and 6PE tunneling [266]. In 6PE tunneling, IPv6 islands are interconnected over a Multi-Protocol Label Switching (MPLS)-enabled IPv4 cloud. In this approach, IPv6 packets are received on IPv6 Provider Edge routers (6PE), which are IPv6/IPv4 dual-stack routers. Afterwards, appropriate encapsulation and decapsulation are performed at the 6PE routers in order to transmit the IPv6 packets.

In order to test the performance of the VTHD network, a testbed as depicted in Figure 14.8 was set up [11]. In the test console, TCL scripts were executed so as to control some parameters like frame loss rate on the test devices. Afterwards, throughput results were measured. As reported in [11], the test was aimed at studying the performance of the three transition mechanisms as mentioned above. Performance results indicated that the dual-stack approach and 6PE tunneling achieved nearly wire-speed performance, and that the IPv6-in-IPv4 tunneling approach had some limitations. Besides investigating the performance of the three transition mechanisms, the performance of IPsec was also studied in the VTHD project. It is reported that IPsec over IPv6 achieved the same performance as IPsec over IPv4 [11].

In the VTHD project, different network devices such as routers and test devices, were intentionally acquired from different manufacturers so as to study the interoperability issues. Through the experiments conducted in the project, it is asserted that network equipment nowadays is IPv6 ready for the transition from IPv4 networks to IPv6 networks.

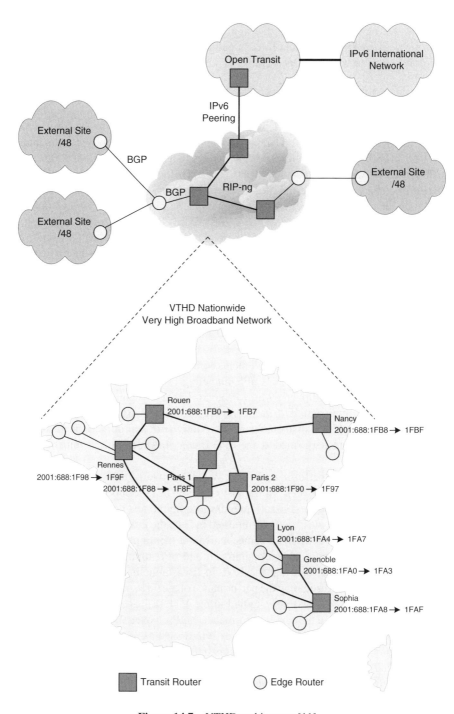

Figure 14.7 VTHD architecture [11].

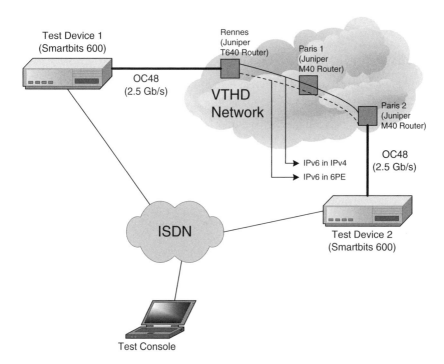

Figure 14.8   VTHD testbed setup [11].

14.6   SUMMARY

In this chapter, we have discussed the key features and design goals of IPv6. Specifically, the addressing architecture, header format, coexisting mechanisms with IPv4, and mobility support (Mobile IPv6) have been introduced. For practical illustrations, we have seen how IPv6 services were deployed in the VTHD network in France. Despite that IPv6-ready hardware (e.g., network equipment) and software (e.g., operating systems) components have been matured and provide optimized performance, IPv6 networks are still being deployed gradually. Nevertheless, IPv6 is inevitable in the long term, and governments or organizations will benefit from earlier introduction of IPv6 services.

PROBLEMS

14.1   What are the features of IPv6 that help to increase packet routing efficiency?

14.2   Briefly compare the advantages and disadvantages of the three transition mechanisms—dual-stack, tunneling, and translation.

14.3   What are the key differences between Mobile IPv4 and Mobile IPv6?

CHAPTER 15

WIRELESS APPLICATION PROTOCOL (WAP)

Translation is at best an echo.

—George Borrow; Lavengro, XXV, 1851

15.1 INTRODUCTION

As the TCP/IP protocol suite was not designed for a wireless environment where bandwidth resource is expensive (consider a cellular network), modifications and enhancements are necessary for adopting it in a wireless network.

Interoperability: WAP aims at bridging the gap between the mobile world and the Internet. It is designed to bring diverse Internet content and other data services to mobile users across different wireless technologies so that mobile users can reach the Internet to have different information services anytime and anywhere. In order to enable mobile devices to access ordinary web servers, WAP tries to utilize existing Internet protocols and standards as much as possible, such as XML, HTML, HTTP, and TLS. The adapted protocols and standards are optimized for deployment in a

Wireless Internet and Mobile Computing. By Yu-Kwong Ricky Kwok and Vincent K. N. Lau
Copyright © 2007 John Wiley & Sons, Inc.

wireless environment where mobile devices are limited in capabilities. To access ordinary web servers, WAP-enabled mobile devices rely on a WAP gateway to provide protocol conversion between the WWW protocol stack and the WAP protocol stack.

Performance: The WAP protocol stack is a lightweight protocol stack that is designed to address the limitations of wireless devices and the wireless network. It is designed with two main goals—to minimize bandwidth requirement and to maximize the number of supported network types (e.g., 9.6 Kbps in GSM). Each layer of the protocol stack is designed to be scalable and efficient. For example, in Wireless Transaction Protocol (WTP), there is no explicit connection setup or teardown, reducing much amount of overhead when compared to TCP. To reduce transmission time, WAP uses binary-coded WML (wireless markup language) pages. In addition, WAP specifies a caching model and user agent profile (UAProf) for efficient delivery of device-specific content.

15.2   WAP SERVICE MODEL

As the Internet provides a wealth of information (and the information provided continues to expand every day), people are becoming more and more reliable on the Internet. Being able to access the Internet, they can retrieve useful information such as maps, weather forecast, etc. They can also handle routine activities electronically, such as e-banking, e-commerce, e-checkin, etc. The Internet has been helping people to make their lives easier. Because of this reason, it would be very convenient if we can reach the Internet to have different information services anytime and anywhere. This is possible with the advancement of wireless technologies such as GSM, CDMA, and TDMA.

Because of the above-mentioned pressing need for providing ubiquitous Internet services (i.e., wireless data services) for mobile users, at the beginning, different service providers developed different proprietary solutions/protocols. The different solutions/protocols developed are characterized by the limitations of mobile phones or mobile terminals and the network that connect them:

Limited Processor Power and Memory: Being portable, mobile devices are usually not equipped with powerful CPU and large memory capacity. Thus, they are not supposed to execute computation-intensive tasks.

Limited Battery Life: Despite the battery technology is evolving, it is still not able to cope with the needs of users (e.g., a mobile device is typically equipped with many functionalities such as camera and music player, which all consume significant power). Thus, the battery life time of mobile devices is still a concern.

Limited Input and Output Facilities: Mobile devices are typically equipped with input devices of limited capabilities, which are not as convenient as keyboards and mouse-pointing devices found in desktop computers and laptops. On the other hand, the display of mobile devices is small, which does not allow substantial information to be displayed on the screen each time.

Low Bandwidth: The bandwidth of wireless networks is relatively low (e.g., 9.6 Kbps in GSM), which constrains the amount of information to be delivered to a mobile user.

High Latency: A relatively high latency is resulted due to the cumulative effects of hardware and software factors when a message goes through a wireless network. The message needs to go through various elements of the wireless network, which consist of various physical network devices and radio transmitters. On the other hand, the message also needs to go through various protocol layers.

Unpredictable Availability and Stability: Wireless networks are much less reliable when compared to wired networks. Mobile users might be frequently disconnected from a wireless network due to channel fading, lost radio coverage, etc., when they roam from place to place.

The above features vary from mobile device to mobile device and from network to network. Moreover, different mobile users have different needs. All these make the proprietary solutions/protocols developed for providing Internet services incompatible to each other. In order to solve the incompatibility issue and to develop a universal and open standard, Wireless Application Protocol (WAP) Forum was founded in 1997 by Ericsson, Motorola, Nokia, and Phone.com. WAP 1.1 was published in 1999 and WAP 2.0 was published in 2001. In 2002, the WAP Forum consolidated into the Open Mobile Alliance (OMA) and the specification work from WAP continues within OMA [262].

WAP aims at bridging the gap between the mobile world and the Internet, by bringing diverse Internet content and other data services or value-added services to mobile users across different wireless technologies. It specifies an application environment and a set of communication protocols so that different manufacturers, network providers, content providers, and developers can collaborate to provide mobile devices with the ability to have technology-independent access to the Internet and telephony services. WAP is designed to meet the following requirements:

Interoperability: Any WAP-enabled mobile device is able to communicate with any network supporting WAP services.

Scalability: Protocols and services should scale with number of customers. It allows the personalization of services in accordance with user needs.

Efficiency: Provide QoS according to the characteristics of wireless networks.

Reliability: Provide a consistent and predictable platform for service deployment.

Security: Ensure user data integrity and provide protection to devices and services.

WAP tries to utilize existing Internet protocols and standards as much as possible (e.g., XML, HTML, HTTP, TLS). Those adapted protocols and standards are optimized for deployment in a wireless environment where mobile devices are limited in capabilities, and the wireless network bears the properties of low bandwidth, high latency and unpredictable availability and stability. The following summarizes the key elements of WAP which specify how to improve existing services and how to develop new applications and value-added services in a quick and interoperable manner:

Programming Model: The WAP programming model is similar to the conventional WWW programming model. Since manufacturers, service providers, and developers are familiar with the conventional WWW programming model, it allows a quick introduction of WAP services.

Markup Language: WAP uses wireless markup language (WML), which is a subset of the extensible markup language (XML). WML is designed to fit small mobile devices, while still being able to fulfill the same functionality of HTML.

Scripting Language: WMLScript, which is similar to JavaScript, is used to to define script-type programs for incorporation of enhanced functionalities to WAP-based services.

Small Browser Specification: It specifies a browser suitable for mobile devices having limited processor power, memory, input capability, and small display.

Optimized Protocol Stack: The lightweight protocol stack is designed to address the limitations of wireless devices and the wireless network. It minimizes the bandwidth required and hence maximizes the number of types of wireless networks that are able to support WAP-based services.

Wireless Telephony Application Interface (WTAI): WTAI specifies an application framework for providing telephony services. It enables a WAP-based mobile device to make calls by calling special WMLScript or accessing special URLs.

When a user goes mobile, his concern about retrieving information from the Internet is different from that when he is at home or office. For example, when he is out of town, he might want to keep in touch with his family or colleagues in a quick and stable way through emails, and have a fast and convenient way to keep an eye on the stock market. On viewing emails and browsing the Internet about the stock market, he does need to have fancy images or a mass of unrelated information, which are simply a waste of the valuable bandwidth and incur a high transmission latency. Therefore, WAP is designed to bring Internet services to mobile users in a fast, accurate, and convenient way.

Wireless Markup Language (WML)

Web pages created using HTML are intended to be displayed in machines with larger screens, such as desktop computers and laptops. In addition, today's HTML pages are often created with fancy images or videos, which require relatively more powerful processors for viewing. Thus, HTML pages are not suitable for use in mobile devices with limited processor power and screen. Moreover, such kind of HTML pages are large in size, which are not suitable for delivery in wireless networks of low bandwidth and high latency. In order to address these challenges in providing web-based information services to mobile devices, WML is designed to describe data and the format that data should be presented.

Like HTML, WML is a tagged language. WML adopts a deck and card metaphor. Each WML document is made up of multiple cards, and cards are grouped into a deck. Similar to an HTML page, a deck is the unit of transmission and is identified by a URL. In order to save bandwidth in a wireless network, WML pages can be encoded in a binary format before transmission. When navigating with a WML browser, a user interacts with cards, and moves forward and back through the deck. Figure 15.1 illustrates an example of a deck with two cards.

 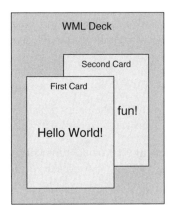

Figure 15.1 A WML deck with two cards.

WMLScript

WMLScript is a scripting langauge which complements WML. WMLScript is designed to work well in a wireless environment, like WML. Scripts written in WMLScript can be compiled into bytecode for efficient transmission over wireless networks. Moreover, the WMLScript bytecode interpreter is compact in size, which allows efficient execution of scripts will less memory and processor requirements.

The key functionalities of using WMLScript are described as follows:

User Interaction and Input Validation: In a WML card, if the user is requested to input some information and then send to a server, then it can save bandwidth and latency if the WML browser can call some WMLScript to verify if the user has input correct type and/or value of information. If there is some input error, error messages can be prompted to tell the user to input the correct type and/or value. In addition, by using WMLScript, the WML browser can gather the results of a series user interaction and then send the results to the server in a batch.

Device Access: Through WMLScript, the user can access the device's hardware facilities and peripherals, as well as the software functions.

Device Software Extension: Since WMLScript can utilize a device hardware and software facilities, it allows the extension of existing software functionalities by downloading new software components from service providers or device manufacturers. This way, new software functionalities can be added without making changes to the firmware of the device.

15.3   WAP SYSTEM ARCHITECTURE

The WAP system architecture is based on the WWW server/clicent architecture of the Internet. Generally, the architecture is composed of three elements (see Figure 15.2):

Client: It is is the WML browser in a wireless device. It issues WAP requests to a server.

Server: It is the entity which provides services and where resources are located. This can be an ordinary Internet-based server or a WAP-capable server.

Gateway: It provides protocol conversion between the WWW protocol stack and the WAP protocol stack, by using content encoders and decoders. Thus, effectively, a gateway acts as a proxy server. When protocol conversion is performed at the gateway, it can minimize wireless communication overhead at the client side. On the other hand, the gateway can also cache frequently-requested contents so as to reduce the request/response time.

Figure 15.2 shows a detailed view of the WAP infrastructure. A mobile user can access to Internet services provided by an ordinary server or a WAP-capable server. If the server is WAP-capable, then it can directly generate WML content. The WML content will then be sent to the WAP gateway using HTTP. After that, the WAP gateway will convert the WML content into a binary form and then transmit to the mobile device using WAP protocols.

On the contrary, when the mobile user needs to access to an ordinary server which generates only HTML pages, the generated HTML pages will go through a HTML filter first. The function of the HTML filter is to strip away contents that not suitable for presenting in a mobile device, and then convert HTML codes into WML codes. The HTML filter can be collocated with the WAP gateway, or located in another

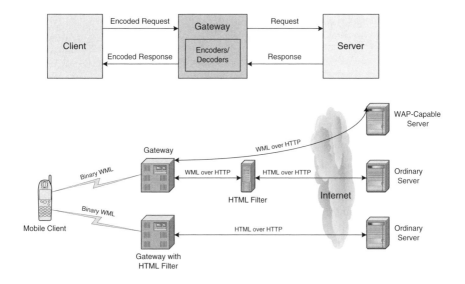

Figure 15.2 WAP infrastructure.

location. If it is located in another location, it will send the filtered WML content to the WAP gateway using HTTP.

Push Architecture

The architecture discussed above is the common pull architecture based on the client/server paradigm, i.e., the client explicitly requests content from the server. The WAP system architecture also specifies a push architecture to enhance the WAP services, in which the server sends messages to the client without explicit request from the client. This push architecture is very useful in delivering messages like instant news, email indication, advertising, etc.

In the push architecture, the server and the gateway are called the push initiator (PI) and the push proxy gateway (PPG), respectively. Figure 15.3 shows the interaction of PI, PPG, and the client. Specifically, the PI sends a push message to the PPG using the push access protocol. The PPG then extracts the client address from the received push message, and checks if the message can be forwarded to the client by finding a mapping of the extracted client address to a format valid in the client's wireless network. If the push message can be forwarded, the PPG then transforms the message to a binary format for efficient transmission to the client using the push over the air (OTA) protocol.

Apart from processing push messages from the PI (i.e., forwarding, cancellation, and replacement of push messages), the PPG also offers other supplementary functionalities to enhance the sending of push messages. These include notifying the PI about the result and status of processing push messages, and telling the PI about the client's capabilities and preferences on content transformation. When a mobile device

Figure 15.3   WAP push architecture.

is not available to accept a push message (e.g., when the device is busy with other tasks or it is at a location of having poor network connection), then the PI can send a short message, namely a service indicator, to the client instead of sending a relatively large push message. The content of the service indicator contains a uniform resource identifier (URI) pointing to a service to be consulted. Consequently, the user can go back to retrieve the push message later by initiating a client request to the server indicating by the URI.

WTA Architecture

To support traditional telephony functionalities in a mobile device, WAP specifies the wireless telephony application (WTA) framework. As illustrated in Figure 15.4, the core of WTA is the WTA user agent. The WTA agent can access to telephony services through a WTA interface (WTAI), providing the user with the ability to make/receive phone calls, access to the phone book, etc. The WTA server provides WTA-related services to the client through the WAP gateway. For example, when there is an incoming voice message, the voice server of the mobile network service provider will indicate the WTA server of the voice message. Then the WTA server may send a push message to the mobile device to notify the user of the new voice message. In the push message, it includes a URL pointing to the server to which the user should contact in order to retrieve the voice message.

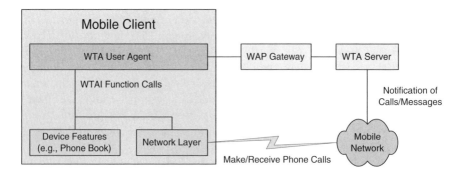

Figure 15.4   WTA architecture.

15.4 WAP PROTOCOL STACK

Overview

Figure 15.5 shows the WAP protocol stack. The lightweight protocol stack is designed with two main goals—to minimize bandwidth requirement and to maximize the number of supported network types. Thus, WAP is designed to be bearer-independent, i.e., it does specify particular bearer services and it tends to use existing data services such as GSM and GPRS. On the other hand, the layered protocol stack is designed to be scalable. Each layer provides a well-defined interface to its upper layer such that each layer can be developed independent of each other. This way, it allows minimum changes to the protocol stack when new bearers or transport protocol stacks are added. It also allows quick addition of new applications and services that are not specified in the WAP specification.

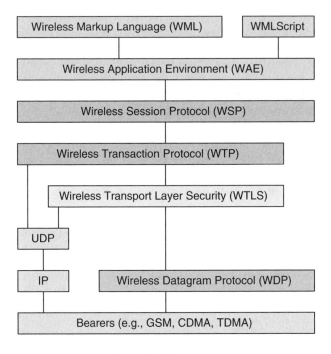

Figure 15.5 The WAP protocol stack.

Wireless Application Environment (WAE)

WAE provides an application framework for mobile devices. It specifies tools and formats for service providers, software developers and hardware manufacturers to develop WAP applications and services. The application environment specified by WAE is based on existing technologies in the WWW paradigm, such as HTML, XML, and JavaScript.

The main elements in WAE are as follows:

WML User Agent: WAE user agents are software executing in mobile devices to support specific functionalities. The WML user agent is a particular user agent that is defined to support WML and WMLScript for content browsing.

WTA User Agent: The WTA user agent is another user agent defined to support telephony services.

Content Generator: It is an application or service residing on ordinary servers. It serves requests from WAP clients and produces standard formats in the response content.

Wireless Session Protocol (WSP)

WSP provides two kinds of session services to its upper layer, namely the connection-oriented session service which operates on top of the transaction service WTP (Wireless Transaction Protocol), and the connectionless session service which operates on top of the datagram service WDP (Wireless Datagram Protocol). Specifically, WSP is based on HTTP and is adapted for optimal use in wireless environments where the data rate is relatively low and there is frequent connection loss. For example, in contrast to HTTP, which is stateless, WSP allows a shared state between a server and a client so as to provide efficient content transfer.

The operation of WSP is based on a request/response paradigm. In addition to this, a server push operation is also defined in WSP to allow a server to send unrequested content to a client. The PDU (protocol data unit) conveyed in either operation consists of a header and a body. In the body, it might contain WML, WMLScript, image data, etc. WSP provides the following functionalities for content transfer:

Session Management: Sessions can be established from a client to a server. Sessions can be long lived, suspended, resumed, and released in an orderly manner.

Capability Negotiation: During session establishment, the client and server can agree on a common level of protocol functionality. Parameters can be negotiated are server/client SDU size, maximum outstanding requests, protocol options, etc.

Content Exchange: WSP defines binary encoding scheme for efficient content exchange between the client and the server in wireless environments.

In WSP, a collection of service primitives are defined to form the interface between WSP and the users of its upper layers in WAE. These primitives have various parameters associated with them, and they form the basis for session management and content exchange. Figure 15.6 shows how session establishment is achieved between the client and the server by using the **S-Connect** primitives. Specifically, the client's WSP user issues an **S-Connect.req** to WSP to request for establishing a session with the WSP user on the server. On issuing the **S-Connect.req**, the parameters passed are server address (SA), client address (CA), client headers (CH), and requested capabilities (RQ). Then at the server side, **S-Connect.ind** is issued to transfer the parameters from the client to the WSP user. If the WSP user at the server side accepts the request for establishing a session, it responds by issuing **S-Connect.res**, with server headers (SH) and negotiated capabilities (NC) being the parameters. Finally, at the client side, **S-Connect.cnf** is issued to the WSP user to inform it about the negotiated capabilities.

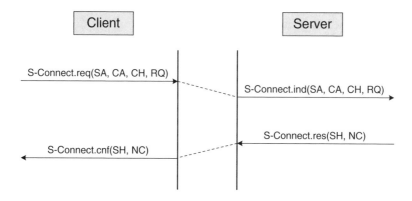

Figure 15.6   WSP session establishment.

Wireless Transaction Protocol (WTP)

WTP is a transaction-oriented protocol, where a transaction is defined to be a request-response pair. Using WTP, there is no explicit connection setup or teardown, which reduces much amount of overhead when compared to TCP. Specifically, WTP provides a reliable connection-less service. Other features of WTP are as follows:

- It achieves reliability by using acknowledgement, retransmission, duplicate removal, and transaction identifier.

- It supports asynchronous transaction, abort of transaction, and report of status of transaction.

- It supports concatenation of messages and delayed acknowledgement so as to reduce the number of messages to be sent out.

- To increase reliability, a WTP user can be requested to confirm the receive of a message.

- There are three classes of transaction services provided, namely Class 0, Class 1, and Class 2. Class 0 provides an unreliable transaction service without the need for a result message, meaning that the transaction is stateless and therefore cannot be aborted. This class of service can be used for providing unreliable push operations. Class 1 provides a reliable transaction service, also without a result message. It can be used for reliable push operations. The responder of this class of service maintains state information for some time after an ACK has been sent out to handle cases where retransmission of the ACK is needed. Finally, Class 2 provides a reliable transaction service of the classic request/response paradigm.

WTP provides the following three service primitives to its upper layer:

- **TR-Invoke**: To initiate a transaction.

- **TR-Result**: To send back the result of a previously initiated transaction.

- **TR-Abort**: To abort an existing transaction.

Wireless Transport Layer Security (WTLS)

WTLS is on top of WDP to provide, when requested by applications, security services between the mobile device and the WAP gateway or WAP server. WTLS is based on the TLS (Transport Layer Security) protocol, and is optimized for use in low-bandwidth and high-latency wireless networks. It also takes into consideration the relatively low processing power and low memory limitations of mobile devices when choosing cryptographic algorithms. WTLS can provide different levels of security measures to mobile devices: privacy, data integrity, authentication, and denial-of-service protection.

Within WTLS, there are four protocols defined:

Handshake Protocol: It is a four-phase protocol to allow the client and the server to authenticate each other and to negotiate the compression and cryptographic algorithms, and other related security settings used to protect data in WTLS records. A four-phase handshaking is performed between the client and the server before any data can be transmitted.

Change Cipher Spec Protocol: The client/server uses this protocol to notify the other of the change of the cryptographic settings.

Alert Protocol: This protocol is used to notify the other peer entity that an error has occurred, by sending an alert message. An alert message can be of the type warning, critical, or fatal. If it is a fatal message, the current connection will be terminated immediately.

Record Protocol: User data from higher layer protocol (e.g., WTP) is encapsulated in a number of steps using the record protocol.

The details of the steps performed in the record protocol are further elaborated in Figure 15.7 First of all, the user data is compressed using a compression algorithm. Then HMAC is used to calculate a message authentication code (MAC) of the compressed data. The MAC is then appended to the compressed data. After that, the concatenated data is encrypted using a symmetric encryption algorithm, and finally a record protocol header is prepended to the encrypted data.

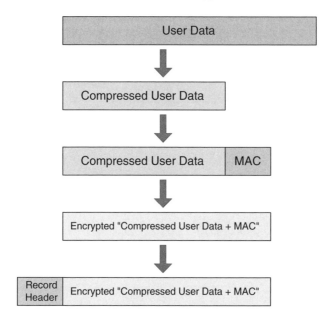

Figure 15.7   Steps performed in the WTLS record protocol.

Wireless Datagram Protocol (WDP)

WDP provides an adaptation on top of many different bearer services to support datagram transport service to higher layers in the WAP protocol stack. WDP is designed to be bearer-independent and thus the adaptation needed depends on the services provided by the target bearer. If the bearer supports IP services, then UDP can be used as WDP, and adaptation is minimal. Otherwise, WDP may need to support functions like data segmentation and reassembly.

During WDP datagram transmission, if errors occur (e.g., destination unreachable), WAP specifies the use of wireless control message protocol (WCMP) for error handling. WCMP is much like the ICMP in IP for diagnostic and informational purposes. In fact, for bearers where IP services are supported, ICMP is used as the WCMP.

15.5 WAP PROFILES AND CACHING

Since different WAP-enabled mobile devices have displays of different screen sizes and color depths, a WAP page might look different in different mobile devices, even the same version of WAP browser is used in the mobile devices. On the other hand, it will be a waste of bandwidth if a color image is downloaded on a mobile device which can only display black and white images. Thus, it is clearly that if WAP contents are to be delivered to a mobile device, it is beneficial to know the capabilities of the mobile device first.

To describe the capabilities of a mobile device, WAP has defined a user agent profile (UAProf). The capabilities of a mobile device are related to software and hardware, including things like processor type, memory capacity, display size, browser type and version, network type, etc. The aim of using UAProf is to allow all elements of the WAP infrastructure (i.e., content servers, application servers, gateways, etc.) to provide mobile devices with device-specific contents.

A user agent profile is basically an XML document containing information about the hardware and software characteristics of a mobile device, and the network to which it will be connected. Therefore, it is not used to specify a user's preference. The user agent profile of a mobile device is stored in its manufacturer's server, called the profile repository. For example, the URL of the user agent profile for the Nokia 6230i cell phone is [333]:

http://nds1.nds.nokia.com/uaprof/N6230ir200.xml

An excerpt of the XML file is shown in Figure 15.8

```
......
<prf:component>
  <rdf:Description rdf:ID="HardwarePlatform">
    <rdf:type rdf:resource="http://www.openmobilealliance.org/
      tech/profiles/UAPROF/ccppschema-20021212#HardwarePlatform"/>
    ......
    <prf:PixelAspectRatio>1x1</prf:PixelAspectRatio>
    <prf:ScreenSize>208x208</prf:ScreenSize>
    <prf:ScreenSizeChar>18x5</prf:ScreenSizeChar>
    <prf:StandardFontProportional>Yes</prf:StandardFontProportional>
    <prf:SoundOutputCapable>Yes</prf:SoundOutputCapable>
    <prf:TextInputCapable>Yes</prf:TextInputCapable>
    <prf:Vendor>Nokia</prf:Vendor>
    <prf:VoiceInputCapable>Yes</prf:VoiceInputCapable>
  </rdf:Description>
</prf:component>
......
```

Figure 15.8   An excerpt of the user agent profile XML for the Nokia 6230i cell phone.

In order to provide mobile devices with device-specific contents, when a mobile device performs a request to a server, the URL of its user agent profile will be included in the header of the request message. A user agent profile contains the following components, each of which contains a number of different attributes:

Hardware Platform: It describes the hardware capabilities of the mobile device, such as color capability, screen size, and text input capability.

Software Platform: It describes the software capabilities or operating environment of the mobile device, such operating system type and version, character sets supported, and audio and video codecs supported.

Network Characteristics: It provides the information about the bearers supported, and their supported services, such as encryption methods supported.

Browser UA: It provides the information about the browser of the mobile device, e.g., browser name and version, and HTML version supported.

WAP Characteristics: It tells about the WAP features supported, such as WAP version and WML deck size.

Push Characteristics: It provides information about the push capabilities of the mobile device such as content types supported, and the maximum size of a push message.

MMS Characteristics: It provides information about the MMS (multimedia messaging service) capabilities, e.g., the maximum size of an MMS message, and the maximum image resolution.

Caching of frequently-accessed data items for answering subsequent requests is a promising technique in a wireless computing environment in order to reduce the response time. This is particularly useful for WAP-enabled mobile devices having limited capabilities. WAP has specified a caching model based on HTTP/1.1 caching as defined in RFC 2068 [118]. Like other adapted technologies like WMLScript and WTLS, in the WAP caching model a number of extensions and clarifications have been specified to facilitate the operation of HTTP/1.1 caching on mobile devices [244].

There are two main types of caching in the WAP architecture, namely caching by WAP devices and caching by WAP gateways. WAP-enabled mobile devices generally cache WML decks in order to reduce the time needed in reloading a page that has been previously loaded, without making a connection to the WAP gateway. This can lead to significant performance improvement in a high latency and low bandwidth wireless environment. However, for dynamic content, caching is inappropriate. The WAP caching model specifies the use of the Cache-control:must-revalidate header in order to signify whether a cached item should be revalidated when it becomes stale.

As for caching by WAP gateways, the WAP caching model specifies that a WAP gateway must faithfully implement the role of an HTTP/1.1 proxy with respect to caching and cache header transmission [244]. Specifically, a WAP gateway acts as a proxy server which caches documents requested by the network. Then the WAP gateway can fulfill subsequent requests from WAP clients, eliminating the round-trip time needed between the WAP gateway and the ordinary server.

15.6   PRACTICAL ILLUSTRATIONS—LOCATION-AWARE ADVERTISING SYSTEM

As mentioned earlier, the WAP push architecture has the nice feature that instead of sending a large push message, the push initiator can just send a short message (i.e., a service indicator) notifying the user of the push message. The short message comprises an embedded URL pointing to the actual content. This way, the user can choose to download the content later, or just simply discard the service indication. Hence, the WAP push architecture is useful for advertising. Suppose you are visiting a city that you are not familiar with and are wandering on the streets, then it will be convenient if there are some advertisements push to your mobile phone showing the information about the shops near to you or details of events in the city. However, one problem arises is how can the advertising system know your location information so as to trigger proper advertisements?

Aalto *et al.* [10] have proposed a permission-based location-aware mobile advertising system based on Bluetooth and WAP technologies. The system is called B-MAD (Bluetooth Mobile Advertising). Specifically, some locations (e.g., in shops) are installed with Bluetooth sensors, which periodically discover Bluetooth-enabled mobile devices and then trigger advertisements. Advertisements, on the other hand, are sent using the WAP push service.

Figure 15.9 shows the architecture of the B-MAD system. As can be seen, there are four elements in the system, namely the end user device, Bluetooth sensor, ad server, and the push sender (i.e., the push initiator in the WAP push architecture). The interaction of the four elements are described as follows:

User Registration: An end user subscribes to the mobile advertisement service through the ad server. During registration, the user needs to input his phone number (MSISDN) and the Bluetooth device address (BD_ADDR). The associated MSISDN and BD_ADDR is stored in a database of the ad server.

End User Bluetooth Device Scanning: The Bluetooth sensor periodically scans for nearby end user Bluetooth devices. If a Bluetooth device is discovered, the BD_ADDR of the device (and also the location identifier of the Bluetooth sensor) is sent to the ad server via a WAP connection.

Advertisement Preparation: When the ad server receives the request from the Bluetooth sensor, it checks in the database to see if there is a matched MSISDN for the received BD_ADDR. If there is a match, the ad server will prepare the required advertisement based on the location identifier of the Bluetooth sensor. Then, the ad server will pass the push sender the MSISDN of the recipient and descriptions of the advertisement (in this case, only short textual description of the advertisement and URL address pointing to the actual advertisement are sent).

WAP Push Message Sending: The push sender sends service indication message of the advertisement to the push proxy gateway using the push access protocol and XML over HTTP. The gateway then encodes the push message to a binary over the air (OTA) format, and uses a Short Message Service Center (SMSC) with SMS as the bearer to deliver the push message [10].

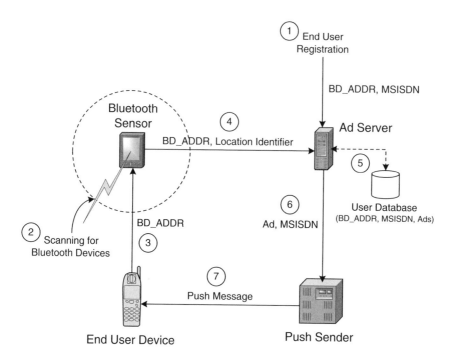

Figure 15.9   B-MAD permission-based location-aware mobile advertising system.

To evaluate the B-MAD system, Aalto *et al.* [10] have conducted quantitative evaluation in a laboratory environment and qualitative user evaluation in a field trial. They reported that there are inherent latencies in discovering end users' Bluetooth devices and delivering advertisements. This will be annoying to users if they receive the advertisements too late but they have already walked quite far away from the Bluetooth sensor. On the other hand, during user registration, the ad server should be modified to accept the user's preference on receiving advertisements so that only personalized advertisements are sent to the user.

15.7 SUMMARY

In this chapter, we have discussed how mobile users are supported with Internet information services through the WAP technology. Specifically, WAP addresses the various limitations of mobile devices, like limited processor power, memory capacity, battery life, input and output facilities. It also address the limitations of wireless networks, such as low bandwidth, high latency, unpredictable available and stability. The lightweight WAP protocol stack, WML, WMLScript, UAProf, and the WAP caching model are designed with the goals to address the limitations of mobile devices and wireless networks. Finally, we have presented how a mobile advertising system can be built with the WAP and Bluetooth technologies.

PROBLEMS

15.1   What is the use of UAProf? How can it enhance the efficiency of delivering WML content to mobile devices?

15.2   How does WAP address the problem of limited battery life of mobile devices?

15.3   Briefly describe the functions of a WAP gateway.

15.4   What are the advantages of using the WAP push service?

CHAPTER 16

TCP OVER WIRELESS

Let every man praise the bridge that carries him over.
> —English Proverb, traced by Smith to 1678

16.1 INTRODUCTION

As we have seen, WAP is a specialized technology designed for adopting TCP/IP in a cellular network. However, the model underlying WAP is inappropriate in many other wireless environments (e.g., a satellite wireless link). Thus, there are many other solutions proposed for enabling TCP over wireless.

Interoperability:

In the past few years, we have witnessed that the Internet has extended its reach to the wireless networking environments. In fact, we are stepping into the ubiquitous mobile computing age. With the help of mobile devices like cell phones and personal digital assistants (PDAs), and 2.5G and 3G mobile data services like GPRS and UMTS, users are provided with ubiquitous mobile access to IP-based applications

Wireless Internet and Mobile Computing. By Yu-Kwong Ricky Kwok and Vincent K. N. Lau
Copyright © 2007 John Wiley & Sons, Inc.

in traditional wired networks. Thus, traditional Internet protocols are also used over wireless links. Since TCP/IP (Transmission Control Protocol/Internet Protocol) is the standard network protocol stack on the Internet, it is certain that TCP is used in different wireless environments. Not only does it leverage a large number of existing applications, but it also provides seamless integration with the wired Internet.

Performance:

The wireless users may commonly find that the provided services in wireless networks are not as good as wired networks because of the limited bandwidth and high bit error rate (BER) in wireless links. On the other hand, a straightforward migration of Internet protocols in traditional wired networks to wireless networks will result in poor performance. In particular, the TCP, which has many salient features that are useful in a wired network, needs significant modifications in order that it can deliver packets over the wireless links efficiently. Indeed, the congestion control mechanisms are particularly problematic in wireless environments. Due to the inherent reliability of wired networks, there is an assumption made by TCP that any packet loss is due to congestion. However, in wireless networks, packet losses are no longer mainly due to network congestion, but due to some wireless specific reasons such as random loss or collision in wireless networks.

Solving the problem of TCP performance degradation in wireless environments is very challenging because TCP is designed and tuned for wired networks. Unfortunately, compared with wired networks, wireless networks have many different characteristics, such as high bit-error-rate (BER), low channel bandwidth and occasional blackout. Recently, many adaptive TCP approaches for various wireless environments have been suggested. There are three main groups of approaches: link layer, split connection and end-to-end approaches. The major objective of these schemes is to make TCP respond more intelligently to the lossy wireless links.

16.2   TCP CONGESTION AND ERROR CONTROL

A Brief Introduction to TCP

Transmission Control Protocol (TCP) is the most commonly used transport layer protocol in Internet for providing reliable end-to-end data transfer. It specifies the procedures for connection setup and connection close, the usage of sequence numbers for reliable and in-order packet delivery, and the timeout retransmission mechanisms.

TCP is a connection oriented protocol and is designed specifically for use over an unreliable network. To establish a connection, a three-way handshake mechanism is adopted in TCP.

In TCP, reliability is achieved through the use of sequence number, cumulative ACKs, and retransmission techniques. Data packets are sent out in order, and each data packet or acknowledgment packet has a unique sequence number. If a packet is received out-of-order, the receiver acknowledges with the highest received in-order sequence number. Thus, each ACK indicates that all the packets whose sequence numbers are less than the sequence number of the currently received ACK have been received. Therefore, ACKs for these out-of-order packets are called duplicate ACKs (DUPACKs). Obviously, TCP uses ACKs as a clocking mechanism to transmit data packets. In the case that no ACK is received during an interval called timeout, TCP uses retransmission to ensure the delivery of data packets.

To prevent sender from delivering too fast, TCP employs a flow control mechanism using advertisement window, which indicates the buffer size currently available at the receiver. The sender gets the advertised window size via the new coming ACK. Using the flow control mechanism, the sender is not allowed to transmit packets more than that the receiver can buffer; otherwise, excessive packets would be discarded.

The original TCP standard mentioned above does not specify any congestion control mechanisms and is not deployed in current Internet, but it is the basis of other TCP variants currently deployed in Internet.

TCP Reno

TCP Reno [165, 327] is widely deployed in Internet. It has a reliable window-based flow control and congestion control protocol. The sender utilizes ACKs to pace the transmission of segments, and interprets triple duplicate ACKs and retransmission timeout as congestion indication. TCP sender reduces the transmission rate by shrinking its window size when it detects a congestion. In Reno, four intertwined algorithms are included in the congestion control protocol. They are slow start, congestion avoidance, fast retransmit, and fast recovery. In addition, retransmission timeout is also employed in the congestion control mechanism.

Two variables are used in Reno, congestion window ($cwnd$) and slow start threshold ($ssth$). The variable $cwnd$ indicates the number of data packets which are allowed to be transmitted currently; $ssth$ is the threshold to determine whether the current congestion control of this connection is in the phase of slow start or congestion avoidance.

The main procedures in Reno's congestion control mechanism are explained in the following:

Slow Start (SS): SS is carried out at the beginning of a TCP connection or after a retransmission timeout. The purpose of SS is to prevent the sender from injecting multiple packets into the network. The SS mechanism prevents the sender from network congestion with burst data and gradually increases the amount of transmitting data to discover the available bandwidth. When a TCP connection is established, $cwnd$ is initialized to one packet. If packet losses are indicated by timeout, $cwnd$ is also set to one packet. Each time an ACK is arrived, $cwnd$ grows by one packet. With reference to Figure 16.1, the increase of $cwnd$ is exponential. The increase will end when $cwnd$ reaches $ssth$, which is given by:

$$ssth = \min\{cwnd, awnd\}, \tag{16.1}$$

where $cwnd$ is based on the estimation of perceived congestion by the sender, $awnd$ is advertised window imposed by the receiver and relative to the available receiver buffer size in this connection. Therefore, SS can effectively avoid packet burst injecting into the network as well as quickly fill up the estimated Internet capacity.

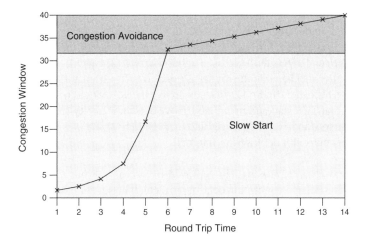

Figure 16.1 Variation of congestion window size due to slow start and congestion avoidance mechanisms.

Congestion Avoidance (CA): When $cwnd$ reaches the value of $ssth$ or receiving several DUPACKs, the congestion control enters the CA phase. In this phase, the TCP tries to slowly probe the network for extra bandwidth. $cwnd$ is increased by one packet per RTT (round trip time). Each time an ACK is arrived, $cwnd$ is increased by:

$$\Delta cwnd = \frac{packet\_size \times packet\_size}{cwnd}, \qquad (16.2)$$

where the packet size and $cwnd$ are maintained in bytes.

Compared with SS, CA leads to linear phase of $cwnd$ growth (see Figure 16.1). In fact, the increase of $cwnd$ is no more than one packet in every RTT.

No matter whether the congestion control is in SS or CA, when packet losses are detected, $ssth$ must be set to half of its original value.

Fast Retransmit and Fast Recovery: When an out-of-order packet reaches the receiver, a DUPACK is sent back to the sender. The purpose of DUPACK is to let the sender know that an out-of-order packet is received and what sequence number is expected, in addition to maintaining ACK clock. However, the TCP sender does not know which packet triggers this DUPACK, and whether the expected packet is lost or out-or-order. Then it waits for a small number of DUPACKs to be received. When the Kth DUPACK is received, it strongly indicates the expected packet has been lost. Then TCP performs retransmission without waiting for timeout. Typically, K is equal to 3. This mechanism is called Fast Retransmission.

Generally, fast retransmit and fast recovery are implemented together. After fast retransmit is triggered, besides the sender retransmits the lost packet, fast recovery algorithm is also employed to govern the transmission until a non-duplicate ACK arrives. Since DUPACK is generated by receiving new and out-of-order packets, there is still data flowing between sender and receiver. Fast recovery is used to avoid reducing the flow dramatically and to maintain flow rate by continuing to send packets according to CA algorithm. When receiving three DUPACKs, the sender retransmits the expected packet, $ssth$ is halved and the $cwnd$ is set to $ssth$ plus three. When another DUPACK is received, the sender transmits one packet and $cwnd$ is increased by one if $cwnd$ is less than $awnd$. After TCP finishes fast recovery, it enters CA with linearly increasing $cwnd$.

Retransmission Timeout: Since TCP is ACK-clocked, round trip time (RTT) which is the time of data packet travels from the sender to receiver plus ACK travels from the receiver to the sender is an important variable to predict packet loss. The value of estimated RTT is applied to another variable called retransmit timeout (RTO), which is the maximum time to wait for an ACK after sending a data packet. RTO governs if retransmission is needed. The estimation of RTO is significant, because if RTO is too small, unnecessary retransmission is triggered; if RTO is too large, the sender will wait without retransmission until the timer expires. The value of RTO is based on the RTT as governed by:

$$S\_RTT_{i+1} = \alpha \times M\_RTT_i + (1 - \alpha) \times S\_RTT_i, \qquad (16.3)$$

where S_RTT is the smoothed RTT, M_RTT is the most recently measured RTT, α is set to 1/8 [164].

Thus,

$$RTO = S\_RTT + 4 \times \Delta RTT, \qquad (16.4)$$

where ΔRTT is the estimate of standard deviation of the difference between S_RTT and M_RTT.

Retransmission timeout mechanism is a supplement when no ACK clock is received for a long time. After the retransmission timer is expired, the value of RTO is doubled. This mechanism is called RTO backoff [183], which can weaken congestion in the network and help the connection leave the congested state.

16.3   DEFICIENCIES AND OVERVIEW OF SOLUTIONS

TCP has been tuned to perform well in traditional networks comprising wired links and stationary hosts. The basic assumption is that congestion is the primary reason for packet losses in wired networks. When packet losses are detected, the TCP sender exponentially throttles down the congestion window size before retransmitting lost packets. It then backs off its retransmission timer, and enters the congestion avoidance phase. All these mechanisms are designed to reduce the load injected into the network. These mechanisms work well in a wired network where the bit error rate (BER) is typically low to allow any packet loss to be treated as an indication of congestion.

However, the situation of wireless networks is much more complicated than that in wired networks. Only congestion control is not enough to deal with so many different sources of TCP performance degradation in wireless networks. We classify the reasons for the deficiencies into five groups. They are high bit error rate, high round trip time, occasional disconnection, the MAC layer and asymmetric bandwidth:

High Bit Error Rate: In particular, high bit error rate is an important characteristic in wireless network. It occurs for different reasons such as path loss, fading, noise and interference.

High Round Trip Time: The high round trip time may reduce the increase rate of congestion window, even in the case of loss of ACK. Since congestion window is a function of number of ACKs, the slow start phase, which is critical time to occupy precious channel, increase slowly.

Occasional Disconnection: Occasional blackout is due to the device mobility. As a mobile host travels between wireless cells, the task of forward data between the mobile host and wired network must be transferred to the new cell's base station. This is called handoff. A long handoff time may result in long latency and packet losses. In ad hoc wireless networks, network partitioning occurs when a give mobile node moves away or is interrupted by the medium, then the neighboring nodes are separated into two parts of the network. Also, the adverse effects of the partitioning on TCP depends on its duration. A long partitioning time may lead to connection abortion.

The MAC Layer: IEEE 802.11 is standard MAC layer protocol for wireless LANs. It can support some ad hoc network architectures. However, this protocol is not designed for such multihop networks. It does not perform well in multihop ad hoc networks [332, 354]. Hidden terminal problem may lead to data collision and exposed terminal problem may result in under-utilization of the available bandwidth (as detailed in Chapter 9.4. And serious unfairness occurs when TCP connections cooperate with IEEE 802.11. Simulations show the TCP performance decreases sharply with the increase of the number of hop counts.

Asymmetric Bandwidth: Asymmetric bandwidth means low bandwidth for the ACK transmission. In some wireless networks, low bandwidth is used for the ACK traverse. Even the ACKs are smaller in size than data packets, the bandwidth of reverse channel cannot afford high rate of ACKs. The result is congestion and loss in ACK channel.

The main reason for the TCP performance degradation in wireless networks is the reaction to non-congestion losses. Generally, if the TCP sender could distinguish different losses, or non-congestion losses are shielded from the sender, the unnecessary congestion control could be avoided. In this context, two methods are used to enhance the TCP performance. The first one is to hide the non-congestion losses from the TCP sender. In this method, the non-congestion losses could be solved locally and thus, most of the losses seen by the TCP sender are indeed caused by congestion. The second on is TCP aware protocol. Using this method, the TCP sender should respond to different losses by different actions.

The existing schemes for enhancing TCP performance in wireless networks can be classified into three groups: link layer, split connection and end-to-end approaches. Simply put, these approaches target networks where only the last hop communicates through a wireless link, as illustrated in Figure 16.2. In this kind of wired-cum-wireless networks, the mobile host communicates with the base station which is connected with a wired backbone. This scenario is common in daily life, such as Internet access through a GPRS phone.

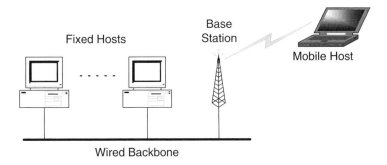

Figure 16.2   Last hop wireless link scenario.

16.4   LINK LAYER APPROACHES

The link layer is governed by unreliable, connectionless and best-effort delivery mechanisms. If the link layer could be reliable and correct the packet errors, TCP sender has less possibility to act with losses. Therefore, the purpose of link layer approaches is to improve the reliability of the wireless network. They attempt to hide the link-related losses from the TCP sender using local remedies, such as local retransmissions, in the wireless link.

Traditional Link Layer Schemes

Traditional approaches for loss recovery at the link layer are FEC (forward error correction) and ARQ (automatic repeat request). The former is to let the receiver recover loss promptly using the redundant correction bits, at the expense of a lower effective throughput even in the case of an error-free transmission. The latter is to let the sender repeat the transmission of any lost packet. These two schemes are generally used in the lossy environments. Many proposed algorithms [19, 25, 253] are based on combined FEC and ARQ.

Asymmetric Reliable Mobile Access In Link-layer (AIRMAIL) [19] is an asymmetric protocol because the base station is obliged for making decisions whenever it transmits or receives packets. The base station sends periodic status messages to allow that the mobile host combines several acknowledgments into a single one. It employs a combination of ARQ and FEC for loss recovery to provide reliable links. It modified the ARQ retransmit strategy: the packets of the whole window are transmitted before a link layer acknowledgment is received by the link layer sender. Furthermore, an adaptive algorithm is employed based on three levels of FEC and the error correction only can be implemented after the whole window. This may cause TCP timeout for the long time latency. AIRMAIL also has mechanism to deal with handoff. Figure 16.3 shows how a mobile terminal (MT) works when handoff occurs using AIRMAIL. It should be noted that AIRMAIL allows the mobile transmitter to continue transmitting data while the handoff is occurring.

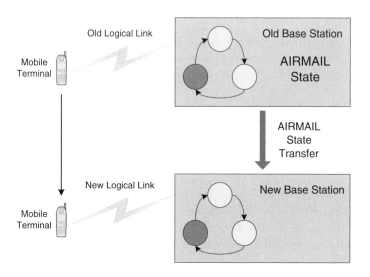

Figure 16.3 Mobility management in AIRMAIL [19].

Radio Link Protocol (RLP) [253] is a link layer based approach designed for local recovery of packet loss by a judicious use of the sequence numbers. It is based on a point-to-point ARQ for radio channels. In this protocol, a packet is retransmitted only if the transmitter is sure that it is lost. This is implemented by using the feedback packets (see Figure 16.4) from the receiver together with the sequence number of packet and a send sequence number at the sender. Thus, it is efficient to make sure the receiver only gets one copy of each packet. An enhanced version of RLP, which piggybacks partial receiver state in the reverse channel on uses data packets, avoids to be idle when all retransmissions are completed.

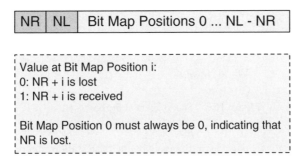

Figure 16.4   Feedback packet [253].

Snoop and Its Enhancements

Snoop [22] is a link layer aware TCP scheme. It is based on a special module called the Snoop agent at the base station (see Figure 16.5). The Snoop agent monitors every packet that travels through the wireless TCP connection in both directions and processes them with different measures according to the types of packets. The most important function of the Snoop agent is to recover the lost packet and retransmit them. In the Snoop protocol, it is assumed that all packet losses in wireless links are not due to congestion. As such, the Snoop agent maintains a cache of TCP packets sent across the link that have not yet been acknowledged by the mobile receiver. When packet loss is detected by either three duplicated ACKs or local time-out, the Snoop agent retransmits the packet if it has been cached and suppresses the duplicate acknowledgments. The fixed TCP sender will not trigger the congestion control or avoidance until time-out. The merit of the Snoop protocol is to retransmit losses locally and suppress the duplicate acknowledgments, thereby avoiding unnecessary fast retransmit and congestion control at the fixed sender.

NewSnoop [155] and FDA [154] are enhancements of Snoop. NewSnoop uses the hierarchical structure to enhance the performance during the handoff. It adds a tag in the packet at base station agent to identify the types of cached packets. Moreover, the Mobile Switch Center also caches the unacknowledged packets to help minimize the latency of retransmission during handoff. Snoop and NewSnoop are efficient when a single link is used to connect the mobile device to the base station. When

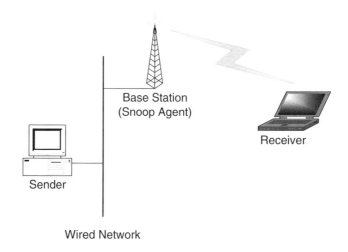

Figure 16.5 Snoop agent at the base station.

multiple connections are used simultaneously, the base station cannot buffer so many unacknowledged packets. FDA is proposed to solve this problem. A predefined buffer threshold indicates the incipient congestion. After the detection of incipient congestion, a special congestion notification, called three forced duplicated ACKs, is generated by the base station instead of the mobile host. This mechanism can potentially reduce the congestion response time for the fixed host. On the other hand, quality guaranteed cache release functions designed to minimize the change of multiple packet dropping at the base station during congestion.

16.5 SPLIT CONNECTION APPROACHES

Split-connection TCP schemes split the logical connection into two parts using the base-station as the break-point—one connection between the sender and the base station, the other between the base station and the receiver. The main idea behind the split connection approaches is to isolate the mobility and wireless related problems from the existing protocols in order to hide the wireless packet loses from the fixed TCP sender. Consequently, the performance of the TCP connection over the fixed part is not influenced by the wireless link and the wired connection does not need any change in existing software on the fixed hosts. The protocol used in the wireless link can be TCP or some other specialized protocols for wireless. Example protocols are I-TCP [20], M-TCP [42], and MTCP [360].

Indirect-TCP (I-TCP) [20] is a split connection approach using TCP for its connection over the wireless link. It attempts to separate loss over the wireless link from what over the wired links and hide packet loss from the TCP sender (see Figure 16.6).

Two components are included in I-TCP: one is on the mobile host and the other on the base station. The component on the base station is used to pump data from one part to other. The other component is similar in functionality and interface to the socket and make the communication transparent to the mobile host. But the TCP resulting does not perform well in wireless links. Furthermore, it violates the end-to-end semantics since the ACKs can reach the fixed sender even before the packets actually reach the mobile device. Therefore, it is not easily deployed in the Internet.

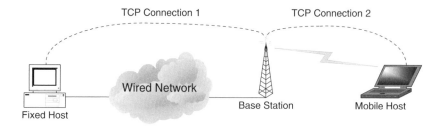

Figure 16.6   Splitting a TCP connection into two connections.

MTCP [360] is based on a very similar idea as in I-TCP. It protects the connection over the wired network from the impact of the short disconnection over the wireless link. The authors [360] proposes a session layer protocol called mobile host protocol (MHP) at the base station and the mobile host in order to compensate for the unreliability of the wireless link. Two implementations are suggested for the session layer. The first one uses TCP and the other uses selective repeat protocol (SRP), which is designed to recover the packet loss quickly over the wireless link.

Another scheme, called M-TCP [42], also splits the TCP connection but preserves the end-toend TCP semantics. This scheme is targeted for the frequent disconnection and low bandwidth. The TCP connection is also split at the base station. The three-layer hierarchical architecture is set up to avoid handoff. As shown in Figure 16.7, the mobile hosts are in the lowest layer, the base stations are in the second layer, and the supervisor hosts connected to the wired networks are at the top level of the hierarchy. The M-TCP protocol is mainly used at the wireless link. The base station adds two TCP clients: one for wired connection; one for wireless connection. When packets or ACKs reach the base station, they are delivered from one client to another. Then the packets are forwarded to the other end point. And a fixed bandwidths are dynamically assigned to the mobile hosts to reduce power consumption.

Another type of splitting TCP connection is called the Remote Socket Architecture (ReSoA) [314]. The main idea is to merge the TCP/IP stacks of mobile host into the base station and leave only user interface on mobile host as depicted in Figure 16.8. The connection between mobile host and base station is implemented by last hop protocol (LHP) which is a reliable protocol. In this way, base station acts as a proxy for the wireless end points.

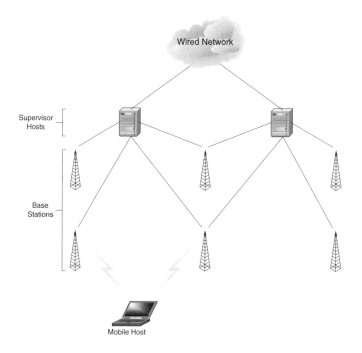

Figure 16.7 Three-level architecture of M-TCP.

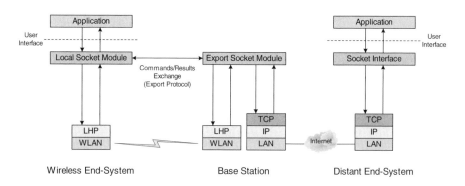

Figure 16.8 Wireless Internet access using ReSoA [314].

However, split-connection schemes must maintain the state of each TCP connection at the base station. As a result, a heavy overhead is incurred, resulting in a low efficiency.

16.6 END-TO-END APPROACHES

End-to-end approaches are designed based on the basic TCP protocol (i.e., Tahoe [164]) to handle packet losses. The first group is TCP Reno's variants, including NewReno, SACK and so on. The second group is based on the help of intermediate routes to predict the congestion and then inform the sender explicitly.

TCP Reno's Variants

Different loss recovery mechanisms are used in the TCP sender to improve the performance. Specifically, the TCP sender uses the cumulative ACKs to determine if packets have reached the receiver and provide reliability by retransmitting lost packets. Packet losses are detected by either three duplicate ACKs or absence of an ACK in a time-out interval. TCP-Tahoe [164] uses the fast retransmit method to recover any loss and enter slow start phase in which it increases the window exponentially.

Fast Recovery is implemented on top of Fast Retransmit in TCP-Reno [165, 327]. Fast Recovery can effectively estimate the amount of packets in flight while recovering from losses. The TCP sender could continue transmitting new packets if this number is below the estimation of the network capacity. The purpose is to preserve the ACK self-clock and avoid the time-out. Then, the congestion window size does not need to slow down dramatically to enter slow start phase, which will result in unnecessary throughput reduction. Instead, it stays in the congestion avoidance phase, where it increases window linearly.

In TCP-Reno, the Fast Recovery algorithm is conservative. Only the first loss packet in the congestion window can be recovered. This is because when an ACK of the first loss packet is received, it leaves Fast recovery phase. This prevents the sender from detecting the other losses through Fast Retransmit. So the other losses in the same congestion window cannot be found until after the long time-out. TCP-NewReno [122, 150] solves this problem by introducing partial ACK—the ACK which covers new data but not all the data outstanding when loss was detected. The objective of doing this is to detect multiple losses in the same congestion window. The TCP sender then does not wait for the duplicate ACK, but retransmits the loss packet immediately when partial ACK is received. The Fast Recovery will end until all packet losses are detected in the same congestion window.

Cumulative ACKs report the number of the packets that have been accumulated. But they do not give the sender much useful information about successfully received packets. If an ACK can carry information about the recently received packets, loss recovery mechanism can be started once receiving such an ACK. SACK (Selective Acknowledgments) [124, 236] are designed to provide such information. They are added as an option to TCP. This option can at most provide three non-contiguous blocks of data which have been received successfully by the receiver. Each block of data is described by its starting and ending sequence number. The earlier is the block, the later have the packets been received. That is, the SACK option always reports the most recently received packets. The redundant appearance of several SACKs can provide enough information about packet losses.

SMART [188] contains not only cumulative ACKs but also the sequence number of the packet which caused the receiver to generate the ACK. The key idea in this algorithm is to build the bitmask of correctly received packets at the sender, instead of carrying it in the ACK header. If the sender detects any gap in the bit-mask, it assumes that the missing packets have been lost and then retransmits the lost packets. However, the out-of-order packets would be misunderstood to be loss.

Explicit Notification

Traditionally, route queue length is set to be a maximum length for each queue. All incoming packets are queued in the router until reaching maximum length. After that, new incoming packets are dropped till old packets are sent out and queue length decreases. This is the well-known "drop tail", because the packet that arrived most recently is dropped when the queue is full. Reference [40] points out two drawbacks of this drop tail technique. The first one is lockout. In some situations, a single or a few connections can monopolize the queue size. The other drawback is full queues. Allowing full queues could result in maintaining queue full for a long time. Based on the above reasons, active queue management is needed. Random early drop (RED) [40, 123] is an active queue management algorithm, which detects incipient congestion by computing the average queue size (avg). RED drops packets depending on the relationship between two thresholds and the average queue size. The avg is a time-averaged length, not an instantaneous one. Besides RED, BLUE [117] is also an active queue management scheme. It aims to remedy the shortcomings of RED.

Floyd and Ramakrishnan [121, 292] first suggested the explicit congestion notification (ECN) mechanism based on RED queue management algorithm. The objective of ECN is to reduce packet drop in the router in order to avoid unnecessary congestion control mechanisms at the TCP sender. Specifically, when a router experience incipient congestion, one randomly selected packet in the queue is marked. The receiver accordingly sends a special ACK packet upon reception of the marked packet. The sender will then reduces the congestion window (which would have been increase) upon reception of the special ACK.

Liu and Jain [218] proposed a strategy called mark front, which can lead to a fast congestion feedback. In the mark front algorithm, a router will mark the packet, which is at the head of the queue, when incipient congestion is detected. Because such a packet is to be sent out first, the feedback time of the mark front strategy is considerably shorter than that of the mark tail method.

Hammann and Walrand [142] suggested an algorithm called New-ECN, which works by preventing a fast connection from opening its congestion window too quickly, but instead enabling a slow connection to open its window more aggressively.

When the marked packet is echoed to the sender, not only the congestion window size is decreased but also the rate of the congestion window increase is modified.

Peng *et al.* [275] proposed the W-ECN algorithm for the wireless environment. In this algorithm, upon noting the first lost packet due to buffer overflow in the router, the headers of passing packets in the queue are marked. In this manner, the sender knows that a packet loss happened due to network congestion but not because of random loss due to degraded channel quality.

Floyd proposed explicit loss notification (ELN) [120]. The explicit information is transmitted to the sender. And then the sender reacts by retransmitting the lost packet without reducing the congestion window. The most difficult task in ELN is to get the information about loss because the corrupted packets have been discarded before they reach TCP.

Bakshi *et al.* proposed explicit bad state notification (EBSN) [21]. Although local recovery by the base station could resolve the packet drop in the wireless link, timeout could happen and then trigger redundant retransmitted packets. The main objective of EBSN is to eliminate the timeout at the fixed host when base station is in the local error recovery. When the wireless link is in the bad state, the base station uses local recovery and the notification is sent to the fixed host to reset the timeout at the fixed host.

16.7 PRACTICAL ILLUSTRATIONS—WIRELESS WAN

On improving the TCP performance over wireless, most research is focused on mathematical analysis and simulations. In [56], Chakravorty *et al.* introduced a detailed experiment-based evaluation and comparison of optimization techniques in a commercial wireless wide-area networks (WWANs). They focused on the performance of web browsing applications in WWAN cellular environments. In their study, they implemented various optimization techniques across different layers (i.e., application, session, transport and link layers) of the protocol stack to investigate the cross-layer interactions of those techniques and their impact on the application performance. The key observation of their work is that even though TCP is relatively well-tuned to perform well in WWAN cellular environments, the performance of HTTP applications is not satisfactory because of the severe mismatch between TCP and default HTTP protocols. In addition, with suitable optimizations implemented in the application and session layers and the use of proxies, the end-user web-browsing experience can be enhanced by at least a factor of two.

Since the focus of this chapter is wireless TCP, the work of Chakravorty *et al.* [56] about performance optimization techniques in the transport layer will be discussed. For optimization techniques in other layers, interested reader is referred to [56].

The experimental testbed is shown in Figure 16.9, which is a GPRS-based WWAN network. Two nodes, SGSN (Serving GPRS Support Node) and GGSN (Gateway GPRS Service Node), are added to the GSM network to support GPRS. SGSN performs functions like signaling, cell selection, routing and handoff between different BSCs (Base Switching Centers). GGSN is the gateway between GPRS network and fixed IP routing network.

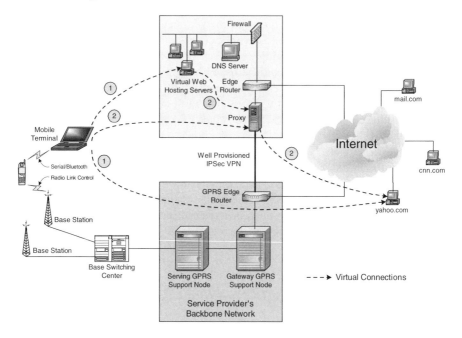

Figure 16.9   WWAN experimental testbed [56].

The network setup is as follows. A mobile terminal (MT) such as a laptop connects to the WWAN network via a GPRS phone. To use the WWAN network, the MT first attaches itself to the GGSN through a signaling procedure and establishes a point-to-point protocol (PPP) connection with the GGSN. Then, the MT successfully connects with the GPRS network, and is dynamically assigned an IP address. Finally, when the MT moves through the network, the WWAN network switches its data to the assigned IP address.

Since the contents of popular websites change frequently, different download attempts may show significant differences in the download content structure, and hence different response time. As such, the comparison of web download from different experiments in the relatively low-bandwidth WWAN network could be meaningless. In order to avoid this problem, in the WWAN experiments conducted in [56], a virtual web hosting system is implemented in a laboratory where the contents of the popular websites are replicated into a set of web servers with public domain names. This way, the MT accesses the virtually hosted web pages using WWAN networks as if they were accessed from actual servers, with the nice property that the access results can be repeated and reproduced.

As mentioned before, the main objective of the WWAN experiments is to evaluate the web-browsing performance over a WWAN network. The MT downloads web contents over the WWAN link via two different paths. The first one is to directly download from the real web servers or virtually hosted seb-servers. This case is shown by Label 1 in Figure 16.9. The second one is to download through a proxy, which is connected to the cellular provider's network via a well provisioned IPSec VPN. This case is shown by Label 2 in Figure 16.9.

In [56], the authors implemented a performance optimization scheme called TCP-WWAN [57] in the transport layer to study the impact of the transport layer to the web-browsing performance. TCP-WWAN is a transparent-proxy solution and uses a 'transparent' proxy located in the cellular provider's network. It has two characteristics in the WWAN network. Firstly, it uses a pre-determined value of bandwidth-delay product to replace the TCP slow start. In addition, it performs aggressive recovery during packet loss and link stalls. Such kind of aggressive behavior makes TCP-WWAN only implemented in the cellular network where appropriate bandwidth sharing mechanism are implemented. Secondly, TCP-WWAN estimates available bandwidth on the wireless link so as to regulate the flow of TCP packets to the MT. Thus, TCP-WWAN can reduce the queue size and prevent spurious timeout, which on the other hand avoid the occurrence of wireless link stalls.

The evaluation of TCP-WWAN was performed using Mozilla browser version 1.4. Instead of using the default setting (HTTP:/1.1, two persistent TCP connections), the browser was modified to use six persistent TCP connections. A 1.4 GHz laptop with Linux installed runs the Mozilla browser, and connects with WWAN GPRS network by using a '3 + 1' GPRS phone. The WWAN operates in the 1.8–1.9 GHz band and uses CS-2 FEC coding scheme. In all experiments, the MT was kept stationary to avoid link variations.

Experiment results show that TCP-WWAN can achieve between 5–13% additional benefit in the browsing of various popular websites. This performance improvement is significant for web-browsing over WWAN links. However, as discussed in [56], the performance improvement can be even more significant when optimizations are done in the application and session layers (about 48–61% improvements), which is not the scope of this chapter.

16.8 SUMMARY

In this chapter, we have discussed the reasons for TCP performance degradation in wireless networks. Particularly, we talk about the reason why the congestion control mechanism can harm the TCP performance significantly in wireless environments. Then, we have introduced a number of representative adaptive TCP algorithms in different wireless environments. These schemes are classified into three groups: link layer, split connection and end-to-end. Finally, as a practical illustration, we have talked about a study of enhancing TCP performance for web-browsing in a commercial wireless wide-area network (WWAN) testbed.

PROBLEMS

16.1   What are the reasons which lead to the TCP performance degradation in wireless networks?

16.2   What are the design objectives of link layer approaches on improving TCP performance in wireless environments?

16.3   For web-browsing in an environment like wireless wide-area network (WWAN), if a web page consists of numerous HTTP objects of small sizes and TCP Reno is adopted, how will be the web-browsing performance? And why?

PART V

WIRELESS RESOURCES MANAGEMENT

CHAPTER 17

WIRELESS PACKET SCHEDULING

Fair as the moon.

—Solomon's Song VI, 10

17.1 INTRODUCTION

In many situations, the wireless medium is a resource shared among all wireless devices (e.g., in a WLAN). Thus, allocation of this shared resource to the users is an important component in the system. Specifically, such a scheduling mechanism is needed for both the uplink (from the devices to the base station or access point) and the downlink (from the base station or access point to the devices) directions.

Interoperability:

Existing wireless packet scheduling algorithms can be classified according to the fairness notions they employ. The key difference is the scheduling set induced by the channel model used. In algorithms with greedy fairness notions, the scheduling set includes only the backlogged flows with "good" channel state. In algorithms with

Wireless Internet and Mobile Computing. By Yu-Kwong Ricky Kwok and Vincent K. N. Lau
Copyright © 2007 John Wiley & Sons, Inc.

realistic fairness notions, it includes all the backlogged flows except the ones with the worst channel state (i.e., cannot transmit any packet). Greedy fairness notions simplify the wireless packet scheduling problem in that the scheduling set is homogeneous with respect to channel state. However, greedy fairness notions are unfair to the flows with channel states between the worst and the best. On the other hand, when realistic fairness notions are used, the scheduling set is larger, but the channel states of the flows in the scheduling set are not the same. This makes the design of fairness notions more complicated.

Performance:
Performance of a scheduler can be characterized by a multitude of metrics. The most widely used metrics are throughput (rate) and fairness. Throughput is obviously important in supporting users' QoS requirements. For instance, many applications required a minimum level of throughput support in order to be viable. By contrast, fairness is a much more subtle issue. First of all, it is difficult to define a universally accepted fairness metric. Indeed, there are many different fairness metrics suggested in the literature, such as proportional fairness, max-min fairness, etc. Secondly, as we will see, fairness is inherently in conflict with traditional metrics such as throughput. Specifically, in order to support a certain level of fairness, the overall system throughput might be reduced. Designing an integrated metric combining throughput and fairness is still a hot research topic. Other useful metrics for evaluating a scheduler include delay, overhead, efficiency, etc.

17.2   THE SCHEDULING PROBLEM

Packet scheduling algorithms have been extensively studied in wired networks. In a packet-switching network, packets coming from different connections arrive at the switching nodes (i.e., the routers) and wait to be delivered. Since each switching node can only deliver a limited number of packets at a time, a certain packet scheduling algorithm is needed at the switching node to decide the order in which the packets are served. On making the scheduling decision, the switching node may need to maintain some QoS guarantees (e.g., bandwidth, delay, fairness) for different connections from different end hosts.

Similar to the scenario in wired networks, a scheduling algorithm is needed at the base station to decide which mobile terminals will be served. For example, in a cellular phone or wireless LAN network, typically there is a central machine (i.e., the base station or the access point), which serves the mobile terminals with heterogeneous traffic demands (e.g., media stream data such as video or voice, and file data). To minimize collisions and delays of data transmissions of the mobile terminals, some coordination schemes, known as MAC (multiple access control) protocols, are employed. A scheduling policy is usually implicitly or explicitly incorporated in a MAC protocol. The goal of the scheduling algorithm is to judiciously handle the MAC problem for the shared wireless communication medium (specifically the uplink—the link from the mobile terminals to the base station/access point, which is the major resource contended by the users) such that the utility of the medium is maximized while satisfying users' demands.

However, scheduling algorithms developed for the wired networks are not applicable to the wireless environment because of the inherently different characteristics of the wireless channel.

The Wireless Channel:

In contrast to the transmission quality of a link in a wired network, which is stationary after it is set, the wireless channel condition is continually affected by short-term fading, long-term shadowing, and depends on the mobility of mobile terminals, surrounding obstacles like buildings, mountains, etc. As these effects are location-dependent and time-varying, the signal strength of the wireless channel can vary widely in amplitude and phase. Thus, only a subset of the waiting flows (commonly called the backlogged flows), which is experiencing good channel conditions, is eligible for consideration by the scheduler. Moreover, this subset is time-varying. As illustrated in Figure 17.1, MT (mobile terminal) A is having a comparatively poorer channel state because it is far away from the base station (BS) and it encounters deep fading. On the other hand, MT B may have a better channel state as it is much nearer to the BS. MT D may have very poor channel state as its signal is obstructed by a large building. Thus, it may turn out that only MTs A, B and C can be scheduled, and they have different channel states.

To simplify the scheduling problem significantly, a two-state Markov chain channel model is commonly used [33, 52, 227]. Under this channel model, the wireless channel has just two states—either "good" or "bad". In a "good" channel state, a flow can transmit using full bandwidth, while in a "bad" channel state, a flow cannot transmit any packet. The model is depicted in Figure 17.2 As shown in the figure, P_g is the probability that the channel state in the next time slot is "good" given that it is "bad" in the current time slot. Likewise, P_e is the probability that the channel state in the next time slot is "bad" given that it is "good" in the current time slot.

The major drawback of the two-state Markov chain channel model is that it is unrealistic because packets can still be transmitted even if the channel is not in the perfect state (the "good" state). In reality, using some channel-adaptive technique like ABICM (Adaptive Bit-by-Bit Interleaved Channel Modulation) [198], the transmitter/receiver in a wireless network can exploit the time-varying nature of the channel and accordingly adjust the effective throughput. Specifically, when the channel state is not good due to multipath fading or shadowing (channel state can be determined by checking the pilot symbols in a feedback channel or reverse link), the amount of protection can be readjusted by choosing a different channel coding and modulation mode [198]. Thus, even in a so-called "bad' channel state, an MT can in fact transmit data and realize a possibly lower effective throughput, instead of being totally unable to transmit.

Fairness:

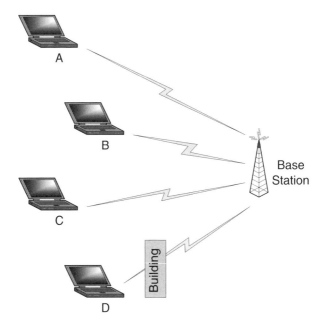

Figure 17.1 A wireless packet scheduling scenario.

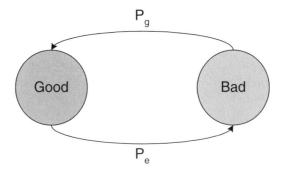

Figure 17.2 The two-state Markov chain channel model.

In a wired network, maintaining fairness and optimizing overall throughput (i.e., optimizing bandwidth utilization) are not conflicting goals. However, in a wireless network, if optimizing throughput is the only principal goal, an ideal method is to schedule among the backlogged flows with the decreasing order of their channel states. For example, consider two backlogged flows i and j, where i belongs to the set of flows with best channel state and j belongs to the set of flows with channel states between the best and worst. If the scheduler always gives the highest priority to the flow with the best channel state, flow i will be served continuously while flow j will be starved. This scheduling philosophy can maximize the overall throughput, but it is obviously unfair.

On the other hand, if fairness is the dominating goal such that it must be maintained at any time instant, an MT should be scheduled to transmit even when it encounters a very bad channel state. In the previous example, flow j will also be scheduled to transmit. Inevitably, such a policy degrades the overall effective throughput since it schedules the flows without the best channel state to transmit. Indeed, if we employ the classical fairness notions in the wired networks (defined with respect to resources only, without regard to utilization) and try to guarantee the same effective throughput between the two flows, and suppose that a flow j happens to have a very bad channel state, then the system will give much more time slots to flow j in order to let it have the same effective throughput as achieved by flow i. It is unfair to flow i because the system allocates much more time slots to flow j with such a poorer channel state. If those time slots were allocated to flow i, the system would have more packets transmitted.

In order to maintain fairness among all flows, when there are backlogged flows with the best channel state, the scheduler should still consider the flows with channel states between the best and the worst to transmit even though such poor flows will realize a lower bandwidth utilization. This is the reason why keeping fairness and maximizing throughput are often conflicting goals. To tackle this challenging scheduling problem, we need to design a fairness notion such that the following two goals are achieved: 1.) flows with different channel states should not be treated as the same; and 2.) the throughput should not be degraded desperately.

17.3 SYSTEM MODEL

Figure 17.3 shows a single cell in a cellular phone network. The cell is managed by a base station (BS), which supports integrated video, voice, and file data services to mobile terminals (MTs) within the cell. Transmissions in the cell involve both the uplink (from a MT to the BS) and the downlink (from the BS to a MT). A scheduling algorithm is running at the BS, responsible for making scheduling decisions for both the uplink and the downlink traffic. In later discussion, we consider the scheduling of packets in a single cell only.

Multiple Access Scenario

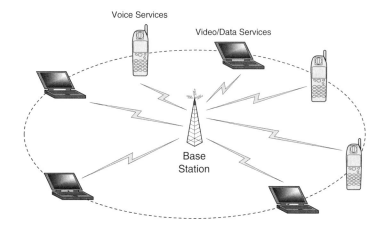

Figure 17.3   Packet scheduling in a cell.

Figure 17.4 shows the multiple access scenario in a wireless environment mentioned before. The system resource at the BS is limited. Different MTs contend for the shared channel, and signals from the admitted MTs are multiplexed and transmitted together by the BS. If $S(t)_i$ stands for the signals from MT i, then the overall received signal at the BS is:

$$y(t) = \sum_{i=1}^{N} a_i \cdot S(t)_i + n(t), \qquad (17.1)$$

where a_i is the channel fading associated with the MT i, $n(t)$ is the thermal white Gaussian noise.

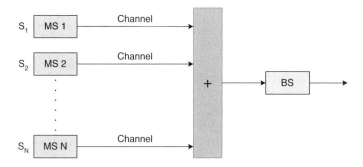

Figure 17.4   Multiple access scenario.

In order to ensure each intended receiver to extract its signal from the received $y(t)$ with multiplexed signals, given a frequency spectrum, the system resource must be divided into orthogonal channels. Each MT is assigned one or more orthogonal channels to transmit its packets, then the intended receiver can identify its signal because its channel is orthogonal to the others.

In general, there are three popular methods to separate the system resource into orthogonal channels: TDMA (Time Division Multiple Access), FDMA (Frequency Division Multiple Access), and CDMA (Code Division Multiple Access). A TDMA scheme divides the time line into slots. MTs can transmit data using the whole bandwidth in their scheduled slots. Different MTs occupy different time slots. A FDMA scheme partitions the whole frequency spectrum into channels. Each MT communicates on one certain channel with bandwidth throughout the communication, and the frequency channel is only reserved for this particular MT and no one else can share it. A CDMA scheme employs another dimension—spreading codes used in spread spectrum communications. Since the codes are orthogonal, channels encoded with different codes are separated from each other. Each MT can transmit using the whole bandwidth. In this chapter, we will mainly focus on discussing packet scheduling in TDMA based networks.

Figure 17.5 shows the TDMA frame structures for the uplink and the downlink in a cellular environment depicted in Figure 17.3. In the uplink, a frame is divided into three subframes. They are the request subframe, information subframe, and the pilot symbol subframe (note that pilot symbols are known reference symbols). Specifically, there are mini-slots in the request subframe for voice requests reservation and data requests contention. Note that a data request is not allowed to make reservation in the sense that even if a data request successfully seizes an opportunity to transmit in the current frame, it has to contend again in the next frame if it has some more data to send. There are information slots in the information subframe for the transmission of voice or data packets. Finally, there are slots in the pilot symbol subframe. A downlink frame is similarly partitioned into four subframes, namely the acknowledgment subframe, poll-for-CSI (Channel State Information) subframe, information subframe, and announcement subframe. The frame duration is 2.5 msec. Such a short frame duration has the advantage of having shorter delay and being practicable in wideband systems.

In TDMA based networks, the BS can transmit one packet at a time. On the contrary, in CDMA based networks, the BS can generally transmit more than one packet at a time because the system capacity is interference limited. As a TDMA based network can only support one MT at a time (at a particular time slot), wireless packet scheduling problem in TDMA based networks can be considered as a single-server problem. Figure 17.6 illustrates a packet scheduling scenario in a TDMA based network.

Due to the short propagation delay in nowadays wireless packet data networks (e.g., cellular networks or wireless LANs), we can normally assume that the MTs can almost immediately know the request result. This assumption can be justified

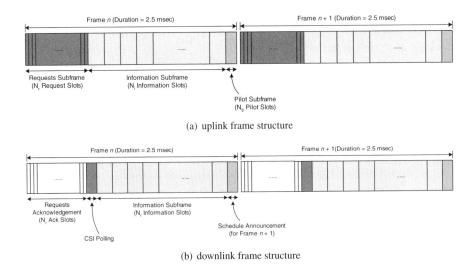

(a) uplink frame structure

(b) downlink frame structure

Figure 17.5 Uplink and downlink frame structure.

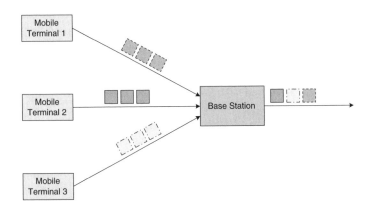

Figure 17.6 Packet scheduling in a TDMA based network.

by the following arguments. First of all, the delay involved from the time when the MTs send the requests to the BS to the time that the MTs realize the result of the contention (by checking the announcement on the downlink) is composed of two parts, namely the processing delay at the BS and the round trip propagation delay between the MTs and the BS. For high speed wireless access systems, the propagation delay is on the order of second (for a 30m separation), which corresponds to less than a mini-slot. The processing time required at the BS is also rather short because at the time of sending the acknowledgment to the MT, the BS does not need to make any decision on resource allocation yet. Thus, the request packet processing delay will be negligible (much shorter than a mini-slot). Considering the worst case, the delay of the acknowledgment is less than two mini-slots. Thus, after transmitting the request to the BS, the MT simply listens for subsequent downlink acknowledgment mini-slots (not necessarily the immediately following acknowledgment slot) with a time-out value (e.g., 5 mini-slots). The MT will retry sending another request when the timer expires.

Source Models:

Video and voice packets are assumed to be delay sensitive while data packets are assumed to be delay insensitive. Thus, voice packets are labeled with deadlines. A voice packet will be dropped by the MT if the deadline expires before being transmitted. Such packet dropping has to be controlled to be within a certain limit (e.g., below 1% as indicated in [137]) in order that the quality of service to the voice MTs is still acceptable. The source and contention models are summarized below.

Voice Source Model: The voice source is assumed to be continuously toggling between the talkspurt and silence states. The duration of a talkspurt and a silence period are assumed to be exponentially distributed with means t_t and t_s seconds, respectively (as indicated by the empirical study in [137], $t_t = 1$, and $t_s = 1.35$). It is assumed that a talkspurt and a silence period start only at a frame boundary.

Video Source Model: We use video teleconferencing as an example video source. Adapting the model used in [197], the number of packets per video frame period (i.e., 40 msec for a 25 fps frame rate) is governed by the DAR(1) model, which is a Markov chain characterized by three parameters: the mean, the variance, and ρ. The transition matrix is computed as: $P = \rho I + (1 - \rho)Q$, where ρ is the autocorrelation coefficient and I is the identity matrix. Furthermore, each row of Q is identical and consists of the negative binomial probabilities: $(f_0, ..., f_K, F_k)$, where $F_k = \sum_{k<K} f_k$, and K is the peak rate. Similar to a voice source, a video source can only tolerate a 1% packet loss rate [197].

Data Source Model: The arrival time of file data generated by an MT is assumed to be exponentially distributed with mean equal to one second. The data size, in terms of number of packets, is also assumed to be exponentially distributed with mean equal to 100 packets. It is assumed that packets arrive at a frame boundary.

Request Contention Model: In a TDMA system, to avoid excessive collisions, even if a voice or data request has some packet waiting to be sent, the MT will attempt to send a request at a request mini-slot only with a certain permission probability.

In a TDMA system, an MT entering a new voice talkspurt or generating a new stream of data packets transmits an appropriate request packet in one of the request slots of the next frame. If there is more than one packet transmitted in the same request slot, collision occurs and none of the requests will be correctly received. At the end of each request slot, the successful or unsuccessful request will be identified and broadcast by the BS. An unsuccessful MT (does not receive the acknowledgment announcement in the downlink frame) can retry in the next request slot. On the other hand, a successful MT then transmits its information packets in the corresponding information slot in the current frame.

Channel Model:

Wireless channel capacities are time-varying and location-dependent. Modeling the wireless channel has historically been one of the most difficult parts of wireless system design.

Let $c(t)$ be the combined channel fading which is given by:

$$c(t) = c_l(t)c_s(t), \tag{17.2}$$

where $c_l(t)$ and $c_s(t)$ are the long-term and short-term fading components, respectively. Both $c_l(t)$ and $c_s(t)$ are random processes with a coherence time (time separation between two uncorrelated fading samples) on the order of a few milliseconds and seconds, respectively.

Short-Term Fading: Without loss of generality, we assume $E[c_s(t)^2] = 1$ where $E[x]$ denotes the expected value of a random variable x. The probability distribution of $c_s(t)$ follows the Rayleigh distribution given by:

$$f_{c_s} = c_s \exp(-\frac{c_s^2}{2}). \tag{17.3}$$

We assume that the mean and maximum speeds of the MT are 50 km/hr and 80 km/hr, respectively. Thus, the Doppler spread [32] is given by $f_d \approx 100 Hz$. It follows that the coherence time, denoted by T_c, is approximately given by:

$$T_c = \frac{1}{f_d}, \tag{17.4}$$

which is about ten milliseconds.

Long-Term Fading: The long-term fading component, $c_l(t)$, is also referred to as the local mean [32] which, as shown by field test measurement, obeys the log-normal distribution:

$$f_{c_l}(c_l) = \frac{4.34}{\sqrt{2\pi}\sigma_l c_l} \exp(-\frac{(c_l(dB) - m_l)^2}{2\sigma_l^2}), \qquad (17.5)$$

where m_l and σ_l are respectively the mean (in dB) and the variance of the log-normal distribution, i.e., $c_l(dB) = 20log(c_l)$. Since $c_l(t)$ is caused by terrain configuration and obstacles, the fluctuation is over a much longer time scale. Again, from field test results, the order of time span for $c_l(t)$ is about one second. Since MTs are scattered geographically across the cell and are moving independently of each other, we assume the channel fading experienced by each MT is independent of each other.

For error protection, redundancy is incorporated into the information packet. To exploit the time-varying nature of the wireless channel, a variable-throughput channel-adaptive physical layer [198] as illustrated in Figure 17.7 can be employed. CSI (Channel State Information), $c(t)$, which is estimated at the receiver, is fed back to the transmitter via a low-capacity feedback channel. Based on the CSI, the level of redundancy and the modulation constellation applied to the information packets are adjusted accordingly by choosing a suitable transmission mode. Thus, the instantaneous throughput is varied according to the instantaneous channel state. The essence of incorporating a channel-adaptive physical layer in the MAC protocol is to capture the notion of multiple channel quality levels such that the scheduler in the MAC protocol can exploit the scheduling diversity to optimize overall system performance. Besides using the above channel-adaptive technique, other channel-adaptive transmission models/techniques [356] can also be employed.

We assume the coherence time of the short-term fading is around ten milliseconds, which is much longer than an information slot duration (which is 2.5 milliseconds). Thus, the CSI remains approximately constant within at least two frames and it follows that the transmission mode for the whole frame is determined only by the current CSI level. Most importantly, scheduling decisions made at the beginning of a frame are based on accurate channel state information about the frame yet to be transmitted. Specifically, transmission mode q is chosen if the feedback CSI, $\hat{c}$, falls within the adaptation thresholds, (ξ_{q-1}, ξ_q). In the case of ABICM, the operation and the performance of the scheme is determined by the set of adaptation thresholds, $\{\xi_0, \xi_1, ...\}$. If the ABICM scheme is operated in the constant BER (bit-error rate) mode, then the adaptation thresholds are set optimally to maintain a target transmission error level over a range of CSI values. When the channel state is good, a higher mode could be used and the system enjoys a higher throughput. On the other hand, when the channel state is poor, a lower mode is used to maintain the target error level at the expense of a lower transmission throughput. Note that when the channel state is very bad (e.g., a very low SINR is perceived), the adaptation range of the ABICM scheme can be exceeded such that the throughput (mode-0) becomes so low, making it impossible to maintain the targeted BER level, and thus, the mobile terminal (or the flow) cannot transmit any packet.

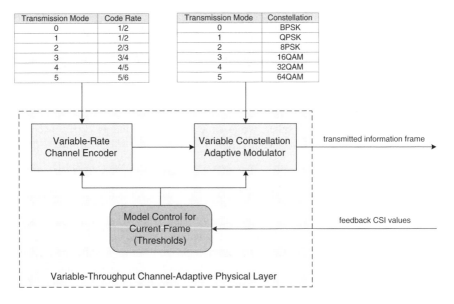

Transmission Mode	Code Rate
0	1/2
1	1/2
2	2/3
3	3/4
4	4/5
5	5/6

Transmission Mode	Constellation
0	BPSK
1	QPSK
2	8PSK
3	16QAM
4	32QAM
5	64QAM

Figure 17.7   A conceptual block diagram of the variable-throughput channel-adaptive physical layer [198].

Given the above considerations about the channel state, the instantaneous throughput offered to the access control layer, denoted by ρ, is also variable and is therefore conceptually a function of the CSI, $c(t)$, and the target BER, P_b, denoted by $\rho = f_\rho(c(t), P_b)$.

17.4   FAIRNESS NOTIONS

In a wired network, all the backlogged flows constitute the scheduling set. The scheduler tries to allocate the bandwidth among all the backlogged flows proportionally to their preallocated service shares. Whenever the system allocates some bandwidth to a certain flow, it will achieve a certain throughput accordingly. This is because in the wired line transmission, the efficiency (the bandwidth utilization) of using the channel is constant among all flows.

However, in the wireless environment, the backlogged flows have different channel states. There are flows which can make full use of the bandwidth, flows which cannot transmit any packet, and flows with channel states in between. The scheduling set is different from that in the wired network's case. For example, the flows which cannot transmit any packet should not be scheduled even they are backlogged.

In a wireless environment, the scheduling set is a subset of a typical scheduling set in a wired network. Moreover, this subset is also variable over time because the channel states vary as time goes by. However, the scheduling set in the wired networks includes all the backlogged flows, and thus, if the wired fairness notions are employed, a scheduler will try to allocate the same throughput level to all the backlogged flows. However, in wireless networks, for the flows which can transmit no packet, the system will simply keep on wasting the resource in vain and possibly starve the other flows. This is inefficient from a resource utilization point of view because those flows suffering from deep fading (i.e., channel quality is not good—with a very low SINR) will not be able to utilize the time slots efficiently (e.g., data loss may occur more frequently). A more intelligent method is to allow the flows having better channel states to proceed first. This is the motivation for designing fairness notions that are adaptive to the wireless environment.

Greedy and Realistic Fairness Notions:

There are several existing fairness notions for the wireless environment, and they can be classified into two categories:

1. For fairness notions which are designed with the assumption of the two-state Markov chain channel model, they are regarded as greedy fairness notions because the scheduler will only schedule the flows with "good" channel state to transmit so that it can make use of the channel greedily;

2. For fairness notions which are designed with the assumption of a realistic channel model, such as the one mentioned in Section 17.3, they are regarded as realistic fairness notions because they are more useful in a realistic wireless network.

As for the greedy fairness notions, the scheduling set includes only the backlogged flows with the best channel state. The common drawbacks of the greedy fairness notions are:

- they are unfair to the flows with channel states between the worst and the best; and

- they are not realistic—flows can spend most of the time in channel states between the best and the worst, and they receive no service, while flows with the best channel state will not relinquish their service to help these unlucky flows catch up.

In the realistic fairness notions, the scheduling set includes all the backlogged flows except those which cannot transmit any packet. Thus, the scheduling set is larger, and more realistic and fair. But the channel states of the flows in the scheduling set are not the same. This makes the design of fairness notions more complicated. For example, we need to decide if flows with different channel states should receive the same normalized amount of time slots or realized throughput. Note that the system will need to allocate much more time slots to the flow with a bad channel state to achieve the same amount of throughput with the flows with a good channel state.

Effort and Outcome Fair:

In a broad sense, fairness can be defined with respect to two aspects: effort and outcome [375]. Intuitively, a policy is called effort fair if the allocation of services to different flows is fair, without regard to the actual amount of data successfully delivered by the flows using the allocated services. Informally, "fair" means a flow gets the service amount that it deserves to get. On the other hand, a policy is called outcome fair if the actual realized data throughput among the flows is fair.

Effort Fair: A scheduler is fair if the system resource (e.g., the bandwidth or time slots in TDMA based network) the system allocates to different flows is proportional to the different service shares of the MTs. Mathematically, that means the difference between the normalized services the system allocates to any two backlogged flows i and j is bounded as follows:

$$\left| \frac{S_i(t_1, t_2)}{r_i} - \frac{S_j(t_1, t_2)}{r_j} \right| < \varepsilon, \tag{17.6}$$

where $S_i(t_1, t_2)$ denotes the allocated service of a certain flow i during time interval $(t_1, t_2]$, r_i is the requested service share, and ε is a finite constant. Such a fair scheduler can be considered as effort fair [293] in the sense that the scheduler only guarantees the effort expended on the flows is fair, without regard to the actual throughput achieved by the different flows.

Outcome Fair: A scheduler is fair if the difference between the normalized amount of realized throughput of any two flows i and j is bounded as follows:

$$\left| \frac{T_i(t_1, t_2)}{r_i} - \frac{T_j(t_1, t_2)}{r_j} \right| < \varepsilon, \tag{17.7}$$

where $T_i(t_1, t_2)$ denotes the actual throughput a flow i achieves during the time interval $(t_1, t_2]$. Such a fair scheduler can be considered as outcome fair [52] in that the scheduler tries to provide a fair actual performance achieved by the flows (rather that the "nominal" performance as in the effort fair definition discussed above).

In a TDMA based network, "effort" refers to the allocated bandwidth (time slots) and "outcome" refers to the realized throughput using the allocated time slots. Note that a variable effective throughput is manifested by the fact that some data may be lost due to poor channel states and thus, inducing retransmissions; or in adaptive FEC schemes such as ABICM, the amount of data protection varies according to the channel states.

Outcome fair and effort fair are equivalent in wired networks. However, in wireless networks, more error protection will be added and the realized throughput will be reduced when a flow does not have a perfect channel state, which results in a discrepancy between "effort fair" and "outcome fair". In other words, there is much difference between allocating the same amount of time slots among MTs and realizing the same amount of throughput among MTs. The question remained is what kind of fairness should be maintained among the MTs with the consideration of their channel states.

The efficiency of using the bandwidth should be considered when trying to allocate time slots "fairly". MTs with different channel states should not be treated as the same (i.e., neither effort fair nor outcome fair should be maintained among MTs with different channel states) because they have different efficiency of using the bandwidth allocated. That is the reason why it is challenging to have a reasonable fairness notion which can be applied in the scheduling set with different channel states, while it greatly simplifies the scheduling problem when the scheduling set includes only the flows with "good" channel state as in greedy fairness notions.

Now, there is a dilemma: 1.) we can make use of the bandwidth greedily to improve efficiency of the system by giving the flow with the best channel state the highest priority but this is unfair to those flows which spend most of the time having poorer channel states; or 2.) we can help to maintain outcome fair at the expense of wasting the bandwidth but this is unfair to those flows with better channel states because they are allocated less time slots.

Another aspect worth noting is that a reasonable fairness notion should not lead to severe degradation of the overall throughput of the system. That is the reason why outcome fair is inapplicable. Indeed, if outcome fair is strictly enforced, some MTs with especially bad channel state will occupy a lot of time slots and have a very low effective throughput. This conflicts with the task of the scheduling algorithm of making use of the system resource intelligently (a system side performance metric) while keeping the fairness property (a user side performance metric).

Comparison of Fairness Notions:

Figure 17.8 shows the classification (according to the fairness notions employed) of a number of existing wireless packet scheduling algorithms: WPS (Wireless Packet Scheduling) [227], IWFQ (Idealized Wireless Fair Queueing Algorithm) [227], CIF-Q (Channel-Condition Independent Fair Queueing) [257], SBFA (Server Based Fairness Algorithm) [293], WFS (Wireless Fair Service) [254], CS-WFQ (Channel State Independent Wireless Fair Queueing) [215], ELF (Effort Limited Fairness) [100], PF (Proportional Fairness) [168], and CAFQ (Channel Adaptive Fair Queueing) [344]. The details of the algorithms are described in Section 17.5.

Table 17.1 gives a qualitative comparison of different fairness notions employed in scheduling algorithms mentioned previously.

In CIF, the need to maximize the total throughput never gives in. Thus, "swap" or "borrowing" takes place to use the channel effectively at the expense of losing the instant fairness among the MTs. But in the long term, outcome fair is achieved. Bias

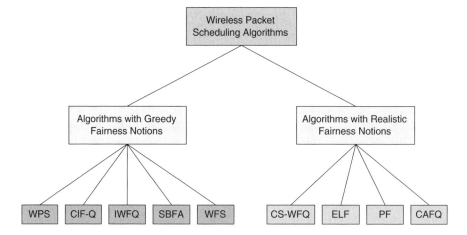

Figure 17.8 Classification of existing wireless packet scheduling algorithms.

Table 17.1 Comparison of different fairness notions.

Fairness	Short-Term	Long-Term
CIF	Short-term fairness is maintained only among the backlogged flows with perfect channel states; neither outcome fair nor effort fair is considered for flows with "not so good" channel states.	Outcome fair
ELF	Outcome fair is maintained among the backlogged flows with channel states better than their predefined thresholds.	Not precisely defined
PF	Short-term fairness is not precisely maintained.	Not precisely defined
CAF	Short-term fairness (normalized by channel states) is maintained among all the backlogged flows except the ones with the worst channel state. A balanced consideration of fairness and system throughput is achieved.	Outcome fair

against the flows with non-perfect channel states exists, and the fairness notion is in favor of the flows with good channel state and maximizing the throughput.

In ELF, the compromise is made between maximizing the throughput and keeping fairness by setting thresholds. When the channel states are within the threshold, output fair is kept at the expense of wasting the bandwidth. There is no any endeavor made to save the throughput of the system unless the channel state is poorer than the threshold. And the worst feature is: there is no action taken to maintain fairness in the long run.

In PF, the adjustment between the two goals is made by putting the channel state element on the numerator and the realized throughput element on the denominator in the priority calculation. However, the fairness notion is not strict and there exists pathological situations.

In CAF, the balance between the two needs can be tuned by a punish factor at any time. The punish factor can help to decide between when to make use of the bandwidth more efficiently and when to treat every flow more "fairly". When the punish factor is set to be very large, CAF will be more like the greedy fairness notions. Flows with bad channel states have little chance to access the bandwidth. Then the system can maximize the throughput. When the punish factor is -1, CAF will behave in the same manner as ELF when the channel state is within the threshold. The system tries to keep output fairness among MTs at the expense of degrading the throughput. For a punish factor larger than -1, the punish factor enables the system to schedule the flows with bad channel states to keep fairness at the same time of punishing them and saving the throughput.

17.5 FAIR QUEUEING APPROACHES

In this section, we discuss several existing packet scheduling algorithms: CIF-Q [257], ELF [100], PF [168], and CAFQ [344].

CIF-Q (Channel-Condition Independent Fair Queueing):

In [257], Ng *et al.* define the notion of Channel-Condition Independent Fair (CIF) such that the scheduler can:

1. provide delay and throughput guarantees for the error-free flows;

2. provide long term fairness: a lagging and continuously backlogged flow which enjoys an error-free channel after some time will be guaranteed to catch up within a certain time bound;

3. provide short-term fairness: the difference between the normalized services received by two error-free flows which are in the same state (lagging, leading, or satisfied/in-sync) during a time interval is bounded; and

4. provide graceful degradation for the leading flows: a leading error-free flow will only give up part of the service it would receive in the companion simulated error-free system to compensate the other lagging flows (i.e., compensation is done in a gradual manner; in some other algorithms [227, 254], once a lagging flow returns to a good channel state, it may get back all its missing services and hence effectively shut down the leading flows).

CIF can be useful in a wireless environment as it takes the non-perfect channel state (i.e., burst errors occur) into consideration. It allows the scheduler to delay the service of a flow if it does not have a good channel state, and the scheduler will try to help this flow to catch up in the long run. However, because the CIF-Q scheduler does not allow a flow having "intermediate" (not so good and not so bad) channel states to transmit, the scheduling set contains only flows with "good" channel state. As we have emphasized earlier, this scheduling philosophy is unsuitable in a practical environment, where we can expect that a significant portion of flows will have intermediate channel states because experiencing moderate levels of burst errors (that can be corrected by appropriate FEC's) is the norm rather than the exception in a wireless network.

The general drawback in CIF-Q is that in the short term, where the system gives overwhelming priority to the flows with good channel state, while all the other error flows (with possibly different channel states that are only slightly worse than the perfect channel state) are just treated as the same: not allowed to transmit. This is unfair among flows and, most importantly, may unduly increase the average delay of many flows. This is unacceptable for supporting isochronous media data.

In fact, a lagging flow can only be guaranteed to catch up when it has an error-free channel afterwards. However, in reality, the time period during which a flow can make full use of the bandwidth is limited. In other words, a flow can spend most of the time in channel states between the perfect and the worst. An efficient scheduling algorithm should provide fairness and performance guarantee even while the channel states of the flows are varying among different quality levels. These drawbacks in CIF-Q can also be found in other scheduling algorithms with greedy fairness notions.

ELF (Effort Limited Fair):

In ELF scheduling [100], it sets the weight of each flow dynamically. By adjusting the weight of each flow according to its channel state, ELF schedules the utilization of time slots among all the backlogged flows except the ones with the worst channel state. The poorer the channel state that a flow perceives, the higher the weight is, and the more time slots the flow is allocated. Thus, ELF scheduling tries to maintain outcome fair among flows with channel states better than their pre-set thresholds. When a flow has a channel state even poorer than its threshold, the system will not go on increasing its weight. Instead, it will be allocated the same weight as when it has a channel state right at the threshold, and thus, the system will not increase the portion of time slots the poor flow occupies. Therefore, there is an effort-limit for each flow, and outcome fair is maintained for flows with channel states better than their threshold. Within the limit, the poorer the channel state a flow has, the more effort is exerted on this flow. But the effort is limited, so no more effort will be paid beyond this limit.

A power factor P_i is used as an administrative control parameter to decide the channel state threshold, and the weight bound. It is used to adjust the weight as follows:

$$\bar{W}_i = \min(\frac{W_i}{1 - E_i}, P_i \cdot W_i), \qquad (17.8)$$

where $\bar{W}_i$ stands for the adjusted weight of flow i, W_i is the original pre-allocated weight of flow i, E_i is the error rate of flow i (e.g., a flow with 30% error rate can make use of 70% of the available bandwidth). For example, if the power factor is set to be 3, then when the error rate is 50%, the new adjusted weight is $2 \cdot W_i$; when the error rate is 66.7%, the new adjusted weight is $3 \cdot W_i$. Thus, as the flow changes from a better channel state to a poorer one, the adjusted weighted will increase so that this flow will be allocated more time slots to achieve outcome fair. However, when the error rate changes from 66.7% to 75%, the weight is bounded by $3 \cdot W_i$ and will not increase. The power factor helps to decide the channel state threshold below which the system will not increase the flow's weight more. The error rate threshold is: $E_i = 1 - \frac{1}{P_i}$.

The power factor is used to separate the outcome region and effort region, or to decide the "effort limit". With a realistic channel model, flows with different channel states will have different efficiency of using the bandwidth. All the flows with channel states better than the worst case have chance to transmit packets. This is more reasonable than the algorithms with greedy fairness notions. Among flows with channel states better than their threshold, outcome fair is maintained at the expense of a low efficiency of using the bandwidth. Effort limit is brought in order to prevent some flows with extremely or continuously bad channel states from monopolizing the bandwidth and demoting the whole performance. Effort limit fair is therefore useful in the realistic channel model and is easy to implement.

However, although there are channel state thresholds to limit the effort, for flows with channel states better than the thresholds, the poorer the flow's channel state is, the more time slots the system gives to the flow. This is unfair to the flows with better channel states. Other problems of the ELF approach include: there is no flow which can take leading and return service to lagging ones later, and there is no pre-allocated bandwidth to help the lagging flows to catch up.

PF (Proportional Fair):

Recently, designed for HDR (high data rate) services in CDMA systems, proportional fair [31, 168] is considered to be a simple yet effective fairness notion. Proportional fairness is formally defined in [187] that a scheduling decision (the bandwidth allocation) is proportional fair if using any other possible allocation will result in a non-positive normalized amount of aggregate allocation. At the scheduling time (time = t), suppose a flow i has an average effective throughput $H_i(\tau)$ over a past time window of length τ (i.e., from time $t - \tau$ to τ; average is calculated using low pass filtering), and the real throughput that can be achieved by flow i at time t is $\lambda_i(t)$, which is the aggregate rate of a certain number of supplemental channels (SCH) determined by the BS according to the interference and power limits. A scheduler can achieve proportional fair if it selects for transmission the flow with the largest value of $\frac{\lambda_i(t)}{H_i(\tau)}$ [168].

Thus, intuitively, a proportional fair scheduler heuristically tries to balance the services (in terms of outcome) of the flows, while implicitly maximizing the system throughput in a greedy manner. Obviously, the proportional fairness notion is a purely outcome fairness metric. Thus, while such a metric is simple to use, proportional fairness does not guarantee fairness in a strict sense. For example, consider the situation where a flow has experienced a prolonged period of poor channel states (hence, has a small value of $H_i(\tau)$), it may not get service even though its channel state improves (e.g., with a moderately large value of $\lambda_i(t)$ if there is a "dominant" flow which has a very good channel state (i.e., a very large value $\lambda_i(t)$).

Another problem is that the lagging flows are not given more consideration. In fact, flows are not classified by leading or lagging in proportional fair queueing. Flows, which have fulfilled their QoS requirements, will not receive less service even when there are other flows which have not fulfilled their QoS requirements. For example, before time t, flow i has received more service than its QoS requirement (we name it as the leading flow) because it has a good channel, however, flow j received much less service than its requirement (we name it as the lagging flow) because it has an especially bad channel before. Between time t and $t + \tau$, flow i has a bad channel and receives less service. At time $t + \tau$, even if the channel state of flow j is better than that of flow i, flow i may still be chosen to occupy the channel as it has a low $H_i(\tau)$ (at this time, it is possible that flow i is seriously leading and flow j is seriously lagging). This is unfair. Furthermore, the delay experienced by flows can also be high in a proportional fair system.

CAFQ (Channel Adaptive Fair Queueing):

CAFQ employs the fairness notion CAF (channel adaptive fairness) [344], and has the following features:

1. contrary to CIF-Q, graceful degradation is not ensured, in order to help the lagging flows more efficiently;

2. a punish factor is used to decide how seriously the scheduler punishes a non-perfect channel state flow that transmits packets; and

3. a virtual compensation flow is incorporated to help the lagging flows to catch up.

From the MT's viewpoint, fairness should be maintained in that so long as a flow can transmit some data, it should be provided some chance to transmit. At the same time, QoS should also be met. However, the BS can transmit one packet at a time in TDMA based networks, so from the system manager's viewpoint, it must improve the efficiency of using the channels to meet the QoS. The goal to keep fairness often conflicts with the goal to improve the efficiency of using the channels in face of the time-varying quality of the channels. Because whenever a flow without a perfect channel state is allowed to transmit, there will be part of the bandwidth wasted, and the wasted bandwidth can never be replenished. It should be noted that this is very different from the idea of swapping flows that are error-free and error-prone, as in existing scheduling algorithms such as CIF-Q. When an error-free flow takes the opportunity of an error-prone flow, it will relinquish the service when the error prone one recovers.

If abundant bandwidth is available or the channel state is most likely to be perfect, we should maintain the graceful degradation, and prevent the non-lagging ones from starving. But in a realistic system in which the channel is not so good, we cannot expect to achieve perfect allocations, but rather we should meet the flows' QoS first. Thus, in CAFQ, graceful degradation is not implemented and the rationale is to compensate the lagging flows as soon as possible. This can help to achieve the QoS requirements of isochronous media data types.

In CAFQ, the virtual compensation flow is incorporated with the following properties:

- swapping of flows is used at the same time of the virtual compensation flow to compensate the lagging flows;

- CAF is strictly maintained in the short term, and the virtual compensation flow helps to keep outcome fair in the long term; and

- the flow which is most seriously lagged has the highest priority of receiving service from the virtual compensation flow.

17.6 PRACTICAL ILLUSTRATIONS—HSDPA

To exploit the advantages of channel adaptation and multiuser selection diversity, the high-speed downlink packet access (HSDPA) scheme is incorporated in a more recent version of the UMTS standard—Rel 5. The HSDPA system adopts a microscopic scheduling approach on the radio resource (timeslots/codes) for its serving active users over very short duty cycles (up to 2ms). To facilitate the microscopic scheduling, the scheduling algorithm resides at the base station instead of the RNC to minimize the potential delay in the execution of scheduling. The base station obtains the instantaneous channel quality estimates from the mobiles (UEs) and selects one mobile (UE) to transmit on the high-data-rate shared-traffic channel at the current time slot, based on a scheduling algorithm (which factors in system throughput, QoS requirements, fairness, or a combination of these). This is fundamentally different from the Rel 99 approach. For example, if there are 10 packet-switched data users in the cell, there will be one high-data-rate traffic channel shared dynamically between these 10 users in the HSDPA systems. In fact, it is shown that the microscopic scheduling approach in the HDSPA systems is theoretically optimal because of the multiuser selection diversity.

17.7 SUMMARY

In this chapter, we have discussed the packet scheduling problem in wireless networks, and we have also talked about the reasons why the extensively explored fair queueing algorithms designed for the wired networks cannot be adopted to the wireless environment. Specifically, we discussed in detail on the properties of the wireless channel. Afterwards, different fairness notions and different wireless packet scheduling algorithms were discussed.

PROBLEMS

17.1   Briefly discuss the difference of packet scheduling in wired and wireless networks.

17.2   As for packet scheduling, what is the problem of using a two-state Markov chain channel model?

17.3   What is the difference between greedy fairness notions and realistic fairness notions? How will the difference affect the packet scheduling problem?

CHAPTER 18

POWER MANAGEMENT

Power is the first good.

—R. W. Emerson: Inspiration, 1876

18.1 INTRODUCTION

Power management is an important issue a wireless system as the devices are usually battery-powered. Indeed, power conservation is sometimes the top priority task in extending the lifetime of the mobile terminals and also that of the whole network.

Interoperability:

We first discuss the power consumption characteristics of the mobile terminals in Section 18.2. With the understanding of power consumption characteristics, it can be concluded that power can actually be conserved in two different situations. The first situation is that the mobile terminal is under active communication and the second situation is that the mobile terminal is under idle time. In the first case, power is generally conserved by adjusting the transmit power levels and selecting

Wireless Internet and Mobile Computing. By Yu-Kwong Ricky Kwok and Vincent K. N. Lau
Copyright © 2007 John Wiley & Sons, Inc.

an appropriate terminals for relaying the packets. These techniques will be briefly discussed in Section 18.3.1. On the other hand, when the terminals are not idle, they can usually be changed into low-power state to further conserve energy. However, they have to wake up periodically to check for packets destined for them. These techniques will be further discussed in Section 18.3.2. Most of the techniques suggested in active communication and that of idle time can actually be used simultaneously to conserve power.

Performance:

Various power conservation schemes are suggested to conserve power during active communication and during idle time. However, as we will see in this chapter, most of the schemes in fact will sacrifice deteriorate the throughput and delay of the network as a tradeoff for power conservation. Even worse, some of the power conservation schemes may actually increase the power consumption. These will be further discussed in Section 18.4.

18.2   CHARACTERIZATION OF POWER CONSUMPTION

In many wireless ad hoc networks, mobile terminals are largely battery powered. Thus, power conservation become an important issue to extending the lifetime of the mobile terminals, and, in return, prolong the lifetime of the network. Before various power conservation schemes are discussed in detail, the power consumption characteristic of mobile terminals will be briefly discussed. Most of the power conservation techniques try to save power based on these characteristics.

As shown in Figure 18.1, mobile terminals spend their power differently in various states and various layers. At the beginning, the mobile terminals in ad hoc networks can be in one of the four states—transmit, receive, idle and sleep. In these states, the energy dissipation is very different. The energy dissipation of mobile terminals in transmit mode is around 1.5 to 3 times that of the receive mode. While the energy dissipation of mobile terminals in transmit mode is around 5 to more than 30 times that of the sleep mode. The energy dissipation of idle mode is somewhere in-between receive mode and sleep mode. However, most specifications do not distinguish energy dissipation between receive mode and idle mode [247].

Moreover, energy is also required for a mobile terminal to actually put information in the wireless medium in the transmit mode and get information from the wireless medium in the receive mode. These constitute the power consumption in physical layer. Generally speaking, it is possible for the mobile terminal to select transmitting power level based on the actual environmental characteristics such as the transmission rate, the received SNR, the distance between the transmitter and receiver, etc. Generally, received power is proportional to the transmitted power and inversely proportional to the square of the distance between transmitter and receiver[1] The higher the

---

[1]Propagation model in free space.

received SNR, the more complex modulation coding can be used to encode more bits into one symbol. In this case, less transmission is needed to transmit the information to the other side [247].

In MAC layer, energy is also required for the sender and receiver to compete for the shared wireless medium channel with other mobile terminals in the network. In DCF mode of IEEE 802.11, four time frame sequence is generally used. In this sequence, the sender terminal needs to send out RTS and the DATA, while the receiver terminal needs to send out CTS and the ACK. On the other hand, power is also required to retransmit the corrupted frames due to propagation error and packets collision with other terminals.

In wireless ad hoc networks, mobile terminals not only need to send out their own packets, but also constantly need to forward packets for other mobile terminals. It is because in such an environment, it is generally impossible for a mobile terminal to directly send a packet to its destination. In this case, it is required for the source mobile terminal to rely on some intermediate terminals to route its packets to destination. The routing of packets use up the power in the intermediate terminals and the selection of different intermediate terminals set will result in different power usages in the network view.

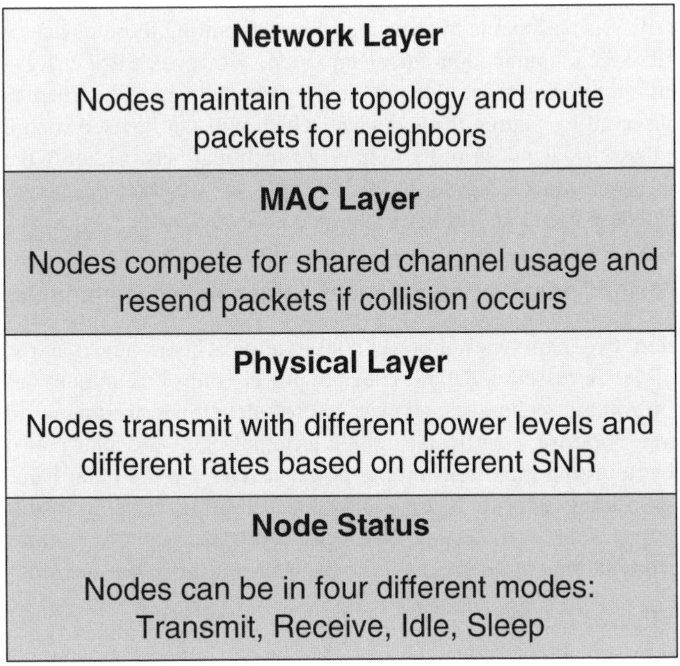

| **Network Layer** |
| Nodes maintain the topology and route packets for neighbors |
| **MAC Layer** |
| Nodes compete for shared channel usage and resend packets if collision occurs |
| **Physical Layer** |
| Nodes transmit with different power levels and different rates based on different SNR |
| **Node Status** |
| Nodes can be in four different modes: Transmit, Receive, Idle, Sleep |

Figure 18.1   Power consumption characteristics.

It is thus obvious that power is consumed in different situations and under different extents. As a result, various power conservation schemes are developed in exploiting power consumption in scheduling of terminals in various states.

18.3 POWER CONSERVATION SCHEMES

We have seen that mobile terminals can be set into multiple states in which different power consumption are resulted. Specifically, during active communication—transmit and receive data, power will be consumed in various layers differently according to the approaches used in these layers. These lead to the development of two main categories of power conservation schemes—conservation under active communication and conservation under idle period [247].

Briefly speaking, power conservation schemes under active communication are based on two main techniques in conserving power for a particular device and/or devices in the whole network. They are power control which mainly target for Physical and MAC layers, and energy-conservation routing which mainly targets for network layer. This will be discussed in Section 18.3.1.

On the other hand, mobile terminals may not be in use all the time. In many practical situations, most of the time, they are idle in the network. Notice that power consumption in sleep mode is much lower than that in transmit, receive or idle mode. Consequently, it is reasonable to conserve power by putting those devices which are not under active communication into sleep mode. However, extra energy and time will be required to bring those sleeping devices into active again. Thus, power can only be conserved by putting those devices which will not be used soon into sleep mode. On the other hand, in order to have a functional network, only parts of the devices can be put into sleep during certain time slots so that the network is not partitioned. These issues are further elaborated in Section 18.3.2.

18.3.1 Power Conservation Schemes under Active Communication

In this section, two main types of power conservation schemes under active communication will be discussed in detail. They are power control techniques and energy-conservation routing techniques. Power control techniques are mainly focused in power conservation for a particular mobile terminal or a particular pair of mobile terminals by adjusting their transmitting power levels. On the other hand, energy-conservation routing algorithms are mainly focused on power conservation for the route paths from sources to destinations. As a result, the network lifetime can also be prolonged with the consideration of the paths of routing in the network [247].

18.3.1.1 Power Control Power control refers to the adjustment of the transmit power level by a mobile terminal which is designed to support transmit power level adjustment to transmit data to the neighbor node. As discussed in the previous section, received signal power level is directly proportional to the transmit power level and inversely proportional to the square of the distance between the communicating nodes (assume in free space propagation model). In this case, for a given threshold received SNR for correctly decoding the messages, the shorter the distance between the communicating nodes, the smaller the transmit power level which is required to achieve this threshold.

There are two main advantages in using this approach. The first is that power can be conserved by using just about right transmit power level to send data to the receiver instead of using the maximum/normal transmit power to send data all the time. The second is that from the network's perspective, spatial reuse can be achieved much more effectively as illustrated in Figure 18.2. Here, nodes A, C, E are planning to transmit to nodes B, D, F, respectively. If normal transmit power is used, only one pair of nodes can transmit data at a time due to the four time frame sequences (i.e., RTS, CTS, DATA, ACK). However, if controlled transmit power is used, these three pairs of nodes can transmit simultaneously. As a result, spatial reuse can be achieved with the use of controlled power.

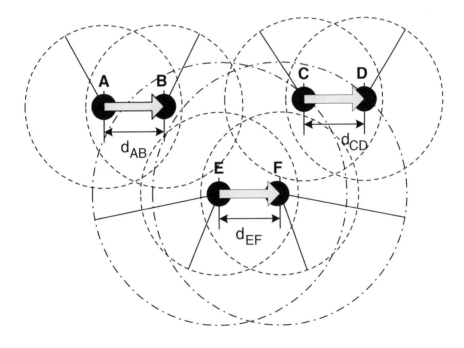

‑ ‑ ‑ ‑ ‑ ‑ ‑ Transmission Range after Power Control

‑ · ‑ · ‑ · ‑ Threshold/Normal Power Transmission Range

Figure 18.2   Power control example.

The main idea behind power control scheme is to use just about right power level to communicate with the neighbor nodes instead of using a standard, fixed, maximum power level to communicate with all the neighbors all the time. As a result, the power needed to communicate with near neighbor nodes can be greatly reduced and this results in power conservation for the network as a whole. Nevertheless, most of the power control schemes also lead to another problem—asymmetrical link problem as shown in Figure 18.3 [247, 177, 216].

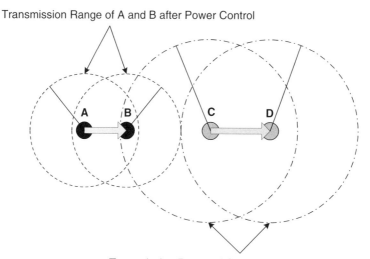

Transmission Range of A and B after Power Control

Transmission Range of C and D after Power Control

Figure 18.3   Asymmetrical link problem.

As can be seen, node A is communicating with node B, and node C is planning to start a communication with node D. Node A sends the data to node B using the about right power level and this results in a transmission range which cannot be detected by node C. At this moment, node C would like to initiate a communication with node D, so it starts by sensing the medium (refers to CSMA/CA in Section 9.4. However, as the transmission range of node A is much reduced by power control scheme and thus, its transmission cannot reach node C. Node C detects a free medium and starts transmitting data to node D using another power level which is high enough to reach node B as shown in Figure 18.3. As a result, a collision is likely to occur at node B if the power level of node C is high enough to corrupt the data of node A at node B. This greatly reduces the effects of power controls and may even *increase* power consumption, as more power will be consumed in retransmission.

In view of this problem, researchers have proposed to transmit RTS and CTS at normal power level, and transmit only DATA and ACK at reduced power level (refer to four-time-frame-sequences in Section 9.4). By doing so, packet collision is greatly reduced. This is because the medium is first reserved by RTS-CTS frames. All the nodes surrounding the communication pair which can successfully decoded[2] the RTS-CTS frames will be able to set the NAV flags to the proper duration as shown in Figure 18.4. As can be seen, nodes C, D and G are within the decoding range of node A and B normal power level. In this case, the RTS and/or CTS can be decoded and detected by these nodes and so they can set the NAV accordingly. For nodes E and F, they can only sense the RTS and/or CTS of node A and B, but cannot decode them. Consequently, after the RTS and CTS transmissions, nodes A and B communicates (DATA and ACK) at a reduced power level. At this moment, if node E or node F would like to transmit packets, it is still possible to cause collision in node A or node B.

Motivated by this observation, a novel scheme called PCMA [248, 216, 247], which is based on the idea from dual bust tone multiple access (DBTMA) [85], is suggested. PCMA relies on the use of a second channel to transmit busy tone to the surrounding senders. Specifically, before the transmission of the frames, the sender can use the busy tone information to determine the transmit power level which will not cause collisions to the existing transmission. Thus, PCMA can effectively reduce data collision at the receiver side. Apart from the use of second channel into the protocol, PCM [177, 247] is based on the use of maximum transmit power level to transmit RTS and CTS. Then, alternately, longer reduced power and shorter maximum power are used to transmit DATA, and reduced power will be used to transmit ACK. This approach can also prevent the senders from missing out the RTS and CTS messages and subsequently consider the medium as free accidentally. However, since the ACK is also transmitted at reduced power in both approaches, ACK collision at sender side is still possible. In fact, it is very difficult to design a perfect MAC protocol which can improve both the capacity and power utilization at the same time.

18.3.1.2 *Energy-Conservation Routing*
In an ad hoc network, it is usually impossible for the source node to directly send a packet to the destination node. As a result, some intermediate nodes have to be selected to help in routing the packets to the destination. Energy-conservation routing can generally conserve power of nodes in the whole network by selecting the minimum power consuming path (minimum energy cost) or selecting the maximum residual power path (maximum battery capacity) or a mix of both (minimum energy cost and maximum battery capacity). There are generally some advantages and disadvantages of these approaches, however. For example, the use of minimum energy path all the time may result in a swift depletion of energy of those mobile terminals on that path, leading to partitioning of the network. By contrast, the use of maximum battery capacity path may also result draining out the battery power of some nodes prematurely since the transmission costs are differ-

---

[2]For the same power level, there are two ranges. The nearer range is the decoding range and the farther range is the carrier sensing range. Nodes within decoding range can decode the messages sent by the originating node, while the nodes within the carrier sensing range and decoding range can only sense the messages sent by the originating node, but cannot decode it correctly.

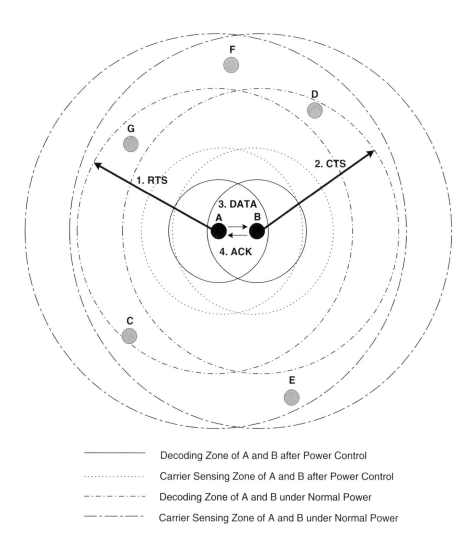

Decoding Zone of A and B after Power Control

Carrier Sensing Zone of A and B after Power Control

Decoding Zone of A and B under Normal Power

Carrier Sensing Zone of A and B under Normal Power

Figure 18.4 Asymmetrical link problem 2.

ent for different mobile terminals due to reasons such as interferences, distances, etc. The mix of minimum energy cost and maximum battery capacity approach generally use a threshold value to determine their favor in minimum energy path or maximum battery capacity path. Generally speaking, when the battery capacity is high, the minimum energy path should be considered. When the battery capacity is low, the maximum battery capacity path should be considered [247]. These considerations will be discussed again in Chapter 19.

18.3.2 Power Conservation Schemes under Idle Period

In an ad hoc network, not all the mobile terminals are in active communications. Some of them are idle most of the time. It is obviously wasting power for those devices to state awake instead of sleeping. As a result, power conservation schemes under idle period usually consider the suspension of some of the mobile terminals in the network [247]. However, additional time and energy are required to bring devices from sleep mode into active mode. These may affect the throughput, delay or may even increase the overall power consumption if the schemes used are not effective enough.

18.3.2.1 *Sleep Mode Scheduling* Although bringing the mobile terminals into sleep mode can effectively reduce the power consumption of the terminals, longer delay, lower throughput and even waste of power maybe resulted. It is because in order to deliver a message from the source terminal to destination terminal nearby, both terminals have to be awake at the same time; otherwise, the message has to be put into the buffer of the source and wait until the destination terminal awake. Consequently, long delay may result due to the waiting for the waking up of the destination terminals. At the same time, the source terminals' buffer queues may be filled up with the messages to destination nodes. Thus, packet dropping may occur during the waiting. Apart from these adverse effects, waste of power will also occur when the destination node awakes but finds that there is no message destined for it. In summary, a good sleep mode scheduling has to be used to coordinate suspension of the nodes in the network.

A straightforward approach to solve the above problems is that every node in the network will be suspended when there is no packet originated from them or destined to them. They will then wake up periodically to check for any packets. This simple idea is generally adapted in most of the approaches. Figure 18.5 illustrates the Power Saving Mode (PSM) of IEEE 802.11 in ad hoc network environment which assumes the existence of synchronized clocks between nearby mobile terminals. With this facility, the mobile terminals will wake up at the same time which is the start of the beacon interval as shown in Figure 18.5. The time period immediately after wake up is called the Ad Hoc Traffic Indication Message (ATIM) period. During this period, a mobile terminal which has packets for another nearby terminal will make an announcement during this period. As shown in Figure 18.5, mobile terminal A has a data packet for mobile terminal B. Now, mobile terminal A will then announce this in the ATIM window. In receipt of the ATIM, mobile terminal will reply with ACK. After that, those mobile terminals which do not have any packets to send nor being announced will again change back into sleep mode. Those mobile terminals which have made announcements or being announced will stay awake for the whole beacon interval period. After the ATIM period, those terminals which have made announcements can then send a packet or several packets to the announced mobile terminals [247, 178].

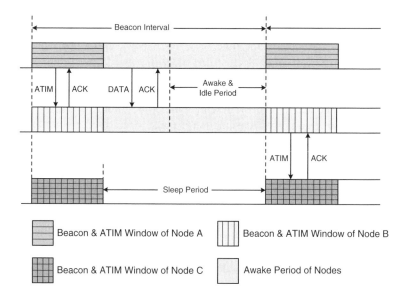

Figure 18.5   IEEE 802.11 PSM mode in ad hoc networks.

The solution suggested by IEEE 802.11 in ad hoc networking environments is based on a fixed ATIM window size for all the nodes in the network. However, some researchers have pointed out that a fixed ATIM window size does not fit all the situations well. It is because if the size of the window is too small, the nodes may not be able to make announcement for all the packets. However, if the window size is too large, more power will be consumed as those nodes are required to be awaked during the ATIM periods and also remaining time for transmitting data in the beacon intervals will be minimal. As a result, low traffic load may favor the use of a small ATIM window and heavy traffic load may favor the use of a large ATIM window.

Consequently, Dynamic Power Saving Mode (DPSM) [178] is suggested. As shown in Figure 18.6, the ATIM window of different nodes are different and will be dynamically adjusted based on the traffic loads in the network. There are two common conditions in which the window size will be increased. The first is that there are remaining packets in the buffer which have not been announced during the original ATIM windows. The second is that a marked packet is received. As shown in Figure 18.6, mobile terminal A tries sending ATIM to mobile terminal B during the first beacon interval. However, mobile terminal B cannot receive it due to short ATIM window size. This packet is then marked. In the second trial, mobile terminal A tries sending this ATIM in the original ATIM window size. In this case, mobile terminal B receives it and finds that the packet has been marked. Mobile terminal B will then increase its window size as it finds that its window size is too small to cope with the network traffic. If all the packets in the buffer can be announced within the same beacon interval, the ATIM window will be decreased in the next interval.

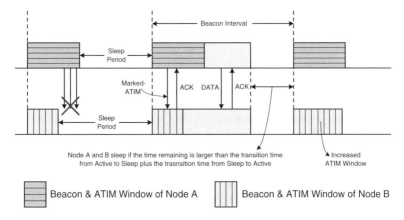

Figure 18.6 IEEE 802.11 DPSM with dynamic ATIM.

Other than an ad hoc networking environment, power saving mode (PSM) is also specified for mobile terminals in infrastructure network. However, there are some differences between PSM in infrastructure network and infrastructureless network. In infrastructure network, the access point will coordinate the access of all the mobile terminals. The mobile terminal will send out PS-Poll to the access point to see if any packet is destined for it. It will change immediately back into sleep mode if there are no pending operations (sending or receiving) that required its attention. However, this power saving mode leads to some problems as shown in Figure 18.7 for some applications. As shown in the figure, the mobile client initiates the first RPC request during the middle of the second beacon interval. It will switch to sleep mode immediately after that. In this case, the first RPC response that arrives during the same beacon interval cannot be sent to the mobile terminal immediately and the access point has to wait for the next beacon interval to do so.

In this kind of applications, power management, paradoxically, leads to higher power consumption and longer delay. This in returns lead to less work to be accomplished before the battery lifetime ends. In view of this, Self Tuning Power Management (STPM) [16] is proposed in which power management will be decided

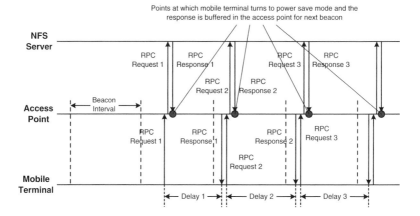

Figure 18.7   IEEE 802.11 PSM mode problem in Infrastructure network.

to be turned on or off based on the hints provided by the applications. For some delay intolerant applications (e.g., interactive applications) and large TCP transfer applications, it may be better to turn off power management.

The above algorithms assumed that all the neighbor nodes are equipped with synchronized clocks. However, this may not be the case in a mobile ad hoc network due to the mobility of the mobile terminals. Since they may not wake up at the same time due to clock skew, they have to stay awake for a longer period of time in order to increase the chance of having overlapping wake up period with their neighbors. However, even if they increase the wake up time, it is still possible for some mobile terminals to be able to send beacon and ATIM to their neighbors, but the reverse may not be true.

Consider IEEE 802.11 PSM in an ad hoc network environment shown in Figure 18.5. Suppose mobile terminal B's clock has a certain time lag compared with that of mobile terminal A and the lagging time is longer than that of ATIM period. In this case, mobile terminal B can never receive any beacon and ATIM from mobile terminal A, but mobile terminal A can receive mobile terminal B's beacon and ATIM. Consequently, various asynchronous approaches have been suggested to ensure at least certain length of overlapping wakeup time between neighbor mobile terminals, so that any pair of neighboring mobile terminals is possible to receive beacon and ATIM from their counterparts. The main reason for this to occur is that the beacon and ATIM period are set at the beginning of each beacon interval. In view of these, researchers have proposed to put beacon and ATIM period at the front in odd interval and put the period at the end in even interval. Furthermore, it is also proposed to put the beacon and ATIM period in both front and end of each beacon interval. This ensures that every neighbors can receive beacon and ATIM signal of the mobile terminal and vice versa [247].

The above solutions can solve the problem of asynchronous clocks, so that every neighbor should be able to receive the ATIM signal. However, the wake up times for the mobile terminals are much longer than that in the case of synchronous clock in order to increase the probability of overlapping between neighbor terminals. To shorten the wake up time, researchers have proposed several methods. Figure 18.8 illustrates the method in which every node will stay awake for the whole beacon interval after every T beacon intervals (e.g., $T = 3$ in the figure). In other intervals, the mobile terminals will just wake up for the ATIM period and will turn back into sleep mode if it has no packet to transmit or is not receiving any ATIM from other mobile terminals. The main drawback of this approach is that, the delay of the packet will be up to the period T.

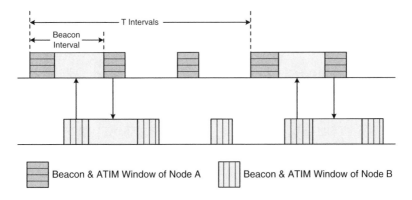

Figure 18.8 Nodes awake for a whole beacon interval every T beacon intervals for asynchronous network.

Another approach suggested by other researchers are the selection of beacon interval slots in which the mobile terminals will wake up for the whole period in the selected slots. The selection algorithm will ensure that every neighbors will be able to overlap for at least one beacon interval in those slots. Figure 18.9 illustrates the idea in which every mobile terminals will select a row and a column of slots in which they will wake up for the whole period in those selected interval slots. We can see that, suppose mobile terminal A selects the yellow row and column, while mobile terminal B selects the blue row and column, these two mobile terminals will have overlap awake period in the green slot. This slot-selection algorithm can ensure every terminals to have at least 2 slots overlap every $n \times n$ slots ($n = 6$ in the figure) [247].

The approaches of selecting rows and columns can ensure every terminals to overlap. However, a variable delay will be introduced, depending on the rows and columns selection of the two mobile terminals. In view of this, another approach is proposed as shown in Figure 18.10. Here, each mobile terminal will select a row of slots. This approach guarantees that every row will have at least one active slot overlapping with

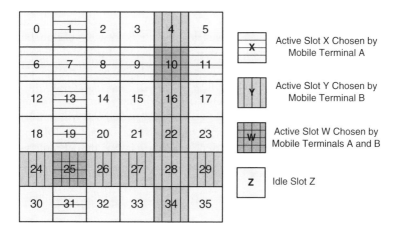

Figure 18.9   Nodes awake for a whole beacon interval during a selected row and a selected column beacon intervals for asynchronous network.

other rows. Consequently, neighbors selecting other rows can communicate during the overlap awake period [247].

18.4   PERFORMANCE ISSUES

In previous sections, various power conservation schemes have been discussed briefly. Generally speaking, it is reasonable to think that the use of power conservation schemes can save power for individual mobile terminals in the network as well as the whole network. However, various researchers have shown that the inappropriate use of the power management schemes in fact will cause throughput dropping, delay increasing, and may even cause higher power consumption due to retransmission or dropping of packets.

Moreover, researchers have found that the use of reduced power in transmission of four frames sequence will in fact cause more packet collisions. This in turns decrease the throughput of the network and increase the end-to-end delay. In view of this, a sensible approach works by using normal power to transmit RTS and CTS frames and reduced power for DATA and ACK. By doing so, the throughput is increased and end-to-end delay is reduced due to fewer packet collisions experienced. However, these two approaches still generate lower throughput and higher delay than the normal IEEE 802.11 transmission schemes without power control. However, a recently proposed scheme, called PCMAC [216] (will be discussed in next section), considers whether the transmission power will cause collision to neighbors transmission and accordingly eliminate the ACK in the sender. As a result, packet collision is reduced with the

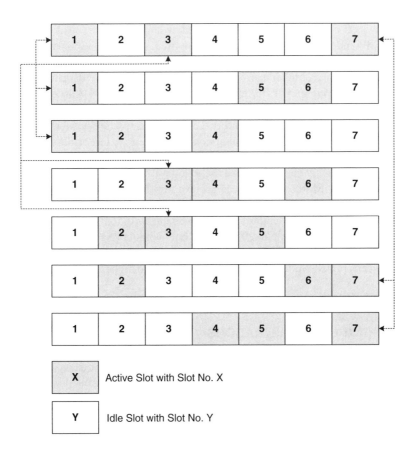

Figure 18.10   Nodes awake for a whole beacon interval during the selected beacon intervals for asynchronous network.

possibility of spatial reuse. In this case, the throughput is higher than that of normal IEEE 802.11 and the end-to-end delay is lower than that of normal IEEE 802.11 scheme without power control [216].

IEEE 802.11 has specified PSM in both infrastructure and infrastructureless networks. As discussed in previous sections, PSM in infrastructureless network makes use of the idea of ATIM window to instruct neighbors to stay awake during the beacon intervals. However, a fixed ATIM window in fact does not fit all the situations in that low traffic load may favor a small ATIM window and high traffic load may favor a high ATIM window. As a result, DPSM is proposed. The results shown that DPSM can achieve the throughput comparable with that of without PSM mode. Furthermore, both of them are higher than that of PSM. At the same time, the power used for achieving such throughput of DPSM is lower than that of both PSM and that without PSM [178].

On the other hand, as discussed earlier, PSM in infrastructure network may cause long delay and increase in power consumption. It has been shown that the performance of PSM in fact depends on applications very much. In many applications, the scheme without power management can achieve the lowest delay, but generally consumes the highest power. Moreover, PSM can achieve lower power consumption than that of without PSM. However, the delay is unacceptably high. STPM, which considers applications' features, can achieve comparable delay with that of without PSM and much smaller power consumption than that of both without PSM and with PSM. One of the interesting result is that in remote X application, PSM scheme in fact consumes more power than that without PSM scheme and achieve highest delay [16].

18.5   PRACTICAL ILLUSTRATIONS—THE PCMAC SCHEME

Various power conservation schemes have been suggested in previous sections. Power conservation schemes included both power conservation schemes under active communication and power conservation schemes under idle. Among the active power conservation schemes, power control has aroused great attention. In this section, a practical scheme known as PCMAC [216] scheme which aims at both power conservation and network throughput improvement by modifying the existing IEEE 802.11 standards will be briefly discussed. As this scheme requires only slight modifications from existing IEEE 802.11 standards, so it can be integrated into existing products easily.

In view of the asymmetric problems suggested in Figure 18.3, PCMAC is proposed to eliminate collision in both the sender (ACK from receiver) and the receiver side (DATA from sender). There are two main modifications in PCMAC. The first is the addition of power control channel with a bandwidth of 500 kbps. After the RTS/CTS exchange through the normal power, the receiver side begins receiving DATA packets. At this moment, it can estimate the noise level it can tolerate during the receipt of the packets from the sender by:

$$\frac{P_r}{SIR_{th}} - P_n \tag{18.1}$$

Where P_r is the received signal power and P_n is the noise power perceived by receiver. SIR_{th} is the threshold signal to noise ratio at the receiver for it to correctly decodes the message.

This information will then be broadcast via the power control channel at normal power level with the frame structure shown in Figure 18.11. With the help of this information, the neighbor nodes can then estimate whether their transmissions will cause corruption in the current transmission of that particular receiver before initiating transmission.

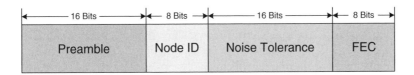

Figure 18.11 Frame structure of power control packet.

The second modification that PCMAC made is the change of four time frame sequence to three time frame sequence in which the ACK frame from the receiver is eliminated. As a result, the ACK frame will not be corrupted at the sender side. In order to facilitate the change, sent-table and received-table will be kept at sender and receiver side, respectively. The sent-table recorded the packets (including the session ID, packet sequence number and a copy of that packet) that have been transmitted to the receiver, while the received-table kept the similar items. Suppose node A is the sender and node B is the receiver. When node B received node A's RTS frame, it will reply with CTS frame with the session ID and packet sequence number of the previously correctly received packet from A. Upon receipt of CTS frame, node A will retransmit previous packet if the session ID and sequence number does not match; otherwise, a new packet will be sent.

With the help of these two modifications, packet collision can be greatly reduced. This is because the sender has to consider whether its transmission will cause collision in its neighbors by using the tolerable noise power level provided by the broadcast message of the neighbors in the power control channel. In this manner, the DATA frame in the receiver side can be efficiently protected. On the other hand, the proposed three frame sequence eliminates the ACK frame. In this case, the ACK collision at the sender side can be efficiently reduced.

18.6 SUMMARY

Since most of the mobile terminals are battery-powered, the lifetime of the devices in a wireless network is constrained by the relatively small amount of energy available. In order to extend the lifetime of individual devices which in turn leads to extension of the lifetime of the whole network, a number of power conservation schemes are proposed. At the beginning of this chapter, the characteristics of power consumption in using the mobile terminals were discussed. Generally, different amounts of power will be consumed in different states. If the mobile terminals need to communicate with other terminals in the network, different considerations at different protocol layers (Physical, MAC and Network) will also cause different power consumptions of that device and other devices in the network.

In view of these power consumption characteristics, various power conservation schemes have been proposed targeting at extending the lifetime of the devices during active communication and extending the lifetime of the devices during idle time. The power conservation schemes targeted at active communication mainly exploits the power consumption pattern in various communication layers. These lead to the use of techniques like power control which mainly targets at power conservation of individual devices in Physical and MAC layers by adjusting the transmit power level and energy-conservation routing which mainly targets at power conservation of all the devices in the network by selecting those intermediate nodes which consume the least power, which have the most abundant remaining power, etc., in participation of routing.

On the other hand, schemes targeted at idle time mainly exploit the different power consumptions of the mobile terminals in different states by putting those devices into low-power consumption states if they are not in active communication in near future. However, extra delay and power will be experienced by changing the devices from low-power state back to active state. In order for the neighboring devices to have ample opportunities to communicate with each other without unexpected and unacceptable delays, scheduling between the power-state traditions have to be considered. Generally, if the mobile terminals are connected to infrastructure or the mobile terminals are near to each other enough (e.g., within one hop), they can be assumed to have synchronized clocks. Synchronous approaches can then be adopted and devices can be assumed to wake up at the same time within each interval. Consequently, the schemes coordinate which mobile terminals can be transited back to low-power state immediately and which mobile terminals have to state awake for certain period in the interval. However, with mobility of mobile terminals, clock synchronization assumption may not be reasonable. In this case, certain clock skew may be experienced by different devices in the network. As a result, neighbors may not wake up at the same time. Thus, asynchronous approaches are proposed to tackle these issues.

Various power conservation schemes have been discussed. Generally speaking, the use of power conservation schemes may deteriorate the performance of other metrics like throughput and delay. In some cases, the power consumption of the power conservation schemes may be even higher than that of without using power conservation schemes due to packet collision and packet dropping.

Finally, a practical power control approach—PCMAC which aims at simple modification of IEEE 802.11 in order to solve various asymmetric links problem caused by the uncoordinated power control of neighbor nodes is discussed. This approach adopts the techniques of additional busy power channel to inform the sender about the presence of current transmission and the modification of four frame sequence into three frame sequence. These techniques effectively reduce the packet collision in the network and result in throughput improvement.

PROBLEMS

18.1 Briefly discuss how power is consumed in mobile terminals.

18.2 Suggest some power conservation schemes during active communications.

18.3 Discuss some power saving strategies during idle time.

a

CHAPTER 19

AD HOC ROUTING

The beaten path is the safest. (Via trita est tutissima.)

—Latin Proverb

19.1 INTRODUCTION

Equipped with powerful communication hardware, many hand-held wireless devices have the capability of forming *ad hoc* connections with nearby devices. However, without infrastructure support, network connectivity is a major resource management problem.

Interoperability:

Ad hoc routing protocols were proposed in view of the additional constraints, e.g., temporarily formation, mobility of nodes, security, energy consumption, etc., that impromptu connectivity brings to the network. There are quite a number of different scenarios which focused on different constraints, some of these application scenarios are discussed in Section 19.2.

Wireless Internet and Mobile Computing. By Yu-Kwong Ricky Kwok and Vincent K. N. Lau
Copyright © 2007 John Wiley & Sons, Inc.

In order to solve the problems induced by different scenarios, at the beginning researchers suggested the adoption of traditional routing protocols—Distance Vector and Link-State used in wired networks to mobile wireless environment. This leads to the table-driven ad hoc routing protocols in which every nodes maintain a complete view of all other nodes in the network. As a result, most table-driven routing protocols are in fact variants of these two traditional routing protocols with some extension techniques to solve certain problems caused by mobility. Notable examples include the addition of sequence numbers, restricted broadcasting by some nodes, different updates for time-triggered and event-triggered updates, the grouping of certain nodes together for management, etc. Combination of different techniques lead to different performance of these table-driven protocol. These routing protocols will be discussed in Section 19.5.1.

Subsequently, researchers started thinking about the problems suggested above in another direction. Specifically, researchers become skeptical about the necessity of having a complete view of the whole networks all the time. In particular, when link breakages or mergings are frequent due to the movement of the nodes in the network. As a result, another main category of routing protocols are suggested, namely, on-demand routing protocols. These protocols do not maintain a very up-to-date view of all destination nodes in the network. They will only try to find routes to the destinations when they have traffics to send to certain destinations. Since they do not maintain a complete view of all the destination nodes, these protocols require a route discovery process and a route maintenance process. Route discovery process involves finding potential routes to a destination when there is no route set up at the time the source node wants to send packets. Route maintenance entails the process of handling link breakages at the time the traffic flows from source to destination are on-going. Such routing protocols are based on techniques like source routing, maintenance of forward and reserve path for certain source-destination pair, etc. These routing protocols will be discussed in Section 19.5.2.

Since both table-driven routing protocols and on-demand routing protocols have their own advantages and disadvantages, hybrid approaches which combines some techniques in table-driven routing protocols and on-demand routing protocols are also proposed. These will also be discussed in Section 19.5.3.

Performance:

There are quite a number of performance metrics which can be considered for routing protocols in ad hoc network. It is because different application scenarios will lead to different importance of certain performance metrics. These scenarios are usually captured by appropriate system models as discussed in Section 19.3 and these performance metrics will be further discussed in Section 19.4. However, generally speaking, average end-to-end delay and throughput, control overhead, etc., are used as performance metrics for comparison between various routing protocols. It is widely agreed that proactive routing protocols generally have higher control overhead than reactive routing protocol, as the former require both periodic and event-triggered route maintenance actions. On the other hand, the data delay performance of proactive routing protocols is usually better than that of reactive routing protocols because proactive routing protocols have up-to-date routes to all destinations while reactive routing protocols may need to look for routes at that time. In terms of throughput, proactive routing protocols can be better or worse than that of reactive routing protocols depending on different scenarios.

19.2   APPLICATION SCENARIOS

Traditionally, ad hoc network came about in response to the need of setting up a network temporarily in hostile environment, e.g., battlefields or disaster sites. Under these situations, fixed and trustworthy telecommunication infrastructure are usually unavailable or are destroyed. As a result, an infrastructureless or semi-infrastructureless set up will be useful to assist military applications or rescue services.

Currently, there are three typical kinds of proposed applications for ad hoc network [28, 216].

- Figure 19.1 illustrates a pure infrasturctureless ad hoc network. In this network, nodes connect with each other to form a temporarily wireless infrastructure without the help of fixed telecommunication infrastructure. Within the network formed, users can share data and/or perform computation cooperatively. Typical example applications are:

  - Business conference meeting: Within the meeting, users may exchange data and business card.

  - Soldiers equipped with wireless devices can communicate with their allies to report the frontline situation to the commander with the help of pure ad hoc network formed temporarily.

  - When disaster situation occurs, telecommunication infrastructure may be destroyed. The rescue personnel equipped with the devices can then go into the sites and report the situations in real time.

- Figure 19.2 illustrates the hybrid infrastructure ad hoc network. In this kind of network, fixed telecommunication infrastructures are usually present. However, these infrastructures may have some deficiencies in supporting and reaching all the users. As a result, hybrid infrastructure is proposed. In those locations where fixed infrastructure may not reach, wireless ad hoc network can be formed to extend the reachability and capability of the fixed network. The main aim of this ad hoc network is to help other nodes which are out of the coverage of fixed infrastructure to reach the infrastructure network with a minimum number of hop count. Consequently, those nodes out of coverage can enjoy services provided by infrastructure network with reasonable throughput and at a smaller cost.

Figure 19.3 illustrates an example of cooperative networks—sensor networks. Sensor networks are usually deployed in a large scale in the target area. The nodes in the network will cooperate with each other to accomplish a target application, e.g., surveillance as shown in the figure. Typically, each sensor node has the following constraints:

- Low Processing Power and Limited Memory: A sensor node is usually embedded with a low processing power CPU in order to achieve low power usage and small in size.

- Battery Powered: A sensor node is usually powered by battery. As a result, energy conversation is important in such kinds of network.

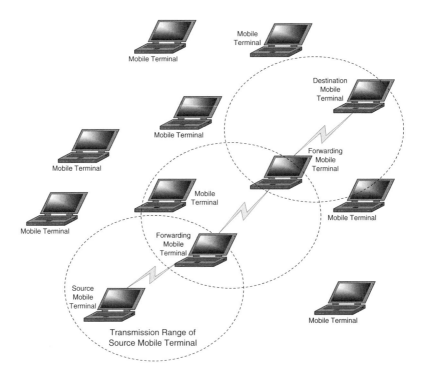

Figure 19.1   Infrastructureless ad hoc network.

19.3   SYSTEM MODEL

Section 19.2 suggested several currently proposed application scenarios, including hybrid-infrastructure, pure infrastructureless, and cooperative infrastructureless ap-

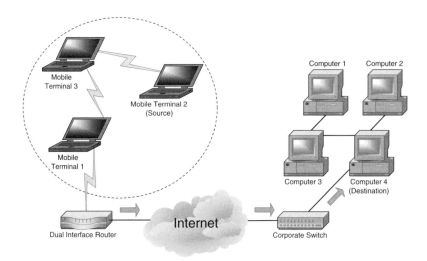

Figure 19.2   Hybrid infrastructure ad hoc network.

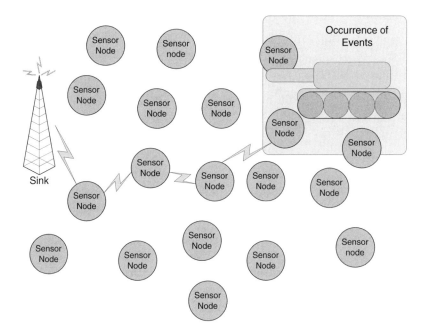

Figure 19.3   Sensor ad hoc network.

plications. However, it is difficult and expensive to deploy and perform real experiments over such a large network in a real scenario. As a result, most researchers usually performed simulations with tools such as NS-2, OPNET, etc. These simulators usually model the scenarios based on some specific system models and some of the parameters specified.

Figure 19.4 shows the system model of a commonly considered kind of ad hoc network. The parameters used depend on the application scenario. They usually include [28, 216]:

- The network is supposed to be bounded by a rectangle with size of width $\times$ height. The size of the rectangle depends on the place where the users expected the scenario to happen, e.g., 200m $\times$ 200m is usually considered a reasonable size for modeling shopping mall or exhibition centre.

- Within the boundary, there are n users equipped with mobile devices. The users may not be stationary. They may move to other location based on some mobility model with a speed denoted by vm/s. One of the commonly used model in representing the movement of users is the Random Way Point Model. Within the model, a user will choose a destination and move to that specific destination with the associated speed. After arrival, the user will pause there for some time and then move to another location following a randomly chosen direction.

- Within the network, a source user will send packets to a destination user. Such source-destination pair will be randomly chosen with variable traffic load, which is usually represented as the number of packets per second.

Based on the setting of the parameters specified above, different ad hoc network scenarios can be modeled with the use of a simulator.

19.4   PERFORMANCE METRICS

In ad hoc routing research, there are quite a large number of protocols proposed. Some of them are proposed in view of the deficiencies of others. However, the improvements are usually made by trading-off various performance considerations. It is, in fact, very difficult to get a all-round protocol which will perform the best in all of the performance metrics. However, it is generally acceptable to considered the following metrics if high speed, high throughput ad hoc network is required as used in [216].

- Average packet end-to-end delay and throughput: These two metric are generally considered as the most important metrics in many research works on routing protocols. It is because the routing protocols developed are mainly used to provide services to an upper layer network protocol, e.g., TCP. Packet

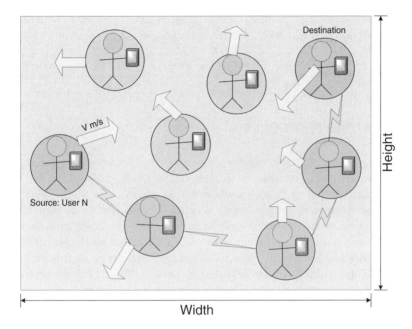

Figure 19.4   Generic system model.

end-to-end delay and throughput can be considered as the quality of services provided by the routing protocols.

- Percentage of packet delivered correctly: The wireless environment and the movement of nodes result in fluctuating wireless channel environment. This increases the difficulties of successfully delivering the packets to the target destinations. On the other hand, some applications allow certain percentage of packet errors, e.g., video streaming, which can be concealed by some error hiding and concealment techniques (this will be discussed in Section 23.4. As a result, percentage of packet delivered correctly is a good indicator on the performance of routing protocols which aim at supporting some error rate sensitive applications in upper layers.

- Control overhead: Due to different approaches on maintenance of the routing table inside the nodes, different routing protocols will induce different control overheads, which manifest as number and size of control packets, the frequency of updates, link break maintenance, etc. Generally, it is considered as unacceptable to have a large control overhead rate with respect to the data packet transmission rate.

Apart from these metrics, energy consumption is generally used as a metric in ad hoc network powered by battery. For instance, percentage of out of order packet is considered a significant metric if TCP is used. Maximum number of nodes supported will be an appropriate concern if an ad hoc network is built for supporting a large group of users (e.g., an army) [28].

19.5 ROUTING PROTOCOLS

Generally speaking, ad hoc routing protocols can be classified into two main categories based on their attitudes towards the maintenance of routing table. They are Table-Driven/Proactive Routing Protocols which will be discussed in Section 19.5.1 and On-Demand/Reactive Routing Protocols which will be discussed in Section 19.5.2[1].

As Proactive routing protocols and Reactive routing protocols have their own advantages and disadvantages. Some other protocols which would like to get the best of both worlds are developed. These protocols can generally be regarded as hybrid routing protocols which will be discussed in Section 19.5.3. Furthermore, some additional routing protocols are proposed in view of some additional constraints in the environment, e.g., Security and Energy Considerations. These will also be discussed in Section 19.5.4.

Figure 19.5 gives a brief example of some of the protocols suggested in these areas [28, 216, 140, 302].

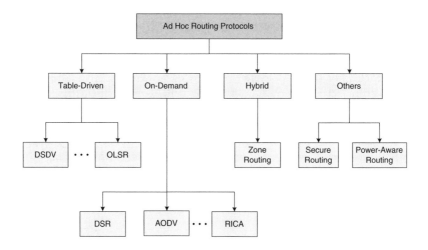

Figure 19.5 Some of the existing ad hoc routing protocols.

---

[1]In later sections, The terms Table-Driven and Proactive will be used interchangeably, while the terms On-Demand and Reactive will also be used interchangeably.

19.5.1   Table-Driven/Proactive Routing Protocols

In this section, representative proactive routing protocols such as DSDV [277] and OLSR [69] are discussed. Proactive routing protocol was proposed based on some traditional routing protocols, e.g., link state and distance vector, to tackle the extra constraints induced by the ad hoc environment. The most notable constraint is the mobility of the wireless devices which can lead to link breakages frequently.

In proactive routing protocol, it is common for all the nodes to know the complete routes to all destination nodes in the network. In order to maintain a consistent, up-to-date view, all the nodes need to actively update their routing tables periodically or when the network topology changes even when there is no traffic to send in the network. However, the main advantage of this kind of protocols is that all the nodes will have a consistent view and will know suitable routes to all destinations. Thus, it is not necessary to have extra delay in finding routes to destination nodes when needed. Indeed, many people consider that these protocols are most suitable to the scenario in which network topology change infrequently and there are large traffics to send in the network.

19.5.1.1   Destination Sequenced and Distance Vector—DSDV   DSDV [277] is a typical example of table-driven routing protocols. Every node in the network will maintain a routing table which has knowledge to all destination nodes in the network. Each entry in the network contains information including destination address, next hop to reach destination, destination sequence number and hop counts to destination [277, 28, 216, 140, 302]. Figure 19.6 illustrates the DSDV scenario.

Destination sequence number is added intentionally in DSDV in order to detect routing loop. It is sometimes possible for a node to learn multiple routes to a destination. For example, in Figure 19.6, it is possible for node D to receive information about getting to node B via node C and node E through completely different route, respectively. In order to decide which path to use, DSDV exploits the use of destination sequence number to resolve the problem. The one with larger destination sequence number is chosen as it represents the more up-to-update information about the destination node. However, if the two routes reaching the same destination are associated with the same destination sequence number, the one will smaller hop count will be chosen.

In order to keep the routing table up-to-date, each node in the network will broadcast their full routing table to other nodes in the network periodically with increment of their own sequence number each time. This is also known as time-triggered update. Apart from periodic update, whenever there is a change in network topology, e.g., link breakage occurs, the neighbor nodes will broadcast the event to other nodes in the network. This is known as event-triggered update. For example, suppose in Figure 19.6, the link CD breaks. Node C and node D, detecting the event, will send update packets to all the neighbors. At the same time, they will set the distance metrics associated with the broken links as ∞.

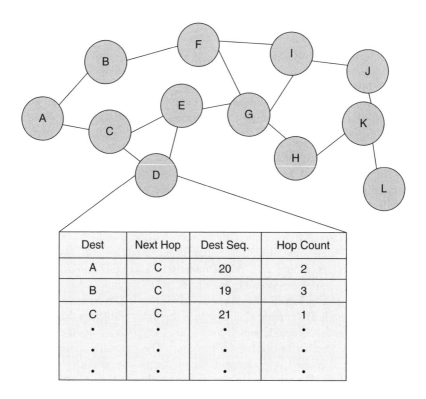

Figure 19.6 DSDV routing scenario.

In order to prevent large amount of traffics generated by the event triggered updates, only the changes from the latest time-triggered update are sent in the event-triggered update. This is known as an incremental update and is usually possible to be packaged into one packet.

19.5.1.2 *Optimized Link State Routing—OLSR*   OLSR [69] is another typical proactive routing protocols. In a traditional link state routing protocol, every node floods the link state information to their neighbor nodes periodically. Subsequently, their neighbor nodes will then flood the information to their own neighbor nodes. This process will continue and the result is that every node in the network will have the complete view of links of all other nodes in the network. This process, in fact, causes a lot of unnecessary overhead as some of the nodes may rebroadcast information that their neighbors know already [69, 28, 216, 140].

Thus, unlike a traditional link state routing protocol, OLSR tries reducing the number of nodes which perform floodings of the link state routing protocols. Specifically, only those nodes selected by the source node (known as Multipoint Relays (MPR)) will perform the flooding actions. The neighbor nodes are selected based on the criteria that all two-hops neighbors of source node will receive the state information of source node. This is done by rebroadcasting the link state information of source node by those selected nodes inside the MPR sets. So those nodes selected as MPR will also keep a list known as MPR Selector set which stores the information about those nodes selected them as MPR, so each source node has to inform their MPRs directly about the selection. At the same time, in order to facilitate the process described above, every node in the network must have information about which nodes are within a two-hop distance. Such information is obtained by broadcasting every node's neighbor list in the beacon (or also known as HELLO message).

Figure 19.7 illustrates the idea of OLSR. In the upper part of the figure, the node E would like to broadcast its own link-state information. Its neighbor nodes C and G which are selected by node E will then rebroadcast the link-state information received by node E. After this step, in fact, all the nodes within two-hop from node E will have known the link-state information of node E as shown in the figure. In the lower part of Figure 19.7, nodes E, F, I, H are multipoint relays of node G. Upon receipt of the broadcast packet from node G for the first time, it will in turn broadcast the packet to their neighbors[2]. After this, those nodes circled in the figure will have received the link-state information of node E.

19.5.1.3 *Other Protocols*   There are many other interesting proactive protocols: CSGR (Cluster Switched Gateway Routing) [65], WRP (Wireless Routing Protocol) [252], GSR (Global State Routing) [63], Fisheye state routing [273], TBRPF (Topology Dissemination Based on Reverse-Path Fowarding) [30], etc. Readers are referred to appropriate references for further information.

---

[2]Node E in this case will not rebroadcast the link-state information as it has broadcast the same information once already.

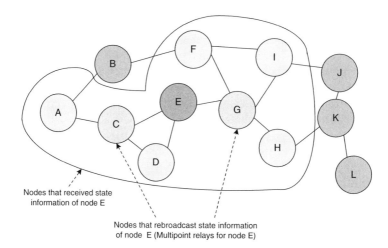

Nodes that received state
information of node E

Nodes that rebroadcast state information
of node E (Multipoint relays for node E)

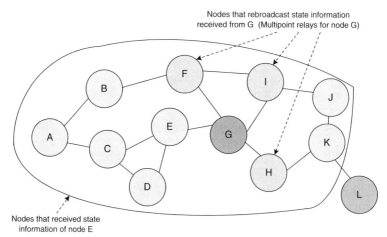

Nodes that rebroadcast state information
received from G (Multipoint relays for node G)

Nodes that received state
information of node E

Figure 19.7 OLSR routing scenario.

19.5.2 On-Demand/Reactive Routing Protocols

In this section, representative reactive routing protocols, namely DSR [173] and AODV [278], are discussed. Reactive routing protocols were proposed in view of a high mobility of wireless devices which may result in frequent updating of routing table. These updates, in fact, are control overhead and may not be necessary in some cases, as there may not have traffic in-between these updates. To prevent this wastage of resources, reactive routing protocols are designed to try finding routes to destination only when there are traffics to destination. The main advantages are that the control overheads are expected to be smaller than that of proactive routing protocols in case of high mobility and less traffics in-between updates.

The main feature of reactive routing protocols is that they will not maintain up-to-date routes to all destinations. There are several common features among reactive routing protocols. Firstly, they usually keep a cache storing the route from a source to certain destinations that they know and maintain expiration times for the routes in order to maintain the freshness of the routes. At the same time, since there are high chances of not finding routes to destinations, they usually employ Route Discovery Process to find the route to destinations. If such destination nodes are reachable, a Route Reply will be generated back to the source with the help of intermediate nodes which will also keep table entries about these routes. In the event that certain intermediate node within a valid route is moved out of range, route maintenance process is invoked.

19.5.2.1 *Dynamic Source Routing—DSR* DSR [173] is a typical example of reactive routing protocols. Every nodes in the network maintains a route cache that contains the source routes to destinations that the mobile devices have learnt before. The entries inside the caches are constantly updated whenever there is more updated information passing through the node. When a source node has packets to send to a certain destination, it would first consult the route cache to see if any entry contains an unexpired route to the destination [173, 28, 216, 140, 302].

If such a route does not exist in the route cache, the source node will initiate a Route Discovery Process as shown in Figure 19.8. Here, the node S generates a RREQ (Route REQuest) packet which includes the source address, the destination address and an unique identification number, for destination D. Node S broadcasts the packet to the neighbor nodes. Upon receipt of the packet, nodes N2, N1, and N5 will check their own route caches to see if any up-to-date routes to the specified destination are known. In case that such a route does not exist, they will append their own addresses into the RREQ packet and broadcast the packet again as shown in the figure. In case that an intermediate node receives multiple broadcast messages with the same source and destination address, it will only rebroadcast the first packet it has received. As shown in the figure, the node N4 receives the RREQ packet from N1 and N6, respectively. Since node N4 received the packet from N1 first, it will only rebroadcast the packet from N1 with its own address appended.

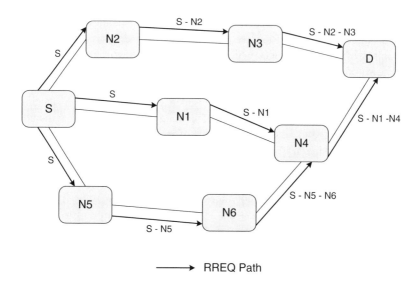

RREQ Path

Figure 19.8   DSR route request propagation.

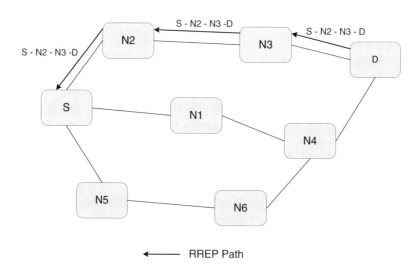

RREP Path

Figure 19.9   DSR route reply propagation.

When the destination node (i.e., Node D) received the RREQ packet or some intermediate node received the RREQ packet which has an up-to-date route (i.e., unexpired route) to destination node D in its route cache, the destination node or the intermediate node will generate a RREP (Route REPly) packet which contains the whole route list from source to the destination node as shown in Figure 19.9. The RREP packet will take the reverse path in the packet if the links are symmetric. The RREP packet may be routed back to source node S through another route discovery process if the links are asymmetric.

One of the unique feature of DSR is that it is a kind of source routing protocol in which every packet is included the whole route to destination inside the packet header.

19.5.2.2 *Ad-hoc On-demand Distance Vector—AODV*   AODV [278] is another typical routing protocol of the reactive type. Its route discovery and route reply process are very similar to that of DSR. Figure 19.10 and Figure 19.11 demonstrate these two processes [278, 28, 216, 140, 302].

When a source node S wants to send a message to destination node D and does not have a valid route to that destination, it initiates a route discovery process to locate the destination by generating a RREQ packet similar to that of DSR. The RREQ packet generated contains source address, source sequence number, broadcast-id, Destination address, Destination Sequence number, and Hop count. The RREQ packet is then broadcast to the neighbors. As shown in Figure 19.10, node S broadcasts the RREP packet to its neighbors. Upon receipt of the broadcast packet, the neighbor nodes N2, N1, and N5 check to see if they have an up-to-date route to destination D. If not, they will rebroadcast the the packet to their neighbors and setup a reverse path entry inside the route table (i.e., the intermediate node will remember the first node from which it has received the RREQ packet). At the same time, the hop count field inside the packet will be incremented by one. Similar to the case in DSR, a neighbor node will only rebroadcast the RREQ packet, which is uniquely identified by the source address and broadcast-id pair which will be incremented by one every time the source address initiates a packet.

The RREP packet eventually reaches the destination node D or an intermediate node which has an up-to-date route to destination. (Whether the intermediate node has an up-to-date route or not is determined by the destination sequence number in the RREQ packet and that known by the intermediate node. Only when the destination sequence number known by intermediate node is larger than or equal to that in the RREQ packet is considered as up-to-date). The RREP packet will be route back through the reverse path setup before back to the source as shown in Figure 19.11. Upon receipt of the RREP packet, the intermediate node will also set up the forward path which points to the node where the RREP packet comes from. This route entry will be associated with a timer to maintain the freshness of the route to the destination.

However, it is still possible for multiple RREP packets to reach certain intermediate nodes or source node. In this case, instead of forwarding the packets back to the source,

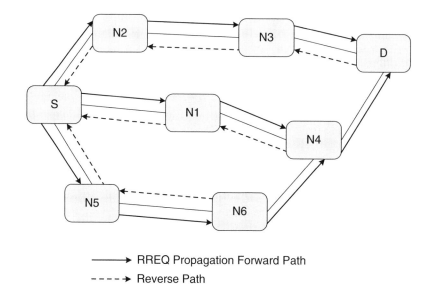

Figure 19.10 AODV route request propagation.

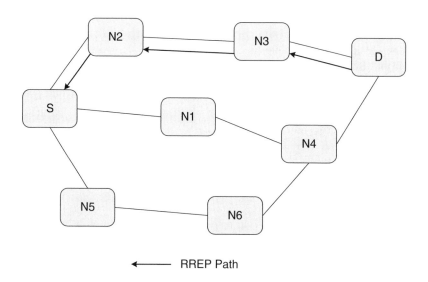

Figure 19.11 AODV route reply propagation.

the intermediate nodes will compare the destination sequence number between the one in record and that newly come RREP packet. The intermediate node will setup the forward path to the node where the newly RREP packet comes and unicast that to its reverse path neighbor only if the destination sequence number of this RREP packet is larger than that in record.

19.5.2.3 Receiver Initiated Channel Adaptive (RICA) Routing Protocol

We describe a variant of reactive routing protocols called RICA in this section. As it has been discussed, it is possible for the routing protocols to be designed to improve certain performance metrics discussed in Section 19.4. The following protocol is designed in view of possibility to improve the throughput by selecting links with high throughput to destination.

In Section 9.5, the physical characteristics of IEEE 802.11b was briefly discussed. It is known that there are 4 different transmission rate supported by IEEE 802.11b—1, 2, 5.5 and 11 Mbps. These transmission rates are supported with the use of different modulation scheme—DBPSK, DQPSK and CCK, respectively. In fact, each mobile terminal will adjust their transmission rate with different modulation scheme based on the perceived SNR or based on the CSI (Channel State Information) of the current channel.

Based on CSI, it is possible to classify each link (or hop) into different classes. In the following example, we refer to IEEE 802.11b. However, the idea can be easily extended to other IEEE 802.11 standards. Different links can be classified into different classes which can then be used to assign different weights to different links or hops. Table 19.1 gives a summary of the classification described above in IEEE 802.11b.

Table 19.1 Classification of different IEEE 802.11b physical transmission rate.

Transmission Rate (Mbps)	Modulation Scheme	Class	Weighted cost
1	DBPSK	D	11
2	DQPSK	C	5.5
5.5	CCK (4-bit)	B	2
11	CCK (8-bit)	A	1

Similar to that of DSR and AODV, RICA [216] is a reactive routing protocol which initiates a route discovery process when it requires to find route to destination. Figure 19.12 illustrates the route discovery and route reply process. At the beginning, the source node S would like to send a packet to destination D, so node S broadcasts a RREQ packet to look for the route to node D. The RREQ packet includes source address, destination address, broadcast id, hop count, CSI hop count and also lists of intermediate addresses.

Upon receipt of the RREQ packet, intermediate nodes will then:

- Append their addresses to the list of intermediate addresses.

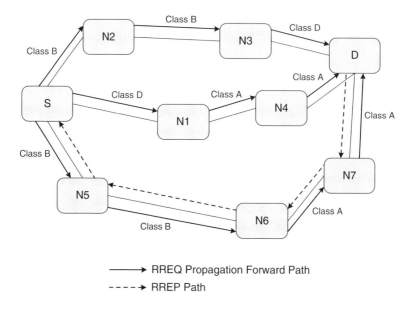

Figure 19.12 RICA route request and route reply propagation.

- Increment the hop count.

- Increment the CSI hop count by the weighted cost suggested in Table 19.1 based on the measurement of CSI of the link from where the RREQ is transmitted.

This process will continue until multiple RREQ reaches the destination node D as shown in the figure. Upon receipt of RREQ, the destination node can learn that there are several routes to destination and will try calculating the CSI hop count. In Figure 19.12, the route S→N2→N3→D has CSI hop count 15, the route S->N1->N4->D has CSI hop count 12 and the route S→N5→N6→N4→D has the CSI hop count 6. The destination node D would then generate a RREP packet which contains the source address, destination address, list of intermediate nodes, hop count, CSI hop count, reply ID, etc. This RREP packet will then be unicast back to the source. When the RREP passes through an intermediate node, a forward path to destination is also set up. At this time, node S can then decide whether to change its route or not, it will then unicast the route update packet through the selected route to the destination.

We have discussed that the channel quality between two mobile devices is time-varying. During the transmission of traffic flow from source to destination, the destination will periodically broadcast CSI-checking packet to gather information about the changing of the channel quality. Figure 19.13 illustrates the idea about the destination initiated CSI checking. The destination will send a CSI checking packet periodically to its neighbor. The packet includes similar items as that of RREQ

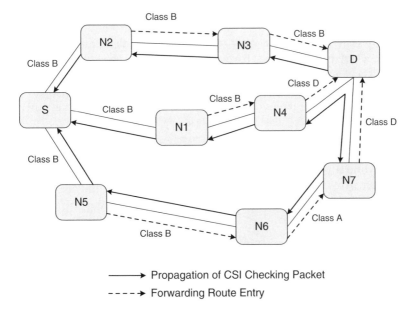

Figure 19.13   CSI checking propagation.

packet described before. However, when an intermediate node receive the packet, it will fill up the hop count field and CSI hop field similar to that suggested in RICA RREQ process. After that, it will set up a forward path to the node in which CSI checking packet arrives and rebroadcast the CSI checking packet with the updated field. In this case, similar to that of AODV and DSR, the intermediate node will only broadcast the first packet received. This process will continue until the source receives the CSI checking packet. At this point, the source can select routes again based on the shortest CSI hop to the destination by unicasting the route update back to the destination similar to that suggested above. As indicated in Figure 19.13, it is easy to see that the source node S will receive 3 copies of CSI checking packet. The source node S will know that the routes to destination via N2, N1 and N5 are of cost values 6, 15, 16, respectively.

19.5.2.4   Other Protocols   There are many interesting reactive routing proto-cols: TORA (Temporally Ordered Routing Algorithm) [269], CBRP (cluster-based routing protocol) [170], SSA (Signal Stability-based adaptive routing) [106], ABR (Associativity-based routing) [302, 334], etc. Readers are referred to appropriate references for further information.

19.5.3 Hybrid Routing Protocols

In previous sections, two main categories of ad hoc routing protocols are briefly discussed. These two approaches have certain drawbacks, however. In table-driven routing protocol, unnecessary overheads are required periodically in order to maintain the routing table of all the nodes in the network. On the other hand, additional delay may result in looking for routes on when there are traffics. In order to get the best of both worlds, researchers proposed some hybrid approaches.

19.5.3.1 *Zone Routing Protocol* Zone routing protocol [272] is a typical example of hybrid approach. Generally speaking, the functions of zone routing protocol can be summarized in Figure 19.14 [28, 140].

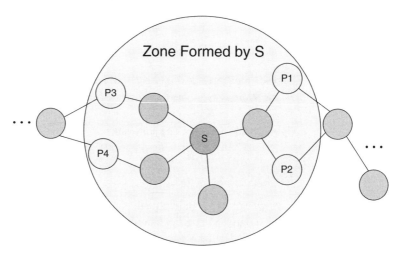

Figure 19.14 Zone routing scenario.

Under this routing protocol, each node will have a zone with certain zone radius which is defined as a predefined number of hops from the node in consideration. In the figure, the zone radius considered is two. Within the zone radius, proactive routing protocols like DSDV are used. In other words, the routing tables of every other nodes within the zone will be maintained actively. As a result, every node will know exactly how to route to other nodes within the zone radius. In Figure 19.14, node S will have direct knowledge on how to route to their neighbor nodes and border nodes, i.e., P1, P2, P3 and P4.

Suppose node S would like to send packets to a destination outside its own zone radius. In this case, node S does not have knowledge on where the destination is located. It will first send the request to their border nodes which are the nodes exactly the same number of hops with that of the zone radius, i.e., the zone radius is two and so node P1, P2, P3, P4 are border nodes in this case. Now, nodes P1, P2, P3, P4 receive the query from node S and thus, they will first look at their own route table to see whether destination node is within their zones. If not, they will continue relay the request to their border nodes. This process will continue until certain node has a route to the destination node which will then send the reply directly back to node S. As a result, this routing protocol is proactive within the zone and reactive outside the zone.

19.5.4   Ad Hoc Routing Protocols with Consideration of Other Constraints

Traditional ad hoc routing protocols suggested mainly aim at solving the problem of mobility of the nodes and temporarily formation of the network with low overhead. However, additional constraints may exist in some specific applications. Example constraints are security and energy efficiency.

19.5.4.1 Secure Routing Protocols   The previously discussed ad hoc routing protocols are generally proposed with the assumption that all the nodes in the network are cooperative. However, this assumption may not be justified. In particular, the network expected to be built in hostile environment in which malicious nodes may exist or the network expected to be built temporarily and without clear identification of each nodes. Consequently, some of them may be selfish in order to save resources. There are quite a number of secure routing protocols proposed. ARAN (Authenticated Routing for Ad Hoc Networks) [309] is one of the typical example.

ARAN is developed based on an extension to AODV. It assumes the existence of a globally trusted certificate authority. Every node can get a certificate associating its public key and IP address. As shown in Figure 19.15, suppose node S would like to find the route to destination D. It will broadcast a RDP (Route Discovery Packet) packet which is a packet with packet identifier i.e., ("RDP"), IP_D, etc. The content is then digitally signed by S's private key. The RDP packet will then be sent together with $cert_S$. Upon receipt of RDP packet and $cert_S$. Neighbor nodes N2, N1 and N5 can then verify the packet by getting the S public key from $cert_S$ and decrypt the packet. After verification, reverse path is setup similar to that of AODV as shown in Figure 19.10[3]. The nodes N2, N1 and N5 will then digitally signed the RDP packet with their own private keys and again broadcast the digitally signed packet. In the next step, N3, N4, and N6 received the broadcast and perform the same actions except that they will extract the RDP packet and signed with their own signature before broadcasting. When the destination node D received the broadcast, it will then verify the source RDP packet and generate a REP (reply) packet similar to that of RDP packet except that the packet identifier is changed to REP and the IP address becomes the source IP address. The REP packet will then be forwarded back through the reverse path using similar approach to that of AODV as shown in Figure 19.11.

---

[3]This part is not shown in Figure 19.15

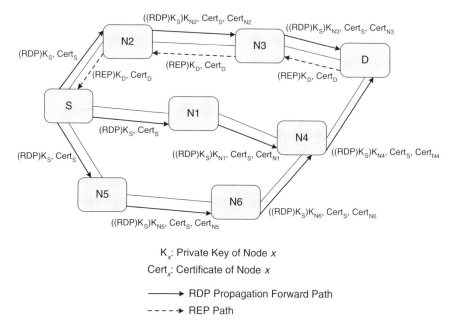

Figure 19.15   ARAN RDP and REP.

Apart from the use of certificates, there are other secure routing protocols proposed. Protocols using MAC (message authentication code) and μTESTLA to provide authenticated ad hoc routing is also suggested. Other than the use of keys and certificates, watchdog mechanisms which involve monitoring the packet forwarding behavior of neighbor nodes and pathrater mechanism which rated the path with the use of information provided by watchdog mechanism are also suggested to be extended from DSR to improve the security by selecting the route with high rate to destination [235]. CONFIDANT [43, 44] is recently suggested to further improve watchdog mechanisms and pathrater by providing trust manager and reputation system concept to evaluate the trust of neighbor nodes by combining the recommendation from other neighbor nodes observation on a certain node.

In addition to the direct security measures suggested, some multipath routing protocols are also proposed. These protocols are designed in view of the possibility of the existence of multiple disjoint routes from source to destinations. These routes can be used as a backup route in case the main route is broken or actively used to increase throughput and/or network resilient in case malicious nodes are present. Thus, the chances of packet delivery can be largely increased even malicious nodes which perform various different kinds of attacks e.g., blackhole attack, modification attack, etc.

19.5.4.2 Power-Aware Routing Protocols Traditionally, ad hoc routing is expected to solve the problem of mobility and temporary connection. Thus, intuitively, traditionally ad hoc routing protocols are expected to use the shortest path to route the packet from source to destination since minimizing the distance could lead to the highest probability of successful delivery. However, this intuition may lead to a path which uses a lot of energy. Moreover, most of the ad hoc network devices nowadays are expected to be powered by battery. The energy-deficient nodes in the network may cause the partitioning of the ad hoc network. As a result, energy consumption becomes an important consideration in ad hoc routing protocols [211].

In fact, there are quite a large number of power-aware routing protocols suggested. They can be mainly classified into two main categories—active conservation and passive conservation. There are also two kinds of objectives expected in active conservation approaches. The first kind of routing protocols aim at minimizing the power consumption of every packet sent. In this kind of routing protocol, they consider the path that can deliver the packet with minimum amount of energy from source to destination. The second kind of routing protocols aim at maximizing the network lifetime. In order words, they consider the path which may not be able to deliver the packet with minimum amount of energy, but possibly may result in maximum remaining energy in all nodes within the path after the delivery of the packet. By doing so, the network lifetime can be extended.

The passive conservation approach takes advantage of the different modes of mobile devices which have different power consumption. These modes are: transmit, receive, idle and sleep. Among them, the transmit mode requires the largest amount of energy while the sleep mode consumes the least. With the appropriate turning off the nodes in the network with minimum connectivity for routing, energy consumption in the network can be greatly reduced. For further information, readers are referred to [211].

19.5.5 Summary

In this section, we have discussed two classical classifications of ad hoc routing protocols. These routing protocols are then extended to various routing protocols with different constraints and different considerations. For instance, considering location information of nodes results in Location-Aided Routing Protocol. There are many new protocols suggested in recent years. Indeed, ad hoc routing continues to be a hot research topic.

19.6 PRACTICAL ILLUSTRATIONS—WLAN BASED AD HOC ROUTING

Various WLAN based Ad Hoc Routing protocols have been discussed in previous sections. In this section, a practical scenario of Ad Hoc Network—Mesh Network which has been proposed recently will be discussed in detail. Some of the routing issues inside such kind of networks will also be discussed.

Figure 19.16 illustrates an example of mesh network. As shown in the figure, it is expected that within a town or a city, not all the buildings will be equipped with Internet Access capability. However, they will have WLAN facilities. In this case, ad hoc network can be formed and the buildings can cooperate with each other to route the packets from other buildings towards the location which has internet access capabilities [256].

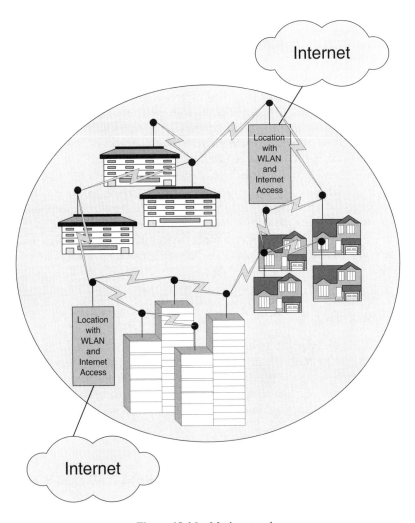

Figure 19.16 Mesh network.

This kind of ad hoc network is relatively static in the sense that the WLAN facilities are located in fixed locations within the buildings. However, it is Ad Hoc in the sense that the owners of the building may decide on their own that whether, when and how long they will turn on the WLAN facilities. As a result, it is possible for the route from source building to the location with Internet access point to be dynamic in nature. At the same time, multiple access points may exist within a town or a city. Thus, the selection of the routes may also be affected by the loads of certain access point usages at that specific period of time.

There are a number of advantages of mesh network.

- Lower cost: The cost of obtaining Internet access in such a network is expected be lower than traditional Internet access methods e.g., DSL which required every unit to subscribe their own services. With the help of mesh network, it only requires several Internet access points to exist in a town or a city. The access points can then be shared by many users through the mesh network.

- Resilient: With the help of mesh network, multi-homing is possible. As a result, the Internet access service will not be affected with the failure of single access point or single internet service provider.

Figure 19.16 shows that there are several routes from the source (e.g., residential unit) to a destination (i.e., Internet Access Point location). In this case, it is required to have an efficient routing algorithm to route the packets from source to destination. Most of the routing protocols suggested in Section 19.5 can be used efficiently here. The most commonly suggested algorithms for mesh network are DSDV, DSR and AODV. However, the main objective of these routing protocols is to find the shortest path (i.e., the smallest number of hops count path). However, it has been suggested that shortest path may not be a good path for routing the packets from source to destination. In particular, in a mesh network, throughput of the path will be more important than the shortest path. As a result, it is suggested that expected number of transmissions (ETX) [76] should be used as a metric to remedy the shortest path problem in ad hoc network routing protocols (e.g., DSDV and DSR) [76, 90]. Moreover, it is expected that the units inside mesh networks may be equipped with more than one WLAN equipment. As a result, Weighted Cumulative Expected Transmission Time (WCETT) [91] is suggested as a metric to utilize multiple channels possibility in mesh network to remedy the shortest path problem.

19.7 SUMMARY

In this chapter, ad hoc routing protocols are briefly discussed. We started our discussion by introducing some of the applications—including the usage in building pure temporarily infrastructureless network, the usage in building hybrid infrastructure and infrastructureless network in which ad hoc network can be viewed as a method to extend the capability of infrastructure network, the usage in sensor networks.

The system model of ad hoc network is briefly discussed. The model generally involves the settings such as the size of location, the number of users, the movement pattern of users, the traffic pattern, the protocols used, etc. These settings are different with different application scenarios.

There are quite a large number of routing protocols proposed in the literature and the list keeps on expanding. Generally, they can be classified into two main categories— proactive and reactive routing protocol or sometime known as table-driven and on-demand routing protocol. In proactive routing protocol, nodes will keep and maintain their routing tables (may be Distance Vector or Link State) actively and periodically. In other words, they will keep track of the routes to other destinations even when there is no traffic. These protocols are known to have relatively larger control overhead due to the periodic update which may be unnecessary or too frequent due to the movement of the nodes. On the other hand, Reactive protocols are proposed to look for the route to the destinations only when there are traffics to destination. The main drawback of these "just-in-time" protocols are that it may have long delay to look for the route. These delays may not be justified when the traffic rates are large and the mobility of the nodes are low. Due to these drawbacks, some hybrid approaches are also suggested in order to strike a trade-off between the traffic loads and mobilities. At the same time, the existence of some additional information such as location information may help in routing. For example, powered by battery or working in hostile environment, energy and security become important considerations which leads to the development of energy conservation routing and secure routing or multipath routing.

Finally, we conclude this chapter with a mesh networking scenario which is expected to be developed in coming future. This can be regarded as hybrid infrastructure and infrastructureless architecture in which residential houses are connected through wireless ad hoc network and the packets are routed to and from the access points through some ad hoc routing protocols.

PROBLEMS

19.1   Identify the reasons for the construction of ad hoc networks.

19.2   Briefly discuss the difference between table-driven and on-demand routing protocols.

19.3   In the practical illustration section, mesh network is suggested. Try suggesting some other practical scenarios in which ad hoc network routing will be useful.

CHAPTER 20

WIRELESS DATA CACHING

Few have too much and fewer too little.

—Norwegian Proverb

20.1 INTRODUCTION

Wireless data access is a typical application level activity in mobile computing. Frequently, wireless devices need to fetch information items from some servers in order to carry out its functions. For instance, a wireless PDA needs to synchronize with some data servers so as to provide updated information (e.g., news, maps, etc.) to its user.

Interoperability:

The ability to obtain on-demand information access in client-server wireless networks would provide valuable support to many interesting mobile computing applications. Researchers have proposed two promising techniques to mitigate the existing bandwidth and energy constraints: data broadcasting and client caching. Data

Wireless Internet and Mobile Computing. By Yu-Kwong Ricky Kwok and Vincent K. N. Lau **501**

broadcasting exploits the sharing nature of the wireless medium, while client caching leverages the skewed access pattern. The major challenge of client caching is to maintain cache consistency. Clients should ensure that the cache is valid before using it to answer any query. Several cache invalidation schemes have been proposed, most of which also take advantage of data broadcasting. The basic scheme is to broadcast periodic invalidation reports (IR). Recently, the addition of updated invalidation reports (UIR) has been proposed. The IR+UIR scheme achieves a much shorter query delay than the basic IR scheme. However, existing invalidation schemes are based on two simplifying and impractical assumptions: 1.) the broadcast channel is error-free; and 2.) no other downlink traffic exists. In reality, a more realistic system model can be employed: 1.) the quality of the wireless channel is time-varying; and 2.) there are other downlink traffic sources in the system. Link adaptation allows more efficient use of bandwidth via dynamically adjusting the transmission rate with channel quality.

Performance:

The performance of various cache invalidation schemes can be evaluated based on three performance metrics: 1.) average query delay; 2.) number of uplink requests per successful query; and 3.) average delay experienced by other downlink traffic (e.g., voice). To study how the different approaches perform under various situations, we can study the performance metrics by varying system parameters like number of clients, mean query generation time, mean update arrival time, access skew (hot access probability), and mean disconnection time.

20.2 APPLICATION SCENARIOS

Scenario 1—At a Conference:

In an academic conference, researchers around the world are gathered at the exhibition center of a city. You are a speaker of a presentation session. In the last few minutes, you have updated the presentation slide, which is made available to the peer researchers via the organizer's wireless network.

You have discovered that there are slight changes in the schedule as printed on the final program booklet. You then download the updated one to your personal digital assistant (PDA) device. In addition, you discover that there is a discussion forum and go through a few interesting topics. The conference organizer also sets up a counter selling latest textbooks of your field. Sitting comfortably in the lobby, you browse the list to decide if there is any suitable one for your next semester's course.

Before departure, you want to take the chance to explore that foreign city. You query for some useful information for planning the adventure. At the end of the day, you check for the up-to-date weather information and confirm the flight schedule before going to bed.

Scenario 2—In a Shopping Mall:

In a weekend, your family is going to spend the evening in a shopping mall. Your favorite restaurant has a full house and you are put in the waiting list. The good news is that you do not need to queue up outside but can check the status via the shopping mall's wireless network. So, you query for the latest events and any special offer using your smart cellular phone. You discover that a famous boutique is having its seasonal sale and download the online catalog to have a look. Then, you decide to visit the shop and make a purchase. There is indeed a promotional campaign. With the receipt, you can participate in a lottery. After finishing the dinner at the restaurant, you query for the winner via the wireless network.

Scenario 3—In the Battlefield:

In military operations, unmanned electronic sensors are used for gathering different types of intelligent information about the enemy, such as taking photos, collecting soil samples, exploring the hostile terrain, etc. Each sensor queries the central control unit about the details of its tasks. To improve reliability, a single task may be duplicated or shared by more than one sensor. For example, sensors 1–200 are responsible for task i, which is to take photos of the missile unit, while sensors 101–200 is responsible for task j, which is to record the ambient temperature. The details of each task may change from time to time. For example, task j is changed to record the temperature only within the time period when the sensor is within a specific region. Each sensor also requests the central control unit for some common information. For example, this may include the current geographical position of itself or its peers.

Some Issues in On-Demand Information Access:

In a wireless environment, one of the distinguishing characteristics is the inherent sharing nature of the communication medium. When the server delivers a piece of information in the common broadcast channel, all the clients, within the communication range, are capable of receiving the broadcast information. Barbará and Imieliński pioneered the use of common broadcast channel to support on-demand information access [26]. This enables more efficient use of the precious wireless bandwidth.

Another promising technique is client caching. A client sends a query request to the server for a particular data item. Server replies with the requested data in the broadcast channel. In addition to answering the query, the client caches the data item in its local storage. If a new query arrives at a later time, requesting for the same data, the cached copy can be used without going through the server as before. This could save wireless bandwidth and battery energy, via reduced amount of uplink traffic. However, the use of client caching also raises another concern—cache consistency. In practice, data items are updated from time to time in an asynchronous manner. A client cannot rely on its cached copies to answer queries forever. Instead, the validity of the caches should be first ensured. The latest value model is a popular consistency model for wireless data access [48, 50, 49, 51, 156, 331], which requires that the most recent version of a data item is used to serve queries. Thus, a cache consistency scheme is needed for proper use of client caching.

20.3 SYSTEM ARCHITECTURE

Figure 20.1 depicts a typical cellular network, which consists of a number of base stations and a set of mobile users. Base stations are connected to a common database via a wired network. Each base station (server) serves the users within its covered area (cell). There are two types of users in the system, namely voice and data users. A voice session requires r_{speech} Kbps. Each data user (client) is making a file transfer session at r_{file} Kbps and also generating a stream of exponentially distributed read-only requests with mean arrival time T_q. The database is divided into hot and cold sets. Updates to each data item follow an exponential distribution with mean arrival time T_u, from which hot items have an update probability of p_u.

Figure 20.2 further illustrates the interaction between the server and the clients.

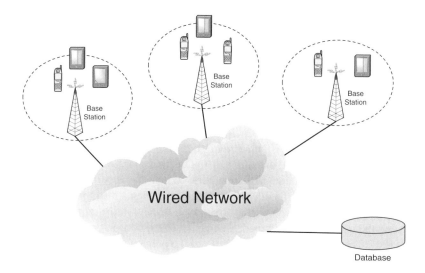

Figure 20.1 A cellular network with data and voice users.

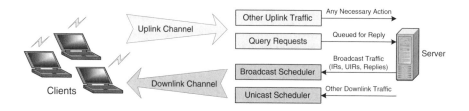

Figure 20.2 Client-server interaction.

Adaptive Physical Layer:

The wireless link between a mobile device and the base station is characterized by two signal propagation components, the fast fading component and the long-term shadowing component, which are discussed in more detail in Section 17.3. Fast fading is caused by the superposition of multipath components and is therefore fluctuating in a very fast manner (on the order of a few milliseconds). Long-term shadowing is caused by terrain configuration or obstacles and is fluctuating only in a relatively much slower manner (on the order of one to two seconds). Since mobile devices are scattered geographically across the system and are moving independently of each other, it is assumed that the channel fading experienced by each mobile device is independent of one another.

To study the channel characteristics and the effect of other downlink traffic on existing cache invalidation strategies, and to exploit the time-varying nature of the wireless channel, we can employ a variable-throughput channel-adaptive physical layer as presented in Section 17.3. The model is based on the concept of exploiting the synergy between physical layer and medium access control, instead of strictly following the traditional layered approach in protocol design.

As depicted in Figure 17.7, the receiver estimates the Channel State Information (CSI) and informs the transmitter via a low-capacity feedback channel, a facility commonly provided in practical cellular systems (e.g., cdma2000). In this system model, the base station obtains the CSI by polling a particular client via the polling subframe, and the client then responds with a known pattern in the pilot subframe (the uplink and downlink frame structure is shown in Figure 17.5). Based on the feedback CSI from the receiver, the level of redundancy and the modulation constellation applied to the information packets are adjusted accordingly by choosing a suitable transmission mode. For example, we can employ a six-mode (mode-0 to mode-5) variable-throughput adaptive bit-interleaved trellis coded modulation (ABICM) [198]. Transmission modes with normalized throughput varying from $1/2$ to $5/6$ are available depending on the channel condition.

The ABICM performs "burst-by-burst" adaptation based on the CSI measurement on the physical layer. Specifically, when the channel condition (as indicated by the CSI) is good (fading attenuation is small), the physical layer employs high order modulation and high rate error correction coding so as to boost the instantaneous throughput. On the other hand, when the channel condition is poor, the physical layer employs low-order modulation and low rate error correction coding so as to better protect the packet transmission at the expense of lower throughput. In fact, due to the salient performance in terms of system utilization, such channel adaptation techniques have been widely employed in practical systems such as 3G1x, UMTS-HSDPA, and wireless LAN systems [46].

Since the coherence time of short-term fading is around 60ms which is much longer than an information slot duration, CSI is approximately constant within a frame and it follows that the transmission mode for the whole frame is determined only by the current CSI level. Specifically, transmission mode q is chosen if the feedback CSI, $\hat{c}$, falls within the adaptation thresholds, (ξ_{q-1}, ξ_q). The adaptation thresholds are set optimally to maintain a target transmission error level over a range of CSI values. When the channel condition is good, a higher mode could be used and the system enjoys a higher throughput. On the other hand, when the channel condition is bad, a lower mode is used to maintain an acceptable error level at the expense of a lower transmission throughput. The performance details of the variable throughput physical layer are shown in Table 20.1.

Table 20.1 Performance of the variable throughput physical layer.

Transmission Mode	Constellation	Transmission Rate (Kbps)	Target Bit Error Rate	Data Payload Capacity (bits)
0	BPSK	160	0.00002	50
1	QPSK	320	0.0001	100
2	8PSK	640	0.0005	200
3	16QAM	960	0.002	300
4	32QAM	1280	0.007	400
5	64QAM	1600	0.05	500

20.4 PERFORMANCE METRICS

The performance of various cache invalidation schemes can be evaluated based on three performance metrics:

1. average query delay;

2. number of uplink requests per successful query; and

3. average delay experienced by other downlink traffic (e.g., voice).

To study how the different approaches perform under various situations, the above performance metrics can be studied by varying six system parameters:

Number of Clients: It measures the scalability of the underlying cache invalidation scheme.

Mean Query Generation Time: It is related to client size, but queries from different clients and the corresponding replies have different corruption probability. Another difference is that larger client size means more downlink traffic in the simulation environment. Using this parameter, we would like to study how different query rates affect the performance metrics.

Mean Update Arrival Time: It aims to study different cache invalidation schemes performance under various update conditions. Update arrival time is directly related to the number of obsolete data items. When data items are updated more frequently, cached items are more likely to be invalidated. Subsequent queries would result in cache misses.

Number of UIRs: If there are more UIRs, clients can invalidate their cache more frequently and answer queries locally with smaller delay. However, more UIRs means larger broadcast overhead, which has a negative effect on other downlink traffic. On the other hand, if the number of UIR replicates is too small, the performance gain from using UIRs is not significant. Thus, this parameter studies the optimal number of UIR replicates between successive IRs such that we can achieve performance improvement without sacrificing other downlink traffic too much.

Access Skew (Hot Access Probability): The effect of access skew on different cache invalidation schemes can be studied by varying the hot data access probability (p_q). The range of p_q value represents different query patterns. If most clients have similar query characteristics, p_q is close to 1; otherwise, pq would be just a small value. Thus, through this parameter, we can study how the performance metrics are affected under different query patterns.

Mean Disconnection Time: It studies the effect of mean disconnection time on different performance metrics. Each client has a mean connection time of, let say, 1000s. Mean disconnection period can be varied from 10s to 10^4s. Short disconnection period, e.g., 10s, is similar to hard handoff.

20.5 DATA CACHING APPROACHES

Existing cache invalidation algorithms can be broadly classified into two types: stateful server and stateless server, as illustrated in Figure 20.3. In a stateful server scheme, the server maintains a set of state information for each of the clients. On the other hand, the server is oblivious to the clients' states in a stateless server scheme. This reduces server's workload and lowers the complexity of the cache invalidation algorithm. Thus, stateless server is more popular than stateful server.

Figure 20.3 shows a simple taxonomy of a number of representative cache invalidation algorithms. The asynchronous and stateful cache invalidation algorithm (AS) belongs to the class of stateful server. Stateless server algorithms include the invalidation report algorithm (IR) and the bit sequence algorithm (BS). Due to its salient characteristics, IR is further extended with the addition of updated invalidation report, i.e., IR+UIR. IR+UIR is again further extended with some channel adaptive techniques by Yeung *et al.* in [362].

Invalidation Report (IR):

Barbará and Imieliński [26] have proposed the basic stateless server approach to maintain cache consistency. The server broadcasts an invalidation report (IR) every L

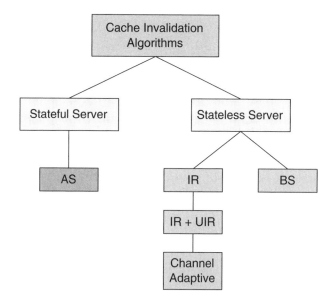

Figure 20.3 Overview of existing cache invalidation algorithms.

seconds. The primary purpose of invalidation reports is to announce a list of recently updated data items. Each client compares its caches with the list in IR and isolates any obsolete ones. Thus, the remaining copies are valid and suitable for serving future queries. Formally, each IR is a list of tuples:

$$IR_i = \{(d_x, t_x)|(T_i - \omega L) < t_x \le T_i\}, \tag{20.1}$$

where d_x is the index of a data item x, t_x is the timestamp of its most recent update, T_i is the current timestamp, and ω is the broadcast window size, which controls the amount of invalidation information to be included in each IR. Clients do not directly use their caches to answer queries. Instead, a client waits for the next IR to determine whether its cache is valid or not. If a valid cached copy exists, the query can be served locally. Otherwise, the client issues an uplink query request to the server and waits for the reply after the next IR broadcast. As illustrated in Figure 20.4, a new query arrives at a time between T_i and T_{i+1}. If the client has a valid copy in its cache, the query can be served locally after the IR broadcast at T_{i+1}. Otherwise, an uplink request will be sent to the server. The client is expected to obtain its requested data item in the following IR broadcast, which is at T_{i+2}. Thus, the expected query delay for cache hits is $\frac{L}{2}$ while that for cache misses is $L + \frac{L}{2} = \frac{3}{2}L$. Furthermore, if a client is disconnected from the server for a period less than ωL, it can still use an IR to invalidate any obsolete data items. Thus, another function of the parameters, ω and L, is to control the longest disconnection time that can be supported by the algorithm.

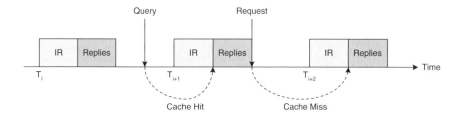

Figure 20.4   The IR cache invalidation algorithm.

Advantages of IR-based approaches include high scalability and energy efficiency: the size of each IR is independent of the number of clients; and IRs are scheduled to be broadcast periodically. As such, clients can switch to doze mode operation between successive IRs to conserve battery power. The major drawbacks, however, are: 1.) clients must flush their entire caches after long disconnection (ωL), even some of the cached items may still be valid. Since each IR contains only invalidation information in the last ωL seconds, a client cannot tell whether its cache is valid or not beyond that period; 2.) client should wait for the next IR before answering a query to ensure consistency, which potentially causes large query delay.

Invalidation Report with Updated Invalidation Report (IR+UIR):

To reduce the long query delay associated with the basic IR cache invalidation approach, Cao [51] has done some pioneering work. He proposed the addition of updated invalidation report (UIR) to the basic IR scheme (IR+UIR). Each UIR contains invalidation information for those data items that have been updated since the last IR. Formally, each UIR is given by:

$$UIR_{i,k} = \{(d_x, t_x)|T_i < t_x\}, k \in [1, m-1], \qquad (20.2)$$

where $(m-1)$ is the UIR replicate times.

In other words, each UIR indicates all the updates since the last IR, which represents the most up-to-date view of the database. Clients use UIRs to invalidate their caches. If there is a valid cache, i.e., cache hit, the query can be served at an earlier time. Otherwise, i.e., cache misses, clients can also send an uplink request at an earlier time. In both cases, the query delay can be reduced.

As shown in Figure 20.5, the same query arrives at a time between T_i and T_{i+1}. For cache hit, the query can be served locally after the next UIR broadcast. Otherwise, the client issues an uplink request to the server and waits for the reply after the next IR at T_{i+1}. In this way, IR+UIR reduces the average query latency at the expense of slightly more broadcast overhead in UIRs [50, 49, 51]. Thus, the expected query delay for cache hits is $\frac{L}{2m}$ while that for cache misses is $\frac{L}{2}$.

However, the use of UIRs at a time between T_i and T_{i+1} assumes that data items in the cache are consistent up to time T_i. This condition is satisfied if the last IR is correctly received, i.e., IR at T_i. Due to disconnection problems or erroneous

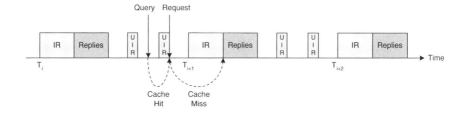

Figure 20.5   The IR+UIR cache invalidation algorithm.

transmissions, some clients may miss this last IR. These clients cannot use any of the successive UIRs before receiving the next correct IR at some later time. Equivalently, the clients are running the basic IR invalidation approach for that period. Therefore, the effectiveness of the IR+UIR scheme largely depends on the integrity of IRs.

Bit Sequence (BS):

Various approaches have been proposed to deal with long disconnection in the basic IR cache invalidation algorithm [156, 171, 331]. Jing *et al.* [171] have proposed the bit-sequence cache invalidation method (BS). Basically, each invalidation report is replaced by a set of bit-sequences. Each bit sequence records the update time of data items, which is broadcast to clients for cache invalidation. However, the BS scheme does not have the time constraint (ωL) as in IR-based schemes.

Each bit-sequence is a fixed-length bit vector. Each bit in a bit-sequence corresponds to a data item in the database. The longest bit-sequence (BS_1) is of length N bits, where N is the number of data items in the database. The bit position indicates the index of a data item in the database. For example, the nth bit of BS_1 represents data item d_n. Each sequence is also associated with a timestamp, T. If the nth data item is updated after T, the nth bit is marked with 1. Otherwise, the bit is 0.

The second longest bit-sequence (BS_2) is of length $N/2$. The length of the next bit-sequence is $N/4$ and so on. The last bit-sequence is a two-bit vector. The kth bit of these bit-sequences represent the position of the kth "1" in the previous bit-sequence. If $BS_1 = 00101011_2$, the 1st and the 2nd bits of BS_2 correspond to the 3rd and the 5th bits of BS_1, respectively. Similarly, a value of 1 (0) represents the data item is (not) updated. If $BS_2 = 0100_2$, this indicates that the 5th bit of BS_1 is updated, which corresponds to data item d_5. For a bit-sequence of length n, the timestamp is the time at which half of the data items have been updated, i.e., $n/2$ bits are marked 1. Initially, the number of 1's is less than $n/2$. The timestamp is then set to zero.

Compared with IR-based cached invalidation schemes (IR and IR+UIR), BS does not have the time constraint (ωL), which allows it to handle clients with arbitrarily long disconnection time. However, the underlying assumption is that the update rates of data items are not too high. Otherwise, as illustrated in the numerical example above, it is possible to result in inconsistency in client caches. Hu and Lee [156] proposed an adaptive scheme in that the server either broadcasts invalidation reports (IRs) or bit-sequences (BSs) based on the present update rates and requests from clients. Their adaptive scheme takes advantage of the two different classes of invalidation schemes, which would perform better in a more dynamic environment.

Asynchronous and Stateful Algorithm (AS):

The previous described schemes (IR, IR+UIR and BS) are based on synchronous or periodic broadcast of different cache invalidation information: invalidation reports, updated invalidation reports or bit-sequences. These schemes assume a stateless server, which is easily scalable to a large number of clients. Kahol *et al.* [180] have proposed an asynchronous and stateful cache invalidation algorithm.

The system architecture is shown in Figure 20.6. Each mobile host (MH) is associated with a home location cache (HLC), which maintains a list of caches kept by the mobile host, i.e., the state information. If a mobile host is roaming to a foreign location, the HLC is duplicated at the foreign mobile switching station. This is to ensure that there is always an MSS (mobile switching station) to maintain the HLC for each MH at any given time. Thus, the server is stateful as opposed to the previous stateless schemes. The state information, kept at HLC, is a list of tuples: $(d_x, t_x, flag)$, where d_x is the id of the data item and t_x is the timestamp of its last invalidation. The flag is set to true when an invalidation has been sent to theMHbut no acknowledgment has been received. The distinguishing characteristic of the AS cache invalidation scheme is that the cache invalidation information is sent asynchronously and also buffered at MSS until an explicit acknowledgment is received.

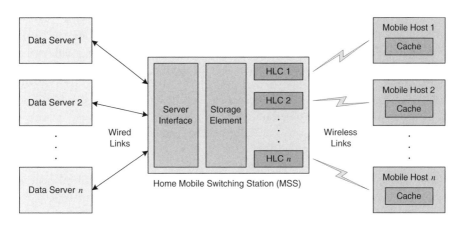

Figure 20.6   System architecture of the AS cache invalidation algorithm [180].

Each MH maintains a set of data items in its local cache. The validity of the cache should be ensured before using it to answer any query. In AS, this is achieved with the aid of the MSS. The data server informs the MSS when a data item is updated. The MSS determines, from HLCs, the set of MHs that caches the data item. Then, the MSS sends the invalidation information to the affected MHs one-by-one. Having received the invalidation information from MSS, the MH invalidates its cached copy.

To support disconnection, each client maintains a cache timestamp, which is the timestamp of the last received message. The client includes the cache timestamp in every message to the MSS. In particular, when an MH resumes after disconnection, it sends a probe message with cache timestamp, t, to the MSS. Upon receiving the message, the MSS consults the corresponding HLC to: 1.) discard all invalidation records with timestamp less than or equal to t; and 2.) send all invalidation records with timestamp greater than t. The MH defers all queries until it receives the acknowledgment from its probe message. In this manner, an MH can determine the set of data items being updated during its disconnection.

We illustrate the operations of the AS cache invalidation algorithm with reference to the scenario shown in Figure 20.7.

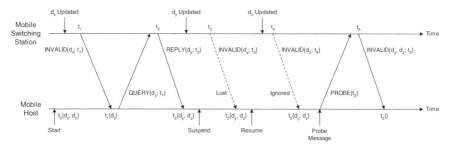

Figure 20.7   An illustration of the AS cache invalidation algorithm [180].

Initially, the mobile host (client) caches two data items, d_x and d_z, with cache timestamp t_0. The data item, d_x, is updated at the mobile switching station (server). At t_1, the server sends an invalidation message, informing the client that d_x has been updated since t_1. Having received the message, the client invalidates its cached copy of d_x and updates the cache timestamp to t_1.

A query for data item d_y arrives at the client. Since there is no cache for d_x, the client sends an uplink query request to the server. The server replies with the requested data item together with the current timestamp, t_2. Using the reply message, the client: 1.) answers the query; 2.) caches the data item d_y; and 3.) updates the cache timestamp to t_2. The client then suspends for a period of time. During its suspension, d_y is updated and the server sends an invalidation message at t_3. However, the message is lost.

Upon its resumption, d_z is also updated, the client receives an invalidation message for d_z. However, this invalidation message is ignored by the client. Instead, it sends a probe message to the server, requesting for all the invalidation records after its cache timestamp, t_2. The server consults the client's HLC and constructs the updated invalidation message. Finally, the client invalidates both dy and dz and updates its cache timestamp to t_5.

The AS cache invalidation is capable of handling arbitrarily long disconnection. However, it relies on the server to maintain a set of state information for each of the clients. Thus, scalability is a major concern. Unlike the stateless server approaches, the server conveys an invalidation message to each client in the unicast channel. If a data item, which is cached by many clients, is updated, this would suddenly generate a large amount of invalidation messages. Furthermore, the server sends invalidation messages in an asynchronous manner. This requires that a client should always in the listening state, which consumes a considerable amount of energy comparable to the receiving state [112, 111].

Reducing the Probability of Corruption in IR:

Existing cache invalidation strategies broadcast each IR in a single packet. However, this is vulnerable to transmission errors. Dividing the IR into a number of segments, with each segment separately transmitted, can reduce both the corruption probability and power consumption.

Server broadcasts invalidation reports (IRs) every L seconds. Each IR contains the current timestamp (T_i) and a list of pairs (d_x, t_x), where d_x is the id of a data item and t_x is the timestamp at which the data item is last updated. To support long disconnection period, data items that are updated in the past ωL seconds are included in each IR, i.e., $t_x > (Ti - \omega L)$, where ω is the broadcast window size. The server also broadcasts ($m - 1$) updated invalidation reports (UIRs) between successive IRs, where ($m - 1$) is the UIR replicate times. Each UIR indicates those data items that are updated since the last IR. Clients make use of IRs and UIRs to keep their caches consistent by discarding obsolete data. From the remaining valid cache, some queries can be served locally. However, the use of UIRs requires that the cache is consistent up to the last IR. If a client misses the last IR, subsequent UIRs before the next IR cannot be used. Thus, the effectiveness of UIRs depends on the integrity of each IR.

Depending on the update arrival rate to the database, an IR may include a large number of id-timestamp pairs. It is crucial to ensure that clients can successfully receive an IR. However, broadcasting the entire IR at once over the error-prone wireless channel is inefficient. This is because such a long transmission is highly vulnerable to channel errors, especially for those clients with particularly unfavorable channel conditions (e.g., with a high mobility or situates in a poor terrain). Since there may be a large number of clients listening to the same IR in the broadcast channel, it is impractical for the server to retransmit an IR to completely avoid corruption. Instead, to reduce the chance of corruption, an original IR is divided into ω segments, IR_j for $j = 1$ to ω. Each IR segment then contains only id-timestamp pairs in a represented broadcast period, i.e., (d_x, t_x) is included in IR_j if and only if $(T_i - jL) < t_x \leq (T_i - (j-1)L)$, for $j = 1$ to ω. This increases the broadcast overhead from one to ω. However, if an id-timestamp pair is included in the original IR, it appears in one and only one IR segment. The total number of id-timestamp pairs in the ω segments remains the same as that in the original IR.

Dividing an IR into ω segments can reduce both the corruption probability and power consumption. Each IR contains cache invalidation information for ωL seconds while an IR segment contains cache invalidation information for L seconds only. Let the size of an IR and an IR segment to be $S_{\omega L}$ and S_L, respectively. Suppose the broadcast channel seen by a particular client has a bit error rate, P_e, and the overhead of each packet is x bits. In IR+UIR, the client should receive the complete IR. Thus, the probability of successfully receiving the IR is $(1 - P_e)^{(S_{\omega L}+x)}$. In the proposed scheme, the last IR segment is required. Thus, the probability of successfully receiving the last IR segment is $(1 - P_e)^{(S_L+x)}$. Since updates to a data item follow exponential distribution with mean update arrival rate of $\frac{1}{T_u}$ (recall that T_u is the mean arrival time), S_L would be smaller than $S_{\omega L}$. The success probability to receive an IR segment would be much higher than a complete IR. Thus, the client has a greater chance to benefit from subsequent UIRs. Although each IR is divided into a number of segments, the number of wake-up per broadcast period is the same as the IR+UIR algorithm. For clients with the previous IR, they are only required to retrieve the latest segment in the immediate IR. Thus, the power consumed in downloading the required cache invalidation information would be reduced.

The idea is further illustrated in Figure 20.8. In both cases, the client correctly receives the IR at T_{i-1} but some contents of IR at T_i are corrupted due to transmission errors. In the IR+UIR scheme, the client should discard the IR at T_i. There are two adverse consequences: 1.) the client cannot answer any query until the next successful IR, which is at least L seconds later; and 2.) the client cannot make use of subsequent UIRs. This increases the query delay and the client is effectively operating in the basic IR scheme. In the proposed scheme, the server divides an IR into ω segments such that each IR segment contains invalidation information for a represented broadcast interval. An id-timestamp pair appearing in the latest segment at T_i indicates the data item is updated at least once during the time period from T_{i-1} to T_i. The client can then use this latest segment, together with the IR at T_{i-1}, to reconstruct the IR at T_i. As a result, the client can use subsequent UIRs to invalidate its local cache and answer queries locally, as if there were no corrupted content in the IR.

Improving Channel Utilization:

In the wireless environment, clients experience the same broadcast channel differently [270]. Some clients perceive a good channel (e.g., the client is near to the server) while others may simultaneously perceive a poor one (e.g., far away from the server, deep fading, etc.). If the traffic is scheduled to broadcast at a fixed low rate in the hope to improve the success probability, the wireless channel is sometimes under-utilized; on the other hand, with a fixed high rate broadcast, some clients may be prone to frequent errors, which leads to increase in average query delay. Thus, dynamically adjusting the transmission rate according to the channel conditions could improve both channel utilization and overall performance.

The work in [195] is mainly about leveraging the time-varying effective channel bandwidth property to improve the performance of individual point-to-point unicast traffic. However, in the cache invalidation scenario, we are concerned about the traffic delivered over the broadcast channel. Specifically, the server cannot optimize the transmission to each particular client by using the "best" transmission mode

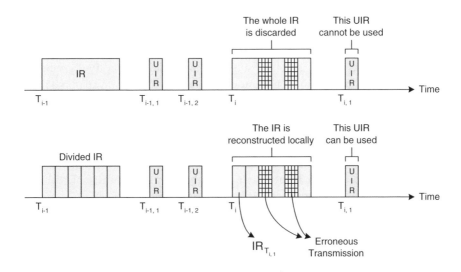

Figure 20.8   Reducing the probability of corruption in IR.

for the client. Instead, the server has to use only a single transmission mode in a broadcast message for all clients. Indeed, if a higher mode is used in the broadcast (i.e., more aggressive), some clients suffering from poor channel conditions may not successfully receive the broadcast message, despite that a shorter delay is achieved through such a high data rate transmission. On the other hand, if a lower mode is used in the broadcast (i.e., more conservative), some clients enjoying good channel conditions may not fully benefit from the good channel quality with such a low data rate transmission, resulting in a wastage of bandwidth resources.

To improve the success probability of broadcast traffic, the optimal transmission rate should depend on: 1.) the current channel status of all clients, and 2.) the importance of the information being delivered. This idea is illustrated as follows: deliver more important information using a low-rate (i.e., higher level of error protection), to improve the success probability; deliver less important information using a high-rate (i.e., lower level of error protection), to better utilize the wireless channel.

For active clients, the latest IR segments (e.g., IR_1, IR_2, IR_3), UIRs and query replies are necessary for correct operations: maintain cache consistency and serve queries. Thus, they are more appropriate for low-rate broadcast to provide better protection against transmission errors. The other IR segments, which contain old cache invalidation information, are only necessary for some clients that have been disconnected for some period of time ($< \omega L$). As such, these old segments are more suitable for high-rate broadcast to better utilize the channel. However, these low-rate and high-rate broadcasts are not kept unchanged for all IR broadcasts. What transmission rates are suitable for low-rate and high-rate broadcast is a function of the channel conditions of all clients. Thus, a broadcast scheduler is introduced at the server side to determine the optimal transmission rates for different broadcast traffic. The broadcast scheduler collects the channel state information (CSI) from all clients with the aid of the polling subframe in the downlink and pilot subframe in the uplink. Since it is not good to use either an aggressive (i.e., high rate) or a conservative (i.e., low rate) broadcast strategy, the average data rate, computed based on all CSIs, is used to indicate the aggregate channel conditions of all clients. The average rate is used to determine the transmission mode in broadcasting the more important information. Simulation results show that the latest three IR segments are more important to mitigate the effect of transmission errors. The less important one is then transmitted with one higher mode (see Table 20.1).

Reducing Delay in Other Downlink Traffic:

In IR-based approaches, server aggregates query requests from all its clients over L seconds and broadcasts the corresponding replies after each IR. The rationale is to let more clients share the same reply via broadcast to improve efficiency, at the expense of increase in query delay among all clients. However, the channel is blocked for the whole broadcast period and other downlink traffic has to be suspended. If query replies are distributed more evenly, the impact of broadcast on other downlink traffic would be reduced.

Depending on the client size and the query generation rate of each client, there may be a long list of query replies following each IR. As discussed before, the size of each IR can be very large in order to support long disconnection period. These two factors cause the broadcast traffic to occupy the channel for a long time, resulting to a large delay in other downlink traffic. The severity of the large delay depends on the nature of other downlink traffic. For elastic traffic, there is little harm of suffering from slightly larger delay. However, large delay is unacceptable for some delay sensitive traffic such as real-time voice or video. To reduce the adverse effect of broadcast on other downlink traffic, in the proposed approach the server broadcasts query replies after both IRs and UIRs instead of just IRs. This can shorten the longest broadcast time, in turn reducing the average waiting time of other downlink traffic spent at the server.

This proposed scheme can also improve the query delay by reducing the penalty due to cache miss. As illustrated in Figure 20.9, after receiving a UIR and invalidating any obsolete data items, the client notices that a query cannot be served locally, i.e., a cache miss. The client then issues an uplink query request to the server. In the IR+UIR scheme, the query reply is scheduled to be broadcast after the next IR at T_i. Equivalently, the expected query delay due to cache miss is half the IR broadcast period, i.e., $\frac{L}{2}$ seconds. In the proposed scheme, however, the expected delay for a cache miss is the same as that for a cache hit, i.e., $\frac{L}{2m}$ seconds.

The tradeoff is the reduction in replies aggregation effect due to the shorter broadcast cycle. On the other hand, a broadcast data item is cached by all clients. When a query for the data item arrives at a later time, depending on the status of the cache, clients will have different actions. This leads to two conflicting effects: if the cached

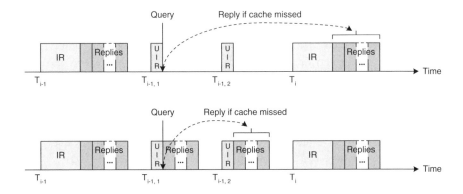

Figure 20.9   Distributing query replies after IRs and UIRs.

copy is still valid, the query can be served locally and the client can conserve battery power in uplink transmission; otherwise, an uplink request will be issued and the resources expended in previous broadcast of the data item are wasted. Which effect outweighs the other depends on the update arrival rate, query distribution in the database, and also the cache size of each client.

20.6   PRACTICAL ILLUSTRATIONS—PROACTIVE KEY CACHING

In a WLAN network, when a user wants to roam from one access point to another, it needs to go through an authentication process to perform a key exchange with the newly associated access point. This process induces a significant delay that can cause data and connections to be lost, which is particular harmful to real-time applications.

In order to solve the above-mentioned problem, Airespace has proposed a proprietary extension to the 802.11i security specification, called proactive key caching (PKC) [186]. Using PKC, when a WLAN user encounters an access point A in the first time, it is issued a key during the authentication process. When the user moves away from access point A and goes to another access point B, it will cache the key that it gets during the authentication process with access point A. Afterwards, when the user roams back to access point A, it can use the cached key to re-authenticate with access point A. This greatly reduces the authentication process. Experiments show that it can reduce 60% of time needed during the authentication process [186]. From this example, we can see that caching is a rather promising technique in wireless networks.

20.7   SUMMARY

In this chapter, we have discussed the on-demand information access problem in client-server wireless networks. Towards the end of this problem, data broadcasting and client caching are two promising techniques to mitigate the existing bandwidth and energy constraints in a wireless network. Data broadcasting exploits the sharing nature of the wireless medium, while client caching leverages the skewed access pattern. The major challenge of client caching is to maintain cache consistency. Clients should ensure that the cache is valid before using it to answer any query. We have discussed several cache invalidation schemes in this chapter.

PROBLEMS

20.1   What are the major drawbacks in invalidation based approaches?

20.2   Discuss the performance gain that we can achieve by adopting data caching in wireless networks.

20.3   What is the main issue involved in using client caching? Briefly describe how we can solve the problem.

CHAPTER 21

SECURITY ISSUES

The way to be safe is never to feel secure.

—H. B. Bohn: Handbook of Proverbs, 1855

21.1 INTRODUCTION

Providing security in wireless communication is a very difficult problem. First of all, security always involves trade-offs. Indeed, the trade-off between performance and integrity is one typical example. Users may not appreciate the high integrity provided if the performance of the communication network is considerably degraded in order to support such integrity.

Interoperability:

There are a number of problems known in current wireless security systems. Section 21.2 in this chapter first gives readers insight on general security considerations in the design of wireless communication. In order to fix some known problems, wireless systems need to make certain tradeoff between the complexity of the solutions and

Wireless Internet and Mobile Computing. By Yu-Kwong Ricky Kwok and Vincent K. N. Lau   **519**
Copyright © 2007 John Wiley & Sons, Inc.

backward compatibility. As a result, in order to remedy certain security problems, a complete upgrade in the wireless system is sometimes required. However, this is often infeasible due to the large scale deployment of the system. In viewer of these, operators or manufacturers usually prefer implementing security remedial solutions in next generation of the technologies. This results in great non-interoperable problems between different versions of the system.

As a case in point, consider that in the GSM cellular wireless system, certain flaws were found in their authentication and encryption processes. Some of them even lead to the possibility of cloning of SIM card of the phone. Certain patch-up types of solutions are incorporated in the algorithms without the need of changing of the handsets in order to provide backward compatibility. The whole authentication and encryption processes of GSM and CDMA will be summarized in Section 21.3.1 and Section 21.3.2. However, some of the possible attacks like faking of base station cannot be solved in these two second generation wireless technologies. Consequently, third generation (3G) wireless security systems are built on top of the security systems in second generation combined with fixes and improvements. Section 21.3.3 explains the authentication and encryption processes in 3G systems.

Meanwhile, Bluetooth WPAN systems come also with a quite complicated security system through a number of key generation processes to produce keys for different functions. The security system level in Bluetooth depends on the application requirements and its default level of the devices.Thus, non-interoperability may result when some of the devices do not presume adequate security support (as in the case of Internet access hotspot) or have fixed PIN code predefined. Consequently, some of the Bluetooth systems may not be able to interoperate among each other if the security levels are different. The authentication and encryption processes in Bluetooth will be discussed in Section 21.4.1, together with a recent flaw found in the pairing processes.

On the other hand, IEEE 802.11 WLANs have been deployed for years. The original design of its security system, called WEP (Wired Equivalent Security), was once considered as secure. It is because the computers at that time were not fast enough to crack such systems. Currently, a typical WEP system can be cracked very easily due to a number of flaws found in the system. In view of this, 802.11i has been proposed. In the mean time, WPA which will become part of 802.11i was released to remedy the situations. However, similar to that of cellular system, all nodes have to be WPA-compatible in order to use this new security measure. This leads to problem of interoperability to the legacy devices. At the same time, the TKIP in WPA developed based on WEP for backward compatibility is optional in the final release of 802.11i. Instead, AES will be the one adopted. However, with the use of AES, legacy devices cannot be supported. All these points will be further discussed in Section 21.4.2.

Performance:

It is notoriously difficult to quantify the performance of a security system. This is partly because a security system usually manifests as a chain of inter-related processes (e.g., key generation, encryption, etc.), among which the "weakest" link determines the overall effectiveness of the system. Indeed, a commonly agreed paradox is that the difficulty of quantifying security undermines the security of the system.

21.2 SECURITY CONSIDERATIONS

21.2.1 Design Goals

In a wired network, malicious physical attachment to the network is difficult as it requires the attacker to intrude into the building to locate the network. However, in a wireless network, the use of a shared and open wireless medium enables a much easier intrusion opportunity. Thus, attackers have a much higher chance to successfully attack the network without being physical close to the network. There are a number of attacks on wireless system. Some of them originally were designed to attack a wired network. Nonetheless, the inherent open nature of wireless networks makes it more easier to succeed. On the other hand, some of them are specifically designed to attack a wireless system. These attacks are summarized in Table 21.1 [97, 242] below.

Table 21.1 Security threat on wireless system.

Threat Type	Description
DOS (Denial-of-Service)	This type of attack is mainly used to prevent valid users from using the network services. There are two main types—sending large amount of bogus packets to the server, e.g., TCP SYN flooding, or preventing a valid user to receive legitimate packets as the intruder performs "man-in-the-middle" attacks.
Replay Attack	The attacker gathers registration requests from a legitimate user and then replay the requests to the server so as to pretend to be a valid user.
Eavesdropping	In this type of attack, the attacker aims at uncover the secret communication data between the communication parties, e.g., username and password of network account, etc.
Session Stealing	The attacker waits for all the authentication processes to complete between the users and server, and then hijacks the session by pretending to be valid users.

In Table 21.1, a number of attacks which can be targeted on both wireless and wired network environment are listed. Since a wireless environment makes these attacks easier to be successful, new wireless security solutions are specifically designed to tackle these attacks and some other wireless specific vulnerabilities.

21.2.2 Components of a Secure Wireless Communication System

In a wireless communication system, it is necessary to ensure two main services—authentication and privacy [97, 242].

Authentication is necessary in a wireless communication system in order to ensure the correctness of identity of the communicating party. In particular, systems that bill the users are required to authenticate the users to ensure that they are valid users and charge their accounts accordingly. Generally speaking, authentication in a wireless environment requires both party to have a shared secret which will then be used to encrypt the identity proof or generate more keys for later use.

On the other hand, since a typical wireless environment is open so that eavesdropping is usually practical. As a result, privacy becomes a serious concern. It is necessary for a wireless system to ensure that the message is only readable by valid target parties, unmodified during the transmission, and remains intact even being intercepted. In a wireless environment, this is generally achieved by encryption of the data using keys generated and only known to the targets parties.

In fact, with the careful handling of authentication processes and privacy issues together with management of the keys. Most of the attacks mentioned above can be effectively reduced. In the following sections, security issues and techniques used in various wireless system will be elaborated.

21.3 CELLULAR WIRELESS SECURITY

Generally speaking, cellular operators have adopted effective measures in authentication and encryption of the messages.

21.3.1 GSM Security

During the development of GSM security techniques, it was decided to make them secret (without publishing for public reviewing). This, in fact, arouse great public and hackers attention. Finally, several flaws and problems were discovered. In the following, the authentication and encryption used are discussed [237].

The security of GSM starts from its SIM (Subscriber Identity Module) card. It is a smart card storing various important information items including IMSI (International Mobile Subscriber Identity), TMSI (Temporary Mobile Subscriber Identity), Ki, $A3/A8$ algorithm, etc. IMSI is used to uniquely identify a specific user globally and TMSI is assigned for subsequent identification within an area after IMSI is used for the first time. The use of TMSI, in fact, provides anonymity to the users as the attackers may not discover the parties under communication easily. Since a SIM card stores much sensitive information, it has to be protected for authorized access only. It is generally protected with a PIN (Personal Identity Number). After a preset number of failed attempts, the card will be locked and can only be unlocked by using the

PUK (Personal Unblocking Key). The card will be locked permanently if the PUK is wrongly input for a number of times.

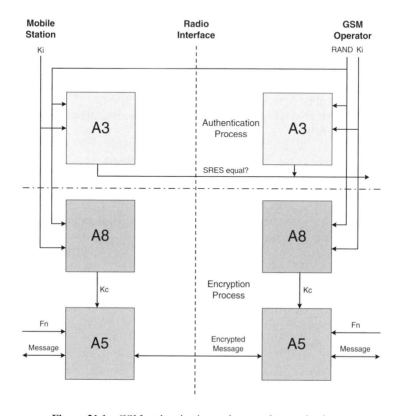

Figure 21.1   GSM authentication and encryption mechanism.

Figure 21.1 illustrates the authentication and encryption processes carried out in a GSM network. Ki in the SIM card plays an important role in the proof of identity and generation of subsequent keys. The key steps are listed as follows.

1. The base station generates a challenge (RAND) and sends it to the phone.

2. The SIM card with the valid Ki encrypts the RAND and produces a SRES (signed response).

3. At the same time, the VLR serving the mobile phone asks for a copy of A3 and Ki from the HLR which in turn obtains *a priori* the information from AuC (Authentication Center).

4. The VLR calculates the SRES from $A3$ and Ki.

5. Phone sends SRES to base station which then forwards it to VLR for comparison. The user is successfully authenticated if the SRES values calculated by both parties match. The user is then allowed to use the network.

The above processes allow a GSM network operator to efficiently authenticate the users. Originally, it was suggested that anonymity is provided with the use of TMSI. However, only anonymity is not enough. Indeed, it is also necessary to protect the user's privacy including the voice and data communication privacy. This is achieved by the encryption of voice and data messages. As shown in Figure 21.1, the encryption processes in GSM are done by the cooperation of $A8$ and $A5$ algorithms, which are briefly described below.

1. The SIM card of the mobile phone receives the RAND at the time of authentication. It will then be processed by the $A8$ algorithm together with Ki to produce the session key Kc.

2. The GSM network operator carries out the same computation to generate the Kc specific to that particular user.

3. The whole communication process can then be encrypted with the use of Kc and message in the $A5$ algorithm.

The above processes provide communication privacy to the GSM users. However, GSM authentication and encryption processes are found to be flawed. Some of the well known problems are listed below.

- Ki in a SIM card can be extracted with physical contact for minutes with the use of side channel attack. As a result, the SIM card can be cloned.

- It was claimed that $A3/A8$ algorithms are based on 64-bit keys Kc. However, the last 10 bits of the keys are fixed as 0 bits. As a result, brute force attacks become much more effective.

- $A5$ was discovered to be flawed also.

- The authentication process only involves mobile phone authenticating itself to the base station, but not vice versa. Thus, a fake base station can be used to capture secret user messages.

21.3.2 CDMA Security

In fact, CDMA is very similar to GSM in the sense that it also uses similar approaches in authentication and encryption processes. However, unlike GSM, user information is stored in the handsets instead of a SIM card. A critical information item is a 64-bit symmetric key known as A-Key [237, 291]. This key is stored into the phone and

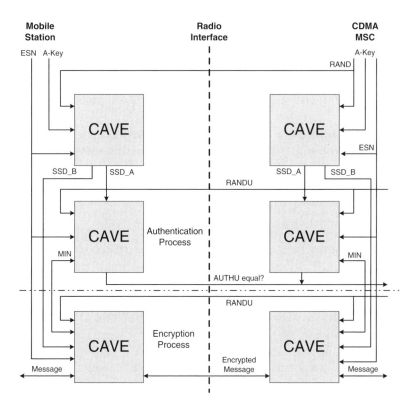

Figure 21.2   CDMA authentication and encryption mechanism.

operator database at the time of purchase. At the same time, the phone itself contains a unique ESN (Electronics Serial Number). These two information items uniquely identify a user.

CDMA authentication and encryption processes are illustrated in Figure 21.2. The authentication steps are summarized below.

1. The mobile phone gets a RAND from the operator and generates two 64-bit keys—SSD_A and SSD_B with the use of the CAVE (Cellular Authentication and Voice Encryption) algorithm, ESN, A-key and RAND. These two keys are also stored inside the database of the operator (generally inside AuC).

2. When the mobile phone indicates to the network that it would like to make a call, MSC in the operator network would try getting the user's information from the HLR and generate a RANDU (challenge) to the user.

3. With the knowledge of SSD_A, ESN, MIN and RANDU, the mobile phone is able to generate a 18-bit AUTHU which is a proof of knowledge of A-Key. At the same time, MSC will calculate the same AUTHU based on the same parameters. (In some case, MSC may not have the ability to calculate CAVE algorithm, then AuC will take over the task).

4. The mobile phone transmits the AUTHU to MSC. If the AUTHU values from both side match, the authentication is successful and the call can be proceeded.

From the above procedures, it can be discovered that the authentication processes in CDMA network is very similar to that of GSM network. The encryption processes are summarized below. (The readers are also referred to Figure 21.2 for illustration of simplified version of encryption processes.)

1. The mobile phone gets the RAND from the operator. With the knowledge of another key—SSD_B, ESN and MIN, a 18-bit VPMASK (Voice Privacy Mask) can be calculated using the same CAVE algorithm.

2. MSC calculates the same VPMASK using the same algorithm and parameters. This VPMASK can then be used to encrypt the voice communication[1].

With the use of CAVE and A-Key, a CDMA operator can provide authentication and encryption to their users. CDMA takes a similar approach as that of GSM in that it keeps the algorithm unpublished. Fortunately, there are fewer flaws found in CDMA.

---

[1]There are other keys generated using similar parameters for signalling message and data encryption.

21.3.3 3G Security

We have briefly outlined second generation cellular security systems. As can be seen, they have several deficiencies including poor encryption algorithm, lack of mutual authentication, difficult upgrading if flaws are discovered, etc. In view of this, 3GPP (Third Generation Partnership Project) which was developed to define standards on 3G (i.e., WCDMA) also established a task group for security [39, 237]. The task group proposed the use of AKA (Authentication and Key Agreement) protocol in 3G network in which the encryption algorithm used is Kasumi, which is widely considered to be effective.

The new authentication and encryption processes are illustrated in Figure 21.3 and the corresponding functions are summarized below.

- The main function of $f1$ is to generate an authentication token (MAC) to authenticate the network operator. This is derived based on the value of K (a root key known only to the operator and stored in the SIM card), SQN (Sequence number) and AMF (implementation specific to network operator).

- The main function of $f2$ is to generate XRES to authenticate the users to the network operator. Consequently, a mutual authentication can be carried out.

- The main function of $f3$ is to derive CK (Ciphering Key) which will be used in $f8$.

- The main function of $f4$ is to derive IK (Integrity Key) which will be used in $f9$.

- Algorithm $f5$ is used to generate AK (Authentication Key). This key is used to encrypt the sequence number to mobile station.

- Algorithm $f8$ is used for encryption of the message. It has a number of new features that are not found in second generation cellular service. This algorithm takes in 128-bit CK, COUNT-C, DIRECTION, BEARER and messages as parameters to produced encrypted ciphers. COUNT-C is mainly used to prevent replay attack and make sure the cipher is updated in every round. The DIRECTION is mainly used to prevent replay attack, while the BEARER is to ensure different values for multiplexed link of a particular user.

- Algorithm $f9$ is used for providing integrity protection to the message. This ensures the message to be unmodified. It makes use of 128-bit IK, COUNT-I (integrity sequence number), fresh and DIRECTION to hash the message. The use of COUNT-I and fresh together prevents the replay attack as COUNT-I and fresh will change frequently. On the other hand, the use of DIRECTION prevents the reflection attack.

From the above description, it can be seen that there are a number of changes made in 3G security system in order to improve security. These include the mutual

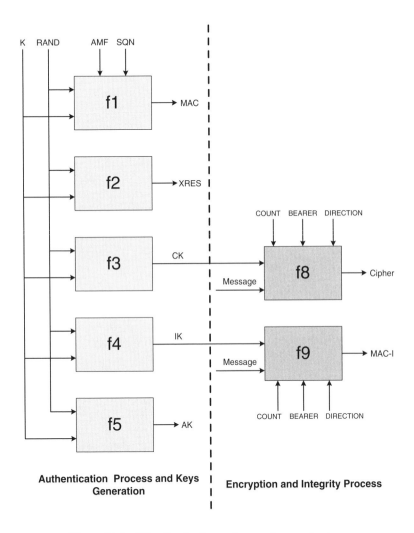

Figure 21.3 3G authentication and encryption mechanism.

authentication between the mobile phone and network operator, the inclusion of integrity check to ensure unmodified messages between mobile phone and operator, the use of longer encryption keys with some well known secure algorithms, the inclusion of several different parameters into the encryption algorithm to prevent attacks like replay attack and reflection attack, etc. Based on these changes, the security system of 3G is much more stronger than that of 2G systems. It is widely expected that this system can withstand most of the known attacks nowadays.

21.4 SHORT-RANGE WIRELESS SECURITY

21.4.1 Bluetooth Security

In this section, the authentication and encryption processes of Bluetooth will be discussed in detail. We also describe a recently discovered flaw in Bluetooth.

21.4.1.1 Bluetooth pairing process   Figure 21.4 illustrates the processes of Bluetooth pairing [130]. Bluetooth pairing starts with the entry of PIN code. There are two kinds of devices. The first kind of devices has predefined (fixed) PIN code pre-installed. The other kind of devices allow entry of a variable size PIN code. The variable size PIN code ranges from 8 bits to 128 bits.

As illustrates in Figure 21.4, after the entry of PIN code at both side, the master generates a 128-bit random number (IN_RAND) and sends it to the slave. They will then make use of $E22$ algorithm to generate the $Kinit$ (initialization key) with the use of PIN code and slave BD_ADDR. At this point, the initialization key generation processes are regarded as completed. Further keys will be generated based on this initialization key.

As both of them possess K_{init}, they will then individually generate a 128-bit random number again (LK_RAND) and transmit this number to the other party after XOR'ing with K_{init}. With the knowledge of K_{init}, both sides' LK_RAND can then be derived. They will then be fed with opposite side's BD_ADDR into $E21$ algorithm to generate LK_Key. Together with self generated LK_RAND, self BD_ADDR and $E21$ algorithm, another set of LK_Key will be derived. The combination of both LK_Key will result in a combination link key K as shown in the middle part of Figure 21.4. The generation of link key K marks the end of link key generation processes.

With the common shared secret (link key K), this two entities can then authenticate each other to perform mutual authentication processes. First of all, one side will act as verifier and the other side will act as claimant. In the case shown in Figure 21.4, slave acts as verifier and master acts as claimant at the beginning. The slave will first generate and then transmit a 128-bit AU_RAND (challenge) to the master. The master will then perform calculation through $E1$ algorithm with the use of its own BD_ADDR, common link key K and the AU_RAND. The slave will carry out the same calculations. The results will consist of a 96-bit ACO (Authentication Ciphering

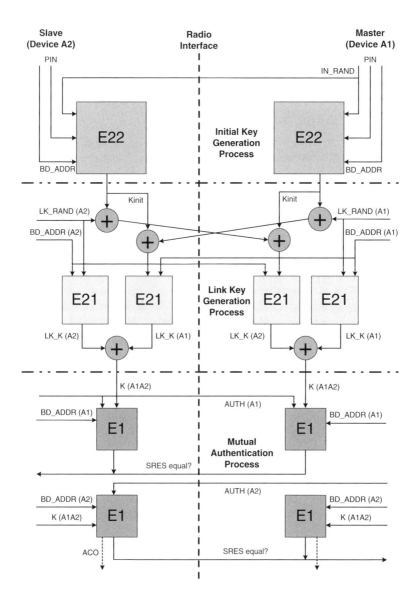

Figure 21.4 Bluetooth pairing process.

Offset) and SRES (Signed Response). The slave authenticates the master if their calculated SRES values match. Afterwards, the roles of verifier and claimant are reversed with the same authentication processes. The ACO of the last authentication process will be kept for the generation of ciphering later.

Up to now the whole authentication process is regarded as completed. The devices can then carry out communication if encryption is unnecessary. Both devices are regarded as paired or bonded at this moment.

21.4.1.2 Encryption Processes
The data encryption process will be discussed in detail in this section. There are two kinds of messages to be considered. The first one is the unicast message that involve two communicating parties—a master and a slave. The second one is the broadcast message that involves a master and multiple slaves. Generally speaking, with the use of different encryption keys for different master-slave pairs, it is impossible to carry data broadcasting. Thus, a single shared link key has to be generated for the piconet for broadcast messages. This involves the generation of master key (called K_{master}).

The generation of K_{master} involves the use of $E22$ algorithm together with use of the two locally generated LK_RAND. This key will then be sent to each slave individually by XOR'ing it with the result of another generated RAND (send to the slave in plaintext) and link key K in algorithm $E21$. For individual slave which has the knowledge of RAND (received from master in plaintext) and the individually shared link key K with master, it can calculate the result and get back the K_{master}.

Now, let us consider the encryption processes [130]. Figure 21.5 illustrates the idea of Bluetooth encryption processes. As can be seen, the link key K, a random number and COF (Ciphering Offset) will be processed by algorithm $E3$ to generate the cipher key Kc. COF will be the exactly the same as the last authenticated ACO in case of encryption between two communicating parties. However, in case of broadcasting, COF will be derived from the master BD_ADDR. Similarly, the link key K here refers to link key between two communicating parties in case of encryption between them. On the other hand, K refers to master key in case of broadcasting.

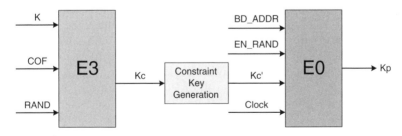

Figure 21.5 Bluetooth encryption processes.

The generated Kc will be 128-bits in size. However, due to some regulations (e.g., export control), sometimes it has to be restricted to be smaller than 128-bits. In this case, it has to pass through constraint key generation algorithm to produce a smaller size constraint cipher key Kc'. This key will then be fed into algorithm $E0$ together with the BD_ADDR, EN_RAND and the clock to produce the payload key Kp. This key will then be the actual key used to encrypt the message to produce cipher at the sender side and decrypt the cipher to produce message at the receiver side.

We can see that the authentication and encryption processes of Bluetooth are quite involved. The keys are generated from the initial PIN code entry, which is then used to construct the initialization key, link key, cipher key, constraint cipher key and finally payload key used in encryption.

21.4.1.3 *Discovered Vulnerability*   It is remarkable that Bluetooth security system was recently found to be flawed [317]. Specifically, the flaw is in the pairing process. The attackers are possible to make use of brute force attack to crack the PIN code if it is possible for them to eavesdrop the whole pairing process. First of all, let us review the messages transmitted over the air during the pairing processes. (The readers are referred to Figure 21.4 also.)

1. The master device sends a 128-bit IN_RAND to slave in plaintext.

2. The master device sends a 128-bit XORed K_{init} and LK_RAND(A1) to slave.

3. The slave device carries out the same actions. That is, it transmits 128-bit XORed K_{init} and LK_RAND(A2) to master.

4. The slave sends a 128-bit AU_RAND (challenge) to master in plaintext.

5. The master responses with a 32-bit SRES in plaintext.

6. The master sends a 128-bit AU_RAND (challenge) to slave in plaintext.

7. The slave responses with a 32-bit SRES in plaintext.

There are in total seven messages expected in the pairing processes. It is assumed that the attackers captured all these messages and know the source and destination of these messages. At this point, the attackers can then, in a brute force manner, try all the combinations of PIN code. With the knowledge of IN_RAND and BD_ADDR, the K_{init} can then be calculated through $E22$ algorithm. The K_{init} can then be used to perform XOR operation to derive two LK_RAND values. They will then be fed into $E21$ algorithm together with their BD_ADDR to derive suspected link key K after combining two LK_Key. Finally AU_RAND and SRES can be used to verify the correctness of the guessed value. If SRES can be derived correctly, the guessed PIN code is correct. With the use of some optimization of the SAFER+ algorithm which is the basis of most of the algorithm used in Bluetooth, 4-digit PIN can be resolved in less than a second and the 7-digit PIN can be resolved in less than an hour.

Some people may argue that this attack only affects the pairing processes. As long as this process is kept secret, this attack can be avoided. Unfortunately, the attackers are possible to spoof the BD_ADDR of one of the paired user and request for the re-pairing processes. This is possible as Bluetooth's design allows either one side of the paired devices to lose the link keys due to whatever reasons. As a result, once the users re-do the pairing processes, there is an opportunity for the attackers to carry out the above procedures.

21.4.2   IEEE 802.11b Security

21.4.2.1   Overview   IEEE 802.11b is widely used nowadays. Early versions of IEEE 802.11 has mandated the use of WEP (Wired Equivalent Privacy) for authentication and encryption of the messages. However, it has been proven flawed and the keys involved can be re-generated relatively easily. IEEE 802.11i is then being developed. In the mean time, some of the components including TKIP and IEEE 802.1x security framework that will be adopted in IEEE 802.11i are used until the final version is released.

21.4.2.2   WEP and its vulnerabilities   Let us first review the mechanisms used in WEP. Afterwards, its vulnerabilities will be elaborated.

Figure 21.6 illustrates the key idea of WEP [27]. As can be seen, the shared secret key will be combined with the initialization vector (IV) to form a block which will then pass through the widely used stream encryption algorithm RC4 to form a keystream. At the same time, the CRC of the plaintext data will be computed and combined with the plaintext data to form a block. The two blocks that have been mentioned will then XOR'ed together to form the ciphertext. After combining with the original IV used in the whole process, the whole block of ciphertext can be sent to the receiver side.

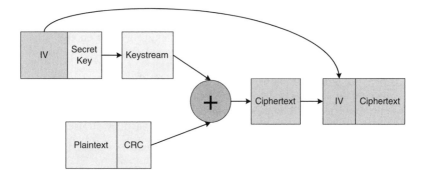

Figure 21.6   WEP mechanism.

At the receiver side, the message can be decrypted by first getting the IV from the received data block and combined it with the secret key to get the same keystream block as that of sender side. This keystream can then be used to XOR with that of cipher to get back the original plaintext message with CRC.

This simple processes form the basis of WEP in encrypting and decrypting the plaintext message block in original IEEE 802.11b design. Parties in communication are required to know the shared secret key beforehand in order to encrypt and decrypt the message. However, this approach, in fact, contains some known flaws and problems. The major ones are listed below.

- WEP involves the use of 24-bit IV which will then be combined with 40-bit or 104-bit key to form a block to feed in RC4 algorithm to form a keystream. However, since the size of IV is small, the keystream will be repeated within hours in heavy traffics environment. It will then be subjected to keystream attack and replay attack. Keystream attack involves the XOR'ing of the two chipertexts using the same keystream. As a result, XOR'ing of two data blocks will be resulted. If one of the data block is known, the other data block can also be recovered. This is often possible due to the existence of some known control message data exchange at upper layers. On the other hand, replay attack is possible because IV out of order usage is possible due to its limited size. Consequently, if certain plaintext and cipertext pair is known, the corresponding keystream can also be recovered which can then be used to create new packet without the knowledge of WEP key.

- Forgery attack is also possible on WEP by simple flipping bits of the packet. It is because WEP ICV (Integrity Check Value) makes use of CRC as a mechanism for integrity check of plaintext. Thus, simply flipping the bits of packets allows the decrypted plaintext to be changed with a valid CRC value.

- FMS attack targets on certain weak keys of WEP no matter it is 40-bit or 104-bit key. In this method, large amount of encrypted packets are captured. During the encryption of these packets, certain concatenation of IV to the key will create weak keys. Consequently, by capturing large enough amount of encrypted packets, the key can then be recovered easily.

- There are also some more problems regarding the key management issues. Since all the devices in the wireless network shares the same key, it is difficult to prevent leakage of the keys. At the same time, it does not have a measure to identify a particular wireless user in the network as they all use the same key with the same authentication method.

21.4.2.3 *TKIP (Temporal Key Integrity Protocol)* In the previous section, some of the known flaws and known problems in WEP currently used in IEEE 802.11 have been discussed. In view of these problems, IEEE 802.11i task group was developed to find a replacement for WEP[2]. However, in the meantime, some of the

---

[2]Originally, this was known as WEP2. However, it is renamed as WPA (Wi-Fi Protected Access).

methods have to be adopted to tackle the serious security issues suggested previously, e.g., some of the attacks that will lead to the leakage of the key. Thus, WPA comprises both TKIP and 802.1x which aims at tackling the encryption key issues and access control issues, respectively [27].

First of all, let us start with TKIP, which mainly targets at solving the problems of the encryption key problems in WEP. It is just an interim solution which aims at providing a patch up solution on existing hardware with simple software and firmware upgrade in order to provide backward compatibility. TKIP was designed as shown in Figure 21.7.

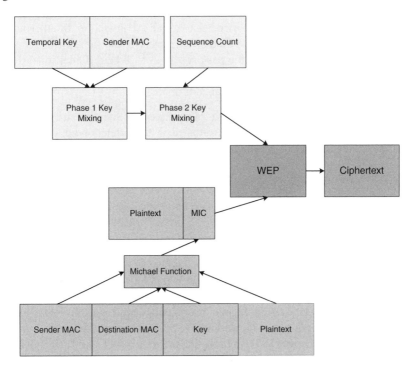

Figure 21.7 TKIP mechanism.

Figure 21.7 illustrates the idea behind TKIP. There are two keys involved. They are 128-bit encryption key (Temporal Key) and 64-bit MIC (Message Integrity Code) key. Their functions are as follows. The temporal key will first be combined with the sender's MAC address to form a intermediate key in phrase 1 which will then be fed into phrase 2 to further combine with the sequence number. Since the sequence number, in fact, depends on the packets transmitted, a 128-bit per-packet key will be generated after phrase 1 and phrase 2 key mixing process. This 128-bit will be broken into 24-bit IV and 104-bit WEP key. Finally, original WEP procedures are adopted as usual. On the other hand, the plaintext will be mixed with several parameters including 64-bit MIC key, Sender and Receiver MAC addresses to form a plaintext and MIC block as shown in Figure 21.7. This block is then fed into the original WEP procedures shown in Figure 21.6.

The above procedures can effectively solve the problems of WEP as supported by the following arguments.

- The packet integrity check no longer makes use of CRC. Instead, a one way hash known as Michael is used. This function makes use a number of items to form the hash value and this effectively solves the integrity attacks.

- FMS attacks, which take advantage on the weak key, are no longer feasible as the per-packet key does not correlate to the IV any more. In other words, the 104-bit key fields will not be the same even for the same 24-bit IV value is used. At the same time, the per-packet key will be different for different wireless users. It is because the two-phase key mixing has taken into account the value of MAC address and sequence number.

- Other than the attacks stated above, with the use of sequence number, replay attack can be avoided. With the avoidance of the correlation between the per-packet key and IV, key stream attacks are much more difficult to be achieved.

It is currently widely considered that TKIP has effectively solved some of them problems in originally WEP idea. In long run, a much more secure encryption idea should be adopted. As a result, IEEE 802.11i proposed the use of AES (Advanced Encryption Standards) for encryption. AES is compulsory in IEEE 802.11i standards, while TKIP is just optional.

21.4.2.4   *802.1X access control*   As it has been explained in previous section, WEP not only contains problems in its encryption problems, but also uses very simple mechanism in authentication process by using the same shared key in the whole network. This also leads to the problem of user identification in the network. In view of this, 802.1x access control is adopted in WPA to tackle these issues [27].

802.1x was adopted from PPP (Point-to-Point Protocol) which is originally used in Modem dial up and later extended to ADSL. It provides point to point connection between two entities with the username and password as authentication measures. In order to provide more authentication options on top of PPP, EAP (Extensible Authentication Protocol) was proposed.

802.1x uses EAP to provide authentication and identification in both wired and wireless environments. The three main components involved and their corresponding functions are summarized below.

- Supplicant is the client who attempts to access the network.

- Authenticator is the one trusted by the network and acts on behalf of the authentication server to authenticate the supplicants. As a result, it can be regarded as a man-in-the-middle and this role is usually played by the access point.

- Authentication Server is the server which contains all the authentication information of the network and makes decision on whether successful authentication occurs. This is usually performed by a RADIUS server.

Figure 21.8 illustrates the idea of 802.1x. First of all, the supplicant sends an EAP start message to authenticator to show its intention to access the network. The authenticator will first put the supplicant in controlled ports in unauthorized state. It sends the EAP Request/Identity message to supplicant. The supplicant then supplies its identity through EAP Request/Identity message back to authenticator which will then forwards the message to authentication server. The communication between the authenticator and authentication server is through an uncontrolled port since they are both trusted in the network. After the EAP-Request(challenge) and the subsequent challenge/response as shown in the figure, the authentication server determines whether the supplicant is authorized to access the network. The authentication server then sends this decision to the authenticator. The authenticator will then put the traffics of supplicant into controlled port authorized state if the authentication is successful.

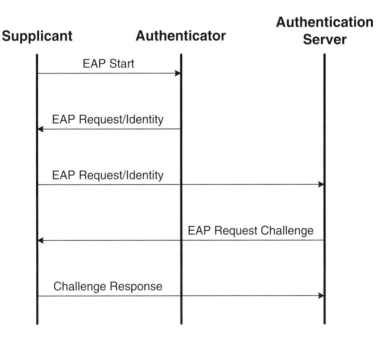

Figure 21.8 802.1x mechanism.

From the processes that have just been discussed, the authentication server authenticates the supplicants through a number of procedures. By doing so, different keys can be used by different supplicants with appropriate EAP algorithms that suit the security needs of the network. At the same time, the identity of the users can be recorded for easily billing later. On the other hand, with the use of 802.1x, periodic rekeying is also possible. Generally speaking, this idea effectively solves the problems that have been suggested.

21.5   PRACTICAL ILLUSTRATIONS—WIRELESS LAN SECURITY AUDITING

In previous sections, different technologies security issues have been discussed. Most of these measures have already been adopted into the products in the market. In this section, a distributed wireless security auditor (DWSA) which has been implemented to detect the rogue access point inside will be discussed to illustrate a practical way to detect security threat by using WLAN [41].

There are a number of attacks in WLAN related to the access point configuration and management. In a corporate environment, even though access point configuration and management processes are usually carefully implemented, they are still susceptible to attacks like insider attack in which rogue access point may be setup by some authorized users whose original intention may just want to provide convenient to other users with wireless access capability. In view of this, DWSA has been implemented to detect the rogue access point. Figure 21.9 illustrates the idea behind DWSA. We can see that the clients become distributed wireless sensors and monitor the existence of any access point located around them. Their functions are summarized below.

- Gather information about the access point searchable around wireless sensor.

- Get wireless client location information by using triangulation which makes use of the received signal strength or using GPS system integrated inside the wireless client

- Periodically report the information gathered to wireless security auditor server through SSL connection.

Moreover, the wireless security auditor server contains much sensitive information including the list of authorized access points and their corresponding physical location inside the organization, the security policy of the organization and the building layout of the organization. This server is used to audit the information gathered from the distributed wireless clients. The main functions of the server are summarized below.

- Authenticate distributed wireless clients that provides information to the server through SSL connection.

- Determine whether rogue access point exists and its corresponding location based on the information provided by various distributed wireless sensors and

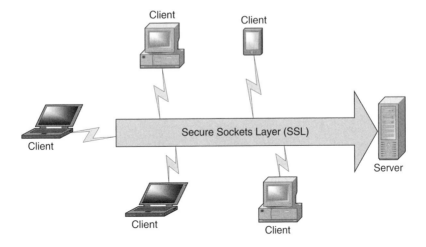

Figure 21.9   DWSA system.

the information originally stored inside the database of the system, e.g., the organization security policy and location of authorized access point.

DWSA architecture relies on the distributed wireless sensors (wireless clients) to gather information about the security inside the corporate network and then send back the information to the processing server through SSL. This allows secure information transfer back to the server. At the same time, only little processing power is required on the clients since all the works are done by the server. The main advantage of this architecture is that it not only detects whether there exist rogue access point, it also tries locating it by first locating the wireless clients itself inside the building.

In general, there are two methods of locating the rogue access point in the network. The first method is by GPS integrated inside the clients. However, this is sometime impossible as not all the clients are integrated with GPS. The second method includes the use of triangulation method. Three-dimensional triangulation environment, just like the case in corporate network with several floors, can be done whenever four fixed access points location and their corresponding distance to the client are known beforehand. The idea behind is similar to find the intersection point of four sphere with the known radii and their center positions. The distance between a fixed access point and a client can be estimated through the received signal strength. With the idea of the current location of wireless client, any other rogue access point location can be estimated by the client through the received signal strength to that access point. Figure 21.10 illustrates an example scenario of how DWSA locates the rogue access point in the network.

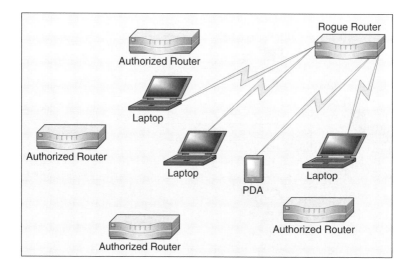

Figure 21.10 DWSA example scenario.

21.6 SUMMARY

In this chapter, different wireless technologies' methods on handling security issues have been introduced. First of all, we discuss different considerations with the main aim at providing authentication, privacy, etc., on wireless medium which is open and public in nature.

Afterwards, specific security issues on cellar network have been suggested with the unpublished security algorithms used in second generation cellular network, e.g., GSM network and CDMA network. However, the unpublished security algorithms are found to be flawed. In view of these, third generation cellular network security has been designed with adoption of the solutions to the known weakness in second generation cellular network security from some published and proven secure algorithms.

After discussing the cellular security issues, short range wireless security issues are described. Short range wireless networks involve quite different characteristics compared with that of cellular network in terms of range, capacity, client portable devices configurations, etc. As a result, they also have different considerations in their security designs. We describe IEEE 802.11 WLAN's and Bluetooth WPAN's methods in tackling security issues. Finally, this chapter is concluded with a practical illustration on a proposed WLAN architecture which is possible to audit fake access point with their corresponding location.

In this chapter, the authors have walked through the security issues in different kinds of cellular and wireless networks. One point in common is that the open and public nature of wireless medium has caused many security problems. Many algorithms have been suggested and adopted to tackle these problems. However, it is generally known that the solutions adopted are hardly perfect due to reasons like implementation difficulties, specific countries export regulation, etc. As a result, there are constantly some flaws found and new types of attack thought of by the attackers. It is obvious that security is an ever on-going battle.

PROBLEMS

21.1 Briefly elaborate the common problems and considerations in wireless communication network security design.

21.2 Briefly explain the security measures adopted in GSM networks and the counter-measures adopted by the attackers in breaking such systems.

21.3 Briefly explain the reasons why WEP is flawed in nature and how WPA has improved it.

21.4 Briefly discuss the security measures adopted by Bluetooth and discuss how it is broken recently.

PART VI

MOBILE COMPUTING APPLICATIONS

CHAPTER 22

VOIP ON WIRELESS

There is no index of character so sure as the voice.

—Benjamin Disraeli:Tancred, II, 1847

22.1   INTRODUCTION

Voice communication is still the dominant form of data transmission in wireless networks. As VoIP over the Internet has proliferated, people naturally expect the same kind of service to be available in wireless environments.

Interoperability:

Indeed, VoIP (Voice over IP) over wireless has become a hot research topic recently. The reason is that it will be much more cost effective to use an IP packet switching network to support voice traffic than a circuit switching network. Specifically, it is not required to have a fixed reservation of resources through the whole path from the origin to the destination. At the same time, some more advanced encoding algorithms can be used to support the voice traffic to further lower the bandwidth required.

Wireless Internet and Mobile Computing. By Yu-Kwong Ricky Kwok and Vincent K. N. Lau
Copyright © 2007 John Wiley & Sons, Inc.

However, there are many problems required to be solved in using wireless VoIP due to interoperability issues. The mainstream wireless and wired voice traffics nowadays are supported by circuit switching network like GSM, CDMA, third generation network and PSTN, in which operators have invested lots of money. As a result, in order to provide real-time voice conversation service to users, wireless VoIP is required to provide interoperability to the circuit switching network, including signal conversion, voice encoding conversion, etc. This is usually done with the addition of a gateway into the infrastructure of the backbone network. Sections 22.2 to 22.4 illustrate the idea of cooperating GSM with H.323 IP network to provides users with a choice of using either technology for telecommunication services.

On the other hand, for the case of using VoIP over WLAN, it may not be a good idea to use the original protocol stack of IEEE 802.11 and/or use it together with other data traffics. However, due to interoperability issues, a large change to IEEE 802.11 especially for VoIP traffics is impossible. As a result, some other methods have to be adopted to mitigate the interoperability issues, while at the same time improve the performance. This is further discussed in Section 22.5.

Performance:

In iGSM, which is a system involving the cooperation of GSM and H.323 VoIP networks, the access delay for call delivery and the number of the occurrence of misrouting due to user mobility are good metrics to evaluate the performance of iGSM, as these are the two problems required to be solved with the use of iGSM. We discuss these two problems in Section 22.5.

On the other hand, using VoIP over WLAN, the number of VoIP streams that can be supported, packet access delay and packet loss are three main metrics for evaluating the performance of VoIP over WLAN. Section 22.5 illustrates some practical ideas on improving the performance related to these three metrics.

22.2 IGSM VOIP APPROACH

iGSM is a kind of value-added service on GSM, allowing users to choose either GSM network or H.323 IP network with the use of handsets or H.323 terminals (e.g., PC), respectively, to access telecommunication services [295]. In other words, users with iGSM accounts are able to make/receive calls using their own handsets, or make/receive calls with the same mobile station ISDN (MSISDN) without their handsets by using H.323 terminals within the IP network. As a result, user mobility can be supported.

Similar to iGSM, there are also some other approaches which propose the integration of GSM network and IP network, such as TIPHON [107] and GSM on the Net [136]. The main difference between these approaches and iGSM is that iGSM targets on providing value-added services to GSM users to support user mobility with the cooperation of H.323 VoIP network, in which no specific wireless capabilities are provided to GSM handsets. However, BSC/BTCs in IP networks and GSM/BTSs in corporate offices are proposed to provide wireless access capabilities to GSM hand-sets to use VoIP services provided within the IP networks in TIPHON and GSM on the Net, respectively.

22.3   IGSM SYSTEM ARCHITECTURE

The architecture of iGSM is illustrated in Figure 22.1 [295]. In the figure, it can be seen that there are two main parts of the backbone network—GSM network and H.323 IP network. Actually, the GSM network portion is unmodified. However, an iGSM gateway is added into the system to act as an interface between the GSM network and H.323 IP network. This gateway is regarded as a member of the IP network and performs several functions similar to that of a traditional H.323 gateway. These functions are summarized below:

- GSM/PSTN/IP Call Control functions including call-setup and call-release.

- GSM MAP protocol and H.225 RAS (Registration, Admission and Status) protocol translation between the GSM network and H.323 IP network.

- Provision of media translation to cope with QoS differences between the IP network and circuit switching network.

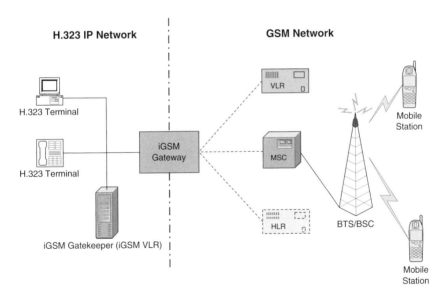

Figure 22.1   iGSM architecture [295].

On the other hand, an iGSM gatekeeper is also present inside the H.323 IP network. In the traditional H.323 network, which is shown in Figure 22.2, the incorporation of a gatekeeper is optional to implement functions like address translation, bandwidth allocation, admission control, zone control, etc. However, in the case of iGSM, the iGSM gatekeeper is implemented inside the H.323 IP network to function as a VLR for users visiting the IP network. A unique ISDN is assigned to each iGSM gatekeeper to let the HLR know that it is a VLR. Being the roles of a gatekeeper and a VLR, the iGSM gatekeeper maintains an iGSM subscriber list and handles appropriate address translation. In other words, it performs functions of the original H.323 gatekeeper and also the functions of the VLR in the iGSM network. We discuss the detailed procedures of registration, deregistration and call-setup in subsequent section.

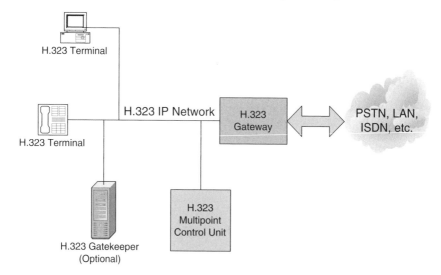

Figure 22.2   H.323 architecture.

22.4   IGSM CALL PROCESSING

In previous sections, the idea of iGSM and its architecture have been discussed. With the additional components, namely the iGSM gateway and iGSM gatekeeper, cooperation services can be provided to iGSM users to use either GSM service or H.323 service. In this section, the detailed procedures involved in the registration of GSM users to the H.323 IP network, explicit deregistration and call setup and delivery are discussed.

22.4.1   Registration Procedures

When iGSM users are within the GSM network, the registration procedures follow exactly the procedures in the traditional GSM network using the GSM MAP protocol.

However, if the users move to the H.323 IP network, the registration procedures are different. Figure 22.3 illustrates the registration procedures required for an iGSM user in the IP network. The details of the procedures are described below:

1. The whole procedures start by the user's moving from the GSM network to the IP network. Within the IP network, the user switches on the H.323 terminal and inputs the MSISDN together with the password. These actions activate the iGSM VoIP service. The H.323 terminal then sends an RRQ (RAS Registration Request) message, which includes the user's MSISDN and password, to the iGSM gatekeeper.

2. On receipt of the MSISDN and password of the user, the iGSM gatekeeper acts as a VLR to verify the password and authenticate the user. After authentication, the gatekeeper then sends an IRQ (Information Request) message to the iGSM gateway. This message includes the information about the user, including operation type (GSM UpdateLoc), VLR's ISDN assigned by the HLR, IMSI of the user, and the MSC number (i.e., the corresponding H.323 terminal address). This information is then translated into standard GSM MAP protocol UPDATE_LOCATION and is sent to the HLR.

3. IMSI is an internationally unique number for a subscriber. With this number, the HLR is able to get the user profile of the iGSM user. This profile is sent back to the iGSM gateway using the standard GSM MAP message INSERT_SUBS_DATA, which is then forwarded to the iGSM gatekeeper using an IRR (Information Request Response) message. Afterwards, the iGSM gatekeeper stores the user profile in its database. In response to the user profile message, the iGSM gatekeeper sends an acknowledgement message using IRQ to the iGSM gateway, which then forwards the acknowledgement message to the HLR.

4. At the same time of forwarding the user profile to the new VLR (i.e., the iGSM gatekeeper), the HLR also updates the old VLR using some deregistration procedures. These include the transmission of CANCEL_LOCATION to the old VLR with the IMSI of the user. The old VLR then responds with an acknowledgement of the CANCEL_LOCATION message to indicate the deregistration of the user from it.

5. At this point, the registration procedures are nearly completed. The HLR then sends an acknowledgement of the UPDATE_LOCATION message with its own HLR number to the iGSM gateway to indicate the success of the registration. This message is then forwarded to the iGSM gatekeeper using an IRR message. Finally, the gatekeeper sends an RCF (Registration Confirmation) message back to the H.323 terminal to indicate the finish of the registration.

After the above procedures, the iGSM user can now use an H.323 terminal device such as a PC to make/receive calls with the same MSISDN.

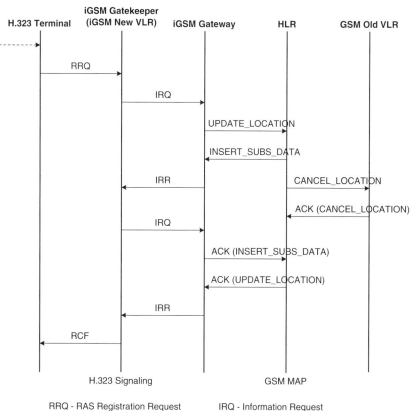

Figure 22.3 iGSM registration procedures.

22.4.2   Deregistration Procedures

As discussed in the previous section on iGSM registration procedures, the registration into the IP network is in fact the deregistration from the GSM network. For iGSM deregistration procedures, the situation is reversed. When the iGSM user moves back to the GSM network, for example, by turning on the handset again, this action will deregister the user from the IP network. Figure 22.4 illustrates the procedures required in the deregistration of the iGSM user from the IP network. The details of the procedures are described below:

1. When the iGSM user turns on the handset again, the registration procedures follow the GSM MAP protocol. The handset registers to the VLR, which then informs the HLR with a UPDATA_LOCATION message. Afterwards, the HLR starts the deregistration procedures. It first locates the iGSM gateway and sends it a CANCEL_LOCATION message with the IMSI of the iGSM user so as to indicate that the user has left the IP network. The iGSM gateway then sends an IRR message to the iGSM gatekeeper with IMSI and CanLoc as the operation type. After that, the iGSM gatekeeper sends a URQ (Unregister Request) to the H.323 terminal.

2. The H.323 terminal responds with a UCF (Unregister Confirmation) message as an acknowledgement to the completion of deregistration back to the iGSM gatekeeper. After knowing that the H.323 terminal has finished the deregistration procedures, the iGSM gatekeeper responds with an IACK (Information Request Acknowledgement) back to the iGSM gateway, which then generates an acknowledgement to the CANCEL_LOCTION message back to the HLR to indicate the completion of deregistration procedures in the IP network.

22.4.3   Call-Setup Procedures

The above procedures describe the registration and deregistration processes of an iGSM user in the H.323 IP network. In this section, we describe the call-setup procedures. If the user is within the GSM network, the call origination and call delivery follow the standard GSM procedures. However, if the user is within the IP network, the call origination and call delivery follow the H.323 procedures. Figure 22.5 illustrates the call-setup procedures while the iGSM user is within the IP network and the call originated from another terminal (e.g., a PSTN or GSM terminal) is delivered to the H.323 terminal of the iGSM user [295]. The whole process is described below:

1. The caller, who may be using another GSM mobile station, dials the MSISDN of the iGSM user. From the MSISDN, the GMSC (Gateway MSC) of the mobile station can be located. As a result, the originating switch sends a SS7 IAM (Initial Address Message) to the GMSC to initiate the call.

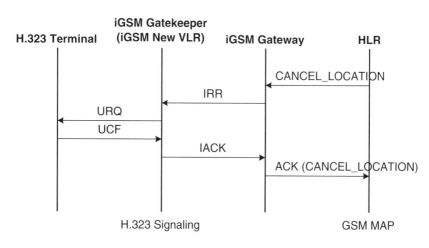

Figure 22.4 iGSM deregistration procedures.

2. On receiving the IAM message, the GMSC needs to find out the current location of the mobile station. Therefore, it needs to query the HLR to ask for the location by using the GSM MAP protocol—SEND_ROUTING_INFORMATION. This message also includes the iGSM user's MSISDN. With this MSISDN, the HLR is able to locate the record of the iGSM user and find out his IMSI. Afterwards, the iGSM user is found to have registered to the HLR through the IP network previously. Then the corresponding iGSM gateway and iGSM gatekeeper can be located. As a result, the PROVIDE_ROAMING_NO message is sent to the iGSM gateway, which is then translated to an RAS LRQ (Location Request) message with RoamNo as the operation type and IMSI of the iGSM user.

3. On receipt of the LRQ message, the iGSM gatekeeper, which is also the current VLR of the H.323 terminal, is able to locate the H.323 terminal (may be a PC) used by the iGSM user. As a result, the MSRN (Mobile Station Routing Number), which is derived from the iGSM gatekeeper and the H.323 terminal address and can be used to uniquely locate the H.323 terminal, is transmitted back together with an acknowledgement in an RAS LCF (Location Confirmation) message to the iGSM gateway. The gateway then sends back the MSRN as a payload in the acknowledgement of the SEND_ROUTING_INFORMATION message to the GMSC, which initiates the query of location of the mobile station at the beginning.

4. The GMSC, which has the unique location information of the H.323 terminal, sends an IAM to the iGSM gateway to show the intention to initiate a call. The IAM message is then translated by the gateway into an ARQ (Admission Request), which is then sent to the iGSM gatekeeper. The gatekeeper responds with an ACF (Admission Confirmation) message to show the possibility of setting up a connection within the IP network. Recall that one of the main functions of the iGSM gatekeeper is bandwidth allocation.

5. The iGSM gateway then sends an H.225 setup message to the H.323 terminal. If the H.323 terminal notifies that there is a call for it and that it would like to accept it, it then replies with a call proceeding message to indicate that it has received enough routing information for the call. Afterwards, the H.323 terminal initiates an ARQ message to iGSM gatekeeper to ask for the permission of accepting the call and to request for bandwidth allocation. Similar to the previous step, the gatekeeper replies with an ACF message.

6. Up to this point, the call-setup is prepared with enough signalling. The H.323 terminal then starts ringing to alert the iGSM user that there is a call for him. At the same time, the H.323 terminal also generates an alerting message to the iGSM gateway, which is then translated into an SS7 ACM (Address Completion Message) message. The ACM message is sent to the GMSC and then to the originating switch.

7. Whenever the iGSM user picks up the call from the H.323 terminal, a connect message is transferred to the iGSM gateway, which is forwarded to the originating switch through the use of SS7 ANM (Answer Message) messages. Finally, conversation between the two communicating entities can be started.

22.4.4   Problems in iGSM

There are two main problems in the use of iGSM. They are the GSM tromboning effect and misrouting due to user mobility [295].

Tromboning effect occurs when the called mobile station and the dialer mobile station are within the same city, but the GMSC is located at other city or country (it is assumed that both GSM MSC and iGSM gateway can function as a GMSC). In this case, it is required to route signals to other city and then transfer back. This kind of calling is expensive.

If the iGSM user is assigned with a GMSC, then the call-setup procedures discussed before are followed by the tromboning effect when the GMSC is in other city or country. However, in the case that the iGSM user is assigned with an iGSM gateway which functions as a GMSC, the following call-delivery procedures are resulted:

1. The originating switch (may be a telephone from a PSTN network) dials the MSISDN of the iGSM user and queries the corresponding iGSM gateway.

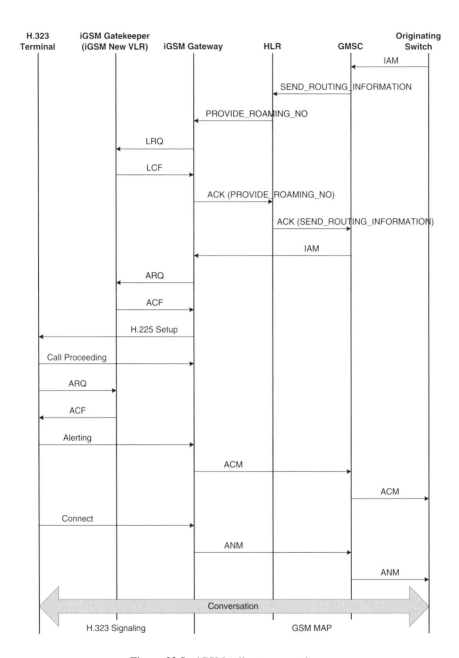

Figure 22.5   iGSM call-setup procedures.

2. The iGSM gateway then looks into its own iGSM gatekeeper (VLR) to see whether the iGSM user has moved from the GSM network to the IP network.

3. If the corresponding MSISDN cannot be found, then the query is forwarded to the HLR and normal GSM call-setup procedures are resulted.

4. If the corresponding MSISDN is found, then normal PSTN-IP call-setup is resulted with the use of normal H.323 call-setup procedures by the iGSM gateway.

Following the above procedures, the tromboning effect is effectively reduced due to the reason that the GSM network is not passed through at all with the iGSM gateway being used as the GMSC. This can result in a much lower call-setup cost than that of the original call-setup procedures. If both communicating entities are from IP networks, then the cost is much lower with the use of a pure VoIP environment. However, this solution may not be adopted by GSM operators if the iGSM gateway is not under their control.

After discussing the tromboning effect, we discuss the misrouting effect due to user mobility. Misrouting effect occurs when the iGSM user moves from the GSM network to the IP network with his mobile station turned on. After moving to the IP network, the registration procedures are carried on for the H.323 terminal which is located within the IP network. With this registration, the HLR remembers that user is in the IP network and the record in the original VLR is modified. After some time, the user leaves the IP network and moves back to the GSM network in the original location. Since his mobile station is still turned on, the mobile station does not explicitly ask for re-registration. In this case, the record in the HLR still indicates that the user is in the IP network, and thus misrouting occurs if someone makes call to the user.

Misrouting can be solved by two measures—implicit registration and explicit registration. Implicit registration occurs when the user dials a call to others. In this case, the VLR will figure out that the record of the user is not present and ask for re-registration. On the other hand, implicit registration also occurs when the user moves to other GSM network or IP network. In this case, registration is forced by the network.

However, the implicit registration scenarios described above actually occur by chance. Therefore, the user should be responsible for using explicit registration to inform the network. This involves turning off of the mobile station within the IP network and then turning it on again after getting back to the GSM network. If this is not possible, then the network can automatically turn off the mobile station if registration from the IP network is detected. With the use of these methods, the misrouting problem can be effectively solved.

22.5 PRACTICAL ILLUSTRATIONS—PROBLEMS AND SOLUTIONS FOR VOIP OVER IEEE 802.11

In previous sections, the integration of GSM network and H.323 VoIP network—iGSM, has been discussed with the detailed procedures of registration, call-setup, deregistration, and the associated problems and solutions of introducing VoIP into the GSM network. In this section, we discuss in detail a practical scenario on the use of VoIP over WLAN. We also discuss how to improve the performance of carrying VoIP traffic in a WLAN environment, and the associated problem when there is also other data traffic [345].

First of all, let us reiterate some of the ideas of WLAN. In IEEE 802.11 WLAN, there are several standards introduced, including a, b, g, etc. IEEE 802.11a/g allows a throughput from 6 Mbps to 54 Mbps, while IEEE 802.11b allows a throughput from 1 Mbps to 11 Mbps. Using GSM 6.10 as the codec, it requires a capacity of 13.2 Kbps, and a frame is generated every 20 ms. As a result, with the use of IEEE 802.11b WLAN and GSM 6.10 codec, it should be able to support around 800 of this kind of VoIP stream. However, it is found that the number of VoIP streams supported is much less than this value (around 12 streams can be actually supported).

There are several reasons behind the discrepancy of the two numbers. One of the main reasons is due to the overhead introduced in packet headers. For example, the use of RTP/UDP/IP introduces 40 bytes of packet overhead, while a packet usually involves 10-30 bytes of payload for VoIP applications. This introduces more than 50% of overhead. At the same time, various MAC control mechanisms (e.g., backoff time, various IFS, etc.) and the PHY layer overhead (e.g., permeable, etc.) can introduce an overhead of more than 800 ms. As a result, the efficiency of VoIP over WLAN is quite low (generally not more than 5%).

In view of the above problems, the voice M-M (Multiplex-Multicast) scheme is proposed [345]. Figure 22.6 illustrates the original idea of VoIP over WLAN, and Figure 22.7 illustrates the modifications made in the M-M scheme. In both cases, a voice gateway is located at the end of each side of the Internet. This gateway is a requirement in the H.323 standard for address translation, call admission control, call signal routing, etc. However, the multiplexer is assumed to be present only in the voice gateway in the M-M scheme.

Originally, without the use of the M-M scheme, VoIP packets are routed from one source to one destination in a unicast manner. However, with the use of the scheme, the packets originated from multiple sources (from S1 to Sn in Figure 22.7) are generated individually and passed to the voice gateway through the access point (AP). In the gateway, multiple packets targeted for the same Basic Service Set (BSS) are combined and multiplexed into one packet with the destination address changed to the multicast address of the BSS. The new packet is then routed through the Internet and arrives at the access point of the designated BSS through another voice gateway. The access point then multicasts the packet within the BSS. In this case, the destination

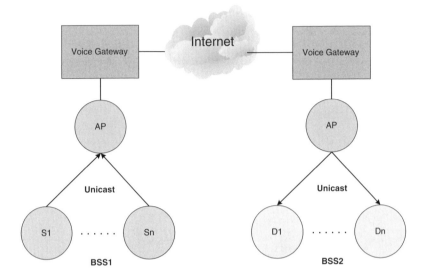

Figure 22.6   VoIP over WLAN.

host receiving the packet looks into the payload to determine whether there is VoIP packet designated for it. The whole process is shown in Figure 22.7

With the use of the M-M scheme, the number of streams supported within one BSS is reduced from $2n$ (n uplink unicast streams and n downlink streams) to $n + 1$ (n uplink unicast streams and 1 downlink stream). As discussed earlier, one of the major reasons for the poor performance of VoIP over WLAN is due to the significant header overhead introduced in various layers. However, the M-M scheme can effectively reduce the headers within each BSS by nearly 50%. As a result, performance improvement of around 90% to 100% is recorded through detailed analysis and simulations [345]. However, one of the limitations of the scheme is that the original security measure used for the unicast stream may not be applicable to the scheme. At the same time, multicast in IEEE 802.11 does not support ARQ. As a result, the performance of the scheme will be varied according to the collision rate.

After discussing the details of the M-M scheme in improving the performance of VoIP over WLAN, we discuss the effect of the presence of TCP data traffic, which is ignored in previous discussion. In fact, TCP traffic can greatly affect the performance of VoIP over WLAN. Figure 22.8 illustrates the scenario that we are going to discus. In the figure, there is an access point in which there are $2n$ VoIP streams and one FTP traffic with uplink and downlink streams. The TCP traffic can affect the VoIP traffic in two ways—the queue in the AP and medium access contention between nodes [345].

As for the queue in the AP, the general implementation of commercial access points involves the deployment of FIFO queue which is shared by all kinds of traffics.

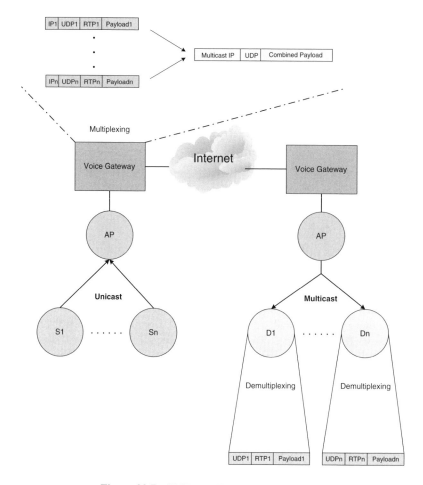

Figure 22.7 VoIP over WLAN M-M scheme.

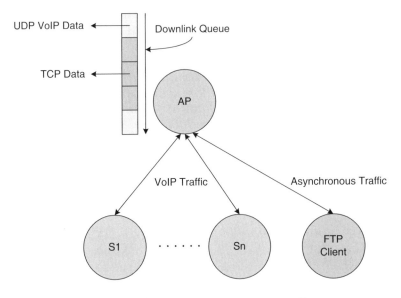

Figure 22.8   VoIP over WLAN with FTP traffic.

As a result, the bursty TCP traffic shares the same queue with UDP (VoIP) traffic. The main behavior of TCP traffic is that it aggressively sends out packets until packet loss occurs. As a result, TCP traffic effectively delays the channel access of VoIP traffic at the access point. A typical solution to this problem is the use of PQ (Priority Queue) in the access point with a higher priority given to the time-sensitive VoIP traffic. Using this method, the access delay can be effectively reduced from around 80ms to less than 5ms in the case of six VoIP streams and one FTP traffic.

On the other hand, the uplink TCP traffic (e.g., the TCP ACK for each TCP packet received) can greatly affect the VoIP traffic as well, especially in the case of the M-M scheme. The reason is that the M-M scheme takes advantage of the multicast nature of the wireless medium. However, multicast does not require any ACK for each packet received. Therefore, there is no mechanism to recover lost packets due to collisions. As a result, a multicast scheme is very sensitive to collisions. On the other hand, the uplink TCP traffic greatly increases the chance of collisions since it waits the same amount of time (DIFS) with other stations. In this case, with a large amount of TCP packets received, there is a large amount of uplink acknowledgement packets, or vice versa if the FTP transmission is in the reverse direction. Subsequently, a large amount of collisions with VoIP streams will be resulted. Using the M-M scheme or M-M scheme with PQ, and in the coexistence of six VoIP streams and one TCP stream, the packet loss rate can be more than 10%.

In view of the above problems, a modification to the MAC layer is proposed [345]. This involves the use of a different length of IFS (Interframe Space). Recall from Section 9.4 that the shorter the length of IFS, the higher the priority to access the medium among different stations. As a result, a MAC-layer multicast priority scheme (MMP) with MIFS, which has a length longer than SIFS but shorter than DIFS, is proposed [345]. With the use of MIFS as the wait time for time-sensitive traffic, priority channel access is given to multicast traffic. As a result, the problem of contention between VoIP traffic and TCP traffic can be solved with the use of different priority wait time. Under the MMP M-M scheme, the loss rate is around zero but the access delay is increased in the case of six VoIP streams and one TCP stream. However, with the use of the M-M scheme together with the two solutions proposed—PQ and MMP, both access delay and loss rate are reduced.

22.6 SUMMARY

In this chapter, the cooperation between GSM network and H.323 VoIP network to support wireless VoIP services (i.e., iGSM) has been discussed. First of all, the architecture of iGSM has been introduced with the additional components that are required to support the services. Afterwards, various important issues in supporting the services have been discussed, including registration procedures, deregistration procedures, call-setup procedures, and some of the problems and their corresponding proposed solutions in using these services such as the tromboning effect and the misrouting problem.

Besides the iGSM system, as practical illustrations, we have also discussed VoIP over WLAN. Various issues on WLAN wireless technologies in supporting VoIP have been discussed, including the low bandwidth capacity of WLAN, poor performance in the presence of other data traffic such as FTP traffic, etc.

PROBLEMS

22.1   Briefly discuss the idea of iGSM in integrating the GSM network and H.323 IP network. Explain why additional components are necessary.

22.2   In iGSM, why there are the tromboning effect and misrouting problem?

22.3   Briefly discuss the idea of the M-M (Multiplex-Multicast) scheme in VoIP over WLAN and explain how it can improve the performance of VoIP over WLAN.

22.4   Why data traffic can greatly affect the performance of VoIP over WLAN? How can this problem be solved with the use of priority queue and MAC-layer multicast priority (MMP) scheme?

CHAPTER 23

WIRELESS VIDEO

Don't part with your illusions. When they are gone you may still exist but you have ceased to live.

—Mark Twain

23.1 INTRODUCTION

Video has become a commodity application in the wired Internet environment. Users routinely play back video files using their computers instead of using traditional audio/video equipment. Naturally, people expect the same to happen in a wireless environment—download video files wirelessly and play them back anywhere anytime.

Interoperability:

A wireless environment is known to have lossy channel with high error rate. This makes it difficult to provide real-time services in a wireless environment, where some QoS requirements have to be met. The real-time services envisioned are discussed in Section 23.2 while the system constraints are discussed in Section 23.3. With

Wireless Internet and Mobile Computing. By Yu-Kwong Ricky Kwok and Vincent K. N. Lau

the system constraints, different error concealment techniques have been suggested in order to tackle the issue of wireless channel. These techniques are introduced in Section 23.4. Error concealment techniques can be divided into three different categories based on the information hiding side, including forward techniques, post-processing techniques and interactive techniques. These techniques are discussed in Sections 23.4.1, 23.4.2 and 23.4.3, respectively. On the other hand, there are techniques which consider the effect of source coding and channel coding based on the varying wireless channel environment. This kind of joint source channel coding techniques are discussed in Section 23.5.

Performance:

On evaluating the performance of a wireless video system, we can adopt metrics like frame rate, bandwidth, delay and corrupted blocks. However, these metrics cannot effectively represent the video quality perceived by the viewer. The reason is that the perception of quality is very subjective, and therefore it is difficult to develop a satisfactory metric to evaluate this. Despite this, it is common to use average PSNR (Peak Signal to Noise Ratio) to evaluate the quality of a decoded video. The main aim of using average PSNR is to identify the maximum signal energy to noise energy ratio, which can be used to approximate the distortion of the decoded video.

23.2   REAL-TIME SERVICE GOALS

Real-time services involve service that requires the sending of data and/or video/audio between two or more parties. This kind of service is usually delay-intolerant. According to the delay-intolerant level accepted by applications, there are two main kinds of real-time applications.

The first kind is real-time interactive multimedia applications [285], such as peer-to-peer phone (VoIP), videoconferencing, etc. The aim of this kind of applications is to enable different users, who may have different communication devices, to enjoy multipoint communication with acceptable visual and audio quality. In addition, it is generally more tolerable to have relatively less quality in visual, but a relatively high quality of audio is preferred. Specifically, the audio service provided should be comparable to that provided by wired telecommunication service like PSTN. Consequently, in order to provide satisfactory quality of interactive service, a number of constraints like delay constraints and bandwidth constraints have to be considered. These system constraints will be further discussed in Section 23.3.

The second kind is streaming live multimedia applications [285], including live television programme, live radio program, etc. Strictly speaking, this kind is not exactly real-time service as it can tolerate a higher level of delay than that of the first kind. This kind of applications usually involves one-way sending of some video or audio from servers to multiple subscribed clients. Since it is relatively more delay-tolerant than interactive service of the first kind, its main aim is to provide clients with acceptable quality of video and audio under loose delay constraints and bandwidth constraints.

23.3   SYSTEM CONSTRAINTS

Under the service goals discussed in the previous section, it is expected that certain minimum levels of network delay and bandwidth are required for real-time services so as to guarantee the quality of audio and video perceived by a user. In the conventional wired circuit-switched network or virtual circuit-switched network (e.g., PSTN), this can be done by resource reservation. However, nowadays, packet-switched network becomes a trend for nearly all kinds of networks, where guarantee on certain minimum levels of network delay and bandwidth is more difficult. The situation is worsen in a wireless network environment, in which an unreliable transmission medium is used. Specifically, the wireless medium is shared by all users, with the properties of having limited bandwidth frequent errors [285, 348].

Another problem is the possibility that different users are using different communication devices of different capabilities, and the devices are connected to different networks. It is expected that users may have real-time interactions through Internet connected by different access networks, for example, wireless networks, cellular networks, and ADSL, etc. The user devices and access networks would have very different capabilities in terms of computational power and bandwidth. Due to this reason, for real-time interactive communication between a 3G mobile phone user and a computer user with broadband Internet access, the heterogeneity between communicating parties have to be solved.

23.4   ERROR CONCEALMENT TECHNIQUES

In order to tackle the system constraints mentioned before, some techniques are required to detect and conceal errors generated. Generally speaking, there are three main types of error concealment techniques, as illustrated in Figure 23.1 [348].

The first type of error concealment techniques, i.e., forward techniques, is mainly initiated by the source. The source intentionally uses some schemes to hide errors from the destination. The most common examples of this type are layered coding techniques and transport level techniques, and the joint source channel coding techniques. In the second type, i.e., postprocessing techniques, the destination tries to hide errors from users by making use of two common techniques, namely spatial and temporal interpolation. The third one usually involves the cooperation between the source and destination. This type is called interactive techniques, and the examples are selective coding and retransmission techniques.

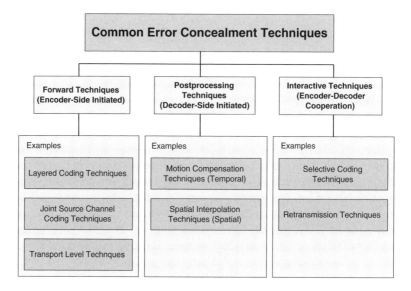

Figure 23.1 Examples of error concealment techniques.

23.4.1 Forward Techniques

In forward techniques, the source encoder tries to use some techniques to hide errors from the destination decoder. Several common techniques are discussed in the following.

23.4.1.1 Layered Coding Techniques Layered coding techniques are one of the most common example of forward error concealment techniques. The idea is summarized in Figure 23.2 [348]. In the figure, the source takes a raw video as the input. Then it encodes the raw video into multiple layers—the base layer (BL) and multiple enhancement layers (ELs). The BL usually involves the minimal encoded information which can be played out individually. Therefore, the BL defines the minimal bandwidth and delay constraints of the video, and is the most important layer. The EL involves the additional encoded information which can be used to refine the quality of the video. In order to provide multiple qualities of the video, several enhancement layers can be built.

Afterwards, the encoded BL and EL bitstreams are packetized and sent to the destination. Based on the bandwidth of the destination, the user can then select the layers that he is planning to receive in order to balance the tradeoff between quality and bandwidth available. As shown in Figure 23.2, the user can enjoy the minimal decoded video quality with the receipt of BL. However, if he can further receive the next enhancement layer, the decoded video quality can be refined by combining the BL and EL.

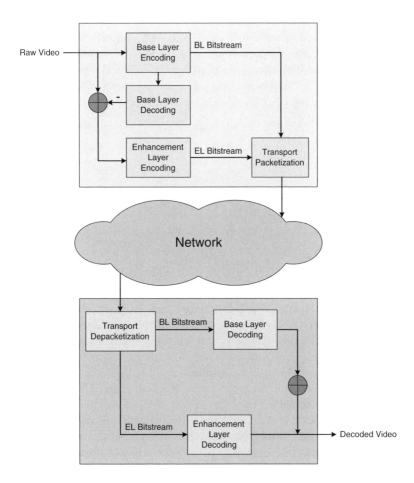

Figure 23.2   Layered coding scenario.

23.4.1.2 Transport Level Techniques There are several kinds of techniques that can be regarded as transport level techniques. The first one introduced in this section, i.e., prioritized packetization of layered coding, is in fact the continuation of the layered coding techniques in the previous section.

Figure 23.3 illustrates the idea of prioritized packetization of layered coding. In layered coding, it comprises the source coding of multiple layers involving the BL and multiple ELs. In fact, the receipt of BL is the most important as it is the basis of the quality of the video, while the receipt of the ELs further enhances the video quality. Due to this reason, the packets of the BL bitstream should be assigned with a higher priority than the EL bitstream, as illustrated in Figure 23.3 In general, packets with a higher priority will experience less packet loss and lower delay if the network in-between the source and destination is QoS-aware, for example, diffserv or intserv [348].

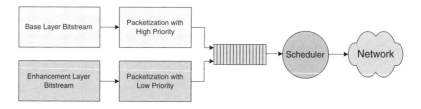

Figure 23.3   Prioritized packetization for layered coding.

Other than prioritized packetization of layered coding, some other techniques like interleaving can also be applied to help hiding errors. In fact, if the loss of packets includes non-successive blocks, the lost blocks can be interpolated with a relatively high accuracy with the help of spatial and temporal interpolation, which will be discussed in Section 23.4.2. Because of this reason, the idea of packetization of odd-even blocks are suggested [373, 348]. As shown in Figure 23.4, the blocks of a frame are marked with odd and even numbers. During packetization, the even blocks from the same slice of a macroblock are filled in first, followed by the odd blocks. By doing so, if there is packet loss, the even and odd blocks in the macroblock of a frame are referred to identify the lost block. Subsequently, by using the correctly-received blocks the lost block can be interpolated with a relatively high accuracy.

In addition to the interleaving techniques mentioned above, cell level forward error correction (FEC) codes like Reed-Solomon codes are commonly applied to conceal errors caused by cell loss [348].

23.4.2   Postprocessing Techniques

The destination side is usually involved in postprocessing techniques to figure out the occurrence of errors due to various reasons and to conceal the errors from the user. If there are missing packets, since the destination side decoder does not have correct

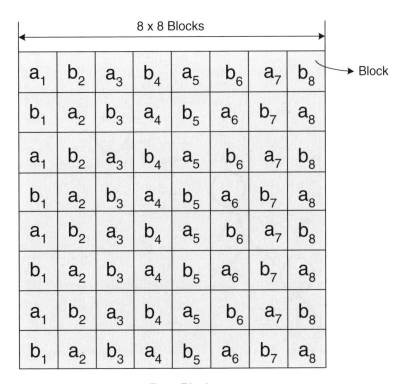

a - Even Block
b - Odd Block

Figure 23.4   Block interleaving.

encoded information, then the corresponding macroblocks will be affected. However, the missing information can be fabricated by referencing the correct information of neighboring blocks in either time or spatial domain. In other words, postprocessing techniques involve the use of interpolation in time domain and spatial domain as illustrated in Figure 23.5 In fact, as for interpolation, some simple techniques can be applied, such as replacing the missing macroblocks from neighboring macroblocks of the same frame or replacing the missing macroblocks from the macroblocks of previous frame. However, these techniques will result in strange images which may be noticeable by the user. This will be further discussed in subsequent sections.

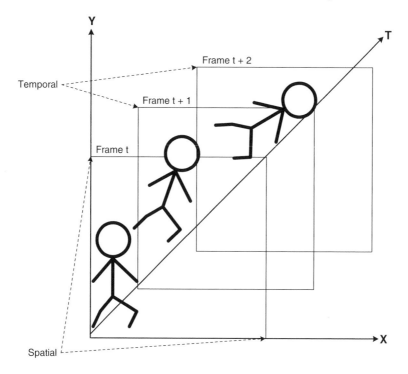

Figure 23.5   The idea of postprocessing techniques.

23.4.2.1 *Spatial Interpolation Techniques*   Spatial interpolation techniques are proposed mainly based on the idea that neighboring macroblocks within the same frame should be very similar to each other. In this section, two spatial interpolation techniques will be discussed.

The first one is block-based spatial interpolation. As depicted in Figure 23.6, a unit square represents a pixel while a square of $8x8$ pixels represents a macroblock. In addition to macroblocks, a square of NxN pixels is defined as a block (in the figure, $N = 4$). Suppose that the macroblock inside the figure is missing, then the pixels inside the missing macroblock are fabricated by interpolating the two nearest pixels from two other correctly-received neighboring blocks. As illustrated in Figure 23.6, the pixel currently in consideration has a distance d_T from the upper block and a distance d_L from the left block. The interpolation is then performed based on these two pixels with distance being a weight [13, 348].

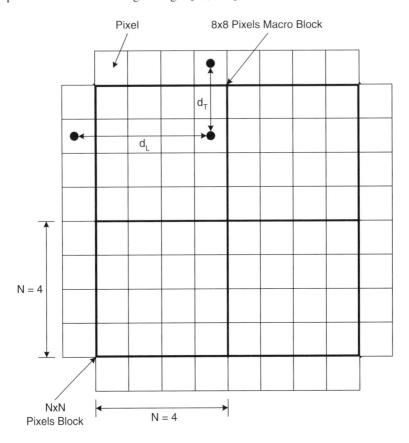

Figure 23.6   Block-based spatial interpolation.

The second one is macroblock-based interpolation. As shown in Figure 23.7, suppose that we are considering the same missing pixel inside the same macroblock as in Figure 23.6, the missing pixel this time is fabricated by interpolating the four nearest pixels from the four neighboring macroblocks. In this case, the pixel in consideration is interpolated from four nearest pixels with distances d_T, d_B, d_L and d_R. Again, the interpolation is performed with distance being a weight.

23.4.2.2 Temporal Interpolation Techniques   Temporal techniques are mainly based on the idea that two macroblocks in the similar region should be very similar

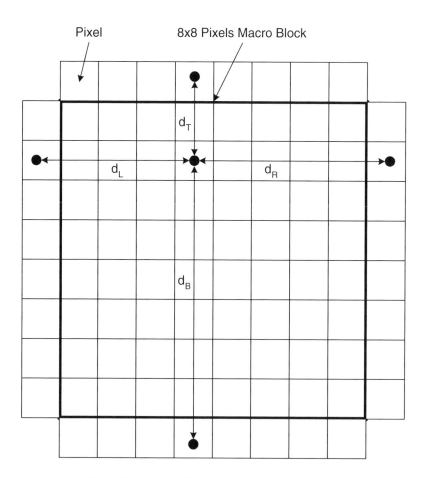

Figure 23.7 Macroblock-based spatial interpolation.

in two consecutive frames. However, this idea is correct for still consecutive frames. If the objects are in movement, there will be differences in two consecutive frames. In this case, directly applying the macroblocks of the previous frame to the missing macroblocks of current frame will result in a strange image noticeable by the user. Consequently, as shown in Figure 23.8, if the object is in motion, motion vectors of the previous frame and current frame are generated and sent to the receiver. During decoding process, the motion vectors are considered in regenerating the images in the frames. If certain macroblocks are missing as shown in the figure, then the motion vector of the missing block is interpolated from the motion vectors of the neighboring macroblocks. This scheme is based on the idea that objects, which usually occupy several macroblocks, will move in the same direction in consecutive frames. With the motion vectors of the missing macroblocks, the missing macroblocks can then be regenerated by either selecting the appropriate macroblocks from the previous frame in replacement of the missing one or interpolating the macroblocks of the previous frame based on the motion vectors. Schemes that make use of this idea are also known as motion compensated concealment schemes [79, 348].

23.4.3   Interactive Techniques

Interactive techniques usually involve the cooperation between the sender and the receiver. Upon discovering the occurrence of some error, the receiver will actively contact the sender, and the sender will respond by sending the original copy of the error packet (retransmission) or changing the encoding of subsequent frames to remove the dependency on the error packet (selective encoding).

23.4.3.1   Retransmission Techniques   When the destination side discovers that there are errors, it will then contact the source about the error region. In some streaming applications as discussed in Section 23.2, it is affordable for the destination to wait for the retransmission packets as there is usually certain amount of play out delay for the buffered packets. However, for real-time applications, retransmission waiting time may not be affordable. In this case, before the retransmitted packets arrive, the destination side can use the error concealment approaches discussed before to conceal the error while waiting for retransmission. At the same time, subsequent frames and blocks depending on the missing blocks are marked. In this case, when the retransmission packets arrive before the play out time of the affected frames and blocks, those affected blocks are then reformed [348].

Besides the stopping of error cascading effect of a missing block, retransmission can also be done based on the priority of the layers as discussed in Section 23.4.1. Recall that the loss of packets in the BL bitstream is much more intolerable than that in the EL bitstream. In this case, multiple copies retransmission can be applied if the high priority packets are lost.

23.4.3.2   Selective Encoding   Specifically, one of the most important effects of retransmission of packets as mentioned above is to stop error propagation due

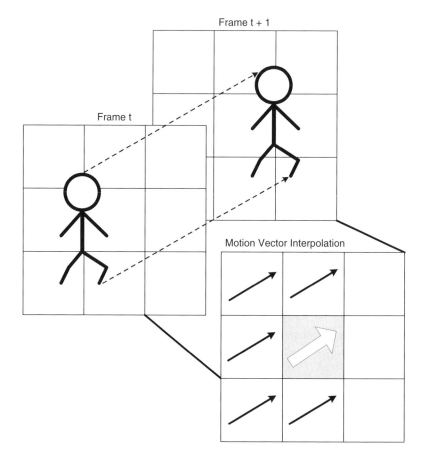

Figure 23.8 Motion compensation temporal interpolation.

to missing or error blocks. In addition to retransmission, there are also some other approaches. The simplest approach is that the receiver informs the sender about which block in which frame is in error. The sender will then encode the next frame into intracode mode. However, under this scheme, there is no dependency on the previous frames, and the compression efficiency will be greatly reduced due to the fact that techniques like motion vector cannot be applied to reduce the information transmitted [348]. To mitigate this problem, selective encoding is introduced. The main idea of selective encoding is that whenever the receiver detects erroneous blocks, it reports the error to the sender. Then the sender can continue to encode the video with dependency of the missing blocks up to current frame and the affected area is also calculated. On the other hand, the next frame is encoded without the dependency of the previous missing blocks or areas.

23.5   JOINT SOURCE CHANNEL CODING TECHNIQUES

In a wireless video scenario as depicted in Figure 23.5, the raw video is first encoded by the source coder, which aims at reducing the redundancy in the raw data, and then further encoded by the channel coder, which aims at protecting data from channel specific errors. Traditionally, source coders and channel coders are designed separately [348, 285].

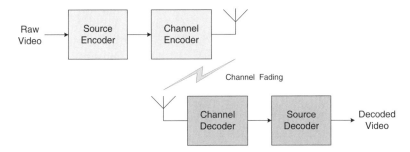

Figure 23.9   Source and channel encoders and decoders.

Shannon suggested that the source coder and channel coder can be designed separately to achieve an optimal performance in the case that the source coder is an optimal one which produces data rates lower than that of channel capacity. At the same time, the channel coder is also an optimal one. However, it has been suggested that Shannon's idea is impractical to be achieved with reasonable design complexity of the coder, especially in a wireless environment where there is channel fading causing various channel errors. In order to solve this problem, joint source channel coding is developed [98].

In fact, under noisy channel, increasing the source coding rate does not necessarily result in a better quality of video due to the reason that less bandwidth will be available for channel coding. As a result, on designing joint source channel coders, a trade-off has to be balanced between the effects of source coder and channel coder under certain fading channel (e.g., Rayleigh fading channel).

Many approaches have been proposed in joint source channel coding. One of the typical examples is the use of convolution code in protecting the MSB (most significant bit) generated by source coder DPCM (differential pulse code modulation) [245] and DCT (discrete cosine transform) [246]. The idea is then extended to the evaluation of the importance of certain bits to the video quality and different coding rates are used to protect the bits from channel errors.

23.6 PRACTICAL ILLUSTRATIONS—RATE ADAPTATION FOR MPEG-4 VIDEO

In previous sections, various error concealment techniques have been introduced. In this section, we present a practical illustration which combines the techniques introduced before to improve the decoded video quality of MPEG-4.

The rate adaptation for MPEG-4 video [319, 370] involves three main components as shown in Figure 23.10. The three components are:

- Rate Adaptation of Scalable Coding

- Prioritized Packetization

- Differentiated Forwarding

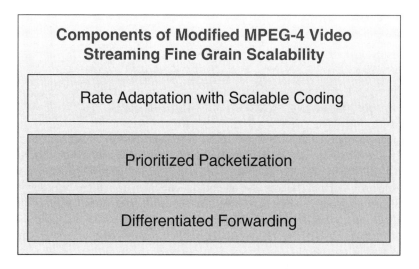

Figure 23.10 Components of rate adaptation for MPEG-4 video streaming.

Figure 23.11 [319, 370] illustrates the framework of the rate adaptation MPEG-4 FGS (fine grain scalability) with prioritized packetization and differential forwarding. In MPEG-4 FGS, the video stream can be encoded into two layers—base layer (BL) and enhancement layer (EL). Unlike the idea of layered coding in Section 23.4.1, the enhancement layer encoding in MPEG-4 FGS is more flexible with the use of bitplane coding in which each bitplane can be adjusted independently based on the bandwidth requirement, i.e., the receipt of certain bitplane of certain rate can increase the video quality in MPEG-4 FGS instead of the need for successful receipt of the whole EL. The BL defines the minimum bandwidth requirements for the decoding of the video and so the bandwidth for the BL bitstream can define the minimum bandwidth requirement of the network. On the other hand, rate-distortion information can be calculated and embedded into each bitplane independently by considering the first several samples because the same bitplane has similar rate-distortion characteristics. The rate for EL bitplanes can then be selected based on the estimation of the current bandwidth so as to maintain a relative constant video quality (minimal distortion variance between different frames of the video). This idea is known as rate adaptation of MPEG-4 FSG.

The rate-adapted BL and EL bitstreams are then packetized and assigned with a priority based on its relative impact on the video quality with the loss of packets (including the impact on current frame and subsequent frames). This idea is known as prioritized packetization.

Afterwards, the prioritized packets are mapped into different QoS queues in the assured forwarding in diffserv, which has different QoS agreements in terms of bandwidth and delay and/or packet dropping rate. Generally speaking, BL packets are assigned to a higher priority queue due to the fact that the BL defines the minimum bandwidth requirements and has relatively less flexible adjustment. The priority assigned inside the packets can then be used to determine which packets to be dropped inside the same queue, so as to minimize the impact on the video quality. This idea is known as differentiated forwarding.

The combination of the three components aims at maintaining a relatively constant video quality even if there are fluctuations in bandwidth availability, with the help of rate adaptation based on the network estimation performed by either the source or destination side. This is illustrated in Figure 23.11

23.7 SUMMARY

This chapter begins by discussing the real-time service goals and system constraints of a wireless video system. There are two main kinds of real-time services expected. The first one is real-time interactive services like videoconferencing and the second one is streaming live multimedia applications. These two kinds of services have different requirements in terms of network bandwidth and delay.

The lossy link and low bandwidth characteristics of wireless medium result in high error rate in packet transmission. As a result, some error concealment techniques are

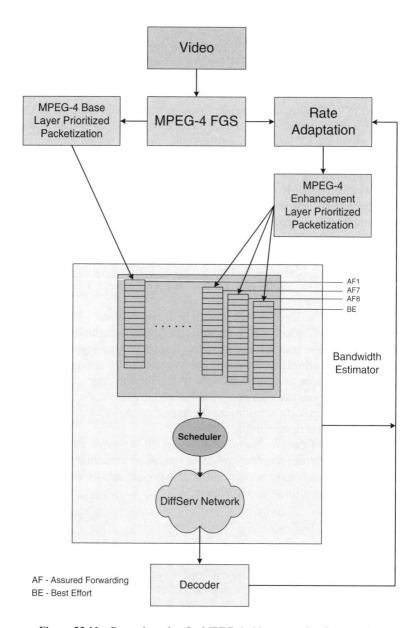

Figure 23.11   Rate adaptation for MPEG-4 video streaming framework.

required to conceal errors from the user. There are three main categories of error concealment schemes—forward, postprocessing and interactive.

Forward techniques usually involve the source which uses some measures to help hiding errors. For example, in layered coding, it encodes the video into BL and EL bitstreams, while the transport level techniques include prioritized packetization of layered coding, block interleaving, and cell-level FEC, etc. Postprocessing techniques usually involve the destination side which uses some measures to help hiding errors from the user. The most common techniques in postprocessing involve spatial and temporal interpolation. In spatial interpolation, it takes advantage of similarities of neighboring macroblocks in the same frame. In temporal interpolation, it takes advantage of similarities of macroblocks of the similar region in consecutive frames. Finally, interactive techniques involve the cooperation between the source and destination. Specifically, the destination side actively informs the source side about the occurrence of errors. Afterwards, the source will try to stop error propagation by either retransmission or selective encoding, so as to remove adverse effects of erroneous macroblocks to later frames.

Lastly, a practical illustration which presents how the error concealment techniques introduced in this chapter are combined in MPEG-4 FGS with rate adaptation, prioritized packetization and differentiated forwarding.

PROBLEMS

23.1 In this chapter, several error concealment approaches are introduced. Briefly discuss these approaches and their pros and cons.

23.2 Briefly discuss the joint source channel coding approach and its pros and cons.

CHAPTER 24

WIRELESS FILE SYSTEMS

Information appears to stew out of me naturally, like the precious ottar of roses out of the otter.

—S. L. Clemens (Mark Twain): Roughing It, pref., 1872

24.1 INTRODUCTION

File system is a fundamental, if not the most basic, service in any computing environment. The "file" abstraction matches human's natural intuition of organizing information. It is mandatory for a wireless system to provide a convenient file system.

Interoperability:

Wireless and wireline file systems have quite a number of differences. Generally speaking, a wireline file system can be assumed to be reachable all the time. As a result, certain modification on the file system can be broadcast to all of the servers at the same time. In order to improve the availability of files, replication of files on multiple servers connected through a wireline environment can be adopted. This can ensure that files accessed by clients are the most updated ones. However, when certain files are being modified by some clients, other clients might not be able to access the files in order to prevent conflicts. This is discussed in Section 24.4.

On the other hand, in a wireless environment, the assumption of always online may not be correct. Due to some bandwidth limitations such as reduced throughput rate and increased data transfer cost, mobile devices might not be able to connect to the servers at all or they only have some limited connections to the servers. To solve the problem, disconnected operations and weakly connected operations are introduced, which are discussed in detail in Section 24.5 and Section 24.6, respectively. These two kinds of operations can help improve the availability of files with the use of some techniques like hoarding, emulating, reintegration, etc.

Both replication method and disconnected and weakly connected operations can help improve the availability of files in a wireless and distributed environment. However, they target on different situations. Replication method targets more on consistency, while disconnected and weakly connected operations target more on availability. This leads to an interoperability problem as the two kinds are considered to be competing with each other in nature. However, the idea of file replication through multiple connected servers can, in fact, be combined with the idea of disconnected and weakly connected operations, so as to get the best of both worlds. As a result, the two kinds can also be complementary to each other, by taking special care of interoperability issues mentioned before. In this chapter, the Coda file system is discussed in detail to demonstrate the possibility of tackling the interoperability issues.

Performance:

File system performance is traditionally quantified by data response time and data transfer throughput. In a distributed file system, the control overhead needed to maintain data consistency is also an important measure. These metrics are also commonly used in assessing the performance of a wireless file system.

24.2 FILE SYSTEM SERVICE MODEL

Coda is a distributed file system designed to support mobile file systems with the support of several novel operations, including optimistic replication in server replications, disconnected and weakly connected operations, etc. With the help of these operations, mobile file system support becomes possible.

Figure 24.1 illustrates the service model of the Coda file system [310]. In the figure, there are two sides involved—client side and server side. In the client side, client $1a$ and client $1b$ represent the client under strongly network-connected situation. In this mode of operation, the client can read files from the server and make updates to the server immediately. On the other hand, client 2 represents the weakly connected mode of operation. In this mode of operation, the client may have limited bandwidth connection to the servers, e.g., a mobile client updates through wireless connections. Lastly, client 3 represents the disconnected mode of operation. In this mode of operation, the client may not be able to reach the servers to get the most updated items.

On the other hand, server $1a$ to server $1c$ represent the servers under normal mode of operation. As a result, any update on one server can be propagated to others immediately. Server 2 represents a server which is isolated from the connected server group. This is possible due to specific server failures or network disconnection between two groups of servers. In this case, the servers need to resolve conflicts after reconnection of the failed server.

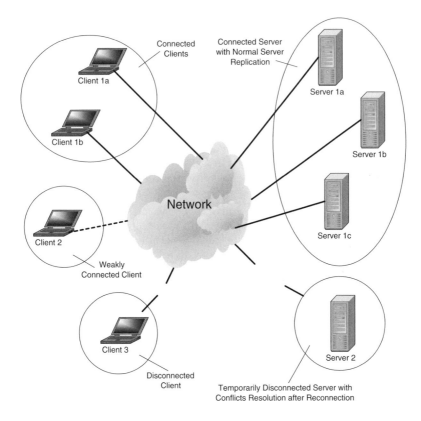

Figure 24.1   Coda file system service model.

With the help of the service model just discussed, mobile clients with different network connection modes (i.e., normal network connection, no network connection and limited network connection) can access to a mobile distributed file system. In the following sections, we discuss how Coda copes with the characteristics of different modes of operation without greatly affecting the services provided to users in other modes.

24.3 GENERAL PRINCIPLES FOR THE DESIGN OF A DISTRIBUTED FILE SYSTEM

In the design of a distributed file system that can support the service model mentioned before, there are a number of considerations. Some of the major ones are summarized in Table 24.1 [310].

Table 24.1 Principles for the design of a distributed file system.

Principle	Description
Client-Server Model	Generally, a distributed file system is expected to have a limited number of trusted servers which can replicate user files and serve user requests, and a relatively large amount of untrusted clients that may access files anywhere.
Transparent File Location	The identities of different servers holding files should be transparent to users. In other words, the users should be able to know the existence of files and access them through local name space as if normal local file access. The contact details of servers are hidden from users.
Local File System Caching	In order to support a large mount of users, and to improve availability and efficiency, local file systems should be used to cache files requested by users, such that operations on the files can be performed locally and modifications are updated back to the servers afterwards.
Volume Based Organization of Data	Data files are grouped into volumes. Each volume refers to a subtree or part of s subtree of the whole name space. This kind of organization can improve administrative efficiency.
Access Control	Authentication mechanisms should be adopted to identify users. Access control mechanisms should also be used to control the access rights of different users.

24.4 REPLICATION SERVICES AND MECHANISMS

The most basic idea for improving the availability of client files is to replicate files. The replication of client files is usually performed by various servers located in different places, and the servers replicate client files based on some predefined rules. As a result, files can be retrieved whenever a client is able to contact either one of the servers with replicated files. This can eventually improve the availability and prevent single point of failure.

However, replication also causes some problems. For example, a write to the distributed file system is required to be propagated to all the servers holding the replica simultaneously. In addition, it is required to handle lots of conflicts like concurrent update by multiple clients on the same file. In order to reduce the complexity of replication, pessimistic replication is widely employed. With this method, access to a file may be blocked with the presence of network partition in order to preserve data consistency. However, this method is not scalable. As a result, optimistic replication is incorporated into the Coda file system [311, 310]. Using optimistic replication, read and write operations are permitted with network partitions as long as the servers within the partition maintain consistency. Conflicts are then left to be resolved until the network partitions are merged.

In Coda, the basic unit of replication is volume. A volume contains a tree structure in which the root directory of the volume, files and subdirectories below the root directory are present. Mobile clients view only a normal file system under the directory /coda. This abstraction is made possible by the Venus cache manager as shown in Figure 24.2. Venus is the cache manager located inside mobile clients and it contacts the appropriate Coda servers holding the required replicas whenever files under /coda are referenced inside the clients.

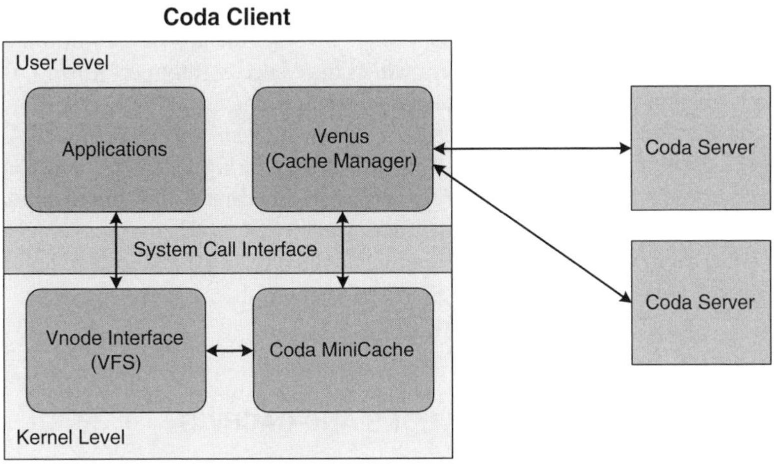

Figure 24.2   Coda client architecture.

The group of servers holding replicas of a particular volume forms the VSG (Volume Storage Group). However, due to network connectivity problems, not all the servers holding the replicas of the particular volume can be contacted. As a result, the group of servers which can be contacted at the time of access forms the AVSG

(Accessible Volume Storage Group). All the accesses including read and write are then be performed on the servers within the AVSG. In this case, it is possible that the size of AVSG is smaller than that of VSG due to some temporary network brokage. However, this optimistic replication approach may result in conflicts which are required to be resolved between servers within the whole VSG. This is in fact a tradeoff between scalability and consistency.

Figure 24.3 illustrates the idea of read operation of the Coda client. When the Coda client references files which are not found within the cache, Venus first randomly selects a server within the AVSG as a PS (Preferred Server). In Figure 24.3, server 1 is selected as PS. Venus then sends data read message with PS to all the servers in the AVSG. The server which is named as PS then returns the status information together with data required by the client back to Venus, while other servers also return the status information back to the client. Afterwards, Venus compares the information obtained by all the servers in AVSG. If there are mismatches in the status information, conflicts occur and have to be resolved. For example, the stale copy of the referenced data inside a server in the AVSG will be updated. After that, the read operation is restarted.

After discussing the read operation, we discuss the update operation. Figure 24.4 illustrates the update operation of the Coda client. When a certain file is modified, the it is sent to all the servers in the AVSG simultaneously. This operation involves two phrases, known as $COP1$ and $COP2$. $COP1$ includes the sending of updated operations such as creating a directory, transferring of the file involved, and/or changing the access lists. On the other hand, $COP2$ involves the process of informing the servers in the AVSG about the servers which have successfully performed $COP1$. This information is summarized in an update set message.

If no failure occurs during the update operation, all the members within the AVSG are synchronized. If a server or network failure occurs during the COP operation, the client times out and assumes that the servers which have not replied are not updated. At the same time, if a server times out, it is assumed that no other servers receive the $COP1$ message. In these scenarios, conflicts will be detected and resolved after the server or network is restored.

24.5   DISCONNECTED OPERATIONS AND CACHING

In the previous section, server replication has been discussed in detail. The growing popularity of portable devices has lead to another view of improving availability of files—disconnected operations [191, 310]. Portable devices usually have limited storage space. Therefore, users usually download files that can be fit into their portable devices and it is highly probable that the files will be used in the near future. Because of these reasons, in disconnected operations, the LRU (Least Recently Used) policy is utilized, which is assisted by the prediction of file access of users.

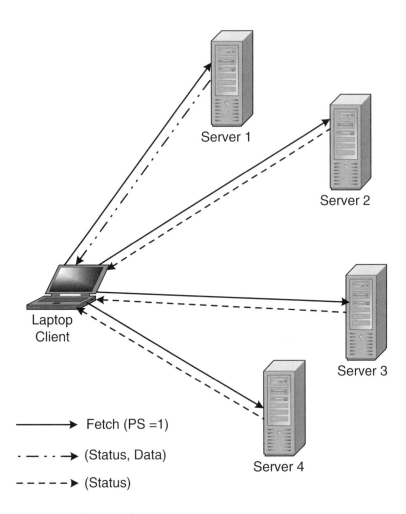

Figure 24.3 Coda server replication read process.

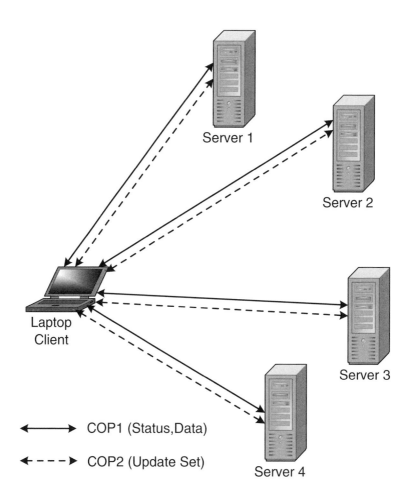

Figure 24.4 Coda server replication update process.

No matter whether it is due to voluntary or involuntary network disconnection, during disconnected operations, the same file system name can still be used. The files which have been downloaded can still be accessed and updated, and possible associated conflicts are automatically resolved after the client reconnects to the network.

Figure 24.5 illustrates the implementation of disconnected operations in Coda. In order to support disconnected operations, three states are defined in the Venus cache manager according to different situations—hoarding, emulating and reintegrating. In normal situation, Venus is in the hoarding state which performs read/write operations normally as discussed in the server replication part previously. However, when disconnection occurs, Venus then transits to the emulating state in which servers are pretended to be present. After reconnection, Venus transits to the integrating state in which all the updates and files are synchronized with the servers. After the synchronization process, Venus returns to the hoarding state. Table 24.2 further describes the characteristics of these states.

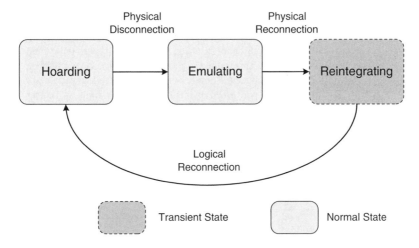

Figure 24.5   Disconnected operations in Coda.

After discussing the server replication method and disconnected operations in Coda, we can see that these two schemes in fact have competing roles in nature. That is, if disconnected operations are possible to solve the problem of availability, then it may not be justified to use server replication, which is generally more expensive. However, it is worthwhile to point out that server replication and disconnected operations also have complementary roles. The replicas in the server replication have higher quality in the sense that they are more complete and accurate. At the same time, users are able to access the files in different geographical areas. On the other hand, the cache in disconnected operations has lower quality, but it provides higher availability of files and performance. Due to the complementary role of these two

Table 24.2   Characteristics of different states in Coda disconnected operations.

States	Description
Hoarding	In this state, the client is always alert of disconnection by caching some files that are important. However, due to the limited storage space in the client device, only certain files can be cached. To choose the appropriate files to be cached, two kinds of information are used—implicit and explicit. Implicit information is based on algorithms like LRU as stated before. On the other hand, explicit information is based on the client hoard database (HDB), which includes the files that are explicitly required to be cached. With these two types of information, Venus periodically checks for the list of caching files to prepare for disconnection.
Emulating	Venus enters this state during disconnection. In this state, read/write requests on the Coda file system are redirected to the cache. However, only those requests which involve files in the cache can be satisfied. On the other hand, any changes to the files are required to logged, so that the updates performed during the period of disconnection can be delivered back to the system without loss. This is done by the client modification log (CML). In order to reduce the number of records on the log, certain optimization is performed on the log file. For example, if the user creates a file, saves it and then deletes it, there will be three actions recorded in the log. By performing log optimization, these three records are figured out and removed.
Reintegrating	During reintegration, with the help of the CML, Venus performs update operations as stated in the server replication part previously. The logged updates are propagated to the AVSG and consistency is checked by the servers in the AVSG.

mechanisms in improving the availability files, Coda is designed to integrate both mechanisms to get the best of both worlds.

24.6   WEAKLY CONNECTED OPERATIONS

As discussed before, disconnected operations are motivated by the popularity of portable devices. Nowadays, most of the portable devices are integrated with certain wireless technologies. For example, mobile phones are integrated with 2G/3G technologies, laptops and PDAs are integrated with IEEE 802.11 a/b/g, etc. On the other hand, since disconnected operations have relatively low quality of files due to the use of caching, researchers have started thinking about the possibility of using temporary wireless connections to improve the quality of the cache. This results in the emergence of weakly connected operations [249, 310].

Weakly connected operations are very similar to disconnected operations. However, as shown in Figure 24.6, there are some fundamental differences. In the design of weakly connected operations, Venus can enter three states—hoarding, emulating and write disconnected. Different from that of disconnected operations, the transient state, i.e., reintegrating, is replaced by the write disconnected state. Upon recovering from disconnection, Venus first enters the write disconnected state. It transits to the hoarding state only after all the updates are performed and a strong connection is detected.

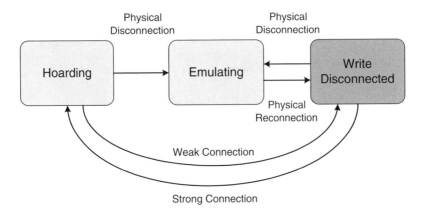

Figure 24.6   Weakly connected operations in Coda.

The write disconnected state is the state in which weakly connected operations differ from disconnected operations. This state is a mixed state between the hoarding, emulating, and reintegrating states. In the write disconnected state, updates are logged similar to that in the emulating state. At the same time, the updates are propagated to the servers using trickle reintegration asynchronously. On the other hand, since the device has limited connection ability in this state, cache misses which cannot be served in the emulating state can now be served. However, if the size of the file requested is too large, user intervention may be required.

As mentioned before, CML (client modification log) is the place for logging updates, and it is very useful for keeping it compact. Due to this reason, optimization techniques are usually applied to the CML as explained in the previous section. However, the use of trickle reintegration may result in the case that some updates are propagated to the servers before optimization can be performed on the CML. To mitigate this situation, an aging window is adopted into the log file to prevent premature updates from reducing the efficiency of log optimization.

Figure 24.7 shows a typical CML log file which is under trickle reintegration. The updates can only be propagated after a certain time governed by an aging window. Specifically, in the figure, only those updates older than time A can be propagated.

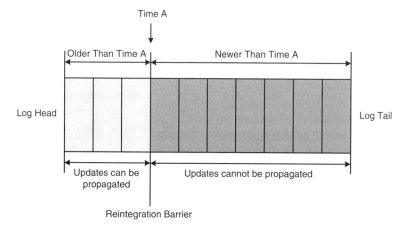

Figure 24.7 Typical CML in trickle reintegration.

Additionally, there is one more technique which can be used in trickle reintegration to improve the performance in the write disconnected state—the choosing of a different reintegration chunk size. This size can be chosen according to the connection link quality and bandwidth. As a result, different connection environments and situations can be covered with this technique.

24.7 PRACTICAL ILLUSTRATIONS—MOBILE DISTRIBUTED DATABASE

In previous sections, we have discussed the Coda file system in detail, especially the details on how it copes with mobile users with different network connection abilities and how it maintains the consistency of files in the system. The discussion focuses on mobile users with local file systems, such as laptop users, handheld device users, mobile phone users, etc. In this section, the practical implementation of a mobile distributed database will be discussed in detail. We discuss how the design of a distributed database can greatly improve the network loading of an existing GSM network [233]. Specifically, we focus on the backbone network of the wireless system.

Figure 24.8 shows the architecture of a GSM network. Within the GSM network, there are two main types of database—HLR and VLR. HLR is the database which stores the user information, including the current locations of users, user profiles, etc. On the other hand, VLR, which is usually located together with an MSC (Mobile Switching Center), serves as a temporary storage of user information within the specific location. Whenever a user moves from one VLR to another VLR, location update is carried out. This update involves the HLR as well. On the other hand, whenever there is a call to a specific user, the user's HLR is queried also. As a result, loading on the HLR is in fact quite heavy.

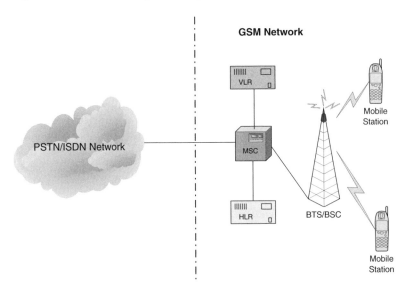

Figure 24.8   GSM database architecture.

A number of methods have been proposed to reduce the loading on the HLR. These methods include:

- Forwarding strategy. A pointer is set up from the old VLR to the new VLR without updating the HLR record.

- Anchoring strategy. A certain location is served as a local anchor and updates only affect this anchor. Moreover, the HLR always points to this anchor.

- Caching strategy. The location of a mobile terminal obtained from the previous call is cached in the signal transfer point (STP). When a call passes through the STP and targets for the specific mobile terminal with an entry inside the STP, the cache is used without querying the HLR.

- Replication strategy. The location of a mobile terminal is replicated in several selected databases. Afterwards, a call originating from the service area of these databases can obtain the required location information.

Obviously, the strategies mentioned above mainly target on some specific situations. For example, the forwarding strategy focuses on the situation where the rate of mobility is higher than the rate of call arrival. In order to further reduce the loading of a centralized root database like HLR, the most effective mechanism is to introduce multilevel database architecture. However, the increase in levels will result in longer delays in location registration and call delivery. In view of such tradeoff, a three-level multitree database architecture has been proposed [233].

Figure 24.9 illustrates the proposed multitree distributed database architecture. In the figure, there are a number of distributed database subsystems (DS). Different DS are probably managed by different service operators. DS can communicate with other DS through the PSTN network, ATM network or other networks with the help of the root database DB0 of their own DS. In this architecture, DB0 and DB2 correspond to the HLR and VLR in the GSM network, respectively. Similar to the GSM network, DB2 is located together with the MSC, which controls an RA (registration area). As a result, when a mobile terminal moves from one RA to another within the same MSC, location update can be avoided. A number of DB2s are grouped together by DB1, which contains a switch called STP to route messages between various databases. Similarly, a number of DB1s are grouped together by the root database DB0. DB0 contains the entries of all the users in the GSM network and service profiles of users within the service area. Each entry in DB0 contains a pointer to either other DS' DB0 when a user is within other DS or to current DS' DB1 when the user is within the same service area. DB1 then contains an entry which has a pointer to DB2 where the user is currently visiting or roaming. The corresponding DB2 has a copy of the service profile of the user.

With the use of the above-mentioned architecture, the loading of the root database can be reduced. This will be explained in the following location update and call delivery procedures. First of all, let us discuss the location update procedures:

1. When a mobile terminal moves from one RA to another RA managed by a different DB2, the registration message is sent to the new DB2.

2. The new DB2 figures out that the user enters the RA from other RA managed by other DB2, it then forwards the registration message to DB1.

3. In DB1, if it can find an entry in its database, it simply updates the pointer to the new DB2 and sends a deregistration message to the old DB2.

4. However, if an entry cannot be found in DB1, then it knows that the user is originally managed by a different DB1. In this case, a new entry is added in the new DB1. The new entry created points to the new DB2. Then a registration message is forwarded to DB0.

5. If an entry is found in DB0, then the user's movement is within the same DS. In this case, DB0 updates the pointer of the user to the new DB1 and then sends a deregistration message to the old DB1. The old DB1 then sends a deregistration message to old DB2 and deletes the entry in its database.

6. However, if an entry is not found in DB0, the user has crossed DS. In this case, the new DB0 requests for the user profile of the user from the old DB0. The

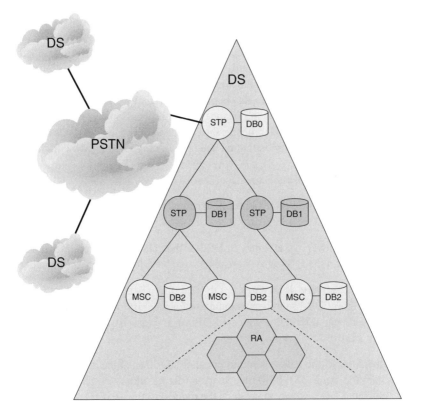

Figure 24.9   Multitree distributed database architecture.

new DB0 then saves the user service profile in its own database and adds an entry to the new DB1, which then points to the new DB2 where the user is present.

7. In the above procedures, if the user's movement is within the same DS, then the copy of a user service profile can be directly sent from one DB2 to another DB2. However, if the movement crosses DS, the service profile of the user is required from the new DB0. After the deployment of the service profile for the user, the registration process is regarded as completed.

The above steps describe the location update procedures with the use of a three-level multitree architecture. With the help of this architecture, location updates on the root database are greatly reduced. The main reason is that the use of multiple levels allows updates to be done in the intermediate level database instead of sending update messages to the root database. In particular, the movement of a user from one RA to another usually involves the change of DB2, occasional change of DB1 and infrequent change of DB0 (this indicates the change of DS). However, one of the main drawbacks of this architecture is that a change involving DB0 may incur changes in DB0 of all DS.

The call delivery procedures are described as follows:

1. When a call is initiated by a mobile terminal and received at caller's MSC, the collocated DB2 then looks for its database entry for the corresponding callee.

2. Without discovery of an entry in DB2 of the caller's MSC, an query is sent to the DB1, and the DB1 also searches its database for an entry of the callee.

3. Without discovery of an entry in DB1, an query is again sent to its associated DB0. Again, DB0 searches its database for an entry of the callee. The search will be repeated in callee's located DB0 if an entry is still cannot be found in caller's DB0.

4. If the entry of the callee is found in caller's DB0 (i.e., the caller and callee are within the same DS), DB0 then sends a routing address request to the callee's associated DB1, which then sends the request to the associated DB2.

5. On discovering the entry of the callee in DB2, a TLDN is assigned to the callee as the routing address and the TLDN is transmitted back to the MSC of the caller. After receiving the TLDN, the caller's MSC can then find the callee and its associated MSC to set up the call.

Similar to that in the location update procedures, intermediate databases can be used to help reduce the loading of the root database whenever the caller and callee are within the same DS.

With the help of the distributed database subsystems architecture discussed above, the signalling load is much reduced. This leads to the possibility to support a larger amount of users. As a result, the system becomes more robust and scalable when compared with the original design of GSM HLR-VLR architecture.

24.8 SUMMARY

In this chapter, the design principles of improving the availability of files in a distributed file system have been discussed. Afterwards, the Coda file system has been discussed in detail. We have discussed the server replication method with use of optimistic replication in Coda. Then, the ideas of disconnected operations and weakly connected operations with the use of hoarding, emulation and reintegration are discussed in detail.

The success of the Coda file system indicates that two very different techniques in improving the availability of files—replication and caching, can actually be combined to cooperatively improve both the quality and availability of files. This is particularly important to the users of mobile portable devices like laptops, PDAs, etc., which may only get limited network access.

Finally, as practical illustrations, the idea of using mobile distributed databases in a distributed multitree database architecture in replacement of the generally adopted GSM HLR-VLR architecture has been discussed in detail. It is found that the use of the multitree architecture can effectively reduce the signalling load when comparing to the conventional GSM HLR-VLR architecture.

PROBLEMS

24.1   Briefly discuss the general concept of a distributed file system.

24.2   What is optimistic replication? How is it different from pessimistic replication?

24.3   Justify the use of server replication and disconnected operations in a distributed file system.

24.4   What are the differences between disconnected operations and weakly connected operations?

24.5   How can a distributed multitree database architecture be used in a cellular network with location updates? Briefly discuss why it can obtain better performance than the GSM HLR-VLR architecture.

CHAPTER 25

LOCATION DEPENDENT SERVICES

The ideal of service is the basis of all worthy enterprise.

—Principles of Rotary, I, 1905

25.1 INTRODUCTION

Anytime, anywhere wireless access does not only enable uninterrupted connectivity to users, but also the possibility of continuous location tracking. Naturally, service providers or other organizations (e.g., government, law enforcement, rescue operations, etc.) would like to exploit such location information so as to provide additional services.

Interoperability:

There are two main types of mobile location services—in-vehicle mobile location service and personal mobile location service. Although their development objectives are different, they require similar components in supporting their service. The components include map database service, spatial analysis service and mobile positioning service. The interoperation of these components contributes to the correct calculation of the location of the user, which is discussed in Section 25.2. Specifically, the spatial analysis service is used for the translation of map information to number and vice versa, while the mobile positioning service is used to estimate the user's current location. The techniques of these two services are discussed in Section 25.3 and Section 25.4, respectively.

Performance:

There are a number of mobile location services envisioned by service providers. These services require different requirements on the accuracy of user's location estimation, response time, and the need for in-door estimation. These requirements are, to some extend, governed by the kind of mobile positioning techniques adopted. This is discussed in detail in Section 25.4.

25.2 MOBILE LOCATION SERVICE MODEL

Generally, mobile location services can be classified into two main types—in-vehicle and personal mobile location services, as illustrated in Figure 25.1 [166].

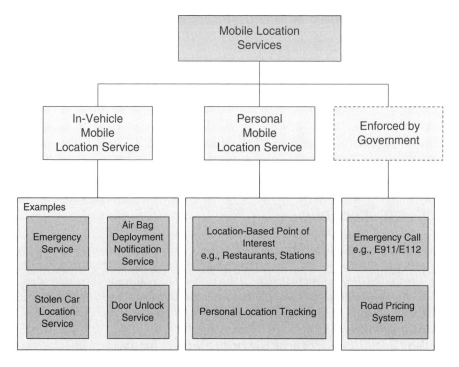

Figure 25.1 Mobile location service model.

25.2.1   In-Vehicle Mobile Location Service

The in-vehicle mobile location service refers to the location service specially designed for automobiles [166]. This service is mainly driven by automobile manufacturers who want to maintain a close relationship with their customers and to generate more revenue by providing services to their customers. For example, in some interactive services, assistance can be provided to drivers when some emergency events occur.

Some of the services provided by OnStar [263], which is a company providing mobile location services for automobiles, are examples of in-vehicle mobile location services:

Emergency Service: In this service, the driver can ask for assistance with the system in the vehicle. Upon receipt of the call, if any further assistance is needed (e.g., ambulance, police, etc.), the service assistant can help by referring the case to the the corresponding emergency department on behalf of the driver.

Air Bag Deployment Notification Service: In this service, the service assistant will contact the driver directly when the system detects that the air bag of the vehicle is deployed. If there is on response from the driver, the emergency department will be contacted.

Stolen Car Location Service: Through this service, the owner of the vehicle can track the location of his vehicle when it is stolen.

Door Unlock Service: When the owner of the vehicle leaves his key inside the vehicle with it locked, he can ask for assistance from the service provider through a phone call.

In fact, besides providing assistance to the driver/owner at the right time, in-vehicle mobile location service also includes some entertainment, Internet access service, weather report service, etc. However, the main focus is to provide a safe driving experience to the customers.

25.2.2   Personal Mobile Location Service

Personal mobile location service is designed for persons, rather than vehicles [166]. The aim is to provide a convenience service to a user located in a particular area. For example, when the user goes to a certain area that he is not familiar with, the map of the area is particularly useful to the user. In addition, a service like personal location tracking and verification is also useful for some emergency scenarios. These services are mainly provided by mobile operators through common communication devices like mobile phones, PDAs installed with GSM cards, etc. Furthermore, these services can be provided with the existing mobile infrastructure coupled with some add-ons. We will discuss this in more detail in Section 25.4.

The following is one of the practical examples of personal mobile location service provided by NTT DoCoMo [89]: When the user arrives at a certain area, through his mobile device he can obtain information like map of the area, locations of restaurants, local weather report, town information, etc.

25.2.3   Government Enforcement

In addition to in-vehicle and personal mobile location services provided by automobile manufacturers and mobile operators for commercial purposes, there are some mobile location services enforced by the government. For example, E911/E112, and road pricing. E911/E112 services[1] are the emergency services suggested by the government such that mobile operators need to handle local 911/112 calls no matter whether the calls are initiated from the mobile phone subscribed to their service or not. On the other hand, road pricing is suggested in some countries in which vehicles are required to be equipped with some location tracking devices. In his way, the road users can be charged with appropriate fares.

25.2.4   Components of Mobile Location Services

In order to support the mobile location services that we have discussed in previous sections, the components as illustrated in Figure 25.2 are required. First of all, because mobile location services require location information of users up to certain degree of accuracy, a location service is necessary. The location service includes three components—map database service, spatial analysis service and mobile positioning service [166]. Map database service is necessary whenever the location of the user is required to be figured out. On the other hand, spatial analysis service is required to map the position represented some keywords to some coding and vice versa. In addition to this, the exact location of the user can be found out by the mobile positioning service. In subsequent sections, we will discuss in detail the spatial analysis and mobile positioning techniques.

25.3   SPATIAL ANALYSIS TECHNIQUES

The spatial analysis service comprises several components, including map database, geocoding, reverse geocoding, routing, map formation, point of interest, map attributes modification, etc.

As Figure 25.3 [166] illustrates, the map database is involved in various operations of the spatial analysis service. There are two types of map model used in the map database for storage of data—entity data model and continuous variation data model. In the entity data model, points, lines and polygons are used for representing real-world objects or entities with some associated attributes predefined. For example, specific locations like a subway station can be represented by points, a road or street

---

[1]E112 services are optional in some European countries.

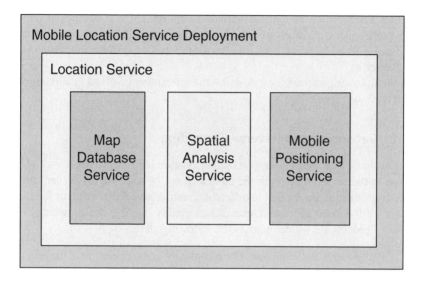

Figure 25.2 Mobile location service components.

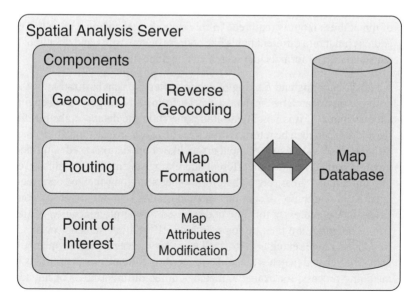

Figure 25.3 Spatial analysis service components [166].

can be represented by lines, a borough can be represented by polygons, etc. On the other hand, in the continuous variation data model, a geographical region is divided into spaces, and each attribute forms a layer with different number of spaces associated with it. In general, vector representation is used in the map database due to its flexibility.

In the following sections, we describe the operations involved in supporting the spatial analysis service.

25.3.1 Geocoding and Reverse Geocoding

One of the basic operations to support mobile location services is to search for a certain location in the digital map, with the address input by the user. However, different addressing schemes are applied in different countries and different input formats are possible for the same address. These factors greatly increase the complexity of the process of geocoding, which includes the following steps:

1. Getting user's address input. The complexity of the operation of the geocoding engine is greatly affected by the items, details and correctness of the address input by the user. Generally, the simpler the user's address input, the higher the complexity of the operation performed by the geocoding engine.

2. Standardizing the address input. Since different countries are using different addressing schemes, in order to perform an accurate search, standardization of the address input is required. In this step, the geocoding engine parses the address input into different fields like address, name, suffix and direction. With the standardized formats, the search can be done more easily.

3. Address searching and ranking matches. After the standardization of the address input, a search on the database is performed by the geocoding engine the address input. If it is possible to find a match in the database, the position of the address can then be returned directly. However, an exact match is generally difficult to occur due to user input errors like abbreviations used in the address input, parsing errors due to different addressing schemes, map database errors like unrecorded areas, etc. Therefore, other searching methods are required to assist in searching. A common method used is soundex. Soundex converts letters in the addresses to some codes based on the pronunciation of the letters. The codes are then be for searching. If multiple matches occurs, then an address range search is performed to verify the validity of the matches. Finally, the verified matches are ranked according to some preset priorities in the matching process, for example, matches in the different fields of the address input.

On the other hand, reverse geocoding is the reverse process of the geocoding that we have just discussed. Given a user's position in terms of latitude and longitude, a map lookup will return the nearest match of the current location of the user on a map. This is useful in figuring out the nearest road segment for directing the driving, and helpful for searching interested points nearby.

25.3.2 Routing

Routing is the process of finding suitable paths from the source location to the destination location based on the criteria given. Basically, the criteria given is turned into costs of links, which are then used in the calculation of the most suitable path by adding the costs of the links involved. With the estimated costs in each link, the routing operations can be captured by one of the following well-known [166]:

Shortest Path Problem: This captures the problem in which the user wants to find the shortest path from his current location to the destination location.

Travelling Salesman Problem: This captures the problem in which the user wants to visit multiple destinations with minimum costs.

Single Depot Multiple Vehicles Routing: This captures the problem in which stocks are stored in a centralized location (i.e., a depot) and multiple vehicles are used to deliver products to the customers with minimum costs.

Multiple Depots Multiple Vehicles Routing: This captures the problem in which multiple stocks are stored in multiple depots and multiple vehicles are used to deliver products to customers with minimum costs.

25.3.3 Other Operations

Other operations of the spatial analysis service include map formation, point of interest search, real-time modification of attributes, etc. These operations are briefly discussed below:

Map Information: There are two main types of map representation, namely vector map and raster map. In vector map representation, a map is represented by points, lines and polygons. This representation method has the advantage of having good modeling of the shape of real objects with smaller amount of data. On the other hand, in raster map representation, a map is represented by pixels. This representation method has the advantage of reducing the complexity of client devices by offloading the complexity to the server. Moreover, the server returns only simple graph images, and the client devices only need to update images after each operation performed on a map (e.g., a zoom-in operation). Currently, raster map representation is commonly used in client devices.

Point of Interest Search: It is a useful operation to help users to identify their targets nearby, so as to direct them to the locations of their targets. There are a number of search types, such as search by name, search by type of facility (e.g., shops), search by distance from a point, search by route/region, etc.

Map Attributes Modification: Basically, map attributes are seldom modified. However, since some attributes like traffic information along a route or weather information can affect the cost calculation process in routing, these kinds of attributes are required to be changed in a short period of time. This is to ensure that optimal operations of routing can be achieved.

25.4   MOBILE POSITIONING TECHNIQUES

Depending on particular requirements needed, different location based applications require positioning information up to certain degree of granularity with certain response time. For example, most in-vehicle mobile location services like air-bag deployment notification are emergency services which require very fine-grained and accurate location positioning information in a very short time, while personal mobile location services like town and weather information of the user's current location require relatively loose accuracy of information and response time.

As a result, there are various techniques for finding the location of the user. Existing techniques can be divided into three main types—cellular location techniques, GPS (Global Positioning System) location techniques, and the hybrid of the two types. Basically, cellular location techniques can achieve an acceptable accuracy of location information with a fast response time. Most of them allow non-line-of-sight operations and therefore in-door operations are possible. However, their accuracy deteriorates in rural areas due to less BTS/BTC deployment. On the other hand, GPS location techniques allow accurate location information to be derived with an acceptable response time. However, they require the antennae to be pointed to the sky, and thus work poorly in in-door environments. In order to further improve the accuracy and speed of GPS location techniques, some hybrid approaches have been proposed.

25.4.1   Cellular Location Techniques

Figure 25.4 [166] shows an example of the GSM architecture with a positioning server for supporting location tracking . In the figure, the Gateway Mobile Location Center (GMLC) is present to accept requests for location information of a user. The HLR of the user is contacted for getting the current Serving Mobile Location Center (SMLC) of the user. The SMLC controls a number of Location Measurement Units (LMUs) which measure radio information from the mobile station (MS). Based on the radio readings from various LMUs, the SMLC returns the result of the estimated location information of the MS.

There are a number of variations of the GSM-based approach. Some of them are summarized below:

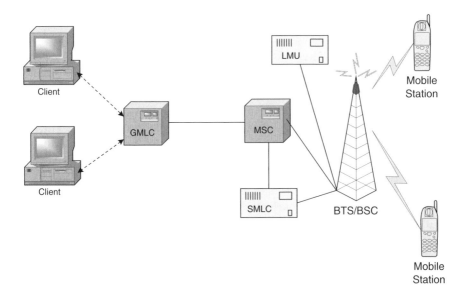

Figure 25.4 GSM architecture for positioning.

Cell of Origin (COO): This approach makes use of the cell concept of the deployment of BTS/BTCs in a mobile system, which usually involves overlapping regions. With the correct position information of each BTS/BTC, the location of the MS can then be estimated through the BTS/BTCs that the MS is currently contacting. The accuracy of this method is on the order of hundred to thousand meters, with a few seconds in response time. This approach is generally referred as a network-based approach since the location is estimated from the cellular network infrastructure.

Time Difference of Arrival (TDOA): This approach makes use of the time differences between the sending signal (uplink) from the MS to at least three BTS/BTCs equipped with LMUs. With the help of the LMUs, the TDOA of the signal from the MS can be estimated. The pairwise time differences and position information of BTS/BTCs can then be used to calculate the position of the MS, with the use of hyperbolic trilateration (discussed in the GPS positioning part in subsequent section). The accuracy of this method is within a hundred meters most of the time. However, it may not be possible to find three BTS/BSCs in some rural areas. Same as COO, this is also a kind of network-based approach.

Angle of Arrival (AOA): Similar to TDOA, this approach makes use of the sending signal (uplink) from the MS to at least three BTS/BTCs to estimate the location of the MS. However, instead of using the arrival time, the angle of arrival to the base station is used. With this AOA information together with the position information of the BTS/BTCs, the position of the MS can be calculated. Same as TDOA, the accuracy of this method is within a hundred meters most of the time. This approach is also a kind of network-based approach.

Enhanced-Observed Time Differences (E-OTD): This approach is very similar to TDOA. However, instead of using uplink signal arrival times, downlink signal arrival times are used, and the location estimation is performed by the MS. However, this requires the MS to have accurate position information of the neighboring BTS/BTCs and receive synchronized signals from the them. Similar to TDOA and AOA, this approach can achieve an accuracy of within a hundred meters most of the time. It is regarded as a mobile-based approach in which the MS is responsible for the location calculation, or as a mobile-assisted approach in which the MS provides information for helping the network to calculate the estimated location.

Location Pattern Matching: In this approach, the frequency used by the MS and the multipath characteristics of the MS are recorded as "signature" for its location, which is stored in a database. Whenever the current location information of the MS is required, the information stored in the database is compared with the current information to estimate the location of the MS. The accuracy of this approach is around a hundred to several hundreds meters. It is regarded as a mobile-assisted approach.

25.4.2   GPS Location Techniques

Figure 25.4 [166] illustrates the idea of GPS location techniques. As shown in the figure, the GPS receiver should receive at least three satellite signals. By using an approach similar to TDOA, the distances between the GPS receiver and the satellites can be estimated. For example, in Figure 25.4 there are three satellites, and their estimated distances to the GPS receiver are $T1$, $T2$ and $T3$, respectively. With this distance information and the position information of the satellites, three circles are drawn as depicted in the figure. The intersection region of the three circles is the region that the GPS receiver is in. This approach is also known as trilateration, and it can achieve an accuracy of within tens of meters.

25.4.3   Hybrid Location Techniques

Figure 25.6 shows a hybrid location technique known as Assisted GPS (A-GPS) [166]. The location information is estimated by the location server in the cellular network. With the use of this information, the location server returns the satellites that are most suitable to the GPS receiver. As a result, the GPS receiver can measure the readings from these satellites directly without searching, which can reduce the initiation time. This approach can achieve an accuracy of within tens of meters.

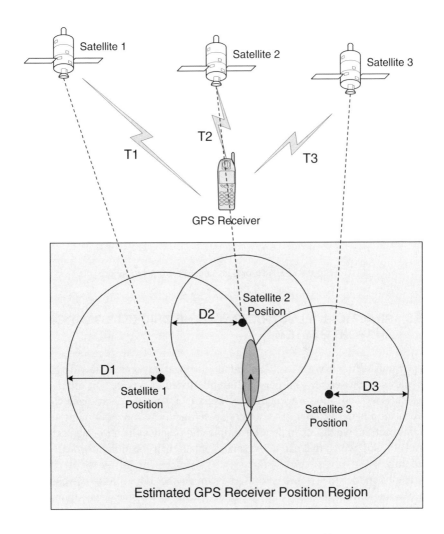

Figure 25.5   GPS architecture for positioning [166].

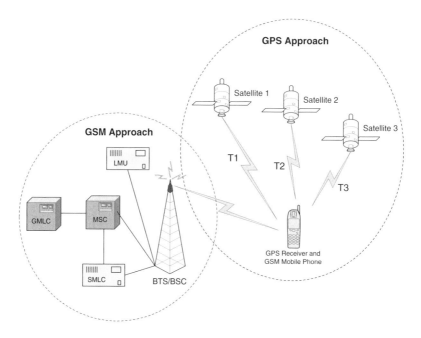

Figure 25.6 Hybrid architecture for positioning.

25.5 PRACTICAL ILLUSTRATIONS—THE CRICKET LOCATION SUPPORT SYSTEM

In previous sections, we have discussed mobile location services including in-vehicle mobile location service and personal mobile location service. In order to support these services, location information of users is always necessary, and the location information is derived with the help of spatial analysis and positioning techniques. In this section, we discus in detail a simple and practical location support system— Cricket [286], which that can help users to obtain their location information within a building.

The basic idea of Cricket is to use beacon devices deployed in various locations within a building to deliver certain information to the listener devices—mobile or static devices attached with a listening interface. The listener devices can then derive their own location from the information broadcast by the beacon devices. The information broadcast by the beacon devices is in fact some strings which can be the identity of the room obtained from the owner, some name servers for other devices to contact with, some authentication messages, etc.

Figure 25.7 illustrates the operation of the Cricket system. In the figure, the listener is located in Room C. It can receive beacon signals from beacon devices A, B and C in Rooms A, B and C, respectively. All the beacon signals carry the corresponding location information of the beacon devices. Using the beacon signals, the listener then estimates its distances from the beacon devices. Together with the information conveyed in the beacon signals, the listener can figure out the beacon device which is the nearest to it, so as to make decision on its current location. For example, in Figure 25.7, the listener should have the shortest distance from beacon device C. Then it can know that it is currently located in Room C after analyzing the information carried by the corresponding beacon signal.

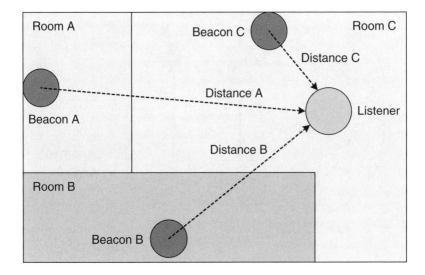

Figure 25.7 Usage scenario of the Cricket location support system.

However, in order to support the above-mentioned operation, a number of issues are required to be handled, e.g., estimation of distances from beacon signals, interferences of beacon devices, nearest beacon device not in current location, etc.

First of all, we will discuss the location estimation process of the Cricket system. Figure 25.8 illustrates the location estimation technique used in the Cricket system. Specifically, each beacon device is equipped with a ultrasonic (US) transmitter and a radio frequency (RF) transmitter. In each transmission, both RF signal and US signal are transmitted simultaneously. However, since the RF signal travels faster than the US signal, with the use of the time differences between the receipt of the first bit encoded in the RF signal and the receipt of the US signal, the distance between the listener and the beacon device can be estimated.

In fact, the absolute distance between the listener and the beacon device is not as important as the relative distances between the listener and other beacon devices. Therefore, the location estimation technique is enough to figure out which beacon device is the nearest from the listener. On the other hand, the coverage area of a

US signal is less than that of an RF signal, and thus with the help of the estimation technique the beacon device can effectively limit its own coverage area.

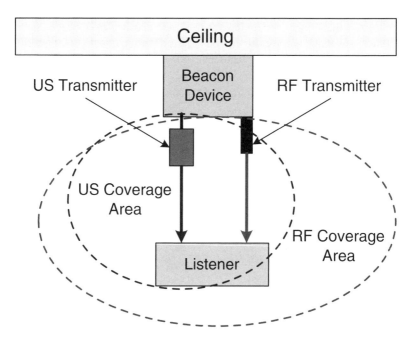

Figure 25.8 Location estimation of the Cricket location support system.

Although the design shown above can limit the coverage area of each beacon device and allow the listener to estimate its distances to beacon devices, it is still possible to have overlapping coverage areas by multiple devices, which can lead to collisions. In order to prevent collisions, a randomization approach is adopted. That is, beacon signals are transmitted randomly within a period of $[R1, R2]$ms. The exact values of $R1$ and $R2$ are chosen based on the situation of proximity, e.g., the bandwidth available, the size of messages delivered by each beacon, etc.

Besides collisions, it is possible for interference to occur between RF and US signals of neighboring beacon nodes. Figure 25.9 shows one of the possible scenarios. In the figure, due to signal attenuation, the listener fails to detect beacon device A's RF signal. However, it can detect beacon device B's RF signal. Before the arrival of beacon device B's US signal, it detects the arrival of beacon device A's US signal. In this case, the distance for beacon device A is induced, instead of that for beacon device B. There are also other possible interferences due to reflection of signals. In order to reduce interferences, proper system parameter values (e.g., size of message, transmission rate, propagation time, etc.) should be chosen. At the same time, some statistical approaches like majority, minmean and minmode, can be applied to further estimate the nearest beacon device.

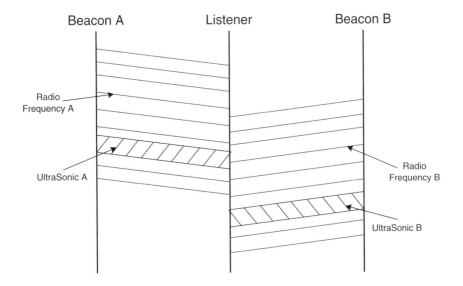

Figure 25.9   Cricket location support system interference problem.

In addition to interference and collision problems, the location of the deployment of beacon devices is also significant to the correctness of the identification of current location of the listener. The main reason is that, in some configuration, the nearest beacon device may not be located in the place where the listener is currently located in. In this case, the listener will get wrong current location information. One possible solution is illustrated in Figure 25.10 The deployment of beacon devices with approximately equal distances between both physical boundary (e.g., different rooms) and virtual boundary (e.g., different areas within the same room) can effectively remedy the problem.

In summary, the Cricket location support system combines several techniques like time differences between two different transmission media, randomized transmission of beacon signals, and certain positioning of beacon devices between the boundary of a location. With the help of these techniques, the system not only can deliver a correct location information up to room-sized level to the user devices with a high accuracy, but also becomes highly practical and can be implemented with low cost. On the other hand, the user devices do not need to transmit any information to the servers. They learn their own location information from the beacon devices instead. As a result, privacy of users can be kept.

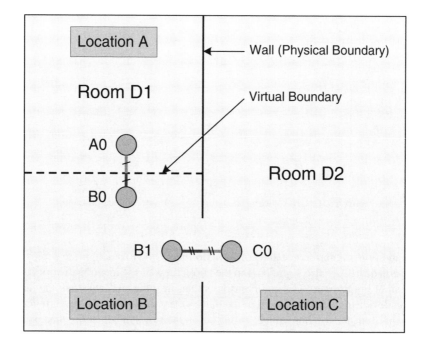

Figure 25.10 Cricket location support system beacon location.

25.6 SUMMARY

In this chapter, the service model of mobile location services has been discussed. Mobile location services can be divided into two main categories—in-vehicle and personal, which are respectively supported by two different operators, namely automobile manufacturers and mobile operators. These services rely on three main components, including the map database service, spatial analysis service and mobile positioning service.

Spatial analysis is required to convert specific locations into some transferable and calculable numbers. This is done by the geocoding process. Sometimes, it is required to have the best route from one location to another, and this requires some routing techniques. Moreover, the routing has to be shown on a map, which requires map generation. Besides some specific locations, users may be interested in some specific point of interests. As a result, point of interest search within a specific location becomes an important component. Finally, the point of interests or the routes within a routing are associated with some attributes. All the above aspects contribute to the spatial analysis service.

The other important issue of mobile location services is the estimation of current location of the user. Existing techniques can be divided into three main types—cellular location techniques, GPS location techniques, and the hybrid of the two types.

Finally, as practical illustrations, we have discussed the Cricket location support system, which is a simple system capable of detecting the location of a user up to room-sized accuracy.

PROBLEMS

25.1   Summarize the types of mobile location services. Discuss briefly which kind of services is more practical to our daily life.

25.2   Briefly discuss the components required in supporting mobile location services and their corresponding functions.

25.3   Discuss the components of the spatial analysis service and their corresponding functions.

25.4   Briefly explain how GPS system works and how GSM system works in tracking the location of a user.

CHAPTER 26

TRUST BOOTSTRAPPING IN WIRELESS SENSOR NETWORKS

If you would wish another to keep your secret, first keep it yourself.

—Seneca: Hippolytus, c. 60

26.1 INTRODUCTION

Establishing pairwise symmetric keys is a critical resource management issue in wireless sensor networks. Usually deployed in a hostile environment where malicious users or adversaries are bound to exist, wireless sensors are subject to a general attack model—a sensor node can be captured, re-programmed, and consequently exhibit arbitrary faulty behaviors. Thus, sensor key management is a challenging research issue, attracting a high level of interests in recent years. In general, a key management system works by first pre-allocating some keys to each sensor before deployment. After deployment, neighboring sensors can undergo a discovery process to set up shared keys for secure communications. An efficient key management scheme has to work under severe system constraints including limited memory storage

Wireless Internet and Mobile Computing. By Yu-Kwong Ricky Kwok and Vincent K. N. Lau
Copyright © 2007 John Wiley & Sons, Inc.

and communication overhead in each sensor. In this chapter we provide a detailed survey of state-of-the-art sensor key management techniques that have demonstrated a high degree of effectiveness.

With the ever increasing advancements in miniaturization of processing and communication hardware, wireless sensors and the networking of them have become popularly used in many applications [14]. However, many research challenges remain to be tackled. One of the most important challenges is the protection of sensor data communications. Indeed, wireless sensors are usually deployed in a hostile environment and the sensors have to operate with little or no infrastructure support. For example, sensors may be dropped from an airplane onto the battlefield for target recognition applications [179]. A key problem is that malicious users (e.g., the enemy) may eavesdrop the sensing data transmitted among the sensors. Furthermore, as the sensors are rarely tamper-resistant, some sensors may be captured by the adversary and re-programmed for eavesdropping or other malicious purposes [60, 318].

Interoperability:

A critical step toward protecting sensor data communications is the establishment of encryption keys among sensors so that they can set up secure communication links [47]. As indicated in by Hu *et al.* [157], setting up pairwise keys for secure communications is a necessary prelude to secure routing [182] in wireless ad hoc networks. However, with a small form factor and limited power supply, it is widely considered that a sensor device cannot employ sophisticated cryptographic technologies such as public key cryptosystems [60, 318]. Most notably, the Berkeley Mica Motes are very commonly used by the research community. In a typical mote, a 8-bit 4MHz Atmel ATmega 128L processor is used. Usually, it is equipped with 128K bytes of program memory and 4K bytes of SRAM. For communications, the sensor uses an ISM band RF module with a peak data rate of 40 Kbps and a maximum transmission range of roughly 30m. The operating system, TinyOS [209], is also small and works in an event driven manner. Finally, the wireless sensors are usually immobile after deployment. With such limited capabilities, symmetric cryptography is considered as the most practicable approach [47].

Setting up pairwise symmetric keys among communicating sensors is an important step for *bootstrapping* [47, 105] the mutual trust required for data exchange. The most obvious approach to enable key setup is to have a centralized key distribution center (KDC) [279, 326] which can allocate keys to sensors that need to communicate among each other. However, in a large scale deployment, it is infeasible to have such a KDC to communicate with all the sensors which may be widely dispersed over the field. Indeed, the KDC may be out of the transmission range of some remote sensors. Moreover, such a KDC is a single point of failure. Indeed, in a hostile environment (e.g., battlefield), it is unrealistic to assume the existence of such a KDC.

As to the topology of a wireless sensor network, there are two possible system architectures: (1) hierarchical sensor networks (see Figure 26.1); and (2) system architecture: fully distributed sensor networks (see Figure 26.2). In a hierarchical sensor network, some sensors are more powerful and thus, act as control nodes

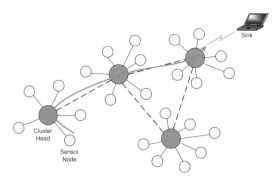

Figure 26.1   A hierarchical sensor network in which nodes are classified as cluster heads and worker sensor nodes.

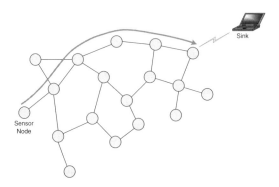

Figure 26.2   A fully distributed sensor network.

(similar to a base station in a cellular network) to provide resource management facilities including encryption keys distribution [47, 174, 374]. These controlling sensor devices are carefully deployed such that they are in the neighborhood of a certain number of "worker" sensors. Indeed, there are many key management schemes designed for hierarchical sensor networks [88, 174, 200, 268, 342, 374] by leveraging the heterogeneity of the sensors. However, in many practical situations [47, 105], it is also infeasible to assume a structured deployment of two types of sensors (e.g., again imagine the situation when a large number of sensors are dropped from the airplane onto a battlefield). In this chapter, we focus on fully distributed sensor networks where the sensors are just randomly deployed and it is very difficult, if not impossible, to control or predict the locations of the sensors after deployment. Hierarchical sensor networks are not considered in this chapter.

In a fully distributed sensor network, the framework that is widely employed in the literature for setting up pairwise encryption keys involve four phases: predistribution, discovery, path-key establishment, and revocation [105]. Predistribution is the process of storing keys in each sensor by a setup server *prior* to deployment. Usually, the key-sets of different sensors are not disjoint. Thus, after two sensors are deployed in nearby locations and they need to communicate, they can set up a secure key by checking whether they have a shared key in their sets. This is the key discovery phase. Obviously, it is possible that two sensors do not have any common key in their sets and they have to rely on other sensors to enable their communication. That is, they may need some intermediate sensors to set up a path in order to communicate. These intermediate sensors need to have shared keys with the two communicating sensors, respectively, so that each link is made secure by using a different key. This is called the path-key establishment phase. In a hostile environment, some sensors may be captured and the assigned keys are revealed to the adversary. These keys may be used in some communication links in the network and thus, have to be revoked. Key revocation is also a critical component in supporting a highly secure and robust network.

Under such a framework, there is large design space for key management schemes [47]. The major design constraints are the memory requirement and communication overhead involved. As detailed in this chapter, many state-of-the-art techniques are designed based on probabilistic key predistribution and discovery. Specifically, random graph theory [103] is widely used for determining the sizes of the predistributed key-sets in order to achieve a certain level of network connectivity.

Performance:

Wireless sensor networks have many challenging limitations compared to traditional Mobile Ad hoc Networks (MANETs). Specifically, the following constraints make the design of a high performance sensor key management scheme challenging.

1. *Energy limitation*: Energy is the main limitation of sensor nodes. Indeed, effective energy conservation scheme is vital for maximizing the lifetime of operation from a single node to the entire network. Typically, a recent battery-powered sensor node can run for a week (under full operation) to several months.

2. *Memory storage limitation*: Sensor device is very tiny and equipped with only a small amount of memory. Apart from storing keying materials, it is necessary to store the key management algorithm and other programs for operation. For example, a popular sensor device, Berkeley MICA2 mote, has 128 KBytes program memory only.

3. *Computational and transmission power limitation*: The small size of device and limited supply of power restrict the processing and communication capabilities of sensor nodes. As a result, complex asymmetric cryptographic algorithms cannot be implemented efficiently in sensor networks without modifications.

4. *Vulnerability to attacks*: As sensor networks may be deployed in hostile environments, nodes are exposed to physical attacks by potential adversary. In

most practical situations, it is possible to compromise and takes full control of the nodes without being detected.

5. *Lack of post-deployment knowledge*: Without deployment knowledge, the distribution of sensor nodes follows random scattering. This makes it hard to determine the network topology *a priori* so as to optimize the distribution of keys.

In summary, a major barrier for enhancing security in sensor networks is its limited computational and communication capabilities. These inherent properties make traditional protocols impractical. As a case in point, public key algorithms such as Diffie-Hellman key agreement as well as RSA signature techniques are too computational intensive so that it is undesirable to be implemented in scarce resource sensor nodes.

Consequently, performance of a key management scheme for wireless sensor networks is usually quantified by the amount of memory storage needed for storing keys in each sensor and the security level. Specifically, security level is measured by determining the number of compromised links when a certain number of sensor nodes are captured.

Moreover, security performance is also a core metric for measuring the degree of success of a key management scheme. Security performance is defined with respect to a certain attack model. In many existing key management schemes, a widely used attack model is that every sensor node is subject to node capturing. Specifically, if a sensor node is captured, keys in the captured node's key ring are assumed to be exposed completely. The compromised keys can then be used to attack other existing links in the network. Thus, to improve the network resilience, most key management schemes are designed to minimize the impact (i.e., the fraction of communication compromised) of node capturing.

This chapter is organized as follows. In Section 26.2, we survey general techniques proposed for the key predistribution phase. In Section 26.3, we describe some predistribution techniques that are specially designed by exploiting deployment knowledge. In Section 26.4, we introduce algorithms designed for the key discovery, establishment, and revocation processes after the sensors are deployed. We discuss some future research directions in Section 26.5. Section 26.6 provides some concluding remarks.

26.2   KEY PREDISTRIBUTION

A major goal of key predistribution [58, 93, 105, 220] is to judiciously allocate keys to the sensor nodes such that a high degree of connectivity of the network can be achieved without requiring a large key storage in each sensor.

Eschenauer and Gligor [105] performed pioneering work in sensor key predistribution in that they suggested a classical approach to allocating keys to sensors using a probabilistic method. Their proposed scheme is simple to implement yet effective in a probabilistic sense. Specifically, before deployment, the centralized sensor controller (also called setup server in this chapter) generates a pool of P distinct keys. Then for each sensor, it is assigned a set of k keys that are randomly selected from the pool. As k is typically much smaller than P (e.g., $k = 75$ while $P = 10000$), the probability that two distinct sensors are assigned exactly the same set of keys is small. On the other hand, as described below, there is also a non-zero probability that two distinct sensors share at least one key.

After this predistribution phase, the sensors can be deployed onto the field. After deployment, if two sensors (presumably located close to each other) need to set up a secure communication link, they need to check whether they have a common key in their predistributed sets of keys. A simple yet practical method is to allow the two sensors to communicate with each other their list of key-IDs (not the keys themselves, of course) in plaintext. Even an eavesdropper can tap such lists, it cannot get to know the keys themselves. If there is no common key, then they need to set up a multi-hop path between them (albeit their physical locations may be close) via some other nearby sensors that share common keys with them.

Mathematical analysis in [105] indicates that the probability p that two sensors share at least one key is given by:

$$p = 1 - \frac{(1 - \frac{k}{P})^{2(P - k + \frac{1}{2})}}{(1 - \frac{2k}{P})^{(P - 2k + \frac{1}{2})}}$$

Suppose $P = 10000$, in order to achieve a probability of $p = 0.5$, only $k = 75$ keys are needed to be stored at each sensor.

It is interesting to note that key revocation and hence rekeying (as triggered by the adversary's capturing of some sensor nodes) can be done by simply executing the key discovery phase again among those nodes that have their shared keys revoked.

Eschenauer and Gligor [105] performed detailed simulations with 10000 sensors. Their simulation results indicated that a value of k smaller than 40 resulted in a disconnected network. Furthermore, with $k = 75$, only half of the nearby sensor nodes were reachable over one single link. A majority of the remaining nodes needed three-hop paths for connections. Finally, out of the 10000 keys generated, only half of them were actually used.

Following up on Eschenauer and Gligor's pioneering work, Chan, Perrig, and Song [58] suggested three clever enhanced schemes which have also been highly cited in the literature. Their first proposed scheme, called q-composite key predistribution, is essentially the same as the original scheme by Eschenauer and Gligor, except that during link setup phase, at least q shared keys are required instead of just one. Specifically, suppose that two sensors have q' shared keys (where $q' \geq q$), then the link key K can be generated by:

$$K = \text{hash}(k_1, k_2, \ldots, k_{q'})$$

where hash is a certain one-way hash function taking all the q' shared keys as input. Analytical results indicated that the q-composite scheme provides greater resilience against node capturing attacks when the number of nodes captured is small. For example, when $q = 2$, suppose 50 nodes have been compromised, the percentage of additional communication links that can compromised is 4.74%. By contrast, this percentage would be 9.52% if the original scheme is used.

The second scheme proposed by Chan *et al.* is based on a multipath transmission of shared keys. The major goal of the multipath key scheme is to update the shared key between two sensors nodes after initial setup. Specifically, suppose nodes A and B want to update their shared key. To do this, they need to find j disjoint paths (where $j \geq 1$) between them. Afterward, node A generates j random values $v_1, \ldots, v_j$, each of which has the same size as the encryption key. Node A then sends these j values along different paths to B. Both nodes A and B can then update their shared key K as follows:

$$K' = K \oplus v_1 \oplus v_2 \oplus \ldots \oplus v_j$$

Consequently, unless a malicious intruder can tap the data on all the j different paths, the updated key can be kept secret.

The third scheme proposed by Chan *et al.* is motivated by the fact that the original Eschenauer-Gligor key-pool scheme cannot provide node-to-node authentication. Indeed, in the original key-pool scheme, two nodes A and B having some shared keys can set up a link between them using a particular shared key. But such a shared key can also be used by one of the two nodes, say A, to set up a different link with another node, say C. Thus, from a data encryption point of view, the two links (and hence the two nodes C and B) are not distinguishable. To address this authentication problem, Chan *et al.* proposed a node-oriented approach (whereas the original key-pool scheme is a key-oriented approach). Suppose that each sensor can store up to m keys and the probability of any two nodes in the network being able to set up a link is set as p (a system parameter). When the key predistribution phase starts, a total of $n = \frac{m}{p}$ unique node IDs are generated. This is the node ID space. Depending on the application, not all IDs will be used up to form a sensor network (i.e., the actual network size may be smaller than n). Each node ID is then associated with m randomly selected distinct node IDs. A distinct pairwise key is generated for each of these m IDs to be shared with the original node ID. Since the keys generated are closely associated with the node ID-pairs, when a link needs to be set up after deployment, node-to-node authentication can be easily carried out.

Liu, Ning, and Li [220, 221, 224, 222] extended the work of both Chan *et al.* [58] and Eschenauer *et al.* [105] by exploiting two new components: location information and bivariate polynomials. Firstly, Liu *et al.* extended the random pairwise key predistribution scheme [58] with the use of location information. Specifically, Liu *et al.* made an important observation in that the setup server, before deployment, should have pretty accurate knowledge about the locations of the sensors to be deployed. Thus, it makes sense for the setup server to arrange the random pairwise predistribution in such a way that each sensor shares pairwise keys with c other sensors whose expected locations are closest to the expected location of c. Here, the magnitude of c is a system parameter depending on the memory constraint in each sensor. Of course, there is always a non-zero per-sensor deployment error, which is the deviation of the actual location from the expected location. Fortunately, Liu *et al.*'s simulation results indicated that with the same storage capacity, the location aware random pairwise predistribution achieved a higher probability to establish pairwise keys than the original version, especially when the deployment error was small.

Secondly, Liu *et al.* [224] made ingenious use of Blundo *et al.*'s polynomial based group key distribution algorithm [38] to design a general framework of key-pool based random key predistribution. Specifically, before deployment, the setup server first generates a bivariate t-degree polynomial $f(x, y)$ over a finite field F_q, where q is a large prime number (as governed by the security requirement in terms of bit length). The bivariate polynomial has a symmetric property, i.e., $f(x, y) = f(y, x)$. For each sensor i (here i is its unique ID), the setup server computes a "polynomial share" of $f(x, y)$, namely $f(i, y)$. Thus, for any other sensor j with $j \neq i$, sensor i can compute a shared key $f(i, j)$ by evaluating $f(i, y)$ at $y = j$. Symmetrically, sensor j can do the same by computing $f(j, i)$.

There are several important merits of this approach [38]. Consequently, each sensor only needs to store a t-degree polynomial, occupying $(t + 1) \log q$ space. A distinct advantage is that there is no need for the two sensors to communicate, apart from knowing each other's unique ID. Furthermore, it is shown [38] that this polynomial pool based technique is t-collusion resistant in that as long as the number of malicious sensors (forming a collusion) is smaller than t, a shared key among two legitimate nodes cannot be broken.

It is interesting to note that if $t = 0$, then the scheme degenerates to the key pool scheme proposed by Eschenauer and Gligor [105].

Observing that the polynomial based predistribution scheme can deter attacks from fewer than t malicious sensors with t limited by the storage capacity of each sensor, Liu *et al.* [224] improved the scheme by including the location awareness concept. Specifically, the target deployment field is partitioned into square cells as shown in Figure 26.3. As can be seen, sensor u is expected to be deployed inside cell $C_{2,2}$. Thus, cells $C_{2,1}$, $C_{1,2}$, $C_{2,3}$, and $C_{3,2}$ are adjacent to u's home cell. With such knowledge, the setup server then allocates u with the polynomial shares: $f_{2,2}(u, y)$, $f_{2,1}(u, y)$, $f_{1,2}(u, y)$, $f_{2,3}(u, y)$, and $f_{3,2}(u, y)$. Subsequently, sensors deployed in these adjacent cells can easily share keys with sensor u.

As reported in [224], the above two-dimensional partitioning of the field into square cells can be extended to n-dimensional hypercube. The advantage of a higher dimension partitioning is the increased security level, at the expense of a higher storage overhead (storing more polynomial shares at each sensor).

Zhou *et al.* [372] proposed a link-layer key establishment scheme by extending Liu, Ning, and Li's bivariate polynomial-pool scheme [224] with a hexagonal-grid partitioning. The resultant extended scheme has lower communication overhead and

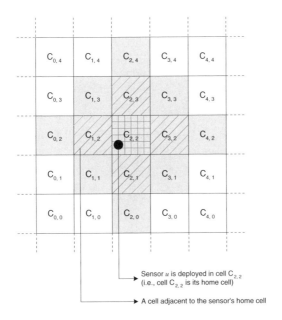

Figure 26.3   The sensor network is partitioned into square shape cells [221].

memory cost. Specifically, suppose D is the inter-cell distance in the grid and ρ is the deployment density, then the number of nodes storing shares of one polynomial in Liu, Ning, and Li's scheme is $5\rho D^2$ whereas in the hexagonal scheme it is only $\sqrt{3}\rho D^2$. Consequently, the memory costs of the two schemes are given by $5(5\rho D^2 + 1)$ and $6(\sqrt{3}\rho D^2 + 1)$, respectively [372].

Du *et al.* [95, 94] exploited the nice properties of a classical key predistribution method proposed by Blom [36]. Specifically, the proposed scheme has an important thresholding feature—as long as no more than λ nodes are compromised, all communication links of uncompromised nodes remain secure.

In Blom's scheme, first of all a $(\lambda + 1) \times N$ matrix G over a finite field GF(q), where N is the size of the network and q is a prime number larger than N. The matrix G is made public. In the key generation process, the setup server produces a random $(\lambda+1) \times (\lambda+1)$ symmetric matrix D over GF(q). It also determines an $N \times (\lambda+1)$ matrix $A = (D \times G)^T$. The matrix D is a secret. Now, as D is symmetric, we have:

$$A \times G = (D \times G)^T \times G = G^T \times D^T \times G = G^T \times D \times G = (A \times G)^T$$

Thus, $A \times G$ is also a symmetric matrix. Now, we denote $A \times G$ by K. Then, we have $K_{ij} = K_{ji}$ (note the similarity between this and the bivariate polynomials used in Liu *et al.*'s scheme), where K_{ij} is an element of the matrix K. Consequently, K_{ij} and K_{ji} can be used as the pairwise key of a pair of sensor nodes. Specifically, the setup server can assign the following items to sensor node k: the k-th row of matrix A and the k-th column of matrix G. Because G is made public, two sensor nodes can exchange their columns of G to identify each other's index. Blom's scheme, just like the bivariate polynomial scheme by Blundo *et al.*, is λ-collusion resilient. This process is illustrated in Figure 26.4. A Vandermonde matrix [36] can be used as the matrix G.

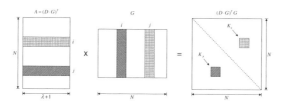

Figure 26.4   Generating sensor pairwise keys based on symmetric matrices [95].

Du *et al.* [95] extended Blom's scheme for sensor networks in the following manner. Du *et al.*'s scheme is called multiple-space key predistribution algorithm. Instead of using one single matrix D, Du *et al.*'s scheme generates ω random symmetric distinct matrices $D_1, \ldots, D_\omega$. Accordingly, there are ω different matrices $A_i = (D_i \times G)^T$. Now, for each sensor, the setup server randomly assigns to it τ (where $2 \leq \tau < \omega$) distinct key spaces from the ω possible choices. Thus, if two sensors share some common key spaces, they can set up shared keys using Blom's scheme.

Du *et al.*'s simulation results indicated that their scheme outperformed Eschenauer-Gligor's and Chan *et al.*'s random key-pool schemes. As a proof of concept, Du *et al.* also implemented their scheme in a real system using MICAz sensor nodes. In their system, $\lambda = 50$ and the matrices are generated over GF(2^{16}). The time for computing a 64-bit key was only 25.67 msec.

Lee and Stinson [201] also proposed a similar scheme as that of Du *et al.* [95] in that the Blom's predistribution method was also used.

Hwang and Kim [161] made an important observation: in practice, the node degree cannot be too large, e.g., 6–8 is optimal for practical considerations (i.e., traffic volume, energy, etc.) instead of the around 20 or more as implied by the random key-pool based or random polynomial-based schemes described above. Indeed, a larger node degree would consume more power in a sensor node. Thus, by using the "giant component" concept in random graph theory [103], Hwang and Kim proposed a slightly modified random key predistribution scheme which is based on judicious adjustment of transmission powers. The resulting network might be partitioned in that a small groups of nodes might be isolated whereas the remaining nodes are connected among each other. But Hwang and Kim argued that this is not a bad result because in practice the isolated nodes can subsequently set up links or paths with the giant component by increasing their transmission powers.

Recently, Traynor *et al.* [335, 335] extended the basic random key-pool predistribution scheme to a hierarchical sensor network. Specifically, they made an observation that the purely peer-to-peer communications among sensors are unrealistic in a practical sensor network. Indeed, they argued that in a real application, usually there are some more powerful sensors acting as control devices. These control devices can then accommodate more keys in the predistribution phase. Thus, accordingly, they proposed an unbalanced random key-pool predistribution scheme, which they found that can give a higher degree of security without increasing the memory cost for the "worker" sensor nodes.

Eltoweissy *et al.* [102] developed a protocol for dynamic re-keying in the post-deployment phase. Under the long life cycle assumption, re-keying is necessary for handling the addition or revocation of nodes. Re-keying can also prevent further compromising of rest of network against node capture. When some nodes are suspected to be compromised, the base station sends re-keying instruction to cluster controllers to trigger the corresponding re-keying operations. The drawback is that their approach requires coordination from a base station and cluster controllers.

Jolly *et al.* [174] proposed a key management protocol based on symmetric keying, primarily for hierarchical sensor networks. It is observed that sensor-to-sensor secure communication is not necessary in many applications. Thus, very few keys (typically two) are enough to establish secure connections between nodes and base station as well as cluster gateways. Their objective is to provide a cost-effective yet sufficient keying infrastructure for security of sensor networks. However, their design requires a centralized key server and the rekeying process is inefficient due to large amount of message exchanges during key renewals.

Table 26.1 gives a qualitative comparison of representative approaches discussed above.

26.3   KEY PREDISTRIBUTION WITH DEPLOYMENT KNOWLEDGE

If the structure of the network after deployment can be predicted or even controlled, predistribution is more efficient because keys can be judiciously allocated to neighboring sensor nodes.

Du *et al.* [92, 94] observed that long distance communications among sensors are not common in many practical applications. Thus, shared keys need only be set up among nodes that are physically close to each other after deployment. Du *et al.* enhanced the original Eschenauer and Gligor's approach by providing a guaranteed probability p of two physically neighboring nodes can set up a shared key. Specifically, N sensor nodes to be deployed are first divided into $t \times n$ groups so that each group $G_{i,j}$ ($i = 1, \ldots, t$ and $j = 1, \ldots, n$) is of equal size. The target deployment location of group $G_{i,j}$ is denoted as (x_i, y_i). For each sensor k in group $G_{i,j}$, its actual deployment location follows the probability density function (pdf) $f(x - x_i, y - y_i)$ where $f(x, y)$ is a Gaussian distribution.

Table 26.1 Comparison of popular key management schemes in wireless sensor networks

	Chan [58]	Du [92]	Eschenauer [105]	Eltoweissy [102]	Jolly [174]	Liu [224]
Description	random pairwise scheme	deployment knowledge scheme	basic scheme	dynamic scheme	low energy scheme	polynomial pool-based scheme
Connectivity	average	high	average	average	N/A (only node to gateways / based station)	average
Resilience	perfect against node capture	impact of attacks is localized	acceptable	good due to controllable rekeying	weak	high due to t-collusion resistant
Scalability	weak as max. network size is limited	simple node addition and revocation	simple node addition and revocation	efficient with help of centralized server	high due to simple rekeying mechanism	simple node addition and revocation
Energy consumption	low	average	average	average	very low	high
Storage cost	high	low	average	low	very low	high
Management	distributed	distributed	distributed	centralized	centralized	distributed
Knowledge assumption	none	location information is available	none	operating in long-lived network	many-to-one communication	none

Now, a set of keys $S_{i,j}$ is allocated to group $G_{i,j}$ during the key predistribution phase. Each set of keys is of the same size, denoted by $|S_c|$. The key feature of Du *et al.*'s scheme is that if two groups' target deployment locations are far away from each other, then the amount of overlap in their respective key sets would be close to zero.

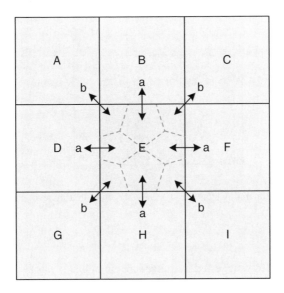

Figure 26.5   A grid based deployment structure illustrating the sharing of keys between neighboring groups [92].

Assuming a grid deployment structure as shown in Figure 26.5, Du *et al.*'s scheme has the following features:

- two horizontally or vertically neighboring groups share exactly $a|S_c|$ keys with $0 \le a \le 0.25$;

- two diagonally neighboring groups share exactly $b|S_c|$ keys with $0 \le b \le 0.25$ and we have: $4a + 4b = 1$;

- two non-neighboring groups would not share any key.

Figure 26.6 shows a simple example with $t = n = 4$.

With these arrangements, it is shown [92] that the key pool size is given by:

$$|S_c| = \frac{|S|}{tn - (2tn - t - n)a - 2(tn - t - n + 1)b}$$

where $|S|$ is the total number of distinct keys in the system. For example, suppose $|S| = 100000$, $t = n = 10$, $a = 0.167$, and $b = 0.083$, then we have $|S_c| = 1770$.

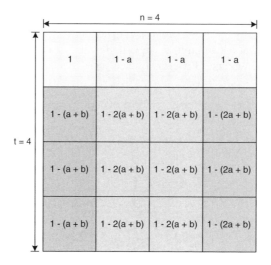

Figure 26.6   An illustrating example ($t = n = 4$) of the sharing among neighboring groups [92].

Huang *et al.* [159] suggested a similar scheme in that it was also based on a grid-group based deployment structure. The major difference is that Huang *et al*'s scheme is based on Blom's key predistribution method [36]. Furthermore, as shown by Huang *et al.*'s detailed analysis, their proposed deployment-knowledge based scheme can effectively tackle some practical malicious attacks such as selective node capturing and node fabrication.

Yu and Guan [367] suggested a predistribution scheme also based on Blom's key allocation method. Their proposed scheme also utilizes the deployment knowledge where the sensor field is divided into hexagonal cells. Their findings indicated that by carefully deciding on the size of the grids, the number of potential neighbors can be significantly reduced. Thus, the resources consumption can be much more efficient.

Ito *et al.* [163] suggested an improved scheme in which they remove Du *et al.*'s assumption [92] that a rectangular grid based deployment structure is used. Essentially, instead of assuming a probability density function for a whole group, Ito *et al.* used a per-node probability density function. Consequently, the key predistribution process is carried on a node by node basis.

Liu, Ning, and Du [223] suggested a scheme similar to that of Du *et al.* [92] in that the sensors are deployed in groups and members of each group are to be deployed to nearby locations. One major distinctive feature is that the target deployment locations (x, y) are not specified as in Du *et al.*'s scheme. As such, Liu *et al.*'s scheme is more flexible. An implication is that the predistribution process cannot leverage on the relationship among the deployment locations, which are unknown. Thus, Liu *et al.* suggested a novel grouping structure with groups G_i and cross-groups G'_i. An example is shown in Figure 26.7. As can be seen, nodes 1, 2, and 3 belong to the group G_1, whereas nodes 1, 4, 7, and 10 belong to the cross-group G'_1.

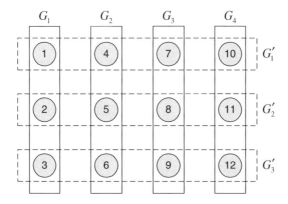

Figure 26.7   Construction of groups and cross-groups [223].

With such a overlapping grouping structure, key predistribution is carried out for both groups and cross-groups. That is, members in each group can set up a key shared key. Similarly, members in each cross-group can also set up a direct shared key. Among nodes that are deployed at physically close locations but do not belong to the same group or cross-group, some "bridge" nodes in a group or cross-group have to be used. For example, if node 1 wants to set up a pairwise key with node 12, the pair could try to use the bridge (1, 10). If this does not work, they can try another bridge (3, 12), and so on.

Liu, Rivera, and Cheng [225] also suggested a deployment knowledge based predistribution method. In their proposed scheme, sensors find out their designated roles and configures themselves automatically based on a pure localized algorithm. Under a hierarchical role-play structure, only service sensors are responsible for key space generation and distribution, with the goal of conserving the energy of other "worker sensors". The core key distribution scheme is, however, also based on bivariate polynomials.

Delgosha and Fekri [84] recently proposed a clever extension to the random bivariate polynomial predistribution scheme. Specifically, each sensor node is assigned a unique ID which is an n-tuple of integers. Each ID then maps to a set of n polynomials that are n-variate to other nodes. As the IDs are organized as the vertices of a hypercube, every two nodes with their IDs differed by a Hamming distance of one can then set up $n - 1$ shared keys. The link key can then be generated by combining these $n - 1$ keys.

26.4   KEY ESTABLISHMENT

Key establishment is a term we use in a narrow sense—the discovery and set up of a shared key between two sensors [338, 160]. As discussed above, after deployment, each sensor has already stored some keys in its memory system. The establishment phase involves the discovery of common keys in the key-sets of two communicating sensors.

Di Pietro *et al.* [281, 280] suggested a key-pool based scheme for setting up pairwise keys among sensors. Specifically, before deployment, a pool of P random keys is generated. Each key is associated with a unique index. Each sensor a is then assigned a set V_a of k keys that are randomly selected from the pool of P keys. The indices of the assigned keys can be deterministically computed by using the ID of the sensor a. Thus, similar to the classical approach suggested by Eschenauer and Gligor [105], there is a non-zero probability that two sensors share some common keys. Precisely, the probability that two distinct sensors a and b share at least one common key is given by:

$$1 - \frac{{}_PC_{P-k}}{{}_PC_k}$$

After deployment, when two sensor nodes a and b want to set up a secure point-to-point communication channel, they independently generate a shared key as follows:

$$\mathcal{K}_{a,b} = \bigoplus_{v_b^i \in V_a} v_b^i$$

That is, the new share key is generated by taking the XOR over all the shared common keys from the key pool.

Referred to as Direct Protocol by Di Pietro *et al.*, it provides only a fixed level of security depending on the system parameters P and k. In view of this, Di Pietro *et al.* suggested an extended version called Cooperative Protocol which works by involving a set of other cooperative sensors, $C = \{c_1, \ldots, c_m\}$ such that $a, b \notin C$, when two sensor nodes a and b want to set up a new shared key. Specifically, the shared key is computed as follows:

$$\mathcal{K}_{a,b}^C = \mathcal{K}_{a,b} \oplus \left(\bigoplus_{\forall c \in C} \text{HMAC}(\text{ID}_a, \mathcal{K}_{c,b}) \right)$$

where $\mathcal{K}_{a,b}$ and $\mathcal{K}_{c,b}$ are generated by using the original Direct Protocol, and HMAC is a one-way hash function. Suppose that sensor a is the one who initiates the key setup procedure, it is responsible for selecting the members of the set C. Then a informs b about the composition of C such that every member $c \in C$ can then set up a pairwise key $\mathcal{K}_{a,c}$ with a and subsequently a pairwise key $\mathcal{K}_{c,b}$ with b. The pairwise key $\mathcal{K}_{a,c}$ is used by a to send the cooperative request to each sensor c for computing the one-way hash function to be sent to b. The security of Cooperative Protocol increases with m.

Chan and Perrig [59] proposed an interesting scheme for key establishment between pairs of sensor nodes. Their proposed scheme is called PIKE (peer intermediaries for key establishment), which works by transmitting a new key via trusted intermediary sensor nodes. The PIKE system critically relies on a regular deployment and logical layout structure, supported by a certain global node-addressing facility. The major distinctive salient feature of PIKE is that the memory cost in each sensor node is $O(\sqrt{n})$, where n is the number of sensor nodes deployed in the system. This represents a significant improvement over many existing schemes which usually have a $O(n)$ memory or communication cost per sensor node.

To illustrate how PIKE works, consider a sensor network with n nodes deployed. We assume that n is a perfect square. Each sensor is then assigned a unique ID (x, y), where $x, y \in \{0, 1, 2, \ldots, \sqrt{n} - 1\}$. Before deployment, each sensor node (x, y) is then allocated $2(\sqrt{n} - 1)$ distinct secret keys which are respectively shared with each member of the two sets of sensor nodes:

$$H = \{(i, y) : i \in [0..\sqrt{n} - 1]\}$$

and

$$V = \{(x, j) : j \in [0..\sqrt{n} - 1]\}$$

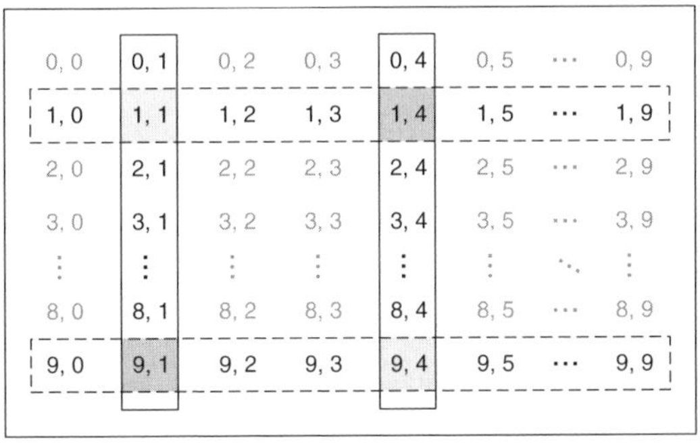

Figure 26.8   Initial key sharing among nodes in both the horizontal and vertical dimensions [59].

Figure 26.8 illustrates the case of $n = 100$. We can see that with the logical layout shown, each pair of nodes in the vertical set $(i, 1)$, where $i = 0, \ldots, 9$, share a distinct secret key which is allocated before deployment. Similarly, each pair of nodes in the vertical set $(i, 4)$ also share a distinct secret key. The same arrangement is carried out for all other vertical sets as well as all horizontal sets, such as $(1, i)$ and $(9, i)$. Consequently, for two nodes that do not have a shared key, they can establish a new shared key via two intermediary nodes. For example, nodes 14 and 91 do not have an *a priori* shared key so that they cannot perform one-to-one secure communications. Now, by transmitting encrypted messages via nodes 11 or 94, nodes 14 and 91 can establish a new key between them.

From the two candidate intermediary nodes, one of them is chosen based on a certain heuristic metric. For example, the candidate that is physically closer to the pair of nodes can be selected. Other metrics such as delay, throughput, or traffic load can also be used.

One critical feature of PIKE is that it relies on a global node-addressing facility which allows a node to locate and reach a node with a shared key. For example, such a facility is required to allow node 14 to locate and reach node 11. This is important because the logical layout has no bearing on the actual physical locations of nodes. Specifically, a geographical routing scheme such as GPSR [184] or a geographic hash table (GHT) based addressing service [297] is needed.

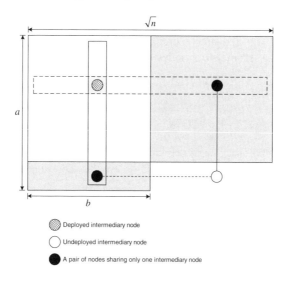

Figure 26.9   A partially deployed sensor network [59].

Furthermore, the sensor nodes need to be deployed in a specific order so that the subsequent establishment of new keys can be carried out with at least one intermediary node to help. Specifically, the sensor nodes should be deployed in the following order: $(0,0), (0,1), (0,2), \ldots, (0, \sqrt{n}), (1,0), (1,1), (1,2), \ldots, (1, \sqrt{n})$, and so on. Figure 26.9 depicts the situation where the last sensor node that has been deployed is $(a - 1, b - 1)$. Here, it can be shown [59] that the probability that only one intermediary node is available for a pair of nodes (i.e., the two nodes in the shaded regions) wanting to establish a new key is given by:

$$\frac{1}{4(a - 1)}$$

Thus, suppose $n = 4900$ and 2000 nodes have already been deployed, then the probability of having only one intermediary node is only around 0.01.

In terms of memory storage requirement, the PIKE scheme is highly efficient. Indeed, based on the description above, each sensor node needs to store $2(\sqrt{n} - 1)$ *a priori* keys. This can be further reduced to $(\sqrt{n} - 1)$ by using a dynamic key generation method [59].

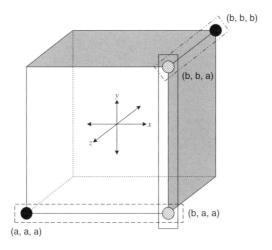

Figure 26.10 A three-dimensional layout structure [59].

There are a couple of useful extensions of the PIKE scheme. Firstly, the 2-D layout structure can be extended to 3-D as shown in Figure 26.10. With three dimensions, each pair of nodes can establish a new key with the help from two intermediary nodes. As each node now shares *a priori* keys with $3(\sqrt[3]{n} - 1)$ other nodes, the memory requirement at each node becomes $O(\sqrt[3]{n})$ instead of $O(\sqrt{n})$. The second extension is that each sensor node can share pairwise keys with other nodes on the diagonals, as shown in Figure 26.11. This arrangement increases the number of possible intermediary nodes for any pair of nodes to choose from.

Chan and Perrig [59] conducted a detailed and extensive simulation study in which two versions of PIKE, namely PIKE-2D (with a 2-D layout) and PIKE-3D (with a 3-D layout), are compared against a KDC based approach and a probabilistic approach [93]. Both versions of PIKE produced performance close to that of the KDC approach which represented as an ideal situation. Furthermore, both PIKE-2D and PIKE-3D required significantly less memory and communication overhead compared with the probabilistic approach. PIKE-2D is more resilient against node-capturing attacks while PIKE-3D is more efficient in terms of memory and communication overhead.

Blaßand Zitterbart [35] suggested a novel approach in establishing a shared key between two sensor nodes that belong to the same data aggregation tree. Motivated by the observation that the key-pool or polynomial-pool based schemes [224, 372] require $O(n)$ memory storage, Blaßand Zitterbart argued that key establishment is needed only between nodes on the same data aggregation tree. For example, as shown

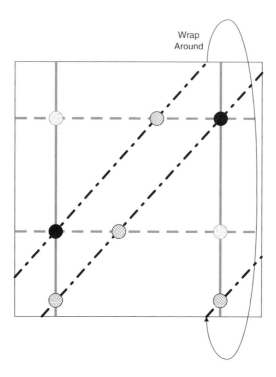

Figure 26.11 Each sensor node shares *a priori* keys with diagonal nodes [59].

in Figure 26.12, the sensor node a only needs to share keys with its ancestors (e.g., x and z) but not other nodes such as b, c, and d. Consequently, the memory cost required is only $O(\log n)$ because the height of the tree is $O(\log n)$.

Specifically, Blaß and Zitterbart argued that for data aggregation applications like the temperature measuring task as depicted in Figure 26.12, sensor devices do not need to communicate in an arbitrary manner to other sensor devices. In this temperature measuring example, it is reasonable to assume that the data sink (i.e., the computer) is interested only in the average temperature of the two rooms. As such, sensor nodes x, y, and z only need to perform the averaging computation. Indeed, there is no need for any communication to happen between a and b, or between c and d.

However, to verify the integrity of the forwarded data (e.g., the average temperature of x and y), an ancestor sensor node in the aggregation tree needs to communicate directly with some nodes on the aggregation paths. For example, the data sink might need to check that the average value as forwarded by z is correct. In order to carry out this checking, the sink needs to communicate with x and y directly. Similarly, z might also need to communicate with a and b directly. The implication of the need

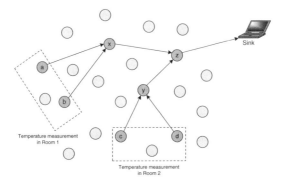

Figure 26.12   Temperature data aggregation tree [35].

for such integrity verification is that a sensor node needs to share a key with each ancestor node on its aggregation path toward the sink.

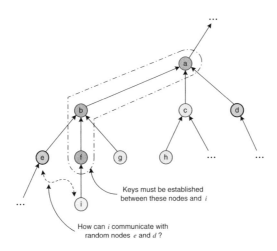

Figure 26.13   A new sensor node i joins the data aggregation tree [35].

The pairwise key establishment protocol proposed by Blaßand Zitterbart can be described *inductively* as follows. Figure 26.13 depicts the situation where a partial aggregation tree has been formed (i.e., the basis of the inductive process). Now, sensor node i is to be added to the aggregation tree with f as its parent. The setup server, called Master Device (MD) in [35], assigns i the following items:

- $K_{\mathrm{MD},i}$: a secret key between i and the MD;

- the IDs of two randomly chosen nodes e and d, together with two secret keys $K_{e,i}$ and $K_{d,i}$ which can be used by i to communicate with these two nodes, respectively; and

- two tickets: $T_e = E_{K_{MD,c}}(i, \text{``is legal player''}, K_{c,i})$ and
$T_d = E_{K_{MD,d}}(i, \text{``is legal player''}, K_{d,i})$.

With the above items, node i can then establish a shared key with node f with the help from e and d, based on an operation called "key splitting". Specifically, a key K can be split into two *key shares* K_1 and K_2 as follows:

$$K_1 = r$$
$$K_2 = K \oplus r$$

where r is a randomly generated number. With key splitting, the original key can be restored ($K = K_1 \oplus K_2$) if and only if both K_1 and K_2 are available.

Now, i generates a shared key $K_{i,f}$ to be used with node f. Obviously, i does not want to reveal this key to any other nodes including e and d. Thus, i splits $K_{i,f}$ into two key shares K_1 and K_2 first. To start communicating with e and d, node i needs to send them the two tickets T_e and T_d, respectively. Given these tickets, nodes e and d trust i and communicate with it. Afterward, i sends the following "request to forward K_1 to f" to node e:

$$C_1 = (i, E_{K_{e,i}}(i, K_1, f))$$

Similarly, i also sends the following forwarding request to d:

$$C_2 = (i, E_{K_{d,i}}(i, K_2, f))$$

Up to this point, if e already possesses a shared key with f (i.e., a key $K_{e,f}$ exists), then e can simply send the following to f:

$$\gamma_1 = (e, E_{K_{e,f}}(i, K_1))$$

Unfortunately, in this example as depicted in Figure 26.13, there is no such shared key between e and f (as well as between d and f) because they are not on the same aggregation path.

In this situation, e (as well as d) iteratively inquires its ancestor nodes until it can find an ancestor node that is also an ancestor node of f. Such an ancestor must exist because ultimately the sink is the ancestor of every node in the tree. Suppose that e finds that b and a to be such suitable ancestors. Similarly, d finds that a and some ancestor of a are such suitable ancestors. Consequently, e splits K_1 (from i) into two key shares K_1^1 and K_1^2 and then sends the following two requests to b and a, respectively:

$$C_1^1 = (e, E_{K_{e,b}}(f, K_1^1))$$
$$C_1^2 = (e, E_{K_{c,a}}(f, K_1^2))$$

Such forwarding actions are illustrated in Figure 26.14.

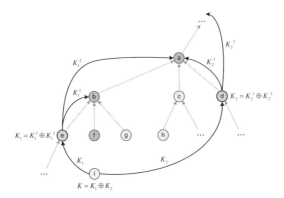

Figure 26.14 Sensor node i sends the key shares to ancestor nodes in the data aggregation tree [35].

Node d performs similar splitting and forwarding actions. Specifically, its share of $K_{i,f}$, K_2, is split into two key shares K_2^1 and K_2^2, which are forwarded to a and some ancestor of a, respectively.

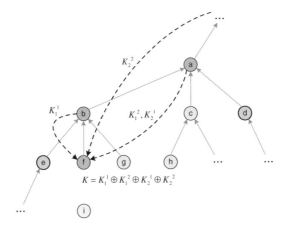

Figure 26.15 Ancestor nodes forward the key shares to the destination node f in the data aggregation tree [35].

As indicated in Figure 26.15, after decrypting the requests from e, a sends the following items to f:

$$\gamma_1^2 = (a, E_{K_{a,f}}(i, K_1^2))$$

Similarly, b sends the following item to f:

$$\gamma_1^1 = (b, E_{K_{b,f}}(i, K_1^1))$$

The ancestor of a and a itself also send similar items to f on behalf of d. Thus, f gets K_1^1, K_1^2, K_2^1, and K_2^2 so that it can restore $K_{i,f}$ by:

$$K_{i,f} = K_1^1 \oplus K_1^2 \oplus K_2^1 \oplus K_2^2$$

This ingenious key establishment protocol can deter attacks from one single malicious node. If there are potentially k malicious nodes ($k \geq 2$), then the protocol can be easily extended by splitting each key into $k + 1$ shares.

Anderson, Chan, and Perrig [17] suggested an interesting idea called "key infection" which is designed based on a practical attacking model. Specifically, Anderson *et al.* argued that it is impractical for the adversary (e.g., enemy in the battlefield) to capture all the traffic flows among all sensors when they are being deployed. The implication of this attacking model is that it becomes more economical and practical to send shared keys among sensors in plaintext when they are being deployed, instead of preloading them with large number of keys from a key pool (as in the classical approach suggested by Eschenauer and Gligor [105]). Their simulation results indicated that the key infection approach worked well for situations where there were up to 3% of malicious sensors in the field.

Miller and Vaidya [241] suggested a clever and comprehensive scheme for practical key establishment by exploiting channel diversity. There are two notable distinctive features in their proposed scheme. First, the sets of keys allocated to the sensors in the predistribution phase are *disjoint*. That is, there is no overlapping between any two sets of keys allocated. Second, in the key material exchange process during key discovery phase, keys are broadcast to neighbor sensors in *plaintext*. The first feature implies that the sensors have to exchange their keys by communicating with each other because they do not share any key implicitly. The second feature implies that during the mandatory keys exchange process, a malicious sensor or an eavesdropper can receive the keys easily because they are sent in plaintext. An important merit of the first feature is that if a sensor is compromised, only its neighborhood would become insecure because the set of keys stored in the compromised sensor would only be possibly used by its neighbors and, as such, other regions of the sensor network are unaffected. By contrast, in a traditional random key-pool predistribution scheme [105], a compromised sensor could possess keys that are used in many regions in the network.

Channel diversity is leveraged to tackle the second issue, namely the broadcasting of plaintext keys during the key discovery phase. Specifically, each sensor is assumed to have an RF module that has multiple channels: a common channel plus a number of other data channels. After the keys predistribution phase, sensors are deployed onto the field and their locations are assumed to be uniformly distributed. Furthermore, the density of the sensors, with respect to the RF transmission range, is assumed to be high enough so that there will be no shortage of one-hop neighbors (e.g., more than ten). When the key discovery phase begins, each sensor u first broadcasts to its neighbors a set data structure S_u, which represents the set of α keys it is allocated by the setup server. The set data structure has the property that it supports checking of membership of a key but it does not reveal the values of the keys. An authentication data structure M_u is also broadcast. Specifically, a neighbor sensor v can use M_u to verify that the received S_u represents a legal set of keys allocated by the trusted setup server. We will describe the details about the data structures S_u and M_u below. After these two data structures are broadcast, each sensor u then switches to a randomly selected channel. The node u sends each of its α keys in plaintext over the channel and listens to the channel for a randomly selected amount of time. Specifically, the sensor u also listens to the broadcasts from its neighbors who happen to use the same channel at the same time (it is assumed that CSMA/CA is used for each channel). Thus, sensor u can record the keys it receives for subsequent key establishment. Note that the key broadcast time is independent of the channel listening time. As such, during the time a channel is used, the sensor u may only be able to broadcast only some of its α keys. The sensor u then switches to another randomly selected channel and repeat the process.

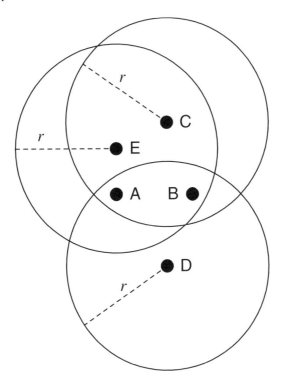

Figure 26.16   Sensor nodes A and B are one-hop neighbors of C, D, and E but nodes D is outside the transmission ranges of C and E. [241].

Channel diversity can probabilistically tackle eavesdropping, as illustrated in Figure 26.16. Suppose each sensor has c data channels. When sensor node C broadcasts a key, the probability that both A and B receive it is equal to $1/c^2$. Furthermore, the probability that E does not eavesdrop the key is equal to $(c-1)/c$. Thus, A and B can possibly set up a secure link even though E is a malicious sensor. Similarly, D may not be able to eavesdrop the key.

After the plaintext key broadcasting phase is over, each sensor u possesses a combined set of $\alpha + \gamma$ keys. That is, contains its own α preallocated keys plus γ keys that it receives from all its neighbors. Note that the total number of received keys from the neighbors may be larger than γ. Nevertheless, only γ of them are randomly chosen as governed by storage limitations. Switching to the common channel again, each sensor u broadcasts a data structure C_u representing the combined set of keys. It is important to note here that the data structure does not reveal the values of the keys themselves (the same is true in the initial broadcasting of S_u). Upon receiving the broadcast from a neighbor, sensor u can then check, for each of its known keys (i.e., the $\alpha + \gamma$ keys), whether it exists in the received combined set of the neighbor. This process is repeated for each neighbor.

Now, the key establishment phase can begin. For each neighbor v of the sensor u, it creates a potential link key k_{uv} by using η of their common keys:

$$k_{uv} = \text{hash}(k_1||k_2||\ldots||k_\eta)$$

Sensor u then sends a Link Request (LREQ) message to v:

$$\text{LREQ} = (v||u||E_{k_{uv}}(RN)||K_{uv})$$

where K_{uv} is again the set data structure representing the set of η common keys, RN is a random nonce (which is encrypted using the potential link key k_{uv}). When v receives this request, it checks its combined set to see if it really has the η common keys. Suppose this is successful and, thus, v can decrypt the nonce. Sensor v then sends u a Link Reply (LREP) message:

$$\text{LREP} = (u||v||E_{k_{uv}}(RN + 1))$$

A link key is then successful set up.

Miller and Vaidya's scheme [241], as described above, critically relies on the set data structure and the authentication of it. Specifically, the set data structure has to support the query of membership without revealing the values of the member keys. Furthermore, the data structure has to be compact so that it does not lead to a heavy storage and communication burden. Miller and Vaidya proposed to use the Bloom filters [37] data structure. Specifically, a Bloom filter is an s-bit vector (initially set to all 0), associated with h one-way has functions: $H_{BF}^1, H_{BF}^2, \ldots, H_{BF}^h$. Given an input value (i.e., a key in our context), each hash function is applied. Every hash function generates an output value between 0 and $s-1$. Suppose hash function H_{BF}^i generates an output value of x, where $0 \leq x \leq s-1$. Then bit x of the vector is set to 1. This is done for all h hash functions. Thus, to check for the membership of a given value, all the hash functions are applied to it and see if the h bits are really set to 1 in the bit vector. The Bloom filter is a compact data structure in that, with constant physical storage (i.e., s bits), it can "store" many input values by just setting the corresponding bit positions to 1 (if they have not already been set to 1 by some earlier members). However, we can see that a Bloom filter can generate false positives because the bits indicating membership of a certain value may actually be set by some other members.

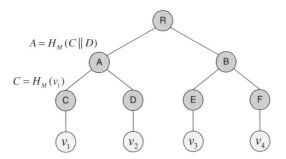

Figure 26.17   An example Merkle tree [241].

For the authentication data structure, Miller and Vaidya [241] used the Merkle trees [240]. Figure 26.17 shows a Merkle tree for authentication of four values: v_1, v_2, v_3, v_4. Notice that the circular nodes represent the tree nodes and leaves, whereas the square nodes represent the objects and are not stored as part of the tree. Each leave node is the output value of applying a one-way hash function to an object. For example, $C = H_M(v_1)$. Each internal node is the output value of applying the hash function to a concatenation of the children values. For example, $B = H_M(E||F)$. Now, each object representing a Bloom filter for a set of keys, can be verified by a receiving sensor with the Merkle tree nodes. For example, given v_2 together with the values of the internal Merkle tree nodes B, C, and R, a receiving sensor can verify the authenticity of v_2 (that is it is an authentic Bloom filter representing the set of keys) by checking the equality:

$$R = H_M(H_M(C||H_M(v_2))||B)$$

Miller and Vaidya [241] performed a detailed simulation study using the NS-2 simulator. With $\alpha = 100$, $\gamma = 100$, and $\eta = 10$, their results indicated that even with just one extra channel (i.e., $c = 2$), the proposed protocol can attain over 90% connectivity among neighboring sensors with link keys that are uncompromised even when 80% of the nodes are malicious.

26.5 DISCUSSIONS AND FUTURE WORK

In the above survey, we can see that techniques proposed for key predistribution and key establishment are practical yet theoretically sound. Researchers have designed their schemes that can cater for various practical constraints such as memory overhead, communication costs, and RF transmission range.

Having reviewed the state-of-the-art techniques proposed for key management in sensor networks, we believe that there are at least two important directions for future research. First of all, while the conventional wisdom is that public key schemes are impractical in a sensor network, Malan *et al.* [231] suggested an ingenious method for practical public key implementation in a resource constrained sensor network. It is likely that further research will make public key schemes more practicable. As such, new predistribution schemes may also be needed for preallocation and authentication of private keys.

Secondly, we consider "post-distribution" also an important topic. Indeed, from a fundamental perspective, the basic problem we need to tackle in key management for sensor networks is the allocation of keys to neighboring nodes so that they can setup secure links. While predistribution followed by discovery is a viable method, it is by no means the only practical approach. We believe that we could also design some post-distribution schemes where keys are allocated or generated *after* the sensor nodes have been deployed and have settled down in the field (i.e., topology of the neighborhood has been identified). The potential merit of a post-distribution approach is that there would be less severe wastage of keys. Toward this research direction, recently Law *et al.* [199] proposed a key redistribution scheme for sensor networks. Their proposed scheme is depicted in Figure 26.18 below. In simple terms, their redistribution scheme works by borrowing keys from neighbors. This is illustrated in Figure 26.19. Obviously, this is just an initial attempt and more work has to be done in this direction. Specifically, the question that needs to be tackled is how a sensor node can compute a key with a neighbor, potentially without predistribution, in a sense that post-deployment computation replaces pre-deployment key allocation.

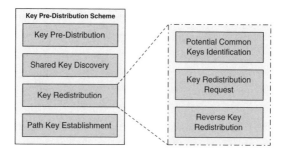

Figure 26.18 Key redistribution approach [199].

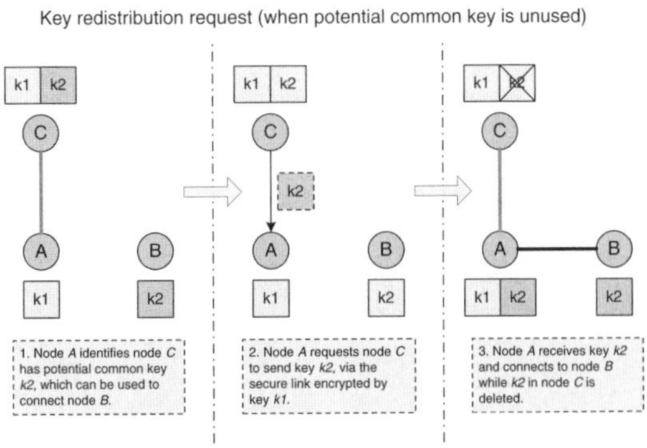

Key redistribution request (when potential common key is unused)

Figure 26.19   Key borrowing concept [199].

26.6 SUMMARY

More often than not, a sensor network is deployed in a hostile environment, where the attack model of an adversary can be very general. Key establishment among sensor nodes is a critical component in protecting the integrity and accuracy of a sensor network application. In this chapter, we have reviewed state-of-the-art efficient techniques proposed for key predistribution and the subsequent phase of key establishment. Many of these schemes are probabilistic in nature and are derived from judicious key space allocation. Despite the demonstrated effectiveness of these existing schemes, sensor network key management still offers a rich space for further research.

PROBLEMS

26.1   Compare and contrast the basic key pool based scheme and the bivariate polynomial based scheme.

26.2   How can deployment knowledge help in optimizing the performance of a key management scheme?

CHAPTER 27

PEER-TO-PEER COMPUTING OVER WIRELESS

A friend in need is a friend indeed.

—English Proverb, traced by Apperson to the XIII century

27.1 INTRODUCTION

Modern computing technologies have decentralized data processing power in an unprecedented manner. An important implication is that user machines, be it a desktop computer or a handheld PDA (personal digital assistant), have data processing power, in terms of instruction processing rate, amount of storage, and reliability, that is inconceivable merely a decade or two ago. Indeed, few people would now argue that computing occurs largely at the "edge" of networks, which have also made tremendous strides thanks to the ever improving communications technologies. Advancements in computing and communication, coupled together, enable a recent unstoppable trend in a new form of distributed processing—peer-to-peer (P2P) computing.

Wireless Internet and Mobile Computing. By Yu-Kwong Ricky Kwok and Vincent K. N. Lau
Copyright © 2007 John Wiley & Sons, Inc.

645

As its name implies, P2P computing involves users (or their machines) on equal footing—there is no designated server or client, at least in a persistent sense. Every participating user can be a server and be a client depending on the context. Some people have referred this to as a "democratic computing environment" [18] because users are free from centralized authorities' control. This new paradigm of distributed computing has spurred many high profile applications, most notably in file-sharing, such as: BitTorrent [70], Freenet [68], Gnutella [134], and, of course, Napster [255].

Interoperability:

In contrast to the traditional client/server file transfer model, peer-to-peer (P2P) systems introduce a new file transfer paradigm—peer-to-peer data transmission. Within this P2P model, there are several types of topology involved, including centralized, decentralized structured and decentralized unstructured. In fact, each kind of topology is developed in view of the disadvantage of others. As a result, they are not interoperable. This is further discussed in Section 27.2 and Section 27.3.

In this chapter, the pioneering idea of wireless P2P systems is our main focus. The wireless medium has actually brought new challenges to the P2P world. Unlike in a traditional wired network where the clients are within a fixed backbone (i.e., the clients connect to ISPs and the ISPs interconnect to form the Internet), in an ad hoc wireless network the mobile clients themselves form the backbone of the system. The formation of the wireless backbone itself forms a topology. As a result, it is required to have a topology control on it. We discuss this in Section 27.4.

There are two topology formations, namely application layer topology (wired P2P systems) and MAC layer topology (wireless ad hoc networks). These two concepts seem to be unrelated and non-interoperable to each other as the former is targeted on wired environments while the latter is targeted on wireless environments. However, the pioneering wireless P2P concept brings these two concepts together. In order to support this concept, it is necessary to integrate these two topology formations together and make them interoperable with each other. The introduction of the TCP2P layer is one of the solutions, and this is further discussed in Section 27.6.

Performance:

It has been suggested that the wireless topology formation can affect the P2P file sharing topology formation. On the other hand, there are some other factors that can severely affect the performance of wireless P2P systems. For example, the file size, the maximum display time of files, the bandwidth required for streaming, the popularity of files, the number of copies replicated and their corresponding locations, etc. In Section 27.5, metrics on file size, delay time and required bandwidth are discussed with the consideration of the access popularity.

27.2   A TAXONOMY OF P2P SYSTEMS

Figure 27.1 illustrates a taxonomy of P2P systems [208]. A P2P system includes several components—peer searching, peer selection, peer membership management and file sharing protocols. On the other hand, as for file sharing protocols, several generations have been developed and they can be classified into three main types, including centralized, decentralized structured and decentralized unstructured. We further discuss this in Section 27.3.

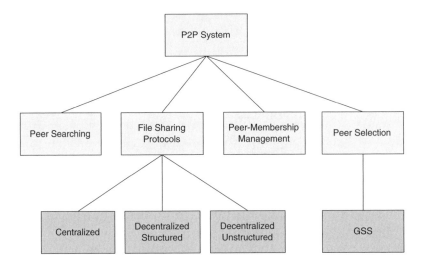

Figure 27.1   Taxonomy of P2P systems.

On the other hand, Figure 27.2 summarizes topology control techniques in the wireless ad hoc network environment. First of all, topology control techniques can be divided into two main categories—graph theoretical approach and non-graph theoretical approach. The graph theoretical approach is not our main focus in this chapter and we will briefly discuss it in Section 27.4. On the other hand, for non-graph theoretical approach, the topology formation is not considered as a graph with edges and vertices. In this case, some techniques are required to control the topology, such as the construction of neighbor sets, transmission power control, nodes sleeping schedule, etc. With the use of these techniques, the topology of a wireless ad hoc network can be restricted to be certain topology. This is further discussed in Section 27.4.

27.3 WIRED P2P SYSTEMS

Wired P2P systems are developed in view of great demand and popularity of file sharing between groups of users. Within a group, users take the responsibility of server and client at the same time. Due to this reason, every user becomes a "peer" in the network. There are many advantages of using this model. The main advantage is that the bandwidth utilization is more evenly distributed among the peers who have replicated the whole copy or chunks of the copy of a shared file. By doing so, no

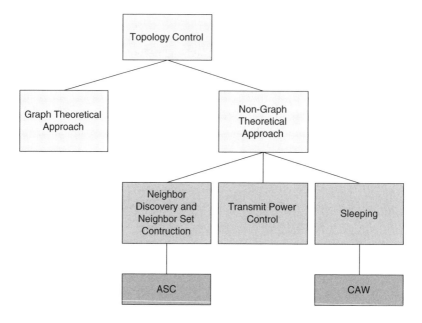

Figure 27.2   Taxonomy of topology control techniques.

single peer will be easily overcrowded since some other peers will also have the same replicated copies.

P2P file sharing systems we encounter in daily life can be classified into three generations [208]. The first generation uses a centralized approach in which a centralized directory is used to store the information of the files that have been shared and the locations of those files. Figure 27.3 illustrates this centralized approach. The clients first update with the server on the list of files that they are willing to share. In the figure, client 1 would like to request an object X. It then sends a query to the centralized directory server. The server then replies with a list of locations where the object X is shared. In this case, client 1 chooses client 3 to download the object and client 3 then sends the required object to client 1 directly.

The centralized approach is adopted by a famous music P2P file sharing company called Napster [255]. Napster keeps a centralized server as a global directory and users share their files using Napster services. However, due its centralized involvement in the file sharing process, Napster has been sued by several music producer companies and was forced to close down its server because of copyright infringement problems.

In view of the legal action against Napster, the second generation of P2P file sharing protocol is developed with the use of decentralized directory approach. Gnutella [134] is one of the most popular examples. In Gnutella, the file query task is distributed to all the nodes within a certain group by flooding the query to neighboring nodes until one of the nodes replies with the presence of the required file or no one replies.

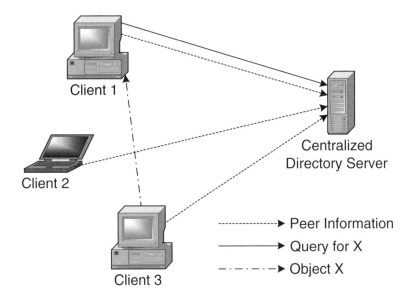

Figure 27.3 A centralized P2P system.

With the use of this idea, there is a bootstrap node to contact with, so that the user can join the group. However, since the bootstrap node itself does not store any file information, it evades the legal action similar to the case of Napster. At the same time, this idea does not guarantee that users must be able to join a group with the files they want, and the users can join and leave the group at any time they want. As a result, the system becomes unstructured as shown in Figure 27.4 and the peer-searching task usually involves flooding.

BitTorrent (BT) [34] can be regarded as the third generation P2P file sharing protocol. From a structural point of view, it is still the same as the second generation file sharing protocol. However, they are different in the sense that BT completely combines the idea of server and client. As a result, a user within a group becomes a "peer". The whole requested file is broken into chunks of smaller size. When a user is requesting chunks that he does not have from other users, other users will also request the chunks that the user has. Due to this reason, a user will download and upload at the same time and the network utilization is highly distributed. With the use of this kind of P2P system, it seems easier to evade from legal actions. However, in Hong Kong, there is a successful legal action against a BT user who had initiated the sharing of movie files.

After the discussion of the evolution of wired P2P systems together with daily life examples, readers may figure out that one type of P2P topology is missing. It is the decentralized structured P2P system. The structure is commonly formed with the use

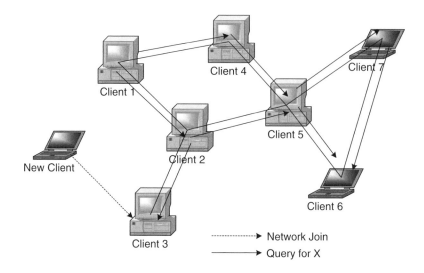

Client 4

Client 7

Client 1

Client 5

New Client

Client 2

Client 6

Client 3

------------▶ Network Join

——————▶ Query for X

Figure 27.4 A decentralized unstructured P2P system.

of distributed hash tables (DHTs). A DHT can be used to search for files and find the location of the peer in a distributed environment. In other words, a DHT can be used for file searching and peer searching. One of the popular examples is Chord [328], which is developed by MIT. In Chord, nodes are conceptually connected as a ring as illustrated in Figure 27.5

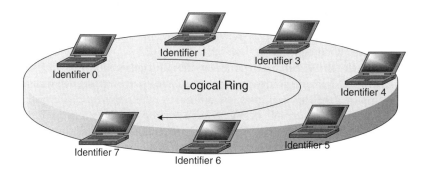

Identifier 1

Identifier 3

Identifier 0

Logical Ring

Identifier 4

Identifier 7

Identifier 5

Identifier 6

Figure 27.5 A decentralized structured P2P system.

The decentralized unstructured topology is more likely to dominate in the wireless P2P environment. It is because its decentralized and unstructured nature is likely to happen in the wireless ad hoc environment. For example, when you are walking with your mobile phone in a shopping mall, there may be some other people doing the same thing. In this decentralized environment, it is unlikely to have a centralized server around nor structured topology built. However, due to the unstructured nature, there are certain difficulties required to be handled, such as selecting which peer to download/upload. On the other hand, the formation of a wireless ad hoc network also brings out new difficulties, as certain topologies will be formed during the formation of the network. Some of the topology formation ideas are discussed in subsequent section. A practical illustration is given in Section 27.6 to demonstrate the possibility of integrating different topologies to form a network dedicated to wireless P2P file sharing.

27.4 TOPOLOGY CONTROL IN WIRELESS AD HOC NETWORKS

After the discussion of the evolution of different wired P2P topologies, we start to discuss topology control techniques that are commonly adopted in wireless ad hoc networks. Recall from Figure 27.2 that there are two main categories of topology control in wireless ad hoc networks, namely graph theoretical approach and non-graph theoretical approach.

27.4.1 Graph Theoretical Approach

Generally speaking, graph theoretical approach involves the consideration of the whole wireless ad hoc network as a graph G. The graph is composed of a set of nodes (i.e., vertices) V. Between vertices, the nodes may be connected with each other, and a set of edges E is formed. However, in a wireless environment, it is not necessary to have a mutual communication between two nodes. In this case, communication is represented as a directed edge to represent one-way communication ability. Based on this information, a topology control algorithm can be applied to the graph G so as to construct a subgraph T of the graph G. After the application of the topology control algorithm, the nodes that communicate as indicated in subgraph T should experience certain performance improvement. Figure 27.6 illustrates the idea discussed.

There are a number of graph theoretical topology control algorithms proposed, such as Minimum-Energy Communication Network (MECN) [300], SMECN (Small MECN) [212], algorithms based on relative neighborhood graph (RNG) [104] and minimum spanning tree (MST) [214], etc. In MECN, it takes into account the energy needed to transmit a packet to the destination node. Specifically, certain nodes are assigned as relay nodes if it is found that less energy is required to deliver the packet to the destination node through relay nodes. SMECN, which is considered as an extended and improved version of MECN, is proposed to enable the support to mobile scenarios. However, Cheng *et al.* [64] discovered that both MECN and SMECN may not be able to find a minimum-energy topology if a node has two duties simultaneously, i.e., being both a relay node and a sink node. As a result, other approaches are required to tackle the problem.

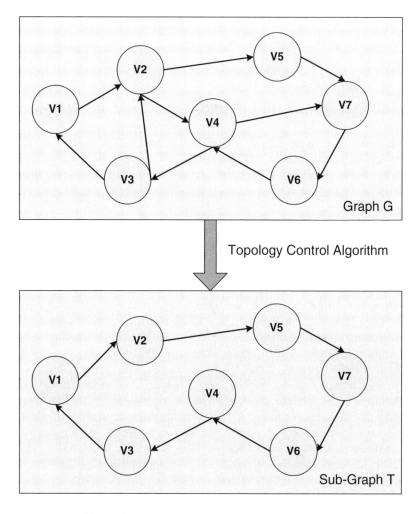

Figure 27.6   Example of graph theoretical approach.

27.4.2 Non-Graph Theoretical Approach

Most of the topology control algorithms of graph theoretical approach may require a node to have all the information about the whole network, so as to calculate the mathematical information required and to form the subgraph. However, this is usually not feasible in a wireless ad hoc network. As a result, it comes the non-graph theoretical approach. In this approach, it considers the nodes' connectivity with their neighbors, which in return controls the topology of the network.

Recall from Figure 27.2 that the non-graph theoretical approach involves the use of three main methods—neighbor discovery, transmit power control and sleeping. These three methods are discussed below:

Neighbor Discovery: The topology of a wireless ad hoc network is greatly varied with the possibility of the discovery of nodes around and the neighbor set construction criteria. For example, the minimum requirement for considering nodes around to be neighbors is that a node should be able to detect the present of other nodes. Generally, radio signals from a node are transmitted omnidirectionally and therefore nodes within certain radius of the node should be able to detect the present of the node. Due to this reason, the coverage area can be changed with the use of directional antennae and this may increase the chance of finding neighbors. On the other hand, certain criteria can be added into the neighbor construction process, e.g., minimum energy level. With the inclusion of different criteria, the topology is greatly varied because the number of potential neighboring nodes may be increased or decreased, depending on the criteria chosen.

Transmit Power Control: The topology of a wireless ad hoc network can also be varied by controlling the transmit power of the nodes. This technique in fact varies the length of the link between two connected nodes. Some examples of topology control with this technique are PCDC [251] and STPM [16].

Sleeping: Sleeping is necessary in some wireless ad hoc networks (e.g., a sensor network) since energy conservation is one of their main consideration and sleeping of a node can effectively save energy and prolong the battery life. With the presence of sleep nodes, the topology of the network keeps on changing. It is because the sleep nodes are not able to respond to any signal transmitted in the proximity. As a result, some of the transmission links will be temporarily disabled, which changes the topology. Due to this reason, sleep nodes must be integrated with some wakeup protocol to wake up from time to time to maintain the basic network connectivity.

27.5 MEDIA OBJECTS REPLICATION TECHNIQUES

In previous sections, various topics related to wired P2P systems and wireless ad hoc network topology control have been discussed. The main purpose of a P2P system is to share files between peers. Obviously, the number of replication of files within the system has a significant effect to the performance of the P2P system. However, typically there are a large amount of files in a P2P system, and it is therefore impossible for every peer to store the whole copies of all files in the P2P system. Due to this reason, a good object replication technique or strategy is crucial to the performance of a P2P system.

In this section, a wireless P2P media objects replication scenario is focused to discuss the impact of replication techniques on the performance of a P2P system [133]. Figure 27.7 illustrates the scenario. In the figure, there is a base station which is supposed to have all the media objects and would like to distribute the objects to various home-to-home on-line (H2O) devices. These devices have the ability of creating or producing media objects to be displayed in other H2O devices, displaying media objects, and relaying media objects data from one H2O device to another H2O device.

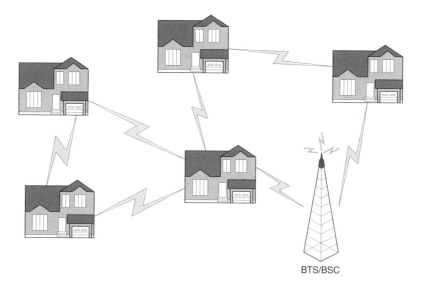

BTS/BSC

Figure 27.7 Architecture home-to-home on-line (H2O) devices.

However, there are several aspects that can greatly affect the performance of the system:

- media objects require continuous bandwidth for steady display;

- the bandwidth between each link depends on the number of transmitting devices in the corresponding region; and

- the topology.

In order to level out the above issues, assumptions on string topology and fixed bandwidth between each link are made. With these assumptions, there are three different kinds of replication strategies. They are based on the size of objects, display time of objects and the bandwidth requirement of objects. These strategies are tested against uniform distribution of objects, proportional to the object access frequency and square root of object access frequency, respectively. With these configurations, the performance comparison is determined by the number of simultaneously displayed media objects that the system can support.

In the configuration of 100 video clips with bandwidth requirement of 4 Mbps and display time of 120 minutes (this is a homogeneous setting in which every clip is the same), the three replication techniques obtained similar performance.

In the heterogeneous setting scenario, over 400 video clips are mixed with over 3500 audio clips. Video clips have four different display times ranging from 30 minutes to 120 minutes and have bandwidth requirement of 4 Mbps. On the other hands, audio clips have three different play times ranging from 2 minutes to 8 minutes and have bandwidth requirement of 340 kbps. Under these configurations, if the bandwidth replication technique is used, video clips are replicated 12 times more than audio clips (this is roughly the ratio between bandwidth requirements of video clips and audio clips). However, when size is used as the replication technique, with the average size of over two gigabytes for video clips and average size of over 10 megabytes for audio clips, video clips are replicated over 100 times more than audio clips. On the other hand, using the display time replication technique, video clips are replicated 14 times more than audio clips.

The results showed that the replication technique which is based on the size of objects obtains the worst performance, as it replicates many video clips and not much space is left for audio clips. The situation is worsen especially in the case that the references of audio clips are much larger than that of video clips. On the other hand, the replication technique which is based on the bandwidth requirement of objects obtains the best results, and it can replicate objects into the nodes with around 1/10 of the space used.

27.6 PRACTICAL ILLUSTRATIONS—LOCALIZED TOPOLOGY CONTROL

Previously, we have discussed various issues that are required to be considered in a wireless P2P system, including wired P2P systems, wireless ad hoc network topology control, and media objects replication techniques. In this section, we discuss a practical wireless P2P system, especially on how the system handles the issues of integrating the topology control of a P2P system with a wireless ad hoc network [208].

First of all, we start by talking about the target application scenario. Figure 27.8 illustrates the target application scenario of the wireless P2P system. As shown in the figure, in a shopping mall there are a number of users with portable devices which have short-distance wireless connection capability. The portable devices may be mobile phones, PDAs or laptops. Under this scenario, it is unlikely that they have a common infrastructure for sharing of files between users. As a result, one of the possible ways of accomplishing the file sharing task is to build up a wireless infrastructure as shown in the figure. In this wireless infrastructure, users can join and leave at any time they want. This concept is similar to that of wireless ad hoc networks. However, the network construction process has to consider several parameters which will affect the network topology formed, including energy constraints, incentive, and fairness.

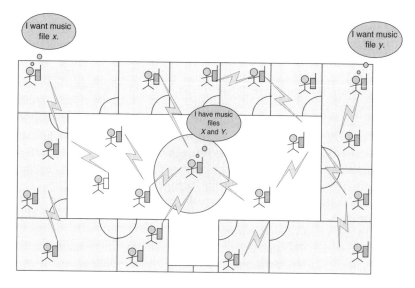

Figure 27.8   Shopping mall scenario of a wireless P2P system.

In view of the differences between wired and wireless P2P file sharing protocols, TCP2P [208] is proposed as illustrated in Figure 27.9 As indicated in the figure, TCP2P can be regarded as a protocol which runs in-between MAC layer and network layer. This protocol layer is mainly used for controlling the topology of the wireless P2P network formed and therefore it is a localized topology control protocol. TCP2P is composed of three main components—Adjacency Set Construction (ASC), Community-Based Asynchronous Wakeup (CAW) and Greedy Server-Peer Selection (GSS).

Recall from Figure 27.2 that one of the main problems in wireless ad hoc network topology control is the neighbor discovery and neighbor set construction process.

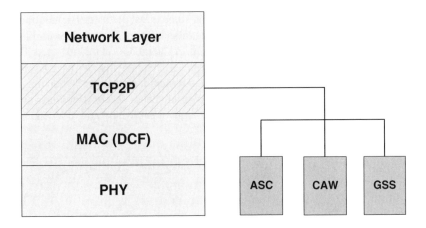

Figure 27.9   TCP2P protocol stack.

ASC is proposed to solve the problem "who is my neighbor?" [208]. In other words, ASC is proposed to construct next-hop set for peers locally, so that the upper layer—network layer can use the neighbor set information to route packets from the source node to the destination node.

As discussed previously, the minimum requirement to construct the neighbor set is that a node should be able to detect the signal transmitted by other peers (i.e., neighbors). After the fulfillment of this minimum requirement, the problem becomes why the neighbors are willing to become next-hop neighbors of the node. There are several factors that a peer needs consider in order to decide whether it should be the next-hop of another peer. The factors include the following:

Fairness: If a peer has been used as a relay node for a long time and it has also contributed a lot in routing control packets for maintaining the network connectivity, it should be exempted from being a next-hop neighbor of other peers.

Aggressiveness: If a peer always asks for some files from other peers without contributing much to the whole P2P system, other peers will have less will to be its neighbor.

Access Popularity: If a peer holds lots of files or a certain amount of very popular files, there is a high chance that other peers will request files from this peer. In this case, the workload will be very high for being a next-hop neighbor of the peer.

Remaining Energy: If the remaining energy of a peer is low, it is unlikely that the peer is willing to be the neighbor of other peers. It is because quite a large amount of energy will be consumed in acting as a relay node.

After quantifying the above factors against some threshold values, a neighbor set for wireless P2P file sharing can then be constructed. There are three possibilities as shown in Figure 27.10 The first case is that peer A's neighbor set includes peer B, but not vice versa. The second case is that peer B's neighbor set includes peer A, but not vice versa. The last case is that both peers A and B include each other in their neighbor sets. The existence of the asymmetric property in adjacency set construction is mainly due to the fact that every peer constructs its neighbor set locally and independently.

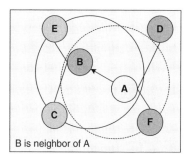

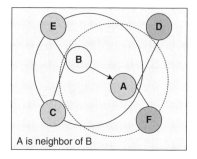

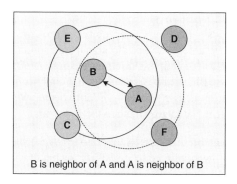

Figure 27.10   TCP2P ASC neighbor set.

The use of ASC in the wireless P2P file sharing scenario has some advantages, including a higher file request success ratio, shorter delays, more even distribution of relay and server nodes, and less deviated energy levels of nodes.

The second component of the TCP2P protocol is Community-Based Asynchronous Wakeup (CAW). One of the expected usages of the TCP2P protocol is to share and exchange some music files. Usually, the fans of a particular singer will have the tendency of downloading music files of this particular singer. As a result, the files stored and shared by fans of this particulars singer are in fact very similar. In this case, these users should have similar interest and they have the potential to form a community [1] as shown in Figure 27.11 However, some measurement techniques have to be adopted in order to measure the similarity between two peers. One of the viable solutions is Information Retrieval (IR) [307]. With the use of techniques like IR, peer nodes can be grouped into different communities.

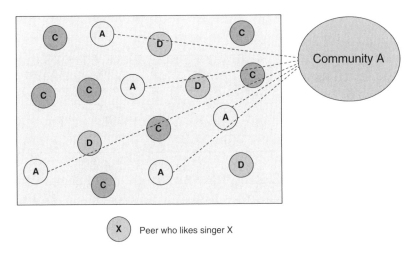

Figure 27.11   Example of community formation.

The problem becomes why should these communities be formed. Recall from Figure 27.2 that another possible technique of controlling the topology of a P2P system is to adopt sleeping in the network. One of the sleeping techniques is to apply asynchronous wakeup scheduling with the use of Wakeup Schedule Function (WSF) [371]. With the use of this function, it is guaranteed that mobile nodes using different schedules will have at least one active time slot overlapping with mobile nodes using other schedules. Figure 27.12 shows an example of WSF design and it is known as "(7,3,1)-design". In the figure, there are seven different schedules in which any schedule will have at least one active time slot overlapped with other schedules.

In the TCP2P protocol, CAW adopts the concept of community into sleeping schedules. Different communities take up different schedules and the peers in the same community, which is defined by the similarity of peers, use the same wakeup schedule. The main advantages of this design is that peers with similar interest can wake up at the same time for file sharing. As a result, those who are not interested in files being shared can be put into sleep mode, which eventually results in a shorter delay for transferring files.

After discussing the questions on "who is my neighbor?" and "when should I sleep and wake up?", we discuss the third component of the TCP2P protocol—Greedy Server-Peer Selection (GSS), which is mainly designed to solve the problem "who

---

[1] A community refers to at least two entities which have particular preferences.

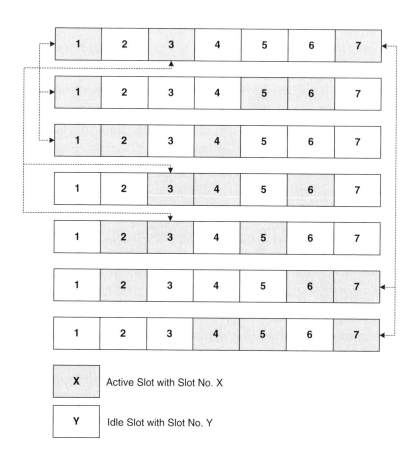

Figure 27.12 TCP2P CAW "(7,3,1)" WSF example.

should I download files from?". Recall from Figure 27.1 that peer selection is one of the main concerns in the P2P system design. It is because a peer usually has not much information on other peers within the system, and therefore it may be difficult to make decision on which peer should be selected to download the file required so that the selection can be beneficial to the whole system. As a result, the idea of Energy-Based Data Availability (EBDA) is proposed [208]. EBDA is a performance metric which takes into account the remaining energy of mobile nodes, the number of files that each node holds and the popularity of files.

In the peer selection process, it is better to select a peer which can make the EBDA metric more positive or at least less negative. Since the EBDA metric takes the remaining energy of peer nodes into consideration, a high value of the the sum of EBDA of all peers is significant to the P2P system. Specifically, the higher the value, the longer the system can survive. The direct consequence is that files can be kept longer in the P2P system.

27.7 SUMMARY

In this chapter, the pioneering concept of wireless P2P systems has been discussed. We have discussed the taxonomy of wired P2P systems and wireless ad hoc networks. Wired P2P systems are very popular nowadays with three different kinds of topology—centralized, decentralized unstructured and decentralized structured. The evolution of these topologies has been discussed.

After that, we have discussed the concept of topology control in wireless ad hoc networks. There are two main types of topology control in wireless ad hoc networks— graph theoretical approach and non-graph theoretical approach. In the non-graph theoretical approach, three main methods can be used to control the topology formation. They are neighbor discovery and neighbor set construction, transmit power control and sleeping.

Besides topology control, replication techniques are also important to the performance of a wireless P2P network sharing media objects. Several parameters are required to be considered, including the size of files, bandwidth required to transmit files at steady rate, display delay of files, and access popularity files. It is found that the performance of the P2P network is sensitive to both bandwidth required and the access popularity.

Lastly, a practical shopping mall scenario of the usage of a wireless P2P network has been illustrated. Specifically, the TCP2P protocol has been discussed. TCP2P includes three components—ASC, CAW, GSS. With the use of TCP2P, a localized topology control on a wireless P2P network can be done.

One major challenging issue in P2P systems is the need to provide incentives so as to entice cooperation among users. We discuss the incentive issues in Chapter 28.

PROBLEMS

27.1   Briefly discuss the wired P2P file sharing protocols.

27.2   Briefly discuss the wireless ad hoc network topology control protocols.

27.3   Summarize the idea of the localized topology control protocol TCP2P and discuss how it can integrate both P2P system topology control and wireless ad hoc network topology control.

CHAPTER 28

INCENTIVES IN PEER-TO-PEER COMPUTING

Even wisdom has to yield to self-interest.

—Pindar: Pythian Odes, III, c. 475 B.C.

It is worse to mistrust a friend than to be deceived by him.

—French Proverb

28.1   INTRODUCTION

As we have introduced in Chapter 27, Peer-to-Peer (P2P) computing has become an important distributed processing paradigm that is pervasive in nature. In a P2P system, users are autonomous and not subject to any central rule. Thus, some users may not be cooperative or altruistic. Indeed, these users usually are just selfishly trying to derive benefits from the P2P community without making any contribution. Such "free-riding" behaviors could render the P2P system ineffective or not viable. To encourage cooperative behaviors, incentives are mandatory. Incentive mechanisms

Wireless Internet and Mobile Computing. By Yu-Kwong Ricky Kwok and Vincent K. N. Lau
Copyright © 2007 John Wiley & Sons, Inc.

manifest themselves as payment based systems, auction based systems, exchange based systems, reciprocity based systems, and reputation based systems. We discuss the usage of various incentive mechanisms in two main P2P applications: file sharing and media streaming. Although most P2P incentive mechanisms designed for the Internet environment can be applied to a wireless system, there is a unique challenge in a wireless environment—the connectivity among peers is by itself a critical sharing problem.

The highly flexible features of P2P computing such as a dynamic population (users come and go asynchronously at will), dynamic topologies (it is impractical, if not impossible, to enforce a fixed communication structure), and anonymity, come at a significant cost—autonomy, by its very nature, is not always in harmony with tight cooperation. Consequently, inefficient or lack of cooperation could lead to undesirable effects in P2P computing. Among them the most critical one is "free-riding" [113] behavior. Loosely speaking, free-riding occurs when some users do not follow the presumed cooperation rules (in order to benefit the whole community) such as sharing files voluntarily, sharing bandwidth voluntarily, or sharing energy voluntarily. Such altruistic actions, presumably, would bring indirect and intangible (and even remote) returns to the users. For instance, if everyone shares files voluntarily, every user would eventually benefit from the high availability of a large and diverse set of selections. Unfortunately, there are always some users that do not believe or buy in to such utopia-like concepts and would, then, "rationally" choose to just enjoy the benefits derived from the community, but not contribute their own resources.

To deter or avoid free-riding behaviors, the P2P community has to provide some *incentives*—returns for resource expenditure that are, more often than not, tangible and immediate. Such incentives would then motivate an otherwise selfish user to rationally choose to cooperate because such cooperation would bring tangible and immediate benefits. Let us consider an analogy: in human society, getting pay from our work is a tangible and immediate incentive to motivate us to devote our energy, which could otherwise be spent on playing, to the work. Indeed, it is important for the incentive to be tangible so that a user can perform a cost-benefit analysis—if benefit outweighs cost, the user would then take a cooperative action [193]. It is also important for the incentive to be immediate (even though this is a relative concept) because any resource is associated with an opportunity cost in that if immediate return cannot be obtained from a cooperative action, then the user might want to save the effort for some other private tasks.

Interoperability:

To provide incentives in a P2P computing system, there are basically five different classes of techniques.

1. **Payment Based Mechanisms:** Users taking cooperative actions (e.g., sharing their files voluntarily) would obtain payments in return. The payment may be real monetary units (in cash) or virtual (i.e., some tokens that can be redeemed for other services). Thus, two important components are needed: (1) currency;

(2) accounting and clearing mechanism. Obviously, if the currency is in the form of real cash, there is a need of a centralized authority, in the form of an electronic bank, that is external to the P2P system. If the currency is in the form of some virtual token, then it might be possible to have a peer-to-peer clearing mechanism. In both cases, the major objective of is to avoid fraud at the expense of usually significant overhead. Proper pricing of cooperative actions is also important—over-priced actions would make the system economically inefficient while under-priced actions would not be able to entice cooperation.

2. **Auction Based Mechanisms:** In some situations, in order to come up with an optimal pricing, auction is an effective mechanism. In simple terms, auction involves bidding from the participating users so that the user with the highest bid get the opportunity to serve (or to be served, depending on the context). An important issue in auction based systems is the valuation problem—how much a user should set in the bid? If every user sets a bid higher than its true cost in providing a service, then the recipient of the service would pay too much that it is deserved. On the other hand, if the bids are too low, the service providers may suffer. Fortunately, in some form of auctions, we can design proper mechanisms to induce bidders to bid at their true costs.

3. **Exchange Based Mechanisms:** Compared to payment based and auction based systems, exchange (or barter) based techniques manifest as a purer P2P interaction. Specifically, in an exchange based environment, a pair of users (or, sometimes, a circular list of users) serve each other in a rendezvous manner. That is, service is exchanged in a synchronous and stateless transaction. For example, a pair of users meet each other and exchange files. After the transaction, the two users can forget about each other in the sense that any future transaction between them is unaffected by the current transaction. Thus, an advantage is that very little overhead is involved. Most importantly, peers can interact with each other without the need of intervention or mediation by a centralized external entity (e.g., a bank). Furthermore, free-riding is impractical. Of course, the downside is that service discovery and peer selection (according to price and/or quality of service) could be difficult.

4. **Reciprocity Based Mechanisms:** While pure barter based interactions are stateless, reciprocity generally refers to stateful and history based interactions. Specifically, a peer A may serve another peer B at time t_1 and does not get an immediate return. However, the transaction is recorded in some history database (centralized in some external entity or distributed in both A and B). At a later time $t_2 > t_1$, peer B serves peer A, possibly because peer B selects peer A as the client due to the earlier favor from A. That is, as peer A has served peer B before, peer B would give a higher preference to serve peer A. A critical problem is how we can tackle a special form of free-riding behavior, namely the "whitewashing" action (i.e., a user leaves the system and rejoins

with a different identity), which enables the free-rider to forget about his/her obligations.

5. **Reputation Based Mechanisms:** A reputation based mechanism is a generalized form of reciprocity. Specifically, while a reciprocity record is induced by a pair of peers (or a circular list of more than two peers), a reputation system records a score for each peer based on the assessments made by many peers. Each service provider (or consumer, depending on the application) can then consult the reputation system in order to judge whether it is worthwhile or safe to provide service to a particular client. Reputation based mechanism is by nature globally accessible and thus, peer selection can be done easily. However, the reputation scores must be securely stored and computed, or otherwise, the scores cannot truly reflect the quality of peers. In some electronic market place such as eBay, the reputation scores are centrally administered. But such an arrangement would again need an external entity and some significant overhead. On the other hand, storing the scores in a distributed manner at the peers would induce problems of fraud. Finally, similar to reciprocity based mechanisms, whitewashing is a low cost technique employed by selfish users to avoid being identified as a low quality users which would be excluded from the system.

Performance:

Different techniques mentioned above are suitable for different applications. Generally speaking, there are two mainstream applications in P2P environments: sharing of discrete data, and sharing of continuous data. Examples of the former include file sharing systems (e.g., Napster), data sharing systems (e.g., sharing of financial or weather reports), etc. A notable example of the latter is P2P video streaming. Indeed, there is an important difference between file sharing and media streaming systems. In the former, a user needs to wait until a file (or a discrete unit of shared information) is completely received before it can be consumed or used. Thus, there could be a significant delay between service request and judgement of service quality. In an extreme case, a user may not discover that a shared file is indeed the one requested or just a piece of junk. By contrast, in a media streaming application, a user would quickly discover if the received information is good enough. The quality of service (QoS) metric used is also different in these two different applications. In a file sharing application, the most important metrics are downloading time and the integrity of the received files. In a media streaming application, the more crucial performance parameters are the various playback quality metrics such as jitter, frame-rate, resolution, etc.

Currently, a majority of P2P systems are implemented over the Internet. However, wireless P2P systems are also proliferating. While most of the incentive techniques designed for a wired environment could be applicable in a wireless system, the wireless connectivity is by itself an important bootstrap sharing problem. Indeed, in the Internet, users seldom pay attention to the connectivity issue because a user can be reached (or can reach) any other Internet user without noticeable effort. The only concern about communication is the uploading or downloading bandwidth consumption. In a wireless environment, however, the mere action of sending a request message from a client peer to a server peer would probably need several intermediate peers to help doing the message forwarding because the server and client peers may be out of each other's transmission range. Consequently, incentives have to be provided to encourage such forwarding actions.

In Section 28.2, we describe approaches designed for providing incentives in Internet based P2P networks. We discuss both file sharing and media streaming applications. In Section 28.3, we describe solutions suggested for wireless P2P systems. We then provide some of our interpretations and suggestions in Section 28.4. We summarize this chapter by providing some remarks in Section 28.5.

28.2 INCENTIVE ISSUES IN P2P SYSTEMS OVER THE INTERNET

28.2.1 File Sharing Systems

In a file sharing system, users would like to retrieve files from other users, and would expect other users to do the same. Thus, each user would need to expend two different forms of resource:

- Storage: Each user has to set aside some storage space to keep files that may be needed by other users, even though such files may not be useful to the user itself;

- Bandwidth: Each user has to devote some of its outbound bandwidth for uploading requested files to other users;

Users usually perform file selection (and hence, peer selection) with the help of some directory system which may or may not be fully distributed. For example, in Napster [255], the directory is centralized.

Under such a sharing model, the most obvious form of free-riding behaviors is that a selfish user just keeps on retrieving files from others but refuses to share its collections (and thus, no need to expend any outbound bandwidth for file uploading). Interestingly enough, in an empirical study using the Maze file sharing system [238] performed by Yang *et al.* [358], it is found that the more direct indicator of free-riding behaviors is the online time of a user. Specifically, the online time of a selfish user in a P2P file sharing network is on average only one-third of that of a cooperative peer.

In this section, we first briefly overview a contemporary file sharing system called BitTorrent [70]. We then survey techniques suggested for various other P2P file sharing networks.

28.2.1.1 BitTorrent   BitTorrent [70] is by far one of the most successful P2P file sharing system. A key feature in BitTorrent is that each shared file is divided into pieces (of size 256KB each), which are usually stored in multiple different peers. Thus, for any peer in need of a shared file, parallel downloading can take place in that the requesting peer can use multiple TCP connections to obtain different pieces of the file from several distinct peers. This feature is highly effective because the uploading burden is shared among multiple peers and the network can scale to a large size. Closely related to this parallel downloading mechanism is the incentive component used in BitTorrent. Specifically, each uploading peer selects up to four requesting peers in making uploading connections. The selection priority is based on descending order of downloading rates from the requesting peers. That is, the uploading peer selects four requesting peers that have the highest downloading rates. Here, downloading rate refers to the data rate that is used by a requesting peer in sending out pieces of some other file. Thus, the rationale of this scheme is to provide incentive for each participating peer to increase the data rate used in sending out file data (i.e., uploading, or, in BitTorrent's term, *unchoking*). There are other related mechanisms (e.g., optimistic unchoking), which are described in detail in [70, 290].

Qiu and Srikant [290] performed an indepth analysis of BitTorrent's incentive mechanism. By using an intricate and accurate model, it is shown that a Nash equilibrium exists in the upload/download game in BitTorrent. At the equilibrium, each peer sets its uploading data rate to be its physical maximum uploading rate (i.e., each peer is fully cooperative). On the other hand, due to the usage of the optimistic unchoking mechanism (a fifth requesting peer is randomly selected in the uploading process, for details, see [70, 290]), a free-rider can potentially achieve 20% of the possible maximum downloading rate. This theoretical result conforms nicely with the simulation findings by Jun and Ahamad [175] who observed that in BitTorrent, free-riders are not penalized adequately while contributors are not rewarded sufficiently.

28.2.1.2 *Hierarchical P2P Systems*

In some situations, the P2P network may be structured in a hierarchical manner so that some specialized machines (called "super-peers") can take up a more important role for handling resource management tasks such as request forwarding and routing, directory listing, etc.

Singh *et al.* [321] studied the impacts of super-peers in a P2P file sharing network. Simply put, a super-peer is a special network node that serves a hub to provide file indexing service to other nodes. The problem is that there is a lack of incentive for a participating node to act as a super-peer because any node can simply join an existing super-peer to obtain good performance. Singh *et al.* observed that some entities external to the P2P system, such as an Internet Service Provider (ISP) or a content publisher, have business driven incentives for designating some nodes to act as super-peers, and for enhancing the capabilities of super-peers. Specifically, a super-peer can cache meta-data only instead of the files themselves. As such, the cost of acting as a super-peer is lower. Furthermore, with properly designed meta-data, a super-peer can support value-added search commands (e.g., topic based search of files).

With such facilities incorporated in each super-peer, a hierarchical P2P file sharing network is then much more efficient that a flat P2P system which relies on request-flooding. Consequently, all nodes in the system have the incentive in maintaining such a P2P file sharing system. Simulation results also indicate that Singh *et al.*'s proposed super-peer scheme is effective.

28.2.1.3 *Payment Based Systems*

Hauscheer *et al.* [145] suggested a token-based accounting system that is generic and can support different pricing schemes for charging peers in file sharing. The proposed system is depicted in Figure 28.1. The system mandates that each user has a permanent ID authenticated by a certification authority. Each peer has a token account keeping track of the current amount of tokens, which are classified as local and foreign. A peer can spend its local tokens for accessing remote files. The file owner treats such tokens as foreign tokens, which cannot be spent but need to be exchanged with super-peers for new local tokens. Each token has a unique ID so that it cannot be spent multiple times.

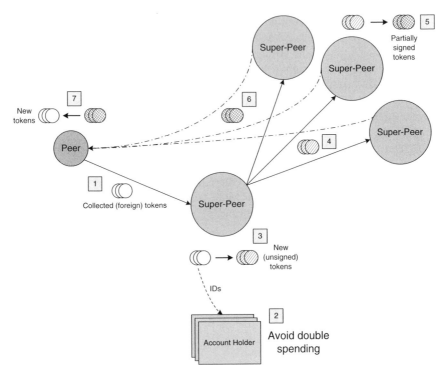

Figure 28.1   A super-peer based token accounting system for P2P file sharing [145].

At the beginning of each file sharing transaction, the file consumer tells the file owner about which tokens it intends to spend. The file owner then checks against file consumer's account kept at the file owner's machine. If the tokens specified are valid (i.e., they have not been spent before), then the file consumer can send the tokens in an unsigned manner to the file owner. Upon receipt of these unsigned tokens, the file owner provides the requested files to the file consumer. When the files are successfully received, the file consumer sends the signed version of the tokens to the file owner. In this manner, Hauscheer *et al.* argued that there is no incentive for the peers to cheat.

Yang and Garcia-Molina [357] proposed the PPay micropayment system in which each peer can buy a coin from a broker. The peer then becomes the "owner" of the coin and can spend it to some other peer. An important feature is that even after the coin is spent, the original owner still has the responsibility to check the subsequent usage of the coin. For example, suppose A is the owner of a coin which is spent to B. If B wants to spend the coin in turn to C, the original owner A needs to check whether such a transaction is valid (e.g., to avoid double spending of the same coin). If A is offline (e.g., temporarily departed the P2P system), then the broker is responsible to perform such checking.

Although the PPay system described above is a useful tool for supporting P2P sharing, Jia *et al.* [169] observed that PPay can be further improved. Specifically, Jia *et al.* proposed a new micropayment system, called CPay (an improved version of PPay), which has one significant new feature. The new feature is that the broker judiciously selects the most appropriate peer to be the owner of a coin. Specifically, the owner of a coin should be one that is expected to stay in the system for a long period of time. Thus, the broker's potential burden of checking coin owners' transactions can be considerably reduced.

Figueiredo *et al.* [119] also considered a payment based system to entice cooperation among peers. Specifically, each peer requiring message forwarding service from other peers needs to pay real money to these peers. Thus, as a peer stays in the P2P network and provides forwarding service to other peers, it can gain money. Indeed, such a monetary gain represents an incentive to enhance the availability of a peer in the network.

Saito [303] proposed an Internet based electronic currency called i-WAT, which can be used by users in a P2P system for "purchasing" services. Each i-WAT message is an electronic ticket signed using OpenPGP. Saito *et al.* [304] then extended the i-WAT system by adding a new feature called "multiplication over time", which means that a requesting peer's debt (in terms of i-WAT units) increases over time. This feature then encourages service providing peers to stay in the system for a longer period of time so as to defer the redemption of i-WAT tickets, thereby increasing the gains from the requesting peers.

28.2.1.4 *Cost of Sharing* Varian [336] reported a simple but insightful analytical study on disincentives in P2P sharing. Table 28.1 lists the notation used in Varian's analysis.

Table 28.1 Notation used in Varian's analysis [336] on disincentives for P2P sharing.

Symbol	Definition
p	unit price
v	value of the item as perceived by each peer
n	number of peers in the system
D	total cost of producing all the items
$d = \frac{D}{n}$	average development cost
k	number of peers in each group that share an item
t	sharing cost incurred by each member of a group
c	cost imposed by the central authority to those peers who participate in sharing
π	profit derived by the central authority

In Varian's model, a single item (e.g., a single music file) is considered and the system is homogeneous in that all peers have the same valuation on the item. There is an external central authority (e.g., the original producer of the music) that has the incentive to discourage sharing among peers in the system. The issue is then how the central authority can introduce proper disincentives into the system. Firstly, observe that for viability in producing the item, we have:

$$v - \frac{p}{k} - t - c \geq 0 \qquad (28.1)$$

$$p\frac{n}{k} \geq D \qquad (28.2)$$

The equilibrium (where the item is just viable to be produced) price and profit are then given by:

$$p = (v - t - c)k \qquad (28.3)$$

$$\pi = (v - t - c)kn - D \qquad (28.4)$$

We can see that profit is, counterintuitively, decreasing in c. An interpretation is that c is not large enough to discourage sharing and price has to be cut for compensation. As an analogy, consider that c represents some copy protection mechanism which merely brings inconvenience to customers but cannot discourage sharing. To compensate for the inconvenience, the price has to be reduced.

On the other hand, for a given value of c, suppose the price is set in such a way that it is marginally unattractive to share. That is, we have:

$$\frac{p}{k} + t + c \geq p \qquad (28.5)$$

This in turn implies that:

$$p = \frac{k}{k-1}(t+c) \qquad (28.6)$$

Now, as the maximum practical value of p is v, we have:

$$c = v - \frac{k-1}{k}t \qquad (28.7)$$

Yu and Singh [366] also investigated the issue of proper pricing in the presence of free-riders in a P2P system. A referral based system is considered in that each peer can either answer a remote query directly (e.g., serving a requested file) or reply with a referral (e.g., pointing to a different peer who may be able to serve the requested file). A requester (i.e., file consumer) needs to pay for both a referral or a direct answer. Each peer keeps track of reserve prices of potential referrals and direct answers from any other peer. These reserve prices are updated dynamically based on transaction experiences in that a satisfactory transaction leads to an increase in the reserve prices while an unsatisfactory one leads to a decrease. From a seller's point of view, these prices are also exponentially decreased as time goes by. Under this model, simulation results indicate that a free-rider will quickly deplete its budget. On the other hand, the price of a direct answer is found to be much higher than that of a referral.

Courcoubetis and Weber [75] recently reported an in-depth analysis of the cost in sharing in a P2P system. In their study, a P2P sharing system is modeled as a community with an excludable public good. Furthermore, the public good is assumed to be nonrivalrous, meaning that a user's consumption of the public good does not decrease the value of the good. Such a model is suitable for a P2P file sharing network, where the excludable public good is the availability of shared files. The model is also considered as suitable for a P2P wireless LAN environment, where the excludable public good is the common wireless channel. With their detailed modeling and analysis, an important conclusion is derived: each peer only needs to pay a fixed contribution, in terms of service provisioning (e.g., a certain fixed number of distinct files to be shared by other peers), in order to make the system viable. Such a fixed contribution is to be computed by some external administrative authority (called a "social planner") by using the statistical distribution of the peers' valuations of the public good.

28.2.1.5 Reciprocity and Reputation Based Systems

Feldman *et al.* [114] suggested an integrated incentive mechanism for effectively deterring (or penalizing) free-riders using a reciprocity based approach. Specifically, the proposed integrated mechanism has three core components: discriminating server selection, maxflow based subjective reputation computation, and adaptive stranger policies.

In the discriminating server selection component, each peer is assumed to have a private history of transactions with other peers. Thus, when a file sharing request is initiated, the peer can select a server (i.e., a file owner) from the private history. However, in any practical P2P sharing network, we can expect a high turnover rate of participation. That is, a peer may only be present in the system for a short time. Thus, when a request needs to be served, such a departed peer would not be able to help if it is selected. To mitigate this problem, a shared history is to be implemented.

That is, each peer is able to select a server from a list of global transactions (i.e., not just restricted to those involved the current requesting peer). A practical method of implementing shared history is to use a distributed hash table (DHT) based overlay networking storage system [329].

A problem in turn induced by the shared history facility is that collusion among non-cooperative users may take place. Specifically, the non-cooperative users may give each other a high reputation value (e.g., possibly by reporting bogus prior transaction records). To tackle this problem, Feldman *et al.* suggested a graph theoretic technique. To illustrate, consider the reputation graph shown in Figure 28.2. Here, each node in the graph represents a peer (C denotes a colluder) and each directed edge represents the perceived reputation value (i.e., the reputation value of the node incident by the edge as perceived by the node originating the edge). We can see that the colluders give each other a high reputation values. On the other hand, a contributing peer (e.g., the top node) gives a reputation value of 0 to each colluder because the contributing peer does not have any prior successful transaction carried out with a colluder. With this graph, we can apply the maxflow algorithm to compute the reputation value of a destination peer as perceived by a source peer. For instance, peer B's (the destination) perceived reputation value with respect to peer A (the source) is 0 despite that many colluders give a high reputation value to B.

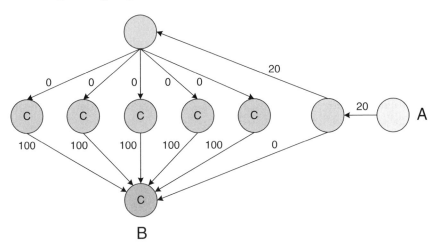

Figure 28.2 A graph depicting the perceived reputation values among peers (C denote a colluder) [114].

Finally, an adaptive stranger policy is proposed to deal with whitewashing. Instead of always penalizing a new user (which would discourage expansion of the P2P network), the proposed policy requires that each existing peer, before deciding whether to do a sharing transaction with a new user, computes a ratio of amount of services provided to amount of services consumed by a new user. If this ratio is great than or equal to 1, then the existing peer will work with the new user. On the other hand, if the ratio is smaller than 1, then the ratio is treated as a probability of working with this new user.

Sun and Garcia-Molina [330] suggested an incentive system called Selfish Link-based InCentive (SLIC), which is based on pairwise reputation values. Specifically, any peer u maintains a reputation value $W(u, v)$ for each of its neighbor peer v, where the reputation value is normalized such that $0 \leq W(u, v) \leq 1$. Here, "neighbor" means a peer v currently having a logical connection with u and thus, such a peer v can potentially request for service from u. With these reputation values, the peer u can then allocate the uploading bandwidth to any requesting neighbor peer v with a value of $W(u, v)/\sum_i W(u, i)$. The reputation value $W(u, v)$ is updated periodically based on an exponential averaging method.

Under this model, Sun and Garcia-Molina [330] observed that each peer has the incentive to do some or all of the following, in order to increase its reputation values as perceived by other peers (and hence, enjoy a better quality of service).

- Sharing out more file data;

- Connecting to more peers (to increase the opportunities for serving others);

- Increasing its total uploading capacity;

28.2.1.6 *Penalty Based Approaches*
Feldman *et al.* [115] also investigated disincentive mechanisms that can discourage free riding. Specifically, they considered various possible penalty schemes in deterring free-riders. A simple model is used. At the core of the model, each user i in the P2P sharing network is characterized by a positive real-valued *type* variable, denoted as t_i. Another key feature of the model is that the cost of contributing is equal to the reciprocal of the current percentage of contributors, which is denoted as x. Thus, for any rational user with type t_i, the user will choose to contribute if $1/x < t_i$ and free-ride if $1/x \geq t_i$.

Furthermore, the benefit each user derived from the P2P network is assumed to be of the form αx^β, where $\beta \leq 1$ and $\alpha > 0$. With this benefit function, the system performance is defined as the difference between the average benefit and the average contribution cost. Specifically, system performance is equal to: $\alpha x^\beta - 1$.

Even with the simplistic model described above, Feldman *et al.* provided several interesting conclusions. Firstly, it is found that excluding low type users can improve system performance only if the average type is low and α is large enough. Unfortunately, exclusion is impractical because a user's type is private and thus, cannot be determined accurately by other peers. It is then assumed that free riding behaviors are observable (i.e., free-riders can be identified). Such free-riders are then subject to a reduction in quality of service. Quantitatively, the benefit received by a free rider is reduced by a factor of $(1 - p)$, where $0 < p \leq 1$. A simple implementation of this penalty is to exclude a free rider with a probability of p. The second interesting conclusion is that the penalty mechanism is effective in deterring free-riders when the penalty is higher than the contribution cost. In quantitative terms, the condition is that $p > 1/\alpha$. Finally, another interesting conclusion is that for a sufficiently heavy penalty, no social cost is incurred because every user will contribute (i.e., choose not to be a free-rider) so that optimal system performance is achieved. In particular, to deal with the whitewashing problem, the analysis suggests that every new user is imposed a fixed penalty. Essentially, this is similar to the case in the eBay system where every new user has a zero reputation and thus, will less likely be selected by other users in commercial transactions. However, this is in sharp contrast to the adaptive stranger policies suggested also by Feldman *et al.* in another study [114] that we have described earlier.

28.2.1.7 *Game Theoretic Modeling* Ranganathan *et al.* [294] proposed and evaluated three schemes induced by the Multi-Person Prisoner's Dilemma (MPD) [265, 312]. Specifically, the key features of the MPD framework can be briefly summarized as follows:

- The MPD game is symmetric in that each of n players has the same actions, payoffs, and preferences.

- Any player's payoff is higher if other players choose some particular actions (e.g., "quiet" instead of "fink").

The MPD framework is used for modeling P2P file sharing as follows. There are n users in the system, each of which has a distinct file that can be either shared or kept only to the owner. The system is homogeneous in that all files have the same size and same degree of popularity. Now, the potential benefit gained by each user is the access of other users' files. The cost involved is the bandwidth used for serving other users' requests. With this simple model, it can be shown that the system has a unique Nash equilibrium in which no user wants to share. Obviously, this equilibrium is sub-optimal (both at the individual level and at a system-wide level) in that each user could obtain a higher payoff (i.e., a higher value of net benefit) if all users choose to share their files.

Motivated by the MPD modeling, Ranganathan *et al.* proposed three incentive schemes:

- **Token Exchange:** This is a payment based scheme because each file consumer has to give a token to the file owner in the sharing process. Each user is given the same number of tokens initially and each file has the same fixed price.

- **Peer-Approved:** This is a reputation based scheme in that each user is associated with a rating which is computed using metrics such as the number of requests successfully served by the user. A user can download files from any owner who has a lower or the same rating. Thus, to gain access to more files in the system, a user has to actively provide service to other users so as to increase the rating.

- **Service Quality:** This is also a reputation based scheme similar to Peer-Approved. The major difference is that a file owner provides differentiated service qualities to users with different ratings.

Theoretical analysis indicates that the Peer-Approved policy with a logarithmic benefit function (in terms of number of accessible files) can lead to the optimal equilibrium where every user contributes fully to the system. Simulation results also suggest that Peer-Approved generates performance (in terms of total number of files shared) comparable to that of Token Exchange, which entails a higher difficulty in practical implementation as it requires a payment system.

Becker and Clement [29] also suggested a simple but insightful analysis of the sharing behaviors using variants of the classical 2-player Prisoner's Dilemma. Specifically, the P2P file sharing process is divided into three different stages: introduction, growth, and settlement. In the introduction stage, the P2P network usually consists of just a few altruistic users who are eager to make the network viable. Thus, sharing of files is a trusted social norm. The payoffs of the two possible actions (supply files or not supply files) are depicted in Figure 28.3. Here, we have the payoffs ranking as: $R > T > S > P$. Consequently, the Nash equilibrium profile is: (Supply, Supply). Notice that the payoffs ranking in the original Prisoner's Dilemma is: $T > R > P > S$, and as such, the Nash equilibrium is the action profile in the lower right corner of the table.

Player 1 \ Player 2	Supply g_2^1		No Supply g_2^2	
Supply g_1^1		R		T
	R		S	
No Supply g_1^2		S		P
	T		P	

Figure 28.3 Payoff table in the introduction stage [29].

In the growth stage, we can expect that more and more non-cooperative users join the network. For these users, the payoffs ranking becomes: $T > R > P > S$, which is the same as the original Prisoner's Dilemma. Thus, the Nash equilibrium for such users occurs at the profile: (No Supply, No Supply). As the P2P network progresses to the mature stage (i.e., the size of the network becomes stabilized), we can expect that a majority of users are neither fully altruistic nor fully non-cooperative. For these users, the payoffs ranking is: $R > T > P > S$. As a result, the payoff matrix is depicted in Figure 28.4. As can be seen, there are two equally probable Nash equilibria: (Supply, Supply) and (No Supply, No Supply). Consequently, whether or not the P2P network is viable or efficient depends on the relative proportions of users in these two equilibria. Results obtained in empirical studies [29] using real P2P networks conform quite well to the simple analysis described above.

Ma *et al.* [228, 229] suggested an analytically sound incentive mechanism based on a fair bandwidth allocation algorithm. Indeed, the key idea is to model the P2P sharing

Player 2 / Player 1	Supply g_2^1	No Supply g_2^2
Supply g_1^1	R R	T S
No Supply g_1^2	S T	P P

Figure 28.4   Payoff table in the settlement stage [29].

as a bandwidth allocation problem. Specifically, the model is shown in Figure 28.5. Here, multiple file requesting peers compete for uploading bandwidth of a source peer. Each requesting peer i sends a bidding message b_i to the source peer N_S. The source peer then divides its total uploading bandwidth W_S into portions of x_i for the peers. However, due to network problems such as congestion, each peer i may receive an actual uploading bandwidth of x'_i which is smaller than x_i.

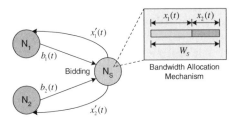

Figure 28.5   Two file requesting peers (N_1 and N_2) compete for uploading bandwidth of a source peer (N_S) [228].

Each bidding message b_i is the requested amount of bandwidth. Thus, we have $x_i \le b_i$. To achieve a fair allocation, the source peer uses the contribution level C_i of each competing peer i to determine an appropriate value of x_i. Ma *et al.* [228, 229] described several allocation algorithms with different complexities and considerations: simplistic equal sharing, max-min fair allocation, incentive based max-min fair allocation, utility based max-min fair allocation, and incentive with utility based max-min fair allocation. The last algorithm is the most comprehensive and effective. It works by solving the following optimization problem:

$$\max \sum_{i=1}^{N} C_i \log(\frac{x_i}{b_i} + 1) \tag{28.8}$$

where:

$$\sum_{i=1}^{N} x_i \le W_S \tag{28.9}$$

Here, the logarithmic function represents the utility as perceived by each peer i.

The above optimization problem can be solved by a progressive filling algorithm that prioritizes competing peers in descending order of the marginal utility $C_i/(b_i + x_i)$.

Given values of b_i and C_i, the source peer can compute the allocations in a deterministic manner. However, from the perspective of a requesting peer, a problem remains as to how it should set its bidding value b_i. Using a game theoretic analysis, it is shown that the action profile in which $b_i = \frac{W_S C_i}{\sum_{j=1}^{N} C_j} \forall i$ is a Nash equilibrium. Furthermore, provided that all cooperative peers use their respective strategies as specified in the Nash equilibrium action profile, collusion among non-cooperative peers can be eliminated. Notice that each requesting peer i needs to know the values of W_S and $\sum_{j=1}^{N} C_j$ in order to determine its own bid b_i. In a practical situation, these two values can be supplied by the source peer to every requesting peer.

28.2.1.8   *Auction Based Approaches*   Gupta and Somani [138] proposed an auction based pricing mechanism for P2P file object lookup services. In their model, each resource (e.g., a file object) is stored in a single node. However, the indices for such a file object are replicated at multiple nodes in the network and these nodes are called *terminal nodes*. When a peer initiates a lookup request for a certain file object, the request is sent through multiple paths toward the terminal nodes, as shown in Figure 28.6. The problem here is that the intermediate nodes need some incentives in order to participate in the request forwarding process.

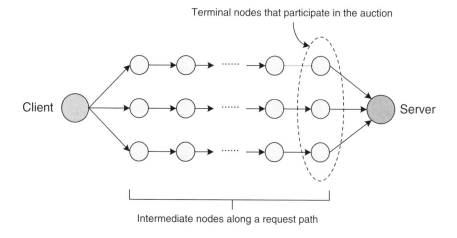

Figure 28.6   The request forwarding process [138].

Gupta and Somani [138] suggested a novel solution to the incentive problem. Specifically, the initiating peer attaches a price in the request message it sends to the first layer of nodes in the request chains. Each intermediate node on the request chains then updates the price by adding its own "forwarding cost". The terminal nodes also do the same updating before sending the request messages to the data source. Upon receiving all the request messages, the data source then performs a second price sealed bid auction (also referred to as Vickrey auction) [265] to select the highest bid among the terminal nodes. The selected terminal node then needs to pay the price equal to the value of the second highest bid. With this auction based approach, all the intermediate nodes on the request chains have the incentive to participate in the forwarding process because they might eventually get paid by the requester should their respective request chain wins the auction.

For example, consider the lookup process shown in Figure 28.7. We can see that the request chain terminated by node T1 wins the auction process and the payoff to the data source node B is 60. The only intermediate node (node 1) then also gets a payoff. Gupta and Somani [138] also showed that a truthful valuation in is the optimal strategy for each intermediate node. Furthermore, based on the requirement that every message is non-repudiable, it is also shown that the proposed mechanism can handle various potential threats such as malicious auctioneer, collusion between data source and a terminal node, forwarding of bogus request message, etc.

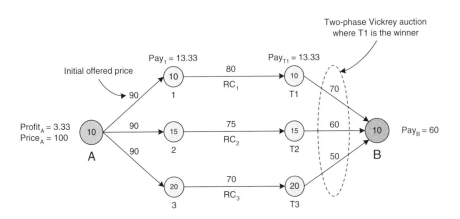

Figure 28.7   An example of the auction process in request forwarding [138].

Wongrujira and Seneviratne [352] also proposed a similar auction based charging scheme for forwarding nodes on a path from a requesting peer to a data source. However, they pointed out an important observation that some potential malicious peers could try to reduce the profits of other truthful peers by dropping the price messages. To mitigate this problem, a reputation system is introduced in that every peer maintains a history of interactions with other peers. The reputation value of a peer is increased every time a message is forwarded by such a peer. On the other hand, if an expected message exhibits a timeout, the responsible peer's reputation value is decreased.

Wang and Li [346] also considered a similar problem in which a peer needs to decide how much to charge for forwarding data. Instead of using auction, a comprehensive utility function is used. The utility function captures many realistic factors: the quantitative benefits of forwarding data, the loss in delivering such data, the cost and the benefit to the whole community. With this utility function, an upstream peer has the incentive to contribute its forwarding bandwidth while a downstream peer is guided toward spending the upstream bandwidth economically. Furthermore, a reinforcement learning component is incorporated so that each peer can dynamically adjust the parameters in its utility function so as to optimally respond to the current market situations.

Sanghavi and Hajek [308] observed that in a typical auction based pricing mechanism as described above, there is a heavy communication burden on the peers. Indeed, the entire set of user preferences has to be communicated from a peer to the auctioneer. Sanghavi and Hajek then analytically derived a class of alternative information mechanisms that can significantly reduce the communication overhead. Specifically, each peer's bid is only a single real number in each case, instead of an entire real-valued function.

Hausheer and Stiller [146] studied a completely decentralized auction approach for electronic P2P pricing of goods in a system called PeerMart (which is built on top of Pastry [301]). The key idea is the usage of a broker set which comprises other peers in the electronic marketplace. Specifically, a broker set consists of peers' whose IDs are closest to the ID of the good in the auction. Each of these peers then potentially acts as the auctioneer in the selling process. The advantage of the broker set based method is that in case a particular peer in the set is faulty (or even malicious in the sense that it does not respond to auction requests), another member in the set can take up the role of auctioneer. An example is shown in Figure 28.8.

28.2.1.9 Exchange Based Systems Motivated by the fact that any payment/credit based system entails a significant transaction and accounting overhead, Anagnostakis and Greenwald [15] proposed an exchange based P2P file sharing system. The fundamental premises is that any peer gives priority to exchange transfers. That is, in simple terms, any peer is willing to send a file to a peer that is able to return a desired file. However, based on this idea, it is incorrect to consider 2-way exchanges only. Indeed, a "ring" of exchange involving two or more peers, as shown in Figure 28.9, is also a proper P2P file transfer.

In the exchange based P2P file sharing system, each peer maintains a data structure called *incoming request queue* (IRQ). Now, a crucial problem is—how each peer can determine whether an incoming request should be entertained, i.e., whether such a request comes from some peer on a ring of exchange requests. It is obviously computationally formidable to determine all the potential multi-peer cycles. Fortunately, Anagnostakis and Greenwald [15] argues that based on simulation results, in practice a peer only needs to check for cycles with up to five peers.

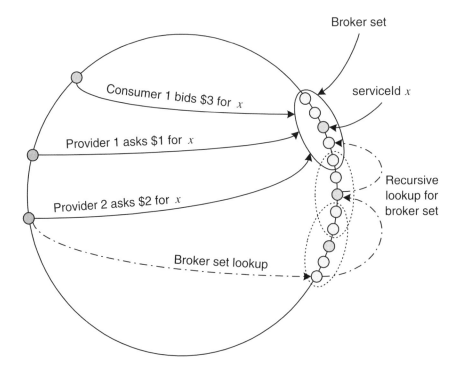

Figure 28.8   An example of fully decentralized auction [146].

Each peer uses a data structure called *request tree* to check for potential request-cycles. For example, as we can see in Figure 28.10, a peer A decides to entertain a request for file object o_2 because A finds that peer P_9 possesses an object that is needed by A. Based on this checking mechanism, the incoming requests are prioritized. Simulation results indicate that the proposed exchange based mechanisms are effective in terms of file object download time.

28.2.2   Media Streaming Systems

In this section, we describe several interesting techniques for providing incentives in a P2P media streaming environment. Broadly speaking, there are two different structures employed in P2P media streaming: asynchronous layered streaming and synchronous multicast streaming.

28.2.2.1   Layered Many-to-One Streaming   Xu *et al.* [353] proposed a fully distributed differentiated admission control protocol called DAC_{p2p}. In their model, each requesting peer needs multiple supplying peers to send different layers of media data. That is, from a topological perspective, each streaming session involves one

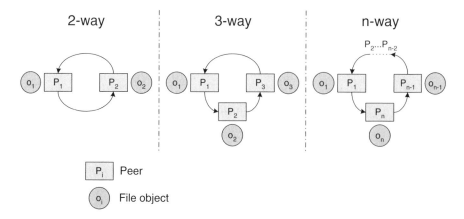

Figure 28.9 Different feasible forms of exchanges [15].

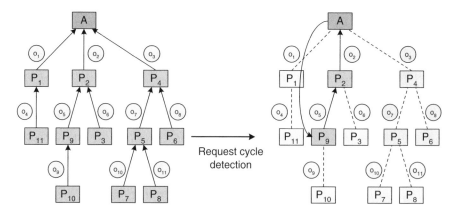

Figure 28.10 Request cycle detection using the request tree data structure maintained at each peer [15].

single recipient and multiple sources, structured as an two-level inverted tree. Of course, each supplying peer may also be a recipient of another logically different streaming session. With this streaming model, each requesting peer needs to actively contact several supplying peers in order to start a session.

This streaming model is based on a practical observation—peer requests are asynchronous and each peer's communication capability is different (i.e., heterogeneous streaming capabilities), as illustrated in Figure 28.11 where peers with different communication supports (e.g., DSL, dial-up, etc.) initiate streaming requests at different times.

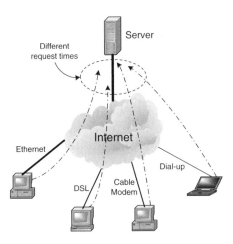

Figure 28.11 Asynchrony and heterogeneity of media streaming peers [80].

A layer-encoded media streaming process is assumed, as shown in Figure 28.12. As can be seen, each peer performs buffered playback so that the received media data are kept in a buffer and thus, can be used for streaming to other later-coming peers. For example, H_1 initiates a streaming session first and thus, it is served solely by the server. Peer H_2 starts its session next and so it can request H_1, which has buffered some media data, together with the server to send it the required data. Similarly, H_3 can stream from H_1 and H_2 without the server as it starts its session just-in-time to use the buffered data from the two earlier peers. On the other hand, H_4, which starts too late, cannot stream from H_1 and H_2. Instead, it receives media data from H_3 and the server.

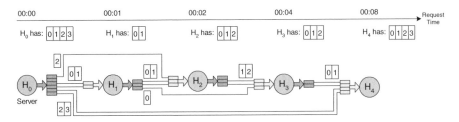

Figure 28.12 Layered streaming with buffering for serving asychronous requests [80].

Each potential supplying peer has only a limited capacity and thus, an admission control mechanism is needed. Peers in the system are classified into N classes according to the different levels of uploading bandwidth available at the peers. Each potential supplying peer P_S maintains an admission probability vector: $(Pr[1], Pr[2], \ldots, Pr[N])$.

Here, a smaller index represents a class with a larger uploading bandwidth. Suppose P_S is itself a class-k peer. Then, its probability vector is initialized as follows:

- For $1 \leq i \leq k$, $Pr[i] = 1.0$;

- For $k < i \leq N$, $Pr[i] = \frac{1}{2^{i-k}}$;

Thus, P_S always grants media streaming requests from a higher class peer (i.e., one that has a larger uploading bandwidth). Notice that this is similar in spirit to the incentive approach used in BitTorrent [70]. For requests from lower class peers, P_S may serve them as governed by the respective probabilities $Pr[i]$ in the vector. If P_S has not served any request during a certain period of time, then the admission probabilities of lower class peers will be increased.

Similar to the case in BitTorrent, each peer has the incentive to report a higher uploading bandwidth because doing so will increase its probability of admission when it needs to initiate a media streaming session. Xue *et al.* [355] extended the DAC$_{\text{p2p}}$ to a wireless environment. The key idea in the extension is to exploit the spatial distribution of mobile devices to form clusters. Users in a cluster interact using the DAC mechanism.

Habib and Chuang [139] also explored a similar idea in providing differentiated peer selection to participating peers. Specifically, a peer has only a limited set of choices (with possibly low media quality) if it behaves selfishly in the system. The degree of selfishness is reflected by a score known to other peers. The score is increased if the peer contributes to other peers, and is decreased if it refuses the requests of other peers. Based on a practical emulation study using the PROMISE [148] streaming system implemented on top of PlanetLab [282], Habib and Chuang [139] found that the proposed incentive scheme is effective in enhancing the performance of the system.

28.2.2.2 *Multicast One-to-Many Streaming*

Ngan *et al.* considered an application level multicast system for video streaming. The system is based on Split-Stream [55] which in turn is built on top of Pastry [301]. The multicast system considered critically relies on a payment based scheme. Specifically, there are five components:

- **Debt Maintenance:** When a peer A forwards video streaming data to a downstream peer B, B owes A a unit of debt.

- **Periodic Tree Reconstruction:** The multicast tree is reconstructed periodically in order to avoid prolonged unfair connections among peers. An unfair connection is one between a well-behaved peer and a selfish peer.

- **Parental Availability:** Any new peer can obtain location and addressing information about any potential parent peers in the multicast tree. Thus, the new peer can identify a potential selfish parent if the latter consistently refuses connection.

- **Reciprocal Requests:** In the system, any two well-behaved peers are expected to have an equal chance of being parent or child in any given multicast tree.

- **Ancestor Rating:** This is a generalization of the Debt Maintenance component. Here, debts are also accounted for all ancestors of a peer. Specifically, all nodes on a path in forwarding data from the source to a peer are credited or debited in cases where expected data is successfully received or not, respectively.

Simulation results indicate that a selfish free-rider is effectively penalized in terms of the amount of video streaming data received.

Chu and Zhang [158] also considered a multicast based streaming environment. The streaming process is synchronous and is supported by a multiple description codec (MDC), in which a server provides several different stripes of video with different quality. A key feature in this streaming environment is that during each streaming session, multiple multicast trees are used, each of which is for sending different stripes of video. This is illustrated in Figure 28.13. With such a streaming structure, a peer can logically join different trees simultaneously at a *different position* in each tree. Specifically, when a peer joins a certain multicast tree at a higher level (e.g., peer A in the tree for stripe II), it needs to provide a larger uploading bandwidth to serve the lower level peers in the same tree. On the other hand, a peer can also join a tree as a leaf so that it becomes a pure recipient in the tree.

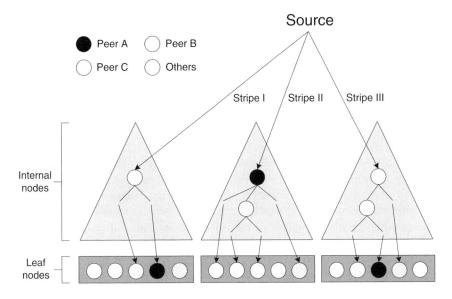

Figure 28.13   Layered video streaming using multiple multicast trees [158].

With this model, Chu and Zhang [158] then studied the effects of different degrees of altruism. They used a parameter $K = f/r$, where f is the total uploading bandwidth provided by a peer and r is its total downloading bandwidth. Thus, a larger value of K indicates a higher degree of altruism for the peer. Similar to many other P2P sharing systems described above in this chapter, a peer with a higher value of K can enjoy a better performance (in terms of media quality in the streaming application). Simulation results indicate that a small average value of K (e.g., 1.5) can already improve the overall performance of the whole system.

Shrivastava and Banerjee [341] demonstrated that streaming based on a multicast structure could be a result of natural selection. The key idea is depicted in Figure 28.14. The left part of the figure illustrates a situation where multiple peers are sharing the capacity of a single server. As a result, each peer can only enjoy a small downloading data rate. However, when peers are organized as a multicast tree, based on strategic *natural selection* (detailed below), each peer can enjoy a much larger downloading rate.

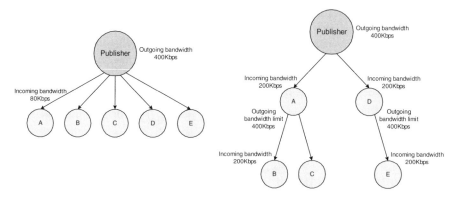

Figure 28.14   A multicast streaming structure is better off for every peer [341].

The natural selection process can be illustrated in Figure 28.15. Here, initially the root is the only source in the system and thus, peer A selects the root as the source, enjoying a downloading rate of 500Kbps. Now, when a new peer B joins the system, peer A has basically two choices: (1) serve peer B; or (2) do not serve peer B. To implement the second choice which seems to be a more favorable one, peer A can declare to the BSE that its uploading rate is 0 or a value smaller than 250Kbps (which is half of the capacity of the root). However, in doing so, peer B has no choice but naturally selects the root to be its streaming source. In that case, peers A and B will share the root's uploading capacity and thus, each obtains only 250Kbps data rate. On the other hand, in anticipation of such an actually unfavorable outcome, peer A should instead strategically declare its uploading bandwidth to be 300Kbps, which is slightly higher than the capacity declared by the root. Consequently, peer A can continue to enjoy a high downloading rate from the root, at the expense of its uploading of data to peer B at a rate of 300Kbps.

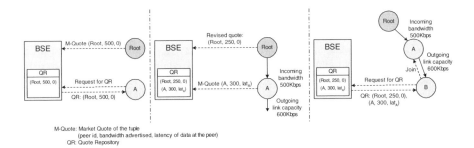

Figure 28.15 An illustration of the strategic natural selection process in connecting streaming sources and destinations (BSE is the bootstrap entity providing service information) [341].

Ye and Makedon [361] proposed a useful detection and penalty scheme to tackle the existence of selfish peers in a multicast streaming session. They observed that a selfish peer may lie to other peers in that it claims its uploading bandwidth is large so that it can enjoy a higher probability of being admitted into a streaming session or enjoy a higher quality of media data. The key of the detection mechanism is that a downstream peer in a multicast tree returns a "streaming certificate" back to its parent peer. For example, as shown in Figure 28.16, peers P_4, P_5, and P_6, send streaming certificates $\mathbf{SCert}(P_i, P_3)$ to the parent peer P_3 ($i = 4, 5, 6$). The certificates are sent periodically and are time-stamped with authentication. Thus, a higher level peer, e.g., P_1, can periodically check whether its children peers (i.e., P_2 and P_3) are selfish by asking for certificates they have received (if any) from their own children peers. If a peer cannot produce such a certificate, the higher level peer can then remove such a potentially selfish peer from the tree. The removal process is manifested as a termination of media data transmission.

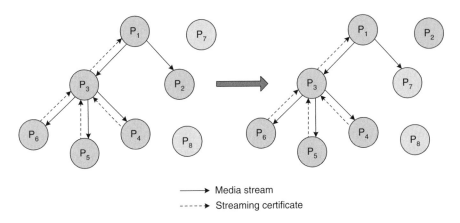

Figure 28.16 Detection and removal of a selfish peer from the streaming multicast tree [361].

Jun *et al.* [176] also explored a similar idea in their proposed Trust-Aware Multicast (TAM) protocol. Targeted for detecting and deterring uncooperative peers which can modify, fabricate, replay, block and delay data, the TAM protocol is based on a message structure that contains four fields: sequence number, timeout period, data payload, and cryptographic signature. The sequence number is used for detecting duplicated or missing data. The timeout period is used for detecting delayed data. Thus, a selfish (or even malicious) peer can be identified by its children peers in the multicast tree. Different from the approach suggested by Ye and Makedon [361], the children peers are responsible for reporting such suspicious selfish peers to the root (or the server) in the tree. Jun *et al.*'s scheme is also more flexible in that even upon receiving such "negative reports", the root may not discard such suspected selfish peers immediately. Instead, the root keeps track of a trust metric for each peer in the tree. A negative report only decreases the trust value. Only when the trust value falls below some threshold, the suspected selfish peer is discarded from the tree and the peers under its sub-tree are relocated.

28.3 INCENTIVE ISSUES IN WIRELESS P2P SYSTEMS

Wireless P2P systems [153] are proliferating in recent years. Thanks to the widely available hot-spot wireless environments, users handheld devices can work with each other in an *ad hoc* and impromptu manner. As will be evident in this section, many techniques designed for wired environments are also applied in a wireless system in a similar manner. Nevertheless, there is a unique challenge in a wireless environment, namely the connectivity issue. Furthermore, there is also one more dimension of cost incurred in each wireless P2P user—the energy expenditure, which is of prime concern to the user as wireless devices are largely powered by batteries.

28.3.1 Routing and Data Forwarding

In a wireless P2P system, the connectivity among peers is itself a bootstrap sharing problem. Indeed, if wireless users are unwilling to cooperate in performing routing and data forwarding, the wireless network can be partitioned so that service providers cannot be reached by potential service consumers. In view of this critical challenge, there has been a plethora of important research results related to incentive issues for ad hoc routing and data forwarding. In the following, we briefly cover several interesting techniques that are based on payment mechanisms, auction mechanisms, reputation systems, and game theoretic modeling.

Ileri2005 *et al.* [162] proposed a payment based scheme for enticing devices to cooperate in forwarding data for other devices in the network. The payment is not in monetary terms but in terms of of bits-per-Joule. Specifically, the utility of a user i in the network is defined as:

$$u_i(p_i) = \frac{T_i(p_i)}{p_i} \tag{28.10}$$

where u_i is the utility, p_i is the transmit power, and T_i is the throughput. That is, the utility is equal to the average amount of data received per unit energy expended, also in bits-per-Joule.

The payment system also involves an access point in the wireless network. Specifically, the access point also tries to maximize its revenue by using two parameters, μ and λ, judiciously. Here, λ is the unit price of service provided by the access point to any device in the network. On the other hand, μ is the unit reimbursement the access point provides to any device which has helped forwarded other devices' traffic. The access point's revenue is therefore given by:

$$\rho = \sum_{\text{all users } i} \lambda T_{\text{in}_i} - \sum_{\text{all forwarders } j} \mu T_{ja}^{\text{eff-for}} \qquad (28.11)$$

where T_{in_i} is the service provided by the access point to user i and $T_{ja}^{\text{eff-for}}$ is the service forwarded by a forwarder j to the access point. The situation is as shown in Figure 28.17.

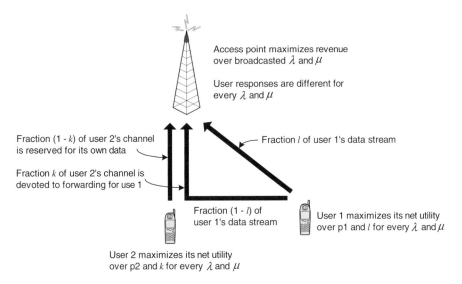

Figure 28.17 Charging of network service and reimbursement of data forwarding [162].

Simulation results indicate that the proposed service reimbursement scheme generally improves the network aggregate utility.

Salem *et al.* [305] also considered a payment based scheme for encouraging cooperation. However, their scheme involves real monetary costs and a payment clearance infrastructure (e.g., a billing account for each user). The charging and rewarding scheme is similar to that we described above—a forwarder will get reimbursed and a normal user using network service will get charged.

Marbach and Qiu [234] investigated a similar problem based on individual device pricing. However, the main differences are that each device is allowed to freely decide how much to charge for forwarding traffic (in previous researches, usually the unit price is the same for all device) and there is no budget constraint on all the devices.

Wang and Li [346] proposed an auction based scheme similar to the work by Gupta and Somani [138] that we described in Section 28.2.1.8. Specifically, each device in a wireless ad hoc network declares its cost for forwarding data when some other device wants to initiate a multihop transmission. After considering all possible paths that are able to reach the destination device, the least cost path is chosen and the devices on the path are paid for their forwarding. Again using a VCG based analysis, it is shown that the dominant action for each device is to report its true cost of forwarding. Wang and Li also showed that no truthful mechanism can avoid collusion between two neighboring devices in the forwarding auction game.

Buchegger and Le Boudec [45] observed that while economic incentives such as payment approaches can entice selfish users to help in routing and forwarding data but may not be able to handle other types of misbehaviors such as packet dropping, modification, fabrication, or timing problems. Thus, they proposed to use a reputation system in which every user provides "opinion" data to the network based on observing the behaviors of neighboring devices. After a user device has gathered such opinions (both from itself or from others, i.e., second-hand information), it can carry out a Bayesian estimation so as to classify the neighboring devices as malicious or normal. A neighboring device that is identified as a malicious user is then isolated from the network by rejecting its routing and forwarding requests.

Felegyhazi *et all.* [116] reported an interesting game theoretic analysis of the forwarding problem in ad hoc networks. Instead of using a payment based strategy, the model employs a purely utility concept in that a device's utility is equal to its payoff when it acts as a data source (i.e., the sender of a multihop traffic), minus the cost when it acts an intermediate device (i.e., a forwarder of other sender's traffic) in any time-slot. Here, both the payoff and cost are defined in terms of data throughput. Thus, an important assumption in this model is that only the sender has a positive payoff, while all the intermediate devices enjoy no payoff but just incur forwarding costs. Specifically, the destination device (i.e., the receiver of the multihop traffic) also enjoys no payoff. This may not conform to a realistic situation. Simulations were done to estimate the probability that the conditions for a cooperative equilibrium hold in randomly generated network scenarios.

28.3.2   Wireless Information Sharing Systems

Wolfson *et al.* [350] investigated an interesting opportunistic wireless information exchange problem in which a moving vehicle transmits the information it has collected to encountered vehicles, thereby obtaining other information from those vehicles in exchange. The incentive mechanisms suggested are based on virtual currency. Specifically, each mobile user carries some virtual currency in the form of a protected counter. Two different mechanisms are considered: producer-paid and consumer-paid. In the former, the producer of information pays while the consumer pays in the latter. The price P of a piece of information item (e.g., availability information about a parking lot) is given by:

$$P = E - t - \frac{d}{v} \qquad (28.12)$$

where E is the gross valuation of the information item, t is the time elapsed since the information item is created, d is the distance to travel before the user of the information can reach the relevant location (e.g., the parking space), and v is the speed of the user. Simulation results under a simple situation where there are only one consumer and two parking spaces indicate that the proposed incentive mechanisms are effective.

Yeung and Kwok [363] considered an interesting scenario in wireless data access: a number of mobile clients are interested in a set of data items kept at a common server. Each client independently sends requests to inform the server of its desired data items and the server replies in the broadcast channel. Yeung and Kwok investigated the energy consumption characteristics in such a scenario.

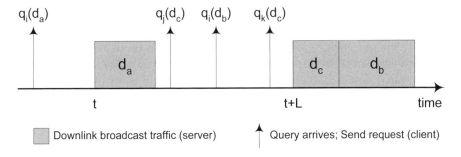

Figure 28.18   System model for wireless data access [363].

Figure 28.18 depicts the system model for wireless data access. It consists of a server and a set of clients, N. The clients are interested in a common set of data items, D, which are kept at the server. To request a specific data item, d_a, client i is required to inform the server by sending an uplink request, represented by $q_i(d_a)$. The server then replies with the content of the requested data item, d_a, in the common broadcast channel. This allows the data item to be shared among different clients. As illustrated in Figure 28.18, both clients j and k request the same data item, d_c, in the second interval. However, the server is required to broadcast the content of d_c only *once* in the next broadcast period, which reduces the bandwidth requirement.

To successfully complete a query, a client expends its energy in two different parts: (1) informing the server of the desired data item, E_{UL}; and (2) downloading the content of the data item from the common broadcast channel, E_{DL}. The energy cost of sending a request to the server is represented by E_s. If a client sends a request for each query, then E_{UL} would be the same as E_s. However, we show that clients need not to send requests for each query even without caching. This implies that E_{UL} does not necessarily equal E_s. In general, the total energy required to complete a query, E_Q, is given by:

$$E_Q = E_{UL} + E_{DL} \qquad (28.13)$$

It is assumed that E_{DL} is the same for all clients, but its value depends on the size of a data item. In practice, E_s is a function of various quantities, including spatial separation, speed, instantaneous channel quality, bit-error rate requirement, etc. For simplicity, however, E_s is also assumed to be a fixed quantity.

Based on this model of wireless data access, a novel utility function for quantifying performance is defined as follows:

$$U_i = \frac{E_{total}}{E_Q^i} \qquad (28.14)$$

The objective is then to reduce the amount of energy consumed in the query process such that *every* client's utility (Equation (28.14)) is increased.

Based on the utility function, Yeung and Kwok formulate the wireless data access scheme as a non-cooperative game—*wireless data access (WDA) game*. Their game theoretic analysis shows that even without using client caching, the clients do not need to send requests to the server. Simulation results also indicate that the proposed scheme, compared with a simple always-request one, increases the utility and lifetime of *every client* while reducing the number of requests sent, at the cost of slightly larger average query delay. They also compared the performance of the proposed scheme with two popular schemes that employ client caching. The simulation results show that caching only benefits clients with high query rates while resulting in both shorter lifetime and smaller utility in other clients.

28.3.3 Network Access Sharing

Efstathiou and Polyzos [101] studied the problem of building a federation of wireless networks using a fully autonomous P2P approach. Specifically, in their system model, there are multiple WLANs, each of which is considered to be completely autonomous. When a user of one WLAN enters the domain of a different nearby WLAN, the latter would also admit the user based on a reciprocity idea. To achieve this, each WLAN is equipped with a domain agent (DA) which is responsible for managing the roaming of foreign users. Each DA maintains a counter of tokens, which is increased when a foreign user is admitted to its WLAN, and is decreased when a local user travels to a foreign WLAN. The DAs of different WLANs interact with each other in a pure P2P fashion. Thus, the advantage of this approach is that there is no need to set up prior *pairwise* administrative agreements among different WLANs.

Based on a prototype built using Cisco WLANs, it is found that the proposed P2P based roaming scheme is efficient.

Kang and Mutka [181] considered an interesting problem in which peers share the access cost of wireless multimedia contents. Specifically, one peer in the network serves as a proxy which pays a network server in order to download some multimedia contents in a wireless fashion. Other peers in the system then share the contents without incurring any cost. This is achieved by having the proxy broadcast the received multimedia data to the peers within its transmission range. For other distant peers, rebroadcasting by the edge peers is employed to serve them. This is illustrated in Figure 28.19 (left side).

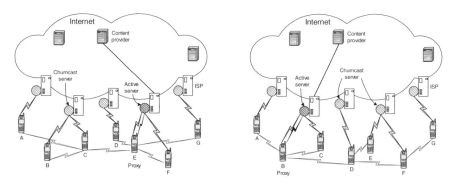

Figure 28.19 Illustration of network access cost sharing (left), and round-robin scheduling of proxy (right) [181].

This idea of cost sharing in wireless data access is called CHUM (cooperating ad hoc networking to support messaging). A key component in such sharing is the cost sharing mechanism. In the CHUM system, the peers take turn, in a round-robin manner, to serve as the proxy. This is illustrated in Figure 28.19 (right side). It is assumed that the peers have the incentive to follow this round-robin rule based on the reciprocity concept. Simulation results indicate that 80% of network access cost is saved even with just six peers in the system.

28.4 DISCUSSION AND FUTURE WORK

We have seen that in both wired and wireless systems, the major techniques for providing incentives are: payment (virtual or real), exchange (barter), reciprocity (pairwise), reputation (global), and game theoretic utility. Payment based systems work by exploiting a user's incentive in increasing or even maximizing its "revenue". However, such an incentive may not be appropriate in some practical situations. For instance, a cellular phone user may not be interested in his or her income (from such sharing) but care more about the quality of service derived from the device. Exchange based systems fit very well in file sharing applications because users have strong incentives for trading interested files (e.g., music files). For other applications such as forwarding of data (e.g., in wireless ad hoc networks), it is debatable as to whether in practice a user would be interested in exchanging data forwarding capabilities. Reciprocity (in a pairwise manner) is similar in spirit as a exchange based mechanism. The major difference is that exchange is usually memoryless in that every exchange transaction is treated as a rendezvous event, while a general reciprocity is achieved when devices help each other during different points in time. Thus, such a difference leads to the requirement that users have to keep memory about the prior transactions so that users can "pay back" each other. Due to this history based feature, reciprocity mechanism suffers from one drawback—a user may not be still in the system when he/she needs to pay back. Such a possible "future loss" may deter a user from genuinely contributing to the community for the fear of not getting deserved pay back. Reputation based systems can be seen as a generalized form of reciprocity. Specifically, while reciprocity is about a particular user pair, a reputation value is a global assessment perceived by all users in the sense that the reputation value is computed by using observations made by many different users. By nature, similar to reciprocity mechanism, reputation systems require substantial memory, centralized or distributed, for recording the reputation values. Thus, it seems that such a system is more suitable for situations where there is a persistent entity, which is logically external to the P2P system, for keeping track of the reputation values. For example, such an entity may be a centralized auctioneer in a electronic auction community, or the base station (access point) in a wireless network. Game theoretic incentive mechanisms are convincing in the sense that the utility function used can usually cover a multitude of important metrics. But the problem is that it is sometimes difficult to achieve an efficient distributed implementation of the resultant protocols.

Perhaps except in a exchange based system which involves "stateless" interactions, all other incentive mechanisms could suffer from tampering or fabrication of the "incentive parameter" used: the money (virtual or real) in a payment based system, the reciprocity metric, and the reputation value. Thus, additional mechanisms, usually based on cryptographic techniques, are needed to guard against such potential malicious attacks to the incentive mechanisms. In particular, whitewashing is widely considered as a very low cost technique for a selfish or malicious user to work around the incentive scheme.

Obviously, there are plenty of room for future research about incentive mechanisms in P2P sharing environments. Most notably, revenue maximizing [364] in a hybrid P2P system (e.g., the so-called converged wireless architecture where an infrastructure based cellular network is tightly coupled with P2P WLANs) is of a high practical interest because we have witnessed more and more cellular subscribers try to share their resources without the intervention of the cellular service provider. In a data sharing environment, server peer selection [204] is another important direction because we believe that users care more about the quality of service achieved than about the revenue or cost they incur in participation. In economics terms, people, especially wired or wireless game players, are quite inelastic about the costs. Nevertheless, energy conservation [206, 205] is still of a prime concern in any wireless P2P sharing network because energy depletion cannot be compensated in any way by increased revenue generated in a payment based sharing system. Thus, perhaps in a game theoretic setting, we should incorporate energy expenditure in the utility function. Finally, topology control [207] in a wired (overlay networks) or wireless (ad hoc networks) is also important in the sense that sharing is usually interest-based, meaning that users naturally form clusters with similar interests, and as such, related users would be more cooperative in following the incentive protocols. Consequently, building an interest-based sharing topology could be helpful in enhancing the effectiveness of sharing.

28.5 SUMMARY

In this chapter, we have presented a detailed survey of incentive techniques for promoting sharing (discrete or continuous data) in a peer-to-peer system. We have considered well known mechanisms that are proposed and have been deployed in the Internet environment. The techniques used can be classified into: payment based, exchange-based, reciprocity, reputation, and game theoretic. These techniques are also applicable in general to a wireless environment. However, a wireless P2P system, while still in its infancy, has a unique challenge—the connectivity among devices is by itself a crucial "sharing" problem (sharing of energy and bandwidth). Much more work needs to be done in related problems such as revenue maximizing (or pricing) in a hybrid wireless P2P system, intelligent peer selection in a data sharing network, energy aware incentive mechanisms, and interest based topology control.

PROBLEMS

28.1   What is the major difference between a barter based incentive scheme and a reputation based incentive scheme?

28.2   What is the key difficulty in implementing a payment based incentive scheme?

28.3   Explain why a wireless P2P system needs incentives in maintaining connectivity among the devices.

APPENDIX A
OPTIMALITY OF MINIMUM DISTANCE
DECODER

In this appendix, we shall prove that minimum distance decoding minimizes the error probability. Consider a demodulator for a M-ary modulator. Let S denote the signal space containing the M constellation points and $\mathcal{R} = \mathcal{R}_1 \bigcup \mathcal{R}_2 \bigcup ... \bigcup \mathcal{R}_M$ be a *decision region* of the demodulator. Note that the decision region $\mathcal{R}$ is in fact a partition of the signal space into M regions where $\mathcal{R}_i \bigcap \mathcal{R}_j = $ (no intersection between regions). Figure A.1 illustrates one example of a decision region.

Given a decision region, the decision rule is to pick $\vec{s}_m$ if the received vector $\vec{y}_n$ falls in region $\mathcal{R}_m$. Hence, any specific decoding scheme can be characterized by a corresponding decision region $\mathcal{R}$. Given that $\vec{s}_1$ is transmitted, the error probability is given by:

$$P_e\left(\vec{s}_1\right) = \int_{R_1^c} p\left(\vec{y}|\vec{s}_1\right) d\vec{y} = 1 - \int_{R_1} p\left(\vec{y}|\vec{s}_1\right) d\vec{y}$$

Wireless Internet and Mobile Computing. By Yu-Kwong Ricky Kwok and Vincent K. N. Lau
Copyright © 2007 John Wiley & Sons, Inc.

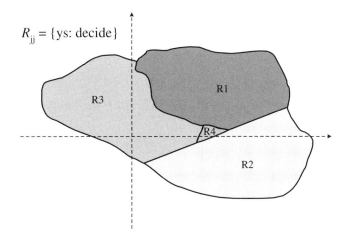

$R_{jj} = \{ys: \text{decide}\}$

Figure A.1 Illustration of decision regions for a digital demodulator with respect to quartenary modulator.

where $\mathcal{R}_1^c = \mathcal{S} \backslash \mathcal{R}_1$. Hence, the average error probability (averaged over all possible transmitted constellation points) is given by:

$$P_e = \sum_{j=1}^{M} p\left(\vec{s}_j\right) \int_{R_j^c} p(\vec{y}|\vec{s}_j) d\vec{y} = 1 - \sum_{j=1}^{M} p\left(\vec{s}_j\right) \int_{R_j} p(\vec{y}|\vec{s}_j) d\vec{y}$$

Since each received vector $\vec{y}$ must belong to one and only one decision region, we should assign $\vec{y}$ to $\mathcal{R}_j$ (for minimizing the average error probability) if $p(\vec{y}|\vec{s}_j)p(\vec{s}_j)$ is maximized. If we assume equi-probable constellations (i.e., $P(\vec{s}_j) = 1/M$), the optimal demodulation rule is given by:

$$\hat{s} = \underset{\vec{s} \in \left\{\vec{s}_1, \dots, \vec{s}_M\right\}}{\arg\max} \; p\left(\vec{y}|\vec{s}\right).$$

This is called the *maximal likelihood demodulation*. Since the channel is AWGN, the received vector is given by:

$$\vec{y} = \vec{s} + \vec{z}$$

and

$$p\left(\vec{y}|\vec{s}\right) = K \exp - \left(\frac{1}{2\sigma_n^2} \left\|\vec{y} - \vec{s}\right\|^2\right)$$

where K is some constant and σ_n^2 is the variance of the channel noise. Hence, the solution of the maximal likelihood decoding becomes:

$$
\begin{aligned}
\hat{s} &= \underset{\vec{s} \in \left\{ \vec{s}_1, \ldots, \vec{s}_M \right\}}{\arg\min} \left\| \vec{y} - \vec{s} \right\|^2 = \underset{\vec{s} \in \left\{ \vec{s}_1, \ldots, \vec{s}_M \right\}}{\arg\max} \left[2\vec{y} \bullet \vec{s} - \left\| \vec{s} \right\|^2 \right] \\
&= \underset{\vec{s} \in \left\{ \vec{s}_1, \ldots, \vec{s}_M \right\}}{\arg\max} \left[2 \int_0^{T_s} y(t)\, s(t)\, dt - \int_0^{T_s} s^2(t)\, dt \right]
\end{aligned}
\tag{A.1}
$$

In other words, the minimum distance decoding is equivalent to the maximal likelihood decoding which is equivalent to the correlator structure as illustrated by the above equation.

REFERENCES

1. Bluetooth Specification Version 1.1. *Specification of the Bluetooth System.* Bluetooth SIG, February 2001.

2. Bluetooth Specification Version 1.2. *Specification of the Bluetooth System.* Bluetooth SIG, November 2003.

3. IrDA Serial Infrared Physical Layer Specification Version 1.4. *Infrared Data Association Serial Infrared Physical Layer Specification.* Infrared Data Association, May 2001.

4. 3GPP. Physical channels and mapping of transport channels onto physical channels (FDD). *3GPP Technical Specifications, TS 25.211*, 2004.

5. 3GPP. Spreading and modulation (FDD). *3GPP Technical Specifications, TS 25.213*, 2004.

6. IEEE Standard 802.11. *Wireless LAN medium access control (MAC) and physical layer (PHY) specifications.* IEEE, 1999.

7. IEEE Standard 802.11a. *Wireless LAN medium access control (MAC) and physical layer (PHY) specifications: High-speed Physical Layer in the 5 GHz Band.* IEEE, April 1999.

8. IEEE Standard 802.11b. *Wireless LAN medium access control (MAC) and physical layer (PHY) specifications: Higher speed physical layer (PHY) extension in the 2.4G Hz band.* IEEE, 1999.

9. IEEE Standard 802.11g. *Wireless LAN medium access control (MAC) and physical layer (PHY) specifications: Further Higher Data Rate Extension in the 2.4 GHz Band*. IEEE, 2003.

10. L. Aalto, N. Göthlin, J. Korhonen, and T. Ojala. Bluetooth and WAP Push Based Location-Aware Mobile Advertising System. In *Proc. MobiSys*, pages 49–58, June 2004.

11. Y. Adam, B. Fillinger, I. Astic, A. Lahmadi, and P. Brigant. Deployment and Test of IPv6 Services in the VTHD Network. *IEEE Communications Magazine*, 42(1):98–104, January 2004.

12. H. Afifi and D. Zeghlache (eds.). *Applications and Services in Wireless Networks*. Kogan Page, 2002.

13. S. Aign and K. Fazel. Temporal & Spatial Error Concealment Techniques for Hierarchical MPEG-2 Video Codec. In *Proc. Globecom*, 1995.

14. I. F. Akyildiz, W. Su, Y. Sankarasubramaniam, and E. Cayirci. Wireless Sensor Networks: A Survey. *Computer Networks*, 38(4):393–422, March 2002.

15. Kostas G. Anagnostakis and Michael B. Greenwald. Exchange-Based Incentive Mechanisms for Peer-to-Peer File Sharing. In *Proc. 24th Int'l Conference on Distributed Computing Systems*, 2004.

16. M. Anand, E. B. Nightingale, and J. Flinn. Self-Tuning Wireless Network Power Management. In *ACM MOBICOM*, 2003.

17. Ross Anderson, Haowen Chan, and Adrian Perrig. Key Infection: Smart Trust for Smart Dust. In *Proc. 12th IEEE Int'l Conf. Network Protocols (ICNP 2004)*.

18. Stephanos Androutsellis-Theotokis and Diomidis Spinellis. A Survey of Peer-to-Peer Content Distribution Technologies. *ACM Computing Surveys*, 36(4):335–371, December 2004.

19. E. Ayanoglu, S. Paul, T. F. LaPorta, K. K. Sabnani, and R. D. Gitlin. AIRMAIL: A link-layer protocol for wireless networks. *ACM Wireless Networks*, 1(1):47–60, February 1995.

20. A. Bakre and B. R. Badrinath. Implementation and Performance Evaluation of Indirect TCP. *IEEE Transactions on Computers*, 46(3):260–278, March 1997.

21. B. Bakshi, P. Krishna, H. H. Vaidya, and D. K. Pradhan. Improving Performance of TCP over Wireless Networks. In *Proc. ICDCS*, pages 365–373, May 1997.

22. H. Balakrishnan, S. Seshan, and R. H. Katz. Improving Reliable Transport and Handoff Performance in Cellular Wireless Networksa. *ACM Wireless Networks*, 1(4):469–481, December 1995.

23. Constantine A. Balanis. *Antenna Theory: Analysis and Design (2nd edition)*. John Wiley and Sons, 1997.

24. J. Bannister, P. Mather, and S. Coope. *Convergence technologies for 3G networks: IP, UMTS, EGPRS, and ATM*. John Wiley & Sons, 2004.

25. C. Barakat and A. Al Fawal. Analysis of Link-level Hybrid FEC/ARQ-SQ for Wireless Links and Long-lived TCP Traffic. *Performance Evaluation*, 57(4):453–476, August 2004.

26. D. Barbará and T. Imieliński. Sleepers and Workaholics: Caching Strategies in Mobile Environments. In *Proc. 1994 SIGMOD International Conference on Management of Data*, pages 1–12, May 1994.

27. L. Barken. *How Secure Is Your Wireless Network? Safeguarding Your Wi-Fi LAN*. Prentice Hall PTR, 2003.

28. S. Basagni, M. Conti, S. Giordano, and I. Stojmenovic. *Mobile Ad Hoc Networking*. John Wiley & Sons, Ltd., 2004.

29. Jan U. Becker and Michel Clement. The Economic Rationale of Offering Media Files in Peer-to-Peer Networks. In *Proc. 37th Hawaii Int'l Conference on System Sciences*, 2004.

30. B. Bellur, R. G. Ogier, and F. L. Templin. Topology Broadcast Based on Reverse-Path Forwarding (TBRPF). *IETF Internet Draft*, 2001.

31. P. Bender, P. Black, M. Grob, R. Padovani, N. Sindhushayana, and A. Viterbi. CDMA/HDR: A Bandwidth Efficient High-Speed Wireless Data Service for Nomadic Users. *IEEE Communications Magazine*, pages 70–88, July 2000.

32. P. Bhagwat, P. Bhattacharya, A. Krishna, and S. K. Tripathi. Using Channel State Dependent Packet Scheduling to Improve TCP Throughput over Wireless LANs. *ACM/Baltzer Wireless Networks*, 3(1):91–102, 1997.

33. V. Bharghavan, S. Lu, and T. Nandagopal. Fair queuing in wireless networks: Issues and approaches. *IEEE Personal Communications*, pages 44–53, February 1999.

34. BitTorrent. http://www.bittorrent.com/, 2006.

35. Erik-Oliver Blaßand Martina Zitterbart. An Efficient Key Establishment Scheme for Secure Aggregating Sensor Networks. In *Proc. ACM ASIACCS 2006*, pages 303–310, March 2006.

36. R. Blom. An Optimal Class of Symmetric Key Generation Systems. In *Proc. Advances in Cryptology (EUROCRYPT'84), Lecture Notes in Computer Science Volume 209*, pages 335–338.

37. B. H. Bloom. Space/Time Trade-offs in Hash Coding with Allowable Errors. *Communications of the ACM*, 13(7), July 1970.

38. C. Blundo, A. De Santis, Amir Herzberg, S. Kutten, U. Vaccaro, , and M. Yung. Perfectly Secure Key Distribution for Dynamic Conferences. In *Proc. Advances in Cryptology (CRYPTO 1992), Lecture Notes in Computer Science Volume 740*, pages 471–486, 1993.

39. K. Boman, G. Horn, P. Howard, and V. Niemi. Umts security. *IEEE Electronics & Communication Engineering Journal*, October 2002.

40. B. Braden, D. Clark, J. Crowcroft, B. Davie, S. Deering, D. Estrin, S. Floyd, V. Jacobson, G. Minshall, C. Partridge, L. Peterson, K. Ramakrishnan, S. Shenker, J. Wroclawski, and L. Zhang. Recommendations on Queue Management and Congestion Avoidance in the Internet. *RFC 2309*, April 1998.

41. J. W. Branch, Jr. N. L. Petroni, L. Van Doorn, and D. Safford. Autonomic 802.11 wireless lan security auditing. *IEEE Security & Privacy Magazine*, May 2004.

42. K. Brown and S. Singh. M-TCP: TCP for Mobile Cellular Networks. *Computer Communications Review*, 27(5):19–43, October 1997.

43. S. Buchegger and J.-Y. Le Boudec. CONFIDANT Protocol: Cooperation of Nodes— Fairness in Dynamic Ad-hoc Networks. In *Proc. IEEE/ACM Symposium on Mobile Ad Hoc Networking and Computing*, 2002.

44. S. Buchegger and J.-Y. Le Boudec. Self-Policing Mobile Ad Hoc Networks by Reputation Systems,. *IEEE Communications Magazine*, 2005.

45. Sonja Buchegger and Jean-Yves Le Boudec. Self-Policing Mobile Ad Hoc Networks by Reputation Systems. *IEEE Communications Magazine*, pages 101–107, July 2005.

46. J. Cai and K. L. Tan. Energy-Efficient Selective Cache Invalidation. *Wireless Networks*, 5(6):489–502, December 1999.

47. Seyit A. Çamtepe and Bülent Yener. Key Distribution Mechanisms for Wireless Sensor Networks: A Survey. In *Technical Report TR-05-07, Department of Computer Science, Rensselaer Polytechnic Institute*, March 2005.

48. G. Cao. A Scalable Low-Latency Cache Invalidation Strategy for Mobile Environments. In *Proc. 6th Annual International Conference on Mobile Computing and Networking*, pages 200–209, August 2000.

49. G. Cao. On Improving the Performance of Cache Invalidation in Mobile Environments. *Mobile Networks and Applications*, 7(4):291–303, August 2002.

50. G. Cao. Proactive Power-Aware Cache Management for Mobile Computing Systems. *IEEE Transactions on Computers*, 51(6):608–621, June 2002.

51. G. Cao. A Scalable Low-Latency Cache Invalidation Strategy for Mobile Environments. *IEEE Transactions on Knowledge and Data Engineering*, 15(5):1251–1265, September 2003.

52. Y. Cao and V. O. K. Li. Scheduling Algorithms in Broadband Wireless Networks. *Proc. IEEE*, 89(1):76–87, January 2001.

53. M. Cardei, I. Cardei, and D. Du (eds.). *Resource Management in Wireless Networks*. Springer, January 2005.

54. Jonathan P. Castro. *All IP in 3G CDMA Networks: The UMTS Infrastructure and Service Platforms for Future Mobile Systems*. John Wiley and Sons, 2004.

55. M. Castro, P. Druschel, A. Kermarrec, A. Nandi, A. Rowstron, and A. Singh. Splitstream: High-bandwidth multicast in cooperative environments. In *Proc. 19th ACM Symposium on Operating Systems Principles*, October 2003.

56. R. Chakravorty, S. Banerjee, P. Rodriguez, J. Chesterfield, and I. Pratt. Performance Optimizations for Wireless Wide-area Networks: Comparative Study and Experimental Evaluation. In *Proc. MOBICOM*, pages 159–173, September 2004.

57. R. Chakravorty, S. Katti, I. Pratt, and J. Crowcroft. Flow Aggregation for Enhanced TCP over Wide-Area Wireless. In *Proc. INFOCOM*, volume 3, pages 1754–1764, March 2003.

58. H. Chan, A. Perrig, and D. Song. Random Key Predistribution Schemes for Sensor Networks. In *Proc. IEEE Sym. Security and Privacy*, May 2003.

59. Haowen Chan and Adrain Perrig. PIKE: Peer Intermediaries for Key Establishment in Sensor Networks. In *Proc. INFOCOM 2005*.

60. Haowen Chan and Adrian Perrig. Security and Privacy in Sensor Networks. *IEEE Computer*, pages 103–105, October 2003.

61. C. H. Chek. *Coexistence Techniques for Heterogeneous Mobile Devices Operating in Uncoordinated Radio Spectrum*. MPhil Thesis, The University of Hong Kong, 2003.

62. Hsiao-Hwa Chen. *Next Generation CDMA Technologies*. John Wiley and Sons, 2007.

63. T.-W. Chen and M. Gerla. Global state routing: a new routing scheme for ad-hoc wireless networks. In *IEEE International Conference on Communications*, 1998.

64. M. X. Cheng, M. Cardei, J. Sun, X. Cheng, L. Wang, Y. Xu, and D. Z. Du. Topology Control of Ad Hoc Wireless Networks for Energy Efficiency. *IEEE Transactions on Computers*, 53(12):1629–1635, December 2004.

65. C. C. Chiang, H. K. Wu, W. Liu, and M. Gerla. Routing in Clustered Multihop Mobile Wireless Networks with Fading Channel. In *Proc. IEEE Singapore International COoference on Networks*, 1997.

66. S. Choi, B. Kim, J. Park, C. Kang, and D. Eom. An implementation of wireless sensor network for security system using bluetooth. *IEEE Transactions on Consumer Electronics*, February 2004.

67. Mooi Choo Chuah and Qinqing Zhang. *Design and Performance of 3G Wireless Networks and Wireless LANs*. Springer Verlag, 2005.

68. I. Clarke, O. Sandberg, B. Wiley, and T. Hong. Freenet: A Distributed Anonymous Information Storage and Retrieval System. In *Proc. Workshop on Design Issues in Anonymous and Unobservability*, July 2000.

69. T. Clausen, P. Jacquet, A. Laouiti, P. Muhlethaler, A. Qayyum, and L. Viennot. Optimized Link State Routing Protocol. In *Proc. IEEE INMIC*, 2001.

70. Bram Cohen. Incentives Build Robustness in BitTorrent. In *Proc. Workshop on Economics of Peer-to-Peer Systems*, June 2003.

71. Atheros Communicatons. 802.11ag: The "clear" choice. *Atheros Communications White Paper*, December 2004.

72. Atheros Communicatons. Atheros extended range xr. *Atheros Communications White Paper*, April 2004.

73. Atheros Communicatons. Super g: Maximizing wireless performance. *Atheros Communications White Paper*, March 2004.

74. C. M. Cordeiro, D. P. Agrawal, and D. H. Sadok. Interference modeling and performance of bluetooth mac protocol. *IEEE Transactions Wireless Communications*, November 2003.

75. Costas Courcoubetis and Richard Weber. Incentives for Large Peer-to-Peer Systems. *IEEE Journal on Selected Areas in Communications*, 24(5):1034–1050, May 2006.

76. D. De Couto, D. Aguayo, J. Bicket, and R. Morris. High-Throughput Path Metric for Multi-Hop Wireless Routing. In *ACM MobiCom*, 2003.

77. Thomas Cover. An achievable rate region for the broadcast channel. *IEEE Transactions on Information Theory*, 31(3):399 – 404, 1975.

78. Thomas Cover and Jay Thomas. *Elements of Information Theory*. John Wiley and Sons, 2006.

79. P. Cuenca, L. Orozco-Barbosa, A. Garrido, D. Quiles, and T. Olivares. A Survey of Error Concealment Schemes for MPEG-2 Video Communications over ATM Network. In *CCECE*, 1997.

80. Yi Cui and Klara Nahrstedt. Layered Peer-to-Peer Streaming. In *Proc. NOSSDAV'03*, pages 162–171, June 2003.

81. G. Davies. *Designing and Developing Scalable IP Networks*. John Wiley and Sons, first edition, 2004.

82. S. Deering and R. Hinden. Internet Protocol, Version 6 (IPv6) Specification. *RFC 1883*, December 1995.

83. S. Deering and R. Hinden. Internet Protocol, Version 6 (IPv6) Specification. *RFC 2460*, December 1998.

84. Farshid Delgosha and Faramarz Fekri. Threshold Key Establishment in Distributed Sensor Networks Using a Multivariate Scheme. In *Proc. IEEE INFOCOM 2006*.

85. J. Deng and Z. J. Hass. Dual Bust Tone Multiple Access (DBTMA). In *ACM MOBICOM*, 2003.

86. K. Dhananjay and N. Guo. Optimal power allocation for the reverse link in a multimedia ds-cdma system. *Proceedings of IEEE Vehicular Technology Conference VTC*, 2:981–984, 1999.

87. S. N. Diggavi, N. Al-Dhahir, A. Stamoulis, and A. R. Calderbank. Great expectations: The value of spatial diversity in wireless networks. *Proceedings of IEEE*, 92(2):219–270, 2004.

88. Tassos Dimitriou and Ioannis Krontiris. A Localized, Distributed Protocol for Secure Information Exchange in Sensor Networks. In *Proc. 19th IEEE Int'l Parallel and Distributed Processing Symposium (IPDPS 2005)*.

89. NTT DoCoMo. http://www.nttdocomo.com/, 2006.

90. R. Draves, J. Padhye, and B. Zill. Comparison of Routing Metrics for Static Multi-Hop Wireless Networks. In *ACM SIGCOMM*, 2004.

91. R. Draves, J. Padhye, and B. Zill. Routing in Multi-radio, Multi-hop Wireless Mesh Networks. In *ACM MobiCom*, 2004.

92. Wenliang Du, Jing Deng, Yunghsiang S. Han, Shigang Chen, and Pramod K. Varshney. A Key Management Scheme for Wireless Sensor Networks Using Deployment Knowledge. In *Proc. INFOCOM 2004*, pages 586–597.

93. Wenliang Du, Jing Deng, Yunghsiang S. Han, and Pramod K. Varshney. A Pairwise Key Predistribution Scheme for Wireless Sensor Networks. In *Proc. 10th ACM Conf. Computer and Communications Security (CCS 2003)*, pages 42–51, October 2003.

94. Wenliang Du, Jing Deng, Yunghsiang S. Han, and Pramod K. Varshney. A Key Predistribution Scheme for Sensor Networks Using Deployment Knowledge. *IEEE Transactions on Dependable and Secure Computing*, 3(1):62–77, Jan.-Mar. 2006.

95. Wenliang Du, Jing Deng, Yunghsiang S. Han, Pramod K. Varshney, Jonathan Katz, and Aram Khalili. A Pairwise Key Predistribution Scheme for Wireless Sensor Networks. *ACM Transactions on Information and System Security*, 8(2):228–258, May 2005.

96. V. A. Dubendorf. *Wireless data technologies*. John Wiley & Sons, 2003.

97. V. A. Dubendorf. *Wireless Data Technologies*. John Wiley & Sons, 2003.

98. R. E. Van Dyck and D. J. Miller. Transport of wireless video using separate, concatenated and joint source-channel coding. *Proc. IEEE*, October 1999.

99. Joerg Eberspaecher, Hans-Joerg Voegel, and Christian Bettstetter. *GSM Switching, Services, and Protocols*. John Wiley and Sons, 2001.

100. D. A. Eckhardt and P. Steenkiste. Effort-Limited Fair (ELF) Scheduling for Wireless Networks. In *Proc. INFOCOM*, pages 1097–1106, 2000.

101. Elias C. Efstathiou and George C. Polyzos. A Peer-to-Peer Approach to Wireless LAN Roaming. In *Proc. WMASH*, pages 10–18, September 2003.

102. M. Eltoweissy, M. Moharrum, and R. Mukkamala. Dynamic Key Management in Sensor Networks. *IEEE Communications Magazine*, 44(4):122–130, April 2006.

103. P. Erdös and A. Rényi. On the Evolution of Random Graphs. In *Institute of Mathematics Hungarian Academy of Sciences*, 1959.

104. O. Escalante, T. Perez, J. Solano, and I. Stojmenovic. RNG-Based Searching and Broadcasting Algorithms over Internet Graphs and Peer-To-Peer Computing Systems. In *Proc. 3rd ACS/IEEE International Conference on Computer Systems and Applications*, pages 17–24, January 2005.

105. L. Eschenauer and V. Gligor. A Key Management Scheme for Distributed Sensor Networks. In *Proc. 9th ACM Conf. Computer and Communications Security (CCS 2002)*, pages 41–47, November 2002.

106. R. Dube et al. Signal Stability based Adaptive Routing (SSA) for Ad Hoc Mobile Networks. *IEEE Personal Communication Magazine*, 1997.

107. ETSI. Telecommunications and Internet Protocol Harmonization over Networks (TIPHON): Description of Technical Issues. *ETSI Technical Report 101 300*, October 1999.

108. ETSI. Broadband Radio Access Networks (BRAN): HIPERLAN Type 2: Data Link Control (DLC) Layer: Part 1: Basic Data Transport Functions. *ETSI Technical Specifications*, 2001.

109. E.V. Eyuboglu and S.U. Qureshi. Reduced state sequence estimation with set partitioning and decision feedback. *IEEE Transactions on Communications*, 38:13–20, 1998.

110. F. LING F, R. LOVE, and H. XU. Behavior and performance of power controlled is-95 reverse link under soft handoff. *IEEE Transactions on Vehicular Technology*, 49(9):1697–1704, 1999.

111. L. M. Feeney. An Energy Consumption Model for Performance Analysis of Routing Protocols for Mobile Ad Hoc Networks. *Mobile Networks and Applications*, 6(3):239–249, June 2001.

REFERENCES

112. L. M. Feeney and M. Nilsson. Investigating the Energy Consumption of a Wireless Network Interface in an Ad Hoc Networking Environment. In *Proc. 20th International Conference on Computer Communications*, volume 3, pages 1548–1557, April 2001.

113. Michal Feldman and John Chuang. Overcoming Free-Riding Behavior in Peer-to-Peer Systems. *ACM SIGccom Exchanges*, 5(4):41–50, July 2005.

114. Michal Feldman, Kevin Lai, Ion Stoica, and John Chuang. Robust Incentive Techniques for Peer-to-Peer Networks. In *Proc. 5th ACM conference on Electronic Commerce*, pages 102–111, May 2004.

115. Michal Feldman, Christos Papadimitriou, John Chuang, and Ion Stoica. Free-Riding and Whitewashing in Peer-to-Peer Systems. In *Proc. 2004 SIGCOMM Workshop on Practice and Theory of Incentives in Networked Systems*, page 228–235, August 2004.

116. Mark Felegyhazi, Jean-Pierre Hubaux, and Levente Buttyan. Nash Equilibria of Packet Forwarding Strategies in Wireless Ad Hoc Networks. *IEEE Transactions on Mobile Computing*, 5(5):463–476, May 2006.

117. W. Feng, K. S. Shin, D. D. Kandlur, and D. Saha. The BLUE Active Queue Management Algorithms. *IEEE/ACM Transactions on Networking*, 10(4):513–528, August 2002.

118. R. Fielding, J. Gettys, J. Mogul, H. Frystyk, and T. Berners-Lee. Hypertext Transfer Protocol—HTTP/1.1. *RFC 2068*, January 1997.

119. Daniel Figueiredo, Jonathan Shapiro, and Don Towsley. Incentives to Promote Availability in Peer-to-Peer Anonymity Systems. In *Proc. 13th IEEE Int'l Conference on Network Protocols*, 2005.

120. S. Floyd. Connections with Multiple Congested Gateways in Packet-Switched Networks Part 1: One-Way Traffic. *ACM Computer Communications Review*, 21(5):30–47, October 1991.

121. S. Floyd. TCP and Explicit Congestion Notification. *ACM Computer Communications Review*, 24(5):10–23, October 1994.

122. S. Floyd and T. Henderson. The NewReno Modification to TCP¡ls Fast Recovery Algorithm. *RFC 2582*, April 1999.

123. S. Floyd and V. Jacobson. Random Early Detection Gateways for Congestion Avoidance. *IEEE/ACM Transactions on Networking*, 1(4):397–413, August 1993.

124. S. Floyd, J. Mahdavi, M. Mathis, and M. Podolsky. An Extension to the Selective Acknowledgement (SACK) Option for TCP. *RFC 2883*, July 2000.

125. A. Ganz, Z. Ganz, and K. Wongthavarawat. *Multimedia Wireless Networks: Technologies, Standards, and QoS*. Prentice Hall, 2003.

126. A. Ganz, Z. Ganz, and K. Wongthavarawat. *Multimedia Wireless Networks: Technologies, Standards, and QoS*. Prentice Hall PTR, 2003.

127. Vijay K. Garg. *IS-95 CDMA and cdma 2000: Cellular/PCS Systems Implementation*. Prentice Hall, 1999.

128. M. S. Gast. *802.11 Wireless Networks*. O'Reilly, 2002.

129. Matthew Gast and Matthew S. Gast. *802.11 Wireless Networks: The Definitive Guide*. O'Reilly, 2002.

130. C. Gehrmann, J. Persson, and B. Smeets. *Bluetooth Security*. Artech House, 2004.

131. P. R. Gejji. Forward-link-power control in cdma cellular systems. *Proceedings of IEEE Vehicular Technology Conference VTC*, 2:981–984, 1992.

132. Wolfgang H. Gerstacker, Frank Obernosterer, Robert Schober, Alexander Lehmann, Alexander Lampe, and Peter Gunreben.   Widely linear receivers for space-time block coded transmission over frequency-selective fading channels.   *url = "citeseer.ist.psu.edu/459853.html*.

133. S. Ghandeharizadeh and T. Helmi. Consistency and Replication: An Evaluation of Alternative Continuous Media Replication Techniques in Wireless Peer-To-Peer Networks. In *Proc. 3rd ACM International Workshop on Data Engineering for Wireless and Mobile Access*, 2003.

134. Gnutella. http://www.gnutella.com/, 2006.

135. Gerardo Gomez and Rafael Sanchez. *End-to-End Quality of Service over Cellular Networks: Data Services Performance Optimization in 2G/3G*. John Wiley and Sons, 2005.

136. O. Granberg. GSM on the Net. *Ericsson Review*, 4:184–191, 1998.

137. J. Gruber and L. Strawczynski. Subjective Effects of Variable Delay and Speech Clipping in Dynamically Managed Voice Systems. *IEEE Transactions on Communications*, 8:801–808, August 1985.

138. Rohit Gupta and Arun K. Somani. A Pricing Strategy For Incentivizing Selfish Nodes To Share Resources In Peer-to-Peer (P2P) Networks. In *Proc. IEEE International Conference on Networks*, November 2004.

139. Ahsan Habib and John Chuang. Service Differentiated Peer Selection: An Incentive Mechanism for Peer-to-Peer Media Streaming. *IEEE Transactions on Multimedia*, 8(3):601–621, June 2006.

140. A. Hac. *Mobile Telecommunications Protocols for Data Networks*. John Wiley & Sons, Ltd., 2003.

141. Timo Halonen, Javier Romero, and Juan Melero. *GSM, GPRS and EDGE Performance: Evolution Towards 3G/UMTS*. John Wiley and Sons, 2003.

142. T. Hamann and J. Walrand. A New Fair Window Algorithm for ECN Capable TCP (New-ECN). In *Proc. INFOCOM*, volume 3, pages 1528–1536, March 2000.

143. Lawrence Harte. *Introduction to GSM: Physical Channels, Logical Channels, Network, and Operation*. Althos, 2004.

144. Lawrence Harte, Morris Hoenig, Daniel McLaughlin, and Roman Kikta. *CDMA IS-95 for Cellular and PCS: Technology, Applications, and Resource Guide*. McGraw Hill, 1999.

145. David Hausheer, Nicolas C. Liebau, Andreas Mauthe, Ralf Steinmetz, and Burkhard Stiller. Token Based Accounting and Distributed Pricing to Introduce Market Mechanisms in a Peer-to-Peer File Sharing Scenario. In *Proc. Third International Conference on Peer-to-Peer Computing*, 2003.

146. David Hausheer and Burkhard Stiller. Decentralized Auction-based Pricing with Peer-Mart. In *Proc. 9th IFIP/IEEE International Symposium on Integrated Network Management*, May 2005.

147. Simon Haykin and Michael Moher. *An Introduction to Digital and Analog Communications*. John Wiley and Sons, 2001.

148. M. Hefeeda, A. Habib, B. Botev, D. Xu, and B. Bhargava. PROMISE: Peer-to-Peer Media Streaming Using CollectCast. In *Proc. ACM Multimedia*, November 2003.

149. R. Hinden and S. Deering. Internet Protocol Version 6 (IPv6) Addressing Architecture. *RFC 3513*, April 2003.

150. J. C. Hoe. Improving the Start-Up Behavior of a Congestion Control Scheme for TCP. In *Proc. ACM SIGCOMM*, pages 270–280, August 1996.

151. H. Holma and A. Toskala (eds.). *WCDMA for UMTS: Radio access for third generation mobile communications*. John Wiley & Sons, third edition, 2004.

152. Harri Holma and Antti Toskala. *WCDMA for UMTS: Radio Access for Third Generation Mobile Communications*. John Wiley and Sons, 2001.

153. Hung-Yu Hsieh and Raghupathy Sivakumar. On Using Peer-to-Peer Communication in Cellular Wireless Data Networks. *IEEE Transactions on Mobile Computing*, 3(1):57–72, Jan.-Mar. 2004.

154. J. H. Hu and K. L. Yeung. FDA: A Novel Base Station Flow Control Scheme for TCP over Heterogeneous Networks. In *Proc. INFOCOM*, volume 1, pages 142–151, April 2001.

155. J. H. Hu, K. L. Yeung, S. C. Kheong, and G. Feng. Hierarchical Cache Design for Enhancing TCP over Heterogeneous Networks with Wired and Wireless Links. In *Proc. GLOBECOM*, volume 1, pages 338–343, November 2000.

156. Q. Hu and D. L. Lee. Adaptive Cache Invalidation Methods in Mobile Environments. In *Proc. 6th IEEE International Symposium on High Performance Distributed Computing*, pages 264–273, August 1997.

157. Yih-Chun Hu, Adrain Perrig, and David B. Johnson. ARIADNE: A Secure On-Demand Routing Protocol for Ad Hoc Networks. In *Proc. 8th Int'l Conf. Mobile Computing and Networking (MOBICOM 2002)*, pages 12–23, 2002.

158. Yang hua Chu and Hui Zhang. Considering Altruism in Peer-to-Peer Internet Streaming Broadcast. In *Proc. NOSSDAV'04*, pages 10–15, June 2004.

159. Dijiang Huang, Manish Mehta, Deep Medhi, and Lein Harn. Location Aware Key Management Scheme for Wireless Sensor Networks. In *Proc. ACM SASN 2004*, pages 29–42, October 2004.

160. Qiang Huang, Johnas Cukier, Hisashi Kobayashi, Bede Liu, and Jinyun Zhang. Fast Authenticated Key Establishment Protocols for Self-Organizing Sensor Networks. In *Proc. WSNA 2003*, pages 141–150, September 2003.

161. Joengmin Hwang and Yongdae Kim. Revisiting Random Key Predistribution Schemes for Wireless Sensor Networks. In *Proc. ACM SASN 2004*, pages 43–52, October 2004.

162. Omer Ileri, Siun-Chuon Mau, and Narayan B. Mandayam. Pricing for Enabling Forwarding in Self-Configuring Ad Hoc Networks. *IEEE Journal on Selected Areas in Communications*, 23(1):151–162, January 2005.

163. Takashi Ito, Hidenori Ohta, Nori Matsuda, and Takeshi Yoneda. A Key Predistribution Scheme for Secure Sensor Networks Using Probability Density Function of Node Deployment. In *Proc. ACM SASN 2005*, pages 69–75, November 2005.

164. V. Jacobson. Congestion Avoidance and Control. In *Proc. ACM SIGCOMM*, pages 314–329, August 1988.

165. V. Jacobson. Modified TCP Congestion Avoidance Algorithm. *Technical Report 30*, April 1990.

166. A. Jagoe. *Mobile Location Servies: The Definitive Guide*. Prentice Hall PTR, December 2002.

167. William C. Jakes. *Microwave Mobile Communications*. John Wiley and Sons, 1975.

168. A. Jalali, R. Padovane, and R. Pankaj. Data Throughput of CDMA-HDR: A High Efficiency High Data Rate Personal Communication Wireless System. In *Proc. VTC*, 2000.

169. Zou Jia, Si Tiange, Huang Liansheng, and Dai Yiqi. A New Micropayment Protocol Based on P2P Networks. In *Proc. 2005 IEEE Int'l Conference on e-Business Engineering*, 2005.

170. M. Jiang, J. Li, and Y. C. Tay. Cluster Based Routing Protocol,. *IETF Internet Draft*, 1999.

171. J. Jing, A. Elmagarmid, A. S. Helal, and R. Alonso. Bit-Sequences: An Adaptive Cache Invalidation Method in Mobile Client/Server Environments. *Mobile Networks and Applications*, 2(2):115–127, October 1997.

172. D. Johnson, C. Perkins, and J. Arkko. Mobility Support in IPv6. *RFC 3775*, June 2004.

173. D. B. Johnson and D. A. Maltz. Dynamic Source Routing in Ad Hoc Wireless Network. *Mobile Computing*, 1996.

174. Gaurav Jolly, Mustafa C. Kuscu, Pallavi Kokate, and Mohamed Younis. A Low Energy Key Management Protocol for Wireless Sensor Networks. In *Proc. 8th IEEE Int'l Symposium on Computers and Communications*, 2003.

175. Seung Jun and Mustaque Ahamad. Incentives in BitTorrent Induce Free Riding. In *Proc. ACM SIGCOMM 2005 Workshop*, August 2005.

176. Seung Jun, Mustaque Ahamad, and Jun (Jim) Xu. Robust Information Dissemination in Uncooperative Environments. In *Proc. 25th IEEE Int'l Conference on Distributed Computing Systems*, 2005.

177. E. Jung and N. H. Vaidya. A Power Control MAC Protocol for Ad Hoc Networks. In *ACM MOBICOM*, 2002.

178. E. Jung and N. H. Vaidya. An Energy Efficient MAC Protocol for Wireless LANs. In *IEEE INFOCOM*, 2002.

179. J. M. Kahn, R. H. Katz, and K. S. Pister. Mobile Networking for Smart Dust. In *Proc. Int'l Conf. Mobile Computing and Networking (MOBICOM 1999)*, August 1999.

180. A. Kahol, S. Khurana, S. K. S. Gupta, and P. K. Srimani. A Strategy to Manage Cache Consistency in a Disconnected Distributed Environment. *IEEE Transactions on Parallel and Distributed Systems*, 12(7):686–700, July 2001.

181. Seung-Seok Kang and Matt W. Mutka. A Mobile Peer-to-Peer Approach for Multimedia Content Sharing Using 3G/WLAN Dual Mode Channels. *Wireless Communications and Mobile Computing*, 5:633–645, 2005.

182. C. Karlof and D. Wagner. Secure Routing in Wireless Sensor Networks: Attacks and Countermeasures. In *Proc. 1st IEEE Int'l Workshop on Sensor Network Protocols and Applications*, pages 113–127, May 2003.

183. P. Karn and C. Partridge. Improving Round-Trip Time Estimates in Reliable Transport Protocols. In *Proc. ACM SIGCOMM*, pages 2–7, August 1987.

184. B. Karp and H. T. Kung. GPSR: Greedy Perimeter Stateless Routing for Wireless Networks. In *Proc. 6th Int'l Conf. Mobile Computing and Networking (MOBICOM 2000)*, pages 243–254, August 2000.

185. H. Kawai, H. Suda, and F. Adachi. Outer-loop control of target sir for fast transmit power control inturbo-coded w-cdma mobile radio. *Electronics Letters*, 35:699–701, 1999.

186. R. Keenan. Caching technique eases wlan roaming, available at: http://www.commsdesign.com/news/tech_beat/showarticle.jhtml?articleid=26806681/, 2006.

187. F. Kelly. Charging and Rate Control for Elastic Traffic. *European Transactions on Telecommunications*, 8:33–37, 1997.

188. S. Keshav and S. P. Morgan. SMART Retransmission: Performance with Overload and Random Losses. In *Proc. INFOCOM*, volume 3, pages 1131–1138, April 1997.

189. H Kim, S Bang, and Y Han. Performance of fast forward power control using coherent sir estimation. *Proceedings of 4th CIC Conference*, 2:981–984, 1999.

190. Kyoung Il Kim. *Handbook of CDMA System Design, Engineering and Optimization*. Prentice Hall, 1999.

191. J. J. Kistler and M. Satyanarayanan. Disconnected operation in the coda file system. *ACM Transactions on Computer Systems*, Feburary 1992.

192. J. Korhonen. *Introduction to 3G mobile communications*. Artech House, second edition, 2003.

193. Ramayya Krishnana, Michael D. Smith, and Rahul Telang. The Economics of Peer-to-Peer Networks. *Journal of Information Technology Theory and Application*, 5(3):31–44, 2003.

194. J. F. Kurose and K. W. Ross. *Computer Networking: A Top-Down Approach Featuring the Internet*. Addison-Wesley, third edition, 2005.

195. Y. K. Kwok and V. K. N. Lau. A Novel Channel-Adaptive Uplink Access Control Protocol for Nomadic Computing. *IEEE Transactions on Parallel and Distributed Systems*, 13(11):1150–1165, November 2002.

196. M.A. Lagunas, A. I. Neira, M. G. Amin, and J. Vidal. Spatial processing for frequency diversity schemes. *IEEE Transactions on Signal Processing*, 48(2):353–362, 2000.

197. T. V. Lakshman, A. Ortega, and A. R. Reibman. VBR Video: Tradeoffs and Potentials. *Proc. IEEE*, 86(5):952–973, May 1998.

198. V. K. N. Lau. Performance of Variable Rate Bit-Interleaved Coding for High Bandwidth Efficiency. In *Proc. VTC*, volume 3, pages 2054–2058s, May 2000.

199. Chun Fai Law, Ka Shun Hung, and Yu-Kwong Kwok. A Key Redistribution Scheme for Distributed Sensor Networks. In *Technical Report HKUEEE-2006-12, The University of Hong Kong*, October 2006.

200. Yee Wei Law, Ricardo Corin, Sandro Etalle, and Pieter H. Hartel. A Formally Verified Decentralized Key Management Architecture for Wireless Sensor Networks. In *Technical Report TR-CTIT-03-07 Centre for Telematics and Information Technology, University of Twente, Enschede*, 2003.

201. Jooyoung Lee and Dou glas R. Stinson. Deterministic Key Predistribution Schemes for Distributed Sensor Networks. In *Proc. SAC 2004, Lecture Notes in Computer Science Volume 3357*, pages 294–307.

202. William C. Y. Lee. *Mobile Cellular Telecommunications: Analog and Digital Systems*. McGraw Hill, 1995.

203. William C. Y. Lee. *Lee's Essentials of Wirelesss Communications*. McGraw Hills, 2000.

204. Andrew Ka Ho Leung and Yu-Kwong Kwok. An Efficient and Practical Greedy Algorithm for Server-Peer Selection in Wireless Peer-to-Peer File Sharing Networks. In *Proceedings of the International Conference on Mobile Ad-hoc and Sensor Networks (MSN'2005), (Lecture Notes in Computer Science, Volume 3794, Xiaohua Jia, Jie Wu, Yanxiang He (eds.), Springer Berlin/Heidelberg)*, pages 1016–1025, December 2005.

205. Andrew Ka Ho Leung and Yu-Kwong Kwok. Community-Based Asynchronous Wakeup Protocol for Wireless Peer-to-Peer File Sharing Networks. In *Proceedings of the IEEE Second Annual International Conference on Mobile and Ubiquitous Systems: Networking and Services (MobiQuitous'2005)*, pages 342–350, July 2005.

206. Andrew Ka Ho Leung and Yu-Kwong Kwok. Energy Conservation by Peer-to-Peer Relaying in Quasi-Ad Hoc Networks. In *Proceedings of IFIP International Conference on Network and Parallel Computing (NPC'2005), (Lecture Notes in Computer Science, Volume 3779, Hai Jin, Daniel Reed, Wenbin Jiang (eds.), Springer Berlin/Heidelberg)*, pages 45l–460, November 2005.

207. Andrew Ka Ho Leung and Yu-Kwong Kwok. On Topology Control of Wireless Peer-to-Peer File Sharing Networks: Energy Efficiency, Fairness and Incentive. In *Proceedings of the IEEE International Symposium on a World of Wireless, Mobile and Multimedia Networks (WoWMoM'2005)*, pages 318–323, June 2005.

208. K. H. Leung. *Localized Topology Control in Wireless Peer-to-Peer File Sharing Networks*. MPhil Thesis, The University of Hong Kong, 2005.

209. P. Levis, S. Madden, J. Polastre, R. Szewczyk, K. Whitehous, A. Woo, D. Gay, J. Hill, M. Welsh, E. Brewer, and D. Culler. TinyOS: An Operating System for Sensor Networks. In *Ambient Intelligence (W. Weber, JM Rabaey, and E. Aarts)*. Springer Verlag, 2005.

210. Chenyang Li and Sumit Roy. Performance of Frequency-Time MMSE Equalizer for MC-CDMA over Multipath Fading Channel. *Wireless Personal Communications*, 18:179–192, 2001.

211. J. Li, D. Cordes, and J. Zhang. Power-aware routing protocols in ad hoc wireless networks. *IEEE Wireless Communications*, December 2005.

212. L. Li and J. Y. Halpern. A Minimum-Energy Path-Preserving Topology Control Algorithm. *IEEE Transactions on Wireless Communications*, 3(3):910–921, May 2004.

213. L. Li, H. B. Li, and Y. D. Yao. Time diversity and equalization for frequency selective fading channels. *Proceedings of IEEE Vehicular Technology Conference (VTC)*, 92(2):1673–1678, 2001.

214. X. Y. Li, Y. Wang, and W.-Z. Song. Applications of k-Local MST for Topology Control and Broadcasting in Wireless Ad Hoc Networks. *IEEE Transactions on Parallel and Distributed Systems*, 15(12):1057–?1069, December 2004.

215. P. Lin, B. Bensaou, Q. L. Ding, and K. C. Chua. A Wireless Fair Scheduling Algorithm for Error-Prone Wireless Channels. In *Proc. WoWMoM*, pages 11–20, 2000.

216. X. Lin, Y. Kwok, and V. K. N. Lau. Power Control for IEEE 802.11 Ad Hoc Networks: Issues and A New Algorithm. In *IEEE ICPP*, 2003.

217. Yi Bing Lin. *Wireless and Mobile Network Architectures*. John Wiley and Sons, 2001.

218. C. Liu and R. Jain. Delivering Faster Congestion Feedback with the Mark-Front Strategy. In *Proc. WCC-ICCT*, volume 1, pages 665–672, August 2000.

219. Chia-Liang Liu and Kamilo Feher. Pilot-symbol aided coherent m-ary psk in frequency-selective fast rayleigh fading channels. *IEEE Transactions on Communications*, 42(1):54–62, 1994.

220. Donggang Liu and Peng Ning. Establishing Pairwise Keys in Distributed Sensor Networks. In *Proc. 10th ACM Conf. Computer and Communications Security (CCS 2003)*, pages 52–61, October 2003.

221. Donggang Liu and Peng Ning. Location Based Pairwise Key Establishments for Static Sensor Networks. In *Proc. 1st ACM Workshop on Security of Ad Hoc and Sensor Networks*, pages 72–82, 2003.

222. Donggang Liu and Peng Ning. Improving Key Predistribution with Deployment Knowledge in Static Sensor Networks. *ACM Transactions on Sensor Networks*, 1(2):204–239, November 2005.

223. Donggang Liu, Peng Ning, and Wenliang Du. Group Based Key Predistribution in Wireless Sensor Networks. In *Proc. ACM WiSE 2005*, pages 11–20, September 2005.

224. Donggang Liu, Peng Ning, and Rongfang Li. Establishing Pairwise Keys in Distributed Sensor Networks. *ACM Transactions on Information and System Security*, 8(1):41–77, February 2005.

225. Fang Liu, Major Jose "Many" Rivera, and Xiuzhen Cheng. Location Aware Key Establishment in Wireless Sensor Networks. In *Proc. ACM IWCMC 2006*, pages 21–26, July 2006.

226. Su-Lin Low and Ron Schneider. *CDMA Internetworking: Deploying the Open A-Interface*. Prentice Hall, 2000.

227. S. Lu, V. Bharghavan, and R. Srikant. Fair Scheduling in Wireless Packet Networks. *ACM Transactions on Networking*, 7(4):473–489, August 1999.

228. Richard T. B. Ma, Sam C. M. Lee, John C. S. Lui, and David K. Y. Yau. A Game Theoretic Approach to Provide Incentive and Service Differentiation in P2P Networks. In *Proc. SIGMETRICS*, pages 189–198, June 2004.

229. Richard T. B. Ma, Sam C. M. Lee, John C. S. Lui, and David K. Y. Yau. An Incentive Mechanism for P2P Networks. In *Proc. 24th Int'l Conference on Distributed Computing Systems*, 2004.

230. W. Ma and Y. Fang. Dynamic Hierarchical Mobility Management Strategy for Mobile IP Networks. *IEEE Journal on Selected Areas in Communications*, 22(4):664–676, May 2004.

231. David J. Malan, Matt Welsh, and Michael D. Smith. A Public-Key Infrastructure for Key Distribution in TinyOS Based on Elliptic Curve Cryptography. In *Proc. First IEEE International Conference on Sensor and Ad Hoc Communications and Networks*, October 2004.

232. Giridhar D. Mandyam and Jersey Lai. *Third Generation CDMA Systems for Enhanced Data Services*. Academic Press, 2006.

233. Z. Mao and C. Douligeris. A distributed database architecture for global roaming in next-generation mobile networks. *IEEE/ACM Transactions on Networking*, Feburary 2004.

234. Peter Marbach and Ying Qiu. Cooperation in Wireless Ad Hoc Networks: A Market Based Approach. *IEEE/ACM Transactions on Networking*, 13(6):1325–1338, December 2005.

235. S. Marti, T.J. Giuli, K. Lai, and M. Baker. Mitigating Routing Misbehavior in Mobile Ad Hoc Networks. In *ACM MobiCom*, 2000.

236. M. Mathis, J. Mahdavi, S. Floyd, and A. Romanow. TCP Selective Acknowledgment Options. *RFC 2018*, October 1996.

237. M. Maxim and D. Pollino. *Wireless Security*. McGraw-Hill, 2002.

238. Maze. http://maze.pku.edu.cn, 2006.

239. Asha Mehrotra. *GSM System Engineering*. Artech House Publishers, 1997.

240. R. C. Merkle. A Certified Digital Signature. In *Advances in Cryptology (CRYPTO 1989)*, August 1989.

241. Mathhew J. Miller and Nitin H. Vaidya. Leveraging Channel Diversity for Key Establishment in Wireless Sensor Networks. In *Proc. IEEE INFOCOM 2006*.

242. S. S. Miller. *WI-FI Security*. McGraw-Hill, 2003.

243. Ajay R. Mishra. *Advanced Cellular Network Planning and Optimisation: 2G/2.5G/3G...Evolution to 4G*. John Wiley and Sons, 2007.

244. WAP User Agent Caching Model. http://www.openmobilealliance.org/release_program/browsing_archive.html, 2006.

245. J. W. Modestino and D. G. Daut. Combined source-channel coding of images. *IEEE Transactions on Communications*, November 1979.

246. J. W. Modestino, D. G. Daut, and A. L. Vickers. Combined source-channel coding of images using the block consine transform. *IEEE Transactions on Communications*, September 1981.

247. P. Mohapatra and S. V. Krishnamurthy. *Ad Hoc Networks Technologies and Protocols*. Springer, 2005.

248. J. P. Monks, V. Bharghavan, W. Mei, and W. Hwu. A Power Controlled Multiple Access Protocol for Wireless Packet Networks. In *IEEE INFOCOM*, 2001.

249. L. B. Mummert, M. R. Ebling, and M. Satyanarayanan. Exploiting Weak Connectivity for Mobile File Access. In *Proc. 15th ACM Symposium on Operating Systems Principles*, December 1995.

250. Y. Mun and H. K. Lee. *Understanding IPv6*. Springer, first edition, 2005.

251. A. Muqattash and M. M. Krun. A Distributed Transmission Power Control Protocol for Mobile Ad Hoc Networks. *IEEE Transactions on Mobile Computing*, 3(2):113–?128, April 2004.

252. S. Murthy and J. J. Garcia-Luna-Aceves. A Highly Adaptive Distributed Routing Algorithm for Mobile Wireless Networks. In *Proc. ACM First International Conference on Mobile Computing & Networking (MobiCom)*, 1995.

253. S. Nanda, R. Ejzak, and B. T. Doshi. A Retransmission Scheme for Circuit-Mode Data on Wireless Links. *IEEE Journal on Selected Areas in Communications*, 12(8):1338–1352, October 1994.

254. T. Nandagopal, S. Lu, and V. Bharghavan. A Unified Architecture for the Design and Evaluation of Wireless Fair Queueing Algorithms. In *Proc. MOBICOM*, pages 132–142, 1999.

255. Napster. http://www.napster.com/, 2006.

256. Mesh Networking. http://research.microsoft.com/mesh/, 2006.

257. T. S. E. Ng, I. Stoica, and H. Zhang. Packet Fair Queueing Algorithms for Wireless Networks with Location-Dependent Errors. In *Proc. INFOCOM*, pages 1103–1111, 1998.

258. NOT-YET-ENTERED2. NOT-YET-ENTERED2. *NOT-YET-ENTERED2*, 2006.

259. NOT-YET-ENTERED3. NOT-YET-ENTERED3. *NOT-YET-ENTERED3*, 2006.

260. T. Ojanperä and R. Prasad. An overview of air interface multiple access for IMT-2000/UMTS. *IEEE Communications Magazine*, 36(9):82–86,91–95, September 1998.

261. Y. Okumura. Field strength and its variability in vhf and uhf land-mobile radio service. *Review of the Electrical Communication Laboratories*, 16(9):825–873, 1968.

262. Open Mobile Alliance (OMA). http://www.openmobilealliance.org/, 2006.

263. OnStar. http://www.onstar.com/us_english/jsp/index.jsp, 2006.

264. L. Ophir and Y. Bitran. 802.11 Over Coax—A Hybrid Coax-Wireless Home Network Using 802.11 Technology . In *Proc. Consumer Communications and Networking Conference*, January 2004.

265. Martin J. Osborne. *An Introduction to Game Theory*. Oxford University Press, 2004.

266. Connecting IPv6 Islands over IPv4 MPLS using IPv6 Provider Edge Routers (6PE). http://www.ietf.org/internet-drafts/draft-ooms-v6ops-bgp-tunnel-06.txt/, 2006.

267. Teredo Overview. http://www.microsoft.com/technet/prodtechnol/winxppro/maintain/teredo.mspx, 2006.

268. Taejoon Park and Kang G. Shin. LiSP: A Lightweight Security Protocol for Wireless Sensor Networks. *ACM Transactions on Embedded Computing Systems*, 3(3):634–660, August 2004.

269. V. Park and M. Scott Corson. A Highly Adaptive Distributed Routing Algorithm for Mobile Wireless Networks. In *Proc. INFOCOM*, 1996.

270. J. D. Parsons. *The Mobile Radio Propagation Channel*. John Wiley & Sons, second edition, 2000.

271. M. Patzold. *Mobile Fading Channels: Modelling, Analysis, and Simulation*. John Wiley and Sons, 2002.

272. M. R. Pearlman and Z. Haas. Determining the Optimal Configuration for the Zone Routing Protocol. *IEEE Journal on Selected Areas in Communications*, August 1999.

273. G. Pei, M. Gerla, and T.-W. Chen. Fisheye State Routing: A Routing Scheme for Ad Hoc Wireless Networks. In *Proc. IEEE International Conference on Communications*, 2000.

274. Jonathan Leary Pejman Roshan. *Wireless Local-Area Network Fundamentals*. Cisco Press, 2003.

275. F. Peng, S. Cheng, and J. Ma. An Effective Way to Improve TCP Performance in Wireless/Mobile Networks. In *Proc. EUROCOMM*, volume 3, pages 250–255, May 2000.

276. C. E. Perkins. Mobile IP. *IEEE Communications Magazine*, pages 84–99, May 1997.

277. C. E. Perkins and P. Bhagwat. Highly Dynamic Destination-Sequenced Distance Vetcor Routing. In *Proc. ACM SIGCOMM*, 1994.

278. C. E. Perkins and E. M. Royer. Ad-Hoc On-Demand Distance Vector Routing. In *Proc. 2nd IEEE Workshop on Mobile Computing Systems and Applications*, 1999.

279. A. Perrig, R. Szewczyk, V. Wen, D. Culler, and J. D. Tygar. SPINS: Security Protocols for Sensor Networks. In *Proc. 7th Annual Int'l Conf. Mobile Computing and Networks (MOBICOM 2001)*, pages 189–199, July 2001.

280. Roberto Di Pietro, Luigi V. Mancini, and Alessandro Mei. Efficient and Resilient Key Discovery Based on Pseudo-Random Key Pre-Deployment. In *Proc. 18th Int'l Parallel and Distributed Processing Symposium (IPDPS 2004)*.

281. Roberto Di Pietro, Luigi V. Mancini, and Alessandro Mei. Random Key Assignment for Secure Wireless Sensor Networks. In *Proc. 1st ACM Workshop on Security of Ad Hoc and Sensor Networks*, pages 62–71, 2003.

282. PlanetLab. http://www.planet-lab.org, 2006.

283. David Poole. *Linear Algebra: A Modern Introduction*. Brooks Cole, 2002.

284. R. Prasad and L. Munoz. *WLANs and WPANs towards 4G Wireless*. Artech House, 2003.

285. R. Prasad and M. Ruggieri. *Technology Trends in Wireless Communications*. Artech House, 2003.

286. N. B. Priyantha, A. Chakraborty, and H. Balakrishnan. The Cricket Location-Support System. In *Proc. 6th Annual International Conference on Mobile Computing and Networking*, 2000.

287. John G. Proakis. *Digital Communications*. McGraw Hill, 2000.

288. M. B. Pursley. Performance evaluation for phase-coded spread-spectrum multiple-access communication–part i: System analysis. *IEEE Transactions on Communications*, 25(8):795 – 799, 1977.

289. Mu Qin. *Space-time-frequency: Diversity for wireless communication*. ProQuest, 2006.

290. Dongyu Qiu and R. Srikant. Modeling and Performance Analysis of BitTorrent-Like Peer-to-Peer Networks. In *Proc. SIGCOMM*, pages 367–377, September 2004.

291. Qualcomm. Cdma 1xrtt security overview. *Qualcomm White Paper*, August 2002.

292. K. Ramakrishnan and S. Floyd. A Proposal to Add Explicit Congestion Notification (ECN) to IP. *RFC 2481*, January 1999.

293. P. Ramanathan and P. Agrawal. Adapting Packet Fair Queueing Algorithms to Wireless Networks. In *Proc. MOBICOM*, pages 1–9, 1998.

294. Kavitha Ranganathan, Matei Ripeanu, Ankur Sarin, and Ian Foster. To Share or not to Share: An Analysis of Incentives to Contribute in Collaborative File-Sharing Environments. In *Proc. Workshop on Economics of Peer-to-Peer systems*, June 2003.

295. H. C. H. Rao, Y.-B. Lin, and S.-L. Cho. iGSM: VoIP Service for Mobile Networks. *IEEE Communications Magazine*, 38(4):62–69, April 2000.

296. Theodore S. Rappaport. *Wireless Communications: Principles and Practice*. Prentice Hall, 2001.

297. S. Ratnasamy, B. Karp, L. Yin, F. Yu, D. Estrin, R. Govindan, and S. Shenker. GHT: A Geographic Hash Table for Data-Centric Storage. In *Proc. 1st ACM Int'l Workshop on Wireless Sensor Networks and Applications (WSNA 2002)*, September 2002.

298. Siegmund Redl, Matthias K. Weber, and Malcolm Oliphant. *An Introduction to GSM*. Artech House Publishers, 1995.

299. C.W. Rhodes and P. Crosby. Measuring peak and average power of digitally modulated advanced television systems. *IEEE Transactions on Broadcasting Technology*, 37(8):885–890, Dec 1992.

300. V. Rodoplu and T. H. Meng. Minimum Energy Mobile Wireless Networks. *IEEE Journal on Selected Areas in Communications*, 17(8):1333?–1344, August 1999.

301. A. Rowstron and P. Druschel. Pastry: Scalable, distributed object address and routing for large-scale peer-to-peer systems. In *Proc. IFIP/ACM Int'l Conf. on Distributed Systems Platforms*, November 2001.

302. E. M. Royer and C. K. Toh. A Review of Current Routing Protocols for Ad-Hoc Mobile Networks. *IEEE Personal Communications*, April 1999.

303. Kenji Saito. Peer-to-Peer Money: Free Currency over the Internet. In *Proc. 2nd Int'l Conference on Human.Society@Internet*, June 2003.

304. Kenji Saito, Eiichi Morino, and Jun Murai. Multiplication Over Time to Facilitate Peer-to-Peer Barter Relationship. In *Proc. 16th Int'l Workshop on Database and Expert Systems Applications*, 2005.

305. Naouel Ben Salem, Levente Buttyan, Jean-Pierre Hubaux, and Markus Jakobsson. Node Cooperation in Hybrid Ad Hoc Networks. *IEEE Transactions on Mobile Computing*, 5(4):365–376, April 2006.

306. J. Eric Salt and Surinder Kumar. Effects of filtering on the performance of qpsk and msk modulation in d-s spread spectrum systems using rake receivers. *IEEE Journal on Selected Areas in Communications*, 12(4):707–715, 1994.

307. G. Salton and M. J. McGill. *Introduction to Modern Information Retrieval*. McGraw-Hill, first edition, 1983.

308. Sujay Sanghavi and Bruce Hajek. A New Mechanism for the Free-Rider Problem. In *Proc. SIGCOMM 2005 Workshop*, August 2005.

309. K. Sanzgiri, B. Dahill, B. N. Levine, C. Shields, and E. M. Belding-Royer. A Secure Routing protocol for Ad Hoc Networks. In *Proc. 10th IEEE International Conference on Network Protocol*, 2002.

310. M. Satyanarayanan. The evolution of coda. *ACM Transactions on Computer Systems*, May 2002.

311. M. Satyanarayanan, J. J. Kistler, P. Kumar, M. E. Okasaki, E. H. Siegel, and D. C. Steere. Coda: A highly available file system for a distributed workstation environment. *IEEE Transactions on Computers*, April 1990.

312. T. C. Schelling. *Micromotives and Macrobehavior*. W. W. Norton & Company, 1978.

313. J. Schiller. *Mobile Communications*. Addison Wesley, 2003.

314. M. Schlager, B. Rathke, S. Bodenstein, and A. Wolisz. Advocating a Remote Socket Architecture for Internet Access using Wireless LANs. *Mobile Networks and Applications*, 6(1):23–42, January 2001.

315. Shunsuke SEO, Tomohiro DOHI, and Fumiyuki ADACHI. Sir-based transmit power control of reverse link for coherent ds-cdma mobile radio. *IEICE TRANSACTIONS on Communications*, E81-B(2):1508–1516, 1998.

316. John S. Seybold. *Introduction to RF propagation*. John Wiley and Sons, 2005.

317. Y. Shaked and A. Wool. Cracking the Bluetooth PIN. In *Proc. 3rd International Conference on Mobile Systems, Applications, and Services*, 2005.

318. Elaine Shi and Adrian Perrig. Designing Secure Sensor Networks. *IEEE Wireless Communications*, pages 38–43, December 2004.

319. J. Shin, D. C. Lee, and C.-C. J. Kuo. *Quality of Service for Internet Multimedia*. Prentice Hall PTR, 2003.

320. Marvin K. Simon and Mohamed-Slim Alouini. *Digital Communication over Fading Channels*. John Wiley and Sons, 2004.

321. Sumeet Singh, Sriram Ramabhadran, Florin Baboescu, and Alex C. Snoeren. The Case for Service Provider Deployment of Super-Peers in Peer-to-Peer Networks. In *Proc. Workshop on Economics of Peer-to-Peer Systems*, June 2003.

322. David Soldani, Man Li, and Renaud Cuny. *QoS and QoE Management in UMTS Cellular Systems*. John Wiley and Sons, 2006.

323. W. Stallings. *Wireless Communications And Networks*. Prentice Hall, 2002.

324. Raymond Steele and Lajos Hanzo. *Mobile Radio Communications*. John Wiley and Sons, 1999.

325. Raymond Steele, Chin Chun Lee, and Peter Gould. *GSM, CDMAOne and 3G Systems*. John Wiley and Sons, 2001.

326. J. Steiner, C. Neuman, and J. Schiller. Kerberos: An Authentication Service for Open Network Systems. In *Proc. USENIX Winter Conf.*, pages 191–202, January 1988.

327. W. R. Stevens. TCP Slow Start, Congestion Avoidance, Fast Retransmit, and Fast Recovery Algorithms. *RFC 2001*, January 1997.

328. I. Stoica, R. Morris, D. Liben-Nowell, D. R. Karger, M. F. Kaashoek, F. Dabek, and H. Balakrishnan. Chord: A Scalable Peer-To-Peer Lookup Protocol for Internet Applications. *IEEE/ACM Transactions on Networking*, 11(1):17–32, February 2003.

329. Ion Stoica, Robert Morris, David Karger, M. Frans Kaashoek, and Hari Balakrishnan. Chord: A Scalable Peer-to-peer Lookup Service for Internet Applications. In *Proc. ACM SIGCOMM 2001*, pages 149–160, August 2001.

330. Qixiang Sun and Hector Garcia-Molina. SLIC: A Selfish Link-based Incentive Mechanism for Unstructured Peer-to-Peer Networks. In *Proc. 24th Int'l Conference on Distributed Computing Systems*, 2004.

331. K. L. Tan. Organization of Invalidation Reports for Energy-Efficient Cache Invalidation in Mobile Environments. *Mobile Networks and Applications*, 6(3):279–290, June 2001.

332. K. Tang, M. correa, and M. Gerla. Isolation of Wireless Ad Hoc Medium Access Mechanisms under TCP. In *Proc. Computer Communications and Networks*, pages 77–82, October 1999.

333. Using UAProf (User Agent Profile) to Detect User Agent Types and Device Capabilities. http://www.developershome.com/wap/detection/detection.asp?page=uaprof, 2006.

334. C. K. Toh. Associativity-Based Routing For Mobile Ad-Hoc Networks. *Journal on Wireless Personal Communications*, 1997.

335. Patrick Traynor, Heesook Choi, Guohong Cao, Sencun Zhu, and Thomas La Porta. Establishing Pairwise Keys in Heterogeneous Sensor Networks. In *Proc. IEEE INFOCOM 2006*.

336. Hal R. Varian. The Social Cost of Sharing. In *Proc. Workshop on Economics of Peer-to-Peer systems*, June 2003.

337. E. P. Vasilakopoulou, G. E. Karastergios, and G. D. Papadopoulos. Design and implementation of the hiperlan/2 protocol. *Mobile Computing and Communications Review*, 2003.

338. Srdjan Čapkun, Jean-Pierre Hubaux, and Levente Buttyán. Mobility Helps Peer-to-Peer Security. *IEEE Transactions on Mobile Computing*, 5(1):43–51, January 2006.

339. A. J. Viterbi. *CDMA: Principles of Spread Spectrum Communication*. Prentice Hall, 1995.

340. A. J. VITERBI, A. M. VITERBI, and K. GILHOUSEN. Soft handoffs extends cdma coverage and increase reverse link capacity. *IEEE Journal of Selected Areas on Communications*, 12(8):1281 – 1288, 1994.

341. vivek Shrivastava and Suman Banerjee. Natural Selection in Peer-to-Peer Streaming: From the Cathedral to the Bazaar. In *Proc. NOSSDAV'05*, June 2005.

342. Ashraf Wadaa, Stephan Olariu, Larry Wilson, and Mohamed Eltoweissy. Scalable Cryptographic Key Management in Wireless Sensor Networks. In *Proc. 24th IEEE Int'l Conf. Distributed Computing Systems Workshops*, 2004.

343. B. Walke, P. Seidenberg, and M. P. Althoff. *UMTS: The Fundamentals*. John Wiley & Sons, 2003.

344. L. Wang, Y. K. Kwok, W. C. Lau, and V. K. N. Lau. Efficient Packet Scheduling Using Channel Adaptive Fair Queueing in Distributed Mobile Computing Systems. *ACM/Kluwer Mobile Networks and Applications*, 9(4):297–309, August 2004.

345. W. Wang, S. C. Liew, and V. O. K. Li. Solutions to Performance Problems in VoIP over a 802.11 Wireless LAN. *IEEE Transactions on Vechicular Technology*, 54(1):366–384, January 2005.

346. Weihong Wang and Baochun Li. Market-driven Bandwidth Allocation in Selfish Overlay Networks. In *Proc. IEEE INFOCOM 2005*, March 2005.

347. Y. Wang, S. Ye, and X. Li. Understanding Current IPv6 Performance: A Measurement Study. In *Proc. 10th IEEE Symposium on Computers and Communications*, pages 71–76, June 2005.

348. Y. Wang and Q.-F. Zhu. Error control and concealment for video communication: A review. *Proc. IEEE*, May 1998.

349. S. Williams. Irda: Past, present and future. *Personal Communications*, Feburary 2000.

350. Ouri Wolfson, Bo Xu, and A. Prasad Sistla. An Economic Model for Resource Exchange in Mobile Peer to Peer Networks. In *Proc. 16th Int'l Conference on Scientific and Statistical Databased Management*, 2004.

351. W. C. Wong, R. Steele, B. Glance, and D. Horn. Time diversity with adaptive error detection to combat rayleigh fading in digital mobile radio. *IEEE Transactions on Communications*, 31(3):378–387, 1983.

352. Krit Wongrujira and Aruna Seneviratne. Monetary Incentive with Reputation for Virtual Market-Place Based P2P. In *Proc. CoNEXT'05*, October 2005.

353. Dongyan Xu, Mohamed Hefeeda, Susanne Hambrusch, and Bharat Bhargava. On Peer-to-Peer Media Streaming. In *Proc. 22nd Int'l Conference on Distributed Computing Systems*, 2002.

354. S. Xu and T. Saadawi. Does the IEEE 802.11 MAC Protocol Work Well in Multihop Wireless Ad Hoc Networks? *IEEE Communications Magazine*, 39(6):130–137, June 2001.

355. Guang-Tao Xue, Ming-Lu Li, Qian-Ni Deng, and Jin-Yuan You. Stable Group Model in Mobile Peer-to-Peer Media Streaming System. In *Proc. IEEE Int'l Conference on Mobile Ad Hoc and Sensor Systems*, pages 334–339, 2004.

356. M. D. Yacoub. *Foundations of Mobile Radio Engineering*. CRC Press, 1993.

357. B. Yang and H. Garcia-Molina. PPay: Micropayments for Peer-to-Peer Systems. In *Proc. 10th ACM Conference on Computer and Communication Security*, pages 300–310, 2003.

358. Mao Yang, Zhengyou Zhang, Xiaoming Li, and Yafei Dai. An Empirical Study of Free-Riding Behavior in the Maze P2P File-Sharing System. In *Proc. IPTPS'05*, February 2005.

359. Y. J. Yang and J. F. Chang. A strength and sir combined adaptive power control for cdma mobile radio channels. *IEEE Transactions on Vehicular Technology*, 48(6):1996–2004, 1999.

360. R. Yavatkar and N. Bhagawat. Improving End-to-end Performance of TCP over Mobile Internetworks. In *Proc. Workshop on Mobile Computing Systems and Applications*, pages 146–152, December 1994.

361. Song Ye and Fillia Makedon. Collaboration-Aware Peer-to-Peer Media Streaming. In *Proc. MM'04*, pages 412–415, October 2004.

362. M. K. H. Yeung and Y. K. Kwok. Wireless Cache Invalidation Schemes with Link Adaptation and Downlink Traffic. *IEEE Transactions on Mobile Computing*, 4(1):68–83, January 2005.

363. Mark Kai Ho Yeung and Yu-Kwong Kwok. A Game Theoretic Approach to Power Aware Wireless Data Access. *IEEE Transactions on Mobile Computing*, 5(8), August 2006.

364. Mark Kai Ho Yeung and Yu-Kwong Kwok. On Maximizing Revenue for Client-Server Based Wireless Data Access in the Presence of Peer-to-Peer Sharing. In *Proc. 17th Annual IEEE International Symposium on Personal, Indoor, and Mobile Radio Communications (PIMRC'2006)*, September 2006.

365. H. K. Yip. *Packet Scheduling Techniques for Coordinating Colocated Bluetooth and IEEE 802.11b in a Linux Machine*. MPhil Thesis, The University of Hong Kong, 2004.

366. Bin Yu and Munindar P. Singh. Incentive Mechanisms for Peer-to-Peer Systems. In *Proc. 2nd International Workshop on Agents and Peer-to-Peer Computing*, July 2003.

367. Zhen Yu and Yong Guan. A Robust Group-Based Key Management Scheme for Wireless Sensor Networks. In *Proc. IEEE WCNC 2005*, pages 1915–1920.

368. J. Zander. Performance of optimum transmitter power control in cellular radio system. *IEEE Journal of Selected Area on Communications*, E81-B(2):51–62, 1992.

369. S. Zeadally, R. Wasseem, and I. Raicu. Comparison of End-System IPv6 Protocol Stacks. *IEE Proceedings—Communications*, 151(3):238–242, June 2004.

370. L. Zhao, J. Shin, and C.-C. J. Kuo J. Kim. http://citeseer.ist.psu.edu/zhao01fgs.html, 2001.

371. R. Zheng, C. Hou, and S. Lui. Asynchronous Wakeup for Ad Hoc Networks. In *Proc. ACM MobiHoc*, pages 35–45, June 2003.

372. Yun Zhou, Yanchao Zhang, and Yuguang Fang. LLK: A Link-Layer Key Establishment Scheme for Wireless Sensor Networks. In *Proc. IEEE WCNC 2005*, pages 1921–1926.

373. Q.-F. Zhu, Y. Wang, and L. Shaw. Coing and cell-loss recovery in dct-based packet video. *IEEE Transactions on Circuits and Systems for Video Technology*, June 1993.

374. Sencun Zhu, Sanjeev Setia, and Sushil Jajodia. LEAP: Efficient Security Mechanisms for Large Scale Distributed Sensor Networks. In *Proc. ACM CCS 2003*, pages 62–72, October 2003.

375. M. Zorzi and R. R. Rao. The Role of Error Correlations in the Design of Protocols for Packet Switched Services. In *Proc. 35th Annual Allerton Conference on Communications, Control, and Computing*, pages 749–758, September 1997.

TOPIC INDEX

3G, 14, 101
See also UMTS
3G1X, 88, 114, 119, 121
CDMA2000, 114, 164, 174
QoS differentiation, 128, 132
ad hoc networks, 287
ad hoc routing
ABR, 493
AODV, 487, 489
CBRP, 493
CSGR, 485
DSDV, 483
DSR, 487
Fisheye state routing, 485
flooding, 485
GSR, 485
OLSR, 485
power aware, 497
proactive, 476, 483, 485, 487, 494–495, 500
reactive, 476, 487, 489, 491, 493, 495
RICA, 491
RREP, 489, 492
RREQ, 487, 489, 491–492

security, 496
SSA, 493
system model, 480
TBRPF, 485
TORA, 493
WRP, 485
AES, 520, 536
AMPS, 39, 44, 113–114, 116–118, 131
IS54, 131
ARP, 389
ARQ, 420
AVSG, 583–584, 588
Barker sequence, 272, 275
binary exponential backoff, 270
Bluetooth, 285–287, 289–290, 292, 295, 304, 330–331, 335, 339, 341, 354–355, 359, 410–412, 520, 529, 531–532, 540–541
ACL, 286, 294–295, 299, 304
baseband, 289, 292
cable replacement, 285, 287, 305, 307
channels, 297
encryption, 531
eSCO, 294, 296

hold mode, 299
IAC, 297
L2CAP, 289
LMP, 289
MAC, 286, 292, 302, 307
master-slave, 293–294, 296–302
OBEX, 286, 290–291
packet types, 294
  DH1, 304
  DH3, 304
  DH5, 304
  DM1, 296
pairing, 529
park mode, 299
piconet, 289, 293–294, 296–299, 301–304, 307, 312, 315–316, 318, 321, 323
  bridge node, 296, 299–300, 302–303
power classes, 292
profiles, 291
protocol stack, 287–288, 292
RFCOMM, 289, 291
scatternet, 296, 299–300, 302, 307
  BlueConstellation, 301
  Bluenet, 300
  BlueRing, 302
  BTCP, 300
  formation, 300–303, 307
  randomized formation, 301
  routing, 302
  scheduling, 303
  search tree, 301
SCO, 286, 294, 296, 299
SDP, 289
SIG, 285–287
sniff mode, 299
specifications, 287
TDD, 302
time slots, 293–294
CDMA, 13, 43–44, 48, 54–55, 57, 59, 61, 63, 66, 81, 83–84, 86, 112, 114, 116–118, 131, 134, 136, 143–144, 154, 164–166, 174, 520, 524, 526, 540, 546
D-CDMA, 61–65, 67, 69, 73–74, 77–78, 81, 84, 86, 125, 133, 136–138, 140, 173
despreading, 59, 62, 66, 68, 84, 101, 103–106, 136, 143
HDR, 451
IS95, 83–85, 88, 101, 114, 116–117, 121, 131, 133–134, 136–140, 144, 146, 148, 154–156, 158, 160–161, 164–165, 167, 173–174, 176
long code, 136–137, 139, 144, 146, 149, 173
power control, 134, 144, 154–156, 158, 160–161, 172–174

RSSI, 156, 161, 173
processing gain, 68–70, 84
R-CDMA, 66–70, 72–73, 76–78, 81, 84, 86, 89, 133, 136–138, 140, 154, 173
RAKE receiver, 63, 88, 101, 103–106, 166, 170
soft handover, 134, 139, 144, 146, 161, 164–168, 170, 172, 174
  active set, 167–168, 170, 172, 174
  candidate set, 167–168, 170, 174
  neighbor set, 167–168, 170, 172, 174
  PSMM, 168, 170, 174
  remaining set, 167–168, 170, 174
spreading, 58–59, 62–63, 65–66, 68–69, 86
spreading factor, 59, 61–62, 64, 69, 74, 84–85, 125, 127, 132
Walsh code, 137–139, 143–144, 148, 156, 173
wideband, 114, 125, 220, 225
cellular location techniques
  AOA, 605
  COO, 605
  E-OTD, 606
  TODA, 605
cellular networks, 111–114, 116–119, 121, 127, 130–131
  cell splitting, 115, 131
  MSC, 111, 130–131
channel, 1–3, 6, 9, 12–14, 16, 19
  assignment, 80
    fixed, 78
  AWGN, 2, 12, 20, 30, 44, 87–89, 96, 106
  coherence bandwidth, 7, 9, 12, 14, 16
  coherence time, 9, 12–13, 96, 107
  Doppler spread, 9, 12–13, 16
  fading, 9, 12–13, 87–88, 90, 93–97, 105, 107
    flat, 87–90, 92, 94, 106
    frequency flat, 12, 14, 16
    frequency selective, 12–14, 16, 87, 89–90, 97–101, 103–104, 106–107
    ISI, 14
    microscopic, 3, 7, 9, 12–14, 16, 155–156, 160
    Rayleigh, 7, 88
  ISI, 57, 63, 90, 98–99, 101, 103, 106
  multiaccess, 50, 52
  multipath, 2–3, 5, 7, 9, 12–14, 16, 143, 156, 166
  multipath dimension, 7
  multipath
    resolvable, 7, 12–14, 16, 89–90, 98, 104, 106
  path loss, 3, 5–7, 13, 16, 155–156
    exponent, 5, 7, 13, 16
    Hata model, 6
    Okumura model, 6
  shadowing, 3, 6–7, 13–14, 16, 155–156
  Shannon's capacity, 49
  spatial, 70, 73